Shock Waves @ Marseille III

Springer
Berlin
Heidelberg
New York
Barcelona
Budapest
Hong Kong
London
Milan
Paris
Tokyo

R. Brun L. Z. Dumitrescu (Eds.)

Shock Waves @ Marseille III

Shock Waves in Condensed Matter and Heterogeneous Media

Proceedings
of the 19th International Symposium
on Shock Waves
Held at Marseille, France, 26-30 July 1993

With 394 Figures

Springer

Professor Dr. Raymond Brun

Professor Dr. Lucien Z. Dumitrescu

Université de Provence, Centre Saint-Jérôme
IUSTI-MHEQ, Case 321, F-13397 Marseille Cedex 20, France

The Preface and lists of the Symposium Committees and of the Sponsoring Organisations are printed in Volume I.

ISBN 3-540-57712-2 Springer-Verlag Berlin Heidelberg New York
ISBN 0-387-57712-2 Springer-Verlag New York Berlin Heidelberg

Set (Volumes I-IV):
ISBN 3-540-57713-0 Springer-Verlag Berlin Heidelberg New York

CIP data applied for

Printed in Germany

Typesetting: Camera ready by editors
SPIN: 10132540 55/3144 - 5 4 3 2 1 0 - Printed on acid-free paper

Contents - Volume III*

* The Contents of the other volumes are given at the end of the book.

Part 2: Shocks in Porous Media

Part 4: Industrial Applications and Environment

Part 5: Biological Aspects

Survey Paper

Plenary Lectures

Volume III: Shock Waves in Condensed Matter and Heterogeneous Media - An Introductory Survey

M.E.H. van Dongen
Faculty of Applied Physics, Eindhoven University of Technology, P.O. Box 513, nl-5600 MB Eindhoven, The Netherlands

1. Introduction

This volume deals with 'Shock Waves in Condensed Matter and Heterogeneous Media'. It comprises 75 papers on a variety of interesting subjects. Shock waves are always related to changes in the state of matter. The range of shock induced changes covered by this volume is immense: from almost acoustical variations to thermonuclear dissociations. The substances under consideration are diverse: shock induced manipulation of single particles, subtle porous media like liquid foams, large scale porous media like real forests, shocked minerals of meteorites, living tissue and many others. The volume is divided into 5 chapters:

Two Phase Flows
Shocks in Porous Media and Liquid Foams
Shocks in Condensed Matter
Industrial Applications and Environment
Biological Aspects

Chapter 4 contains some contributions that do not strictly belong to the class of problems corresponding to the title of this volume. However, they certainly belong to the category 'Industrial Applications'.

2. Two-phase flows

Several interesting new results, ideas and developments are reported concerning the investigation of two-phase flows by means of shock waves, shock tubes, supersonic nozzles and supersonic free jets. In a majority of the contributions the continuous medium is a gas; some papers are related to bubbly mixtures. Important topics are: the acceleration of single solid or liquid particles by shock induced flow, shock wave propagation and diffraction in particle-laden gas flows, droplet formation, nucleation, evaporation, break-up and coalescence, shock induced film condensation, shocks and solitons in bubbly mixtures.

The study of propagation, reflection and refraction of shock waves in dusty gases is relevant for the understanding of the physical mechanisms of particle drag, dust explosions and suspensions like fluidized beds. An experimental shock tube facility designed to study the head-on interaction of a shock wave with a semi-infinite dusty gas was recently described by Geng et al.(1994). In the bottom section of a vertical shock tube a suspension of dust particles is maintained by an upward gas flow. A sharp interface between dusty gas and particle-free gas above it is realized. When a shock wave interacts with the interface, interesting phenomena occur, in particular when the shock-induced flow is supersonic. A numerical analysis of this problem is presented by Yu et al., who find an intriguing peak-shaped temperature distribution in the dusty gas suspension, when the incident shock wave is sufficiently strong.

Some new results are reported concerning the numerical analysis of shock wave propagation, reflection and diffraction in dusty gases. Shock wave diffraction around a cylinder in a dusty gas is described by Eidelman et al., a detailed numerical analysis of shock wave diffraction around a 90° corner is provided by Wang et al. The very complex problem of the entrainment of dust particles in shock-induced turbulent flows is successfully tackled numerically by Chien et al. The

Shock Waves @ Marseille III
Editors: R. Brun, L. Z. Dumitrescu

emphasis of their work is on the unstable behaviour of particle-laden boundary layers and on the formation of large coherent structures.

In almost all studies of shock-induced dusty gas flows, it is assumed that the acceleration of the dust particles can be described in terms of a quasi-steady drag coefficient. In general, however, the drag interaction between accelerating particles in a non-uniform unsteady flow field is much more complicated with explicit contributions of the pressure gradient, of the virtual mass related to the relative acceleration of particles and carrier gas, and of the Basset history term (Clift, Grace and Weber 1978). The importance of the Basset term for the description of the motion of particles induced by a plane shock wave is discussed by Sommerfeld and Decker for shock Mach numbers around 1.5 and particle diameters ranging from 10 to 50 μm. They find that the largest contribution of the Basset force is found directly behind the shock front and is of the order of 5%. When the velocity relaxation proceeds, and the net forces are vanishing, the relative contribution of the Basset term increases. Based upon a combined numerical and experimental study concerning the attenuation of shock waves is dusty gases by Igra et al., it is concluded that it is not possible to explain the experimental findings by assuming the drag interaction to be quasi-steady. A large range of Mach numbers, dust loading ratios and particle diameters was investigated.

A new and interesting experimental facility to study shock wave induced drag of individual particles is described by Rodriguez et al. It is basically a vertical shock tube with the end plate on top. A 'train' of particles with diameters of the order of 1 mm is released from the top end plate. The trajectories of the particles, affected by the upward propagating shock wave are observed by means of high speed photography. From the particle trajectories and from a numerical analysis of the shock-induced flow, accurate data can be deduced concerning the unsteady drag interaction. A complicated interaction process occurs when an oblique reflecting shock wave, moving along a flat plate, hits a solid dust particle. Such a particle can be swept away from the wall depending on the nature of the reflecting shock wave and the subsequent induced flow. First results of a numerical simulation are shown by Olim et al. The work is still in progress and opens interesting possibilities regarding the small-scale manipulation of single particles adjacent to solid walls.

Shock induced flow phenomena in dusty gases are governed by momentum and heat exchange between the continuous medium and the solid particles. When the particulate medium consists of droplets, two more relaxation processes may become important: mass transfer due to evaporation or condensation and mass transfer due to droplet break-up. Important parameters are the slip Reynolds number, and the Weber number. For Weber numbers sufficiently small, droplet break-up does not occur and the mass transfer is restricted to evaporation and condenstion processes. For a more detailed discussion of the structure of the relaxation zone behind a shock wave in a cloud, the reader is referred to recent papers by Young and Guha (1991) and by Smolders and Van Dongen (1992). Hirahara et al. use a combined shock-expansion tube, in order to produce a homogeneous sub-micron cloud. A shock wave is propagating through this cloud and diffracts from a wedge. The relaxation zone following the shock wave strongly affects the Mach reflection process at the wedge. The evaporation relaxation in an argon-steam mixture is studied both experimentally and numerically by Maršík et al.

A refined experimental set up to determine with excellent accuracy shock induced evaporation rates of droplets is described by Peters and Paikert. A pulse-expansion tube is used to form a homogeneous cloud of mono-dispersed particles, after which a shock wave is generated that causes re-evaporation. By utilizing 90 degree Mie-scattering, extremely accurate growth curves are obtained. The technique has potential possibilities for studying heat and mass transfer processes in the transition regime and for determining accurate diffusion data (Roth 1991a).

The investigation of two-phase flows in nozzles with the formation and evaporation of droplets is important because of possible implications in turbo-machinery, rocket engines, cryogenic wind-tunnels, for the control of acoustic noise, and for the control and manipulation of droplet sizes in supersonic droplet generators. Utheza et al. from the Université de Provence, present results of a numerical analysis of supersonic nozzle flow in which a steady shock wave is present. They discuss the effect of the injection of droplets from different injection sites upstream or downstream the shock wave. Itkin et al. present a numerical study of the effect of an oblique shock wave on the homogeneous nucleation and condensation process based on their own monomolecular condensation theory.

The interaction of droplets with shock waves also occurs in underexpanded free-jet atomizers. An experimental study of such a system is provided by Kurian and Sommerfeld who inject liquid in the settling chamber of a supersonic free-jet and who study the velocity and mean droplet diameter along the free-jet center line. One of the interesting observations is that the average droplet size increases (from 34 to 38 μm) when the droplets pass the Mach disc. This could be due to shock-induced coagulation, first described by Roth (1991b), a process that is associated with the size dependence of velocity relaxation.

Droplet break-up, either gradual or catastrophic, is the subject of a contribution by Widdecke et al. They describe a special shock tube in which a 'train' of isopropanol droplets is injected, perpendicular to the shock tube axis. In contrast to the work by Rodriguez et al. with their vertical shock tube, the droplets are quite small: 50 - 200 μm, and a special micro-projection equipment is used to study droplet break-up by optical imaging.

Two more contributions are related to homogeneous nucleation phenomena in shock tubes. Delale and Schnerr apply an asymptotic analysis of the nucleation equations to the unsteady rarefaction fan in a gas-vapour mixture. Their paper is an extension of earlier work of Delale et al. (1993) on the structure of condensation shocks in supersonic nozzles. A very special nucleation problem arises when a shock wave propagates in Perfluoro-n-hexane, a gas that shows retrograde thermodynamic behaviour. It is shown by Herrmann and Obermeier that nucleation occurs some distance behind the shock front, but that the shock thickness itself is not affected. Usually, nucleation occurs in expanding vapours, in which the processes are relatively slow. A shock-induced nucleation process could give valuable information on the validity of the assumption of "steady-state" nucleation, which is possibly not justified under retrograde circumstances.

A shock wave propagating in a vapour can cause film-condensation along walls. If the shock strength is chosen properly, then only condensation occurs in the reflected shock region. Teske and Obermeier compared the growth of the liquid film at the end-wall with that at the side-wall, and confirm that non-condensable impurities accumulate in the boundary layer adjacent to the liquid film.

A hypothetical 'new shock tube' is considered theoretically on the basis of a gas-kinetic model equation, the BGK equation, by Onishi and Miura. The 'driver' is a liquid interface at the end of the tube, that is assumed to have experienced an instantaneous temperature increase. As a result evaporation starts and a shock wave is formed.

Only three papers are related to shock wave phenomena in bubbly mixtures, which is not in agreement with the importance of this subject for two-phase flow. The reason might be that shock wave conferences traditionally have a somewhat strong gas-dynamic character. Matsumoto et al. present numerical and experimental results about shock waves in bubbly mixtures. The unsteady thermal problem of individual bubbles is treated numerically and the results are implemented in macroscopic wave theory, whithout making use of the polytropic assumption. Experiments are in progress. Malykh describes an experimental study of resonance solitons, formed when a bubbly mixture is subjected to a short duration high amplitude pressure pulse. Lugovstov presents some analytical techniques to describe the interaction of high-amplitude sonic beams with bubbly mixtures.

3. Shocks in porous media and in liquid foams

The investigations of wave propagations in porous media is related to a variety of motives. Traditionally, porous media are used for the damping of acoustic waves in internal flows. It is quite natural to extend the knowledge on the damping properties of such materials to the non-linear regime. Other applications are related to the effects of explosions and blast waves and the corresponding risk analysis and to problems in soil mechanics and petroleum engineering.

In the last 25 years interesting research has been performed concerning wave propagation in porous media in general and in flexible and permeable (polyurethane) foam materials in particular. An interesting review of this work is given by Skews. Polyurethane foams are characterized by a high porosity and by a strong non-linear elastic behaviour. In a typical experiment a slab of foam is placed in contact with the solid end plate of a shock tube. When the foam is subjected to an incident shock wave the peak end-wall pressure is higher than without the presence of foam. Interesting new results were published by Skews et al. (1993)

In recent years much progress was made concerning the physical and numerical modelling of wave propagation in foams. Examples are the numerical study of Baer on the basis of mixture theory and an analytical study by Van Dongen et al. who apply a linear theory due to Biot, which is generally accepted in soil mechanics. In fact, both theories make use of the concept of the so-called effective stress, which describes the forces exerted by the solid particles in a cross-section in excess to the (pore) pressure. In the most simple constitutive model, the effective stress is uniquely related to the porosity. Baer is able to simulate with good accuracy the experiments by Skews et al. (1993). A typical result of the analytical linear study by Van Dongen et al. is that in general two wave modes are found with their own frequency dependent wave speeds and damping factors. It is quite interesting that the pressure traces of the Skews experiments can be qualitatively explained by linear theory! In a numerical study, Olim et al. (1994) show that it is also possible to explain the results of weak-shock reflection experiments in foams with a dusty gas model. Van Dongen et al. also describe a method to measure the end wall gas pressure and the effective foam stress separately. New experimental data on wave propagation and foam deformation are given by Gvozdeva et al., who also discuss the effect of the incident wave form on the measured end wall peak pressures. They simulate the pressure history of a blast wave. It is interesting to compare their results with those of Lee and Lee, who investigate several deformable materials, including energy absorbing ones, both in a shock tube and in a blast chamber. They determine the effect of the porous material on peak pressure and on the total impulse. Energy absorbing materials appear to be effective for the reduction of peak pressures and impulses.

Zaretsky and Igra present the results of a preliminary study about the reflection of a shock wave from a foam containing water. Peak pressures decrease with increasing water content. The problem of the propagation of very strong shock waves in rigid low porosity porous media is treated analytically by Levy et al. A parameter study of shock wave reflections from flexible foams is given by Ben-Dor et al.

A very special problem of shock wave formation and propagation is porous media arises when the pure gas is a chemically reactive mixture. The problem is analysed by Gol'dstein et al., who determine the conditions for the formation of self-sustained pressure shocks and give the structure of the shock waves formed.

Foams of a rather different nature are liquid foams, also characterized by large porosity and gas volume fraction but with "pores" that are closed. Liquid foams form the counterpart of bubbly mixtures. Experiments in liquid foams consisting of sulphanol solutions are described by Goldfarb et al. The foam cells are small: 0.1 mm. The foam is studied in a vertical shock tube. The wave speed of the disturbance is in between isothermal and isentropic; neither oscillatory behaviour is observed nor steepening of the wave front. This is in agreement with their own theoretical model and with that of Elperin et al. The reason is that strong dissipative processes are present due to thermal damping and to viscous damping asssociated with flow in the liquid channels separating the gas bubbles, the so called Plateau-Gibbs channels. In a second contribution, Elperin et al. analyse the formation of shock waves in gas-liquid foams both for low and high volumetric gas content. They are able to explain why in liquid foams shock wave formation is not observed, while under similar conditions in bubbly mixtures shock waves are easily detected.

4. Shocks in condensed matter

There is a variety of problems associated with waves and shock waves in condensed matter. Shock waves generated by laser focusing in liquids or solids may modify the properties of materials. Strong shock waves in solids reveal information on the equation of state under rather extreme circumstances. Researchers are looking for ways to utilize shock waves for the consolidation of powders and for obtaining materials with new physical and mechanical properties. Shock wave phenomena occur when dust particles collide with the protective screens of spacecrafts.

Two contributions are treating the complicated problem of the explosive boiling of a liquid which is exposed to a sudden pressure release. This is realized by subjecting a liquid to an intense rarefaction wave. Several processes can be identified: the formation and dynamics of clusters of bubbles, wave phenomena in a cavitating liquid, the transformation of a liquid to a wet foam and then to a cloud of droplets. In real systems micro-inhomogeneities and walls play an important role. A treatment of the various physical properties, of experimental techniques and of experimental results is given by Kedrinskii and by Karpov.

Laser focusing in liquids and solids in a second important topic. When a laser pulse of sufficient strength is focused in a liquid, energy absorbtion leads to evaporation, the formation of a plasma, bubble collapse and the formation of a shock wave. Alloncle et al. were able to measure the pressure profiles in water by means of interferometry with time intervals of 50 ns. They show that even for a laser pulse energy of only 3 mJ, the peak pressure is 100 bar when the shock has moved away from the focus point over a distance of 1mm. Similar results are described by Emmony et al., who report a measured shock front thickness in water of 30 μm.

Laser-induced shock waves can modify the local physical and mechanical properties of surfaces of solid materials. A high-power pulse laser focused at a solid surface can heat, melt and vaporize the solid material. The shock wave propagating into the solid as a reaction to the mechanical impulse of the evaporating material, can be so strong that the surface and in-depth mechanical properties can be permanently modified. If the solid to be treated is covered by a so-called thermo-absorbing overlay, it is possible to induce compressive residual stresses in the surface area of a material (up to 1mm), without changing the shape of the treated surface, which is shown by Masse and Barreau. For such experiments a laser energy of 80 J and a high power density (5 GW/cm^2) is necessary. Quantitative information concerning the relation between the applied laser power density and the maximum shock pressure in the solid is obtained by Pratt et al. by means of several piezo-electric and piezo-resistive gauges. Maximum pressures are reported as high as 130 MPa. Baumung et al. have used a rather powerful alternative device for studying the response of matter to high power beams. With KALIF, which is an acronym for the Karlsruhe Light-Ion Facility, they are able to produce a pulsed (50 ns) proton beam with a peak power density of 10^{12} W/cm^2. When such a beam hits a target, a very high density plasma is formed at the ablating surface. Baumung et al. report on a variety of experimental findings, such as the acceleration of an Al foil of 10 μm with 10^{11} g, and a rise-time of 600 ps for a 25 GPa shock in aluminium!

Very strong shock waves can be applied for the consolidation and compaction of powder materials. Chiba et al. describe a device that is used for the consolidation of TiAl powder, with particles originally shaped like a Japanese fan. The shock wave is formed by an explosive, is focused in a converging water channel and causes a pressure load of the order of 10 GPa. A similar technique in an ellipsoidal water vessel is described by Itoh et al., who report about investigations to find optimum conditions to achieve shock induced consolidation of powder and to process metal materials.

The experimental study of shock velocities and shock-induced velocities in solids yields information on the EOS properties of such material. Shorokov and Litvinov present Hugoniot adiabate data for a special material (TATB) in the pressure range of 0.1 GPa to 40 GPa. Krupnikov and Kupnikova discuss the shock adiabates of epoxy-resins, which shows a region of elevated compressibility in the pressure range of 18-35 GPa. They argue that this is possibly related to a first order phase transition, associated with a proceeding chemical reaction.

The mechanical properties of composite materials are discussed by Resnyanski and by Patuelli. The first author proposes a phenomenological method of averaging the physical parameters of the different materials and analyses shock wave propagation in such material. Patuelli presents results of a dynamic shock wave loading of composites made of Aluminium and a ceramic material. The shocks are produced by means of explosively accelerated flyer plates.

The effects of really strong shock waves are outlined by Litvinov et al., who give a theoretical treatment of the effect of thermonuclear dissociation on the limiting shock compression ratios. The work refers to very strong shock waves indeed with temperatures of the order of 100 keV (10^9 K).

An adaptive mesh method to speed-up the numerical simulation of an underwater explosion is presented by Boice and Hocquellet. The presence of three different time and length scales corresponding with the initial detonation wave, the water shock wave, and the bubble resonance, respectively, makes such a method attractive. The focusing of cylindrical shock and detonation waves is treated numerically by Hiroe et al. The shock wave focusing occurs in a solid which is described with a Grüneisen EOS. As a numerical technique the Random Choice Method is applied.

Strong shocks in multicomponent media are analysed numerically by Kryukov and Fortov, applying parallel computing. Several examples are given such as the high-speed collisions of dust particles with the protective screen of the Vega spacecraft, the optimization of anti-meteorite protection and the collision of asteroids with the Earth surface.

5. Industrial applications and environment

There is not always a very sharp distinction between the topics belonging to the different categories. In the present chapter we shall meet a spectrum of problems: shock-induced ablation, shock waves in excimer lasers, shock waves in railway tunnels, in automobile exhaust systems, volcano eruptions, shock waves related to meteorite impact, shock waves as forest fire extinguishers and a shock tube study of coal pyrolysis.

The heat transfer between a pulsed high power CO_2 laser and several metallic samples is studied experimentally and theoretically by Dumas et al. They investigate the dependence of the ablation process on the pressure of the ambient gas. In high power excimer lasers, shock waves are formed in the laser cavity as a result of the discharge. At specific pulse rate frequencies the laser output decreases. Hirai et al. convincingly show that this is due to the presence of shock waves in the discharge area. They connect two excimer lasers. One is used as a shock wave generator. By varying the delay time between the two discharges, the effect of shock wave position on laser performance is determined. A detailed analysis of the wave phenomena in the laser tube is presented by Truong et al., and a comparison is made with experimental pressure traces. Transverse wave modes appear to give important contributions to the wave field. Both papers propose technical solutions for the damping of the disturbing waves.

Two Japanese groups are studying the wave phenomena associated with high-velocity trains passing tunnels. The formation of shock waves in tunnels is a typical Japanese problem because of the high 'tunnel density' of the Japanese Shinkansen high-velocity train network. Special measures have to be taken when train-velocities are increasing. Both studies present experimental and numerical data. The group from Kyushu (Aoki et al.) describes an experimental open-end shock tube facility with a special fast-opening gate. The experimental results are compared with a numerical model in which unsteady friction is taken into account. The Sendai group (Saito et al.) describes a different facility. A long cylinder, representing the train, is sliding with a constant velocity under the action of gravity along a tube (the tunnel), tilted with respect to the horizontal. Experimental results are compared with results of a numerical simulation. Both studies are in progress.

In modern motor engines weak shock waves appear in the exhaust pipes, which makes it worthwile to look for silencer designs with a better performance. Sekine et al. present results of a combined experimental/numerical study of different silencer configurations. Matsumura et al. present numerical solutions of the full Navier-Stokes equations for exhaust pipe systems and present measurements of internal and external pressure levels.

Phan discusses the design and performance of an intelligent muzzle brake. Kailasanath et al. have performed a numerical study on the effects of shock waves on the stability and noise generation of under-expanded free jets.

The potential of numerical techniques for studying shock-induced flow phenomena in very complex geometries is illustrated by Kinoshita et al. They present a numerical simulation of a blast wave caused by the eruption of the Japanese volcano Mount Unzen. The topography of the Unzen area is represented with a spatial resolution of 25 meters.

An important technical problem concerning the protection of space stations from the penetration of space debris and micro-meteoroids is tackled by Schmidt et al. They present laboratory results of spherical impacters on shield materials, looking for similarity rules in order to translate the restricted laboratory conditions to conditions actually met in space.

Miura, Kato, Imai and Okamoto have analysed the physical properties of quartz, silica and carbons, found in meteorites and impact craters and found deviating physical properties which is associated with the effects of very strong shock waves during impact. New results on properties of artificially shocked minerals by means of the impact of projectiles in the laboratory, are presented by Miura, Takayama and Kato.

A very special experimental and numerical study concerning shock waves in 'porous' media is presented by Grishin. He discusses the possibility of utilizing shock waves or blast waves to extinguish a forest fire. Small scale experiments are reported with a burning 'model forest' and an identical non-burning 'reference forest'. Making use of chord charges a blast wave was generated, and its effect on both forests was investigated. Also results of a numerical analysis are presented, modelling the forest as a porous reactive medium. Chang describes a special shock tube facility designed to study shock wave-induced pyrolysis of coal powder. The reaction products are analysed for various residence times and various pressure levels.

6. Biological aspects

Shock wave generators have become important tools in clinical practice. In his plenary lecture Prat reports on the present status of an important new development: cytotoxic effects of shock waves and of micro-cavitation on cancerous tumours. *In vivo* experiments with rats showed that a combination of an intra-vascular infusion of micro-bubbles, shock wave exposure and chemotherapy was promising. Prat describes the state of the art of the next step in the development: the design of a special device that causes a negative pressure pulse first, which generates cavitation bubbles, followed by a compression shock, which causes bubble collapse. The results of experiments with a rabbit liver *in vivo* are discussed. The importance of such experiments for a better understanding of the limits and perspectives of the new method is emphasized.

The effects of single and multiple blast exposures on animals is discussed by Vassout et al. The effect of pulse duration, pulse amplitude and number of exposures is specified for the trachea, the lungs, the intestines. The mortality limits for living animals are investigated and estimates are made of the consequences for human beings. More insight is needed in effects of shock waves on living tissue, such as the formation of lesions. To that end, a special shock tube device has been built. The shock wave induced displacement of living tissue is compared with results of numerical simulations. Quantitative physical parameters are specified (Tamagawa et al.).

Ise et al. report on new results of the treatment of gallstone patients with Extracorporeal Shock Wave Gallstone Lithotripsy (ESWGL). If the patients are selected carefully, gallstones can be fragmented without important side effects with a reasonable 'success ratio'.

In order to make progress in the physical understanding of shock wave interaction with living tissue and biological substances, much work has to be done to characterise the relevant physical properties of such materials. Nagoya et al. are presenting EOS data for human blood and pig liver, obtained by determining isothermal pressure-density relations. They also simulate the interaction of a focused shock wave with human tissue. Štuka et al. have experimentally studied the effect of de-gassing the water on the amplitudes and profiles of the pressure in the point of shock wave focusing. The background of their work is that they are looking for ways to affect the waveforms.

Cavitation bubble collapse is most probably an important factor in causing wanted or unwanted damage of human tissue. In order to be able to study these phenomena, a dedicated soft flash X-ray source for shock wave research was developed by Sato et al. It has a pulse width of 50 - 200 ns. Several applications of the system are discussed. Šunka et al. specify a new discharge circuit for the efficient generation of shock waves and detailed measurements of the various characteristics of the generator are presented. Veroman et al. describe a super-powerful generator for shock wave focusing. Several shock wave generators are placed within a torus with an ellipsoidal cross section. While the generators are placed on one line of focus, the shape of the cross section is chosen such that all shock waves converge in the torus center.

Acknowledgment

The author is much indebted to Mrs Brigitte van de Wijdeven for her most efficient help in preparing this manuscript.

References

Clift R, Grace JR, Weber ME (1978) Bubbles, Drops, and Particles. Academic Press, New York

Delale CF, Schnerr, GH and Zierep J (1993) Asymptotic solutions of transonic nozzle flows with homogeneous condensation I. Subcritical flows. Phys. Fluids A 5:2969, II. Supercritical flows. Phys. Fluids A 5:2982

Geng JH, Van de Ven Q, Zhang F, Grönig H (1994) Interaction of a shock wave with a two- phase interface. Shock Waves 3:223

Olim M, Van Dongen MEH, Kitamura T, Takayama K (1994) On numerical simulation of propagation of shock waves in compressible open cell porous foams. Intl. J. MultiPhase Flows, to be published

Smolders HJ, Van Dongen MEH (1992) Shock wave structure in a mixture of gas, vapour and droplets. Shock Waves 2:255

Skews BW, Atkins MD, Seitz MW (1993) The impact of a shock wave on porous compressible foams. J. Fluid Mech. 253:245

Roth P (1991a) The shock tube technique applied to study aerosol rate processes. In: Takayama K (ed) Proc. 18th Intl. Symp. on Shock Waves, Sendai. Springer, pp 73-82

Roth P (1991b) Private communication

Young JB, Guha A (1991) Normal shock wave structure in two-phase vapour-droplet flows. J. Fluid Mech. 228:243

Shock Wave Impact on Porous Materials

B.W. Skews
University of the Witwatersrand, Johannesburg, South Africa

Abstract. This review will concentrate on shock impact studies in media where the pores are interconnected within a solid matrix which may itself deform under load, as well as allowing flow through it, and particularly the case where the matrix compressibility is somewhat lower than that of the fluid. Such materials, when placed adjacent to a rigid wall exhibit the interesting property of amplifying the reflected pressure above that that would have been achieved with no foam present. This paper reviews aspects of this field of research over the past twenty years, including studies of both rigid and compressible porous materials, with both liquid and gaseous pore fluid, and in more than one dimension.

Key words: Foam, Porous material, Refraction

1. Introduction

Early studies of the interaction of shock waves with porous materials were largely motivated by the need to explore methods to attenuate the effects of a wave. These studies form part of the broader field of shock wave propagation in two-phase media, which include areas such as propagation in dusty gases, bubbly liquids, and liquid foams, as well as permeable ground, snow, and forests.

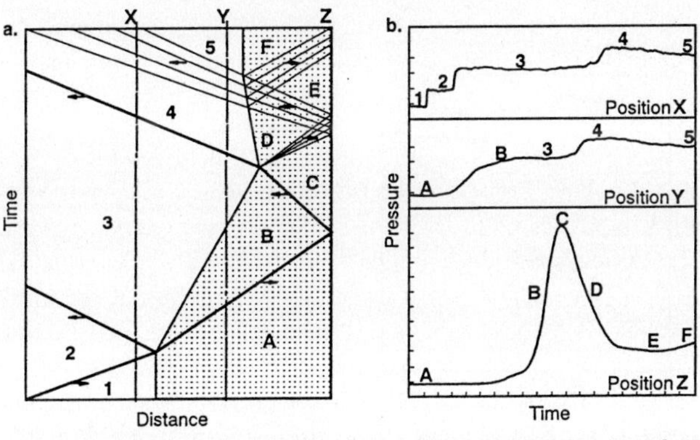

Fig. 1. Simple refraction model and pressure traces (Source: Skews et al. 1991)

2. One-dimensional studies

One of the most commonly conducted experiments in this field is the case of a block of compressible porous material placed against a rigid wall. On the assumption that this interaction may be represented by the one-dimensional refraction of a wave at an interface between two homogeneous materials, the wave diagram is that shown in Fig.1a. When the incident wave (1-2) strikes the material, a reflected wave (2-3), and a transmitted wave (A-B), are formed, resulting in compression of the foam material in region B, and the movement of its front face towards the rigid wall. The transmitted wave is then reflected off the end wall compressing the foam still further (region C), and when reaching the gas/foam interface gives rise to a transmitted shock wave propagating

Shock Waves @ Marseille III
Editors: R. Brun, L. Z. Dumitrescu

out into the gas, and an expansion wave reflected back into the foam material. This latter wave will in turn reflect off the back wall, relieving the pressure there. On reaching the interface an expansion wave will be transmitted into the gas, and a compression wave back into the foam material. Typical pressure measurements obtained from tests in a shock tube using polyurethane foam, with transducers positioned as indicated in Fig.1a, are given in Fig.1b. The pressure traces generally follow the expected pattern, although they differ in a number of respects from what would have been obtained if the simple refraction wave diagram were an accurate description of the phenomena. These differences have been the focus of much of the recent research.

The earliest theoretical and experimental study using porous foam, is that of Monti (1970). He was concerned with the general behaviour of a shock wave impacting an arbitrary, deformable, solid material. In order to assess the theory against experiment, Monti sought a material that was sufficiently soft to give measurable differences in reflected wave pressure. He settled for expanded foam material. These materials are highly non-linear, displaying a stress-strain characteristic which may be divided into three main regions: a small linearly elastic region, followed by a plateau in which the stress level hardly changes, followed by a densification region where the stress increases very rapidly. This approach was to become the main characteristic of investigations of the process for some considerable time, namely the assumption of simple refraction, with its associated implication of no gas flow across the interface, as well as the treatment of the foam as a single phase material, with its implication of a single sharp-fronted transmitted wave.

The first measurements of pressure alongside and behind compressible porous material mounted on the back wall of a shock tube were conducted by Gelfand et al. (1975). These were the first experiments to show that the peak pressure on the back wall is considerably higher than what would be obtained with no foam present. The authors interpreted this result as being due to the momentum associated with the solid phase being transferred to the back wall.

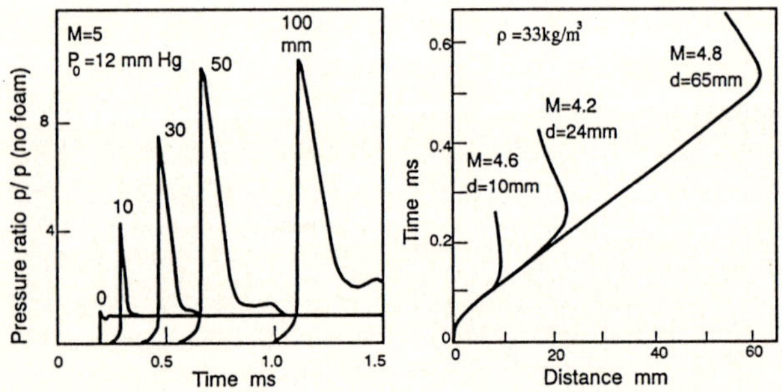

Fig. 2. Typical pressure and deflection characteristics (Source: Gvozdeva et al. 1985)

2.1. The early eighties - research with soft skeletons

A number of significant papers appeared in the eighties from two groups of researchers led respectively by Gelfand and Gvozdeva. In 1983 Gelfand et al. published their test results with no backing to the foam. The significant spreading out of the transmitted wave propagating through the foam was noted. An interesting theoretical analysis was presented, which was the first attempt to take into account the two-phase nature of the material. This was based on the concept of a pseudo-gas. In this model the foam/gas combination is treated as a single phase medium with thermodynamic properties reflecting that of the mixture. Results of the theoretical calculations using this model give reasonable predictions for the wave strengths.

In 1985 an equally interesting paper by Gvozdeva and Faresov appeared. They presented an analytical model based on the assumption that the elasticity of the medium was largely determined by the elasticity of the gas and that the elasticity of the solid phase contributed little to the material response. Two sets of calculations were done, the one assuming isothermal and the other adiabatic conditions. The adiabatic calculation predicted the pressure on the back wall satisfactorily over a Mach number range of 1.3 to 1.8. The authors proposed that the discrepancy at higher Mach numbers is due to percolation of gas remote from the shock wave through the permeable skeleton. A companion paper (Gvozdeva et al. 1985b) notes that the value of the maximum pressure on the back wall for a given foam, increases as the thickness of the foam increases up to a certain thickness whereafter it remains essentially constant. (Fig. 2a). The reasons for this behaviour have not yet been satisfactorily explained. In addition, tests showed that the front face of the foam moves at constant velocity during most of the compression phase, as indicated in Fig.2b.

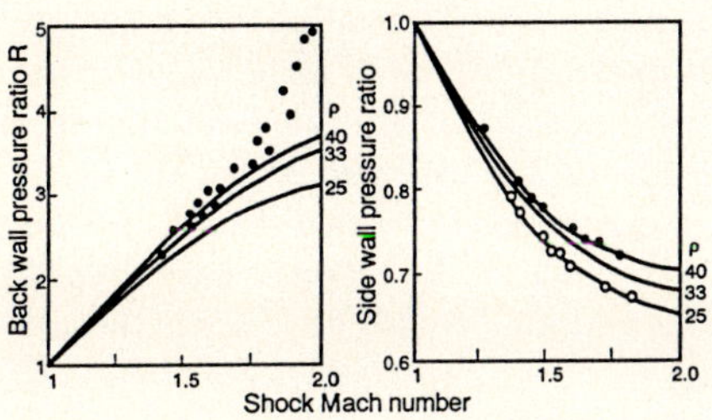

Fig. 3. Pressure on back and side walls (Source: Gvozdeva et al. 1987)

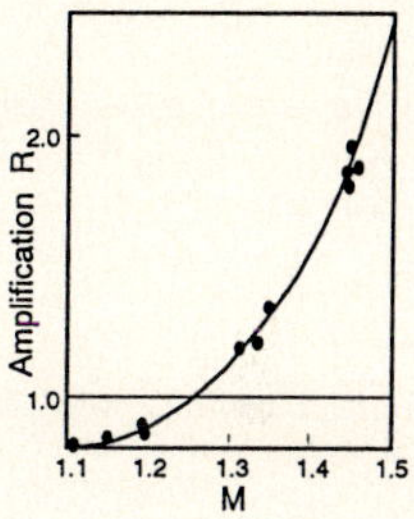

Fig. 4. Blast wave effects (Rayevsky et al. 1989)

This work was extended by Gvozdeva et al. (1986). It was assumed that the material would behave in a similar fashion to that of a liquid containing gaseous bubbles, or as a dusty gas. It was argued that during rapid compression of the material, air does not have time to filter through the pores and that the elasticity is predominantly determined by the gaseous phase. It was also noted that the peak pressure on the back wall was followed by a number of oscillations which were found to be proportional to the length of the material. This is consistent with the wave diagram of Fig.1a extended to longer times. Subsequently Gvozdeva et al. (1987) conducted one-dimensional numerical calculations. Satisfactory agreement was obtained during the loading phase of the foam, but agreement deteriorated during the recovery phase. Fig.3 shows the experimental and calculated results for the non-dimensionalised back wall and side wall pressures, as a function of Mach number and foam density (porosity 0.975). The authors ascribe the differences to the neglect of interphase motion inherent in the pseudo-gas model. The above numerical model was extended by Rayevsky et al. (1989). They assumed a one-fluid two-temperature model with the skeleton being treated as an aggregate of absolutely hard particles, and included both the elasticity of the skeleton as well as heat flux from the gaseous phase to the skeleton material. However, the assumption that interphase momentum is negligible was retained. Good agreement was obtained between the experimental and theoretical results up to the point where the rarefaction wave reaches the back wall and the peak pressure is terminated. Calculations also confirmed that the contribution of the elasticity of the skeleton was negligible. An interesting aspect of this work was the extension to the case of blast wave reflection, which was simulated experimentally by allowing a shock wave to expand out of the open end of the shock tube before impinging on the foam. Of particular significance here is that on the axis of symmetry amplification at the rigid back wall occurs for Mach numbers greater than about 1.25. (Fig.4). This effect was interpreted as being due to the reflection of the expansion phase of the blast, and filtration effects. This work clearly identifies the need for more two-dimensional studies.

2.2. Recent experimental work

Henderson et al. (1989) used specimens very much longer than those used previously, some 1600 mm, in a circular tube, using polyurethane. It was also noted that the front of the transmitted wave became less steep as the wave propagated down the tube. Of particular interest was the finding that the speed of the wave front decreases with incident wave strength up to a Mach number of about 2.4, and then increases again as shown in Fig.5. As far as is known this behaviour has not been tested by other workers and the reasons would be of interest. The sonic velocity in the material was estimated by extrapolating the data to a pressure ratio of unity.

A detailed study of the reflected wave field following impact of a weak shock wave on a range of foams was conducted by Skews (1991). It was shown (Fig.6) that the first reflected shock wave strength is accurately predicted by the pseudo-gas model of Gelfand et al. (1983). This agreement is somewhat surprising because there is evidence of gas entering the foam during the early stages of the process, and the wave processes within the foam are clearly much more complex than allowed by the simple refraction model; the transmitted wave spreads out in time, and the head of the wave moves faster than that predicted from the theory. The net result is that the second reflected wave emerges from the foam face, back into the gas, very much earlier than the pseudo-gas model predicts. Nevertheless the reasons for this agreement in first reflected wave strength require further clarification. This may well come from the numerical models discussed in the next section.

At the 18th Symposium Skews et al. (1991) reported on measurements of the skeletal motion, as well as the more usual pressure measurement. In particular it was shown that a compaction wave moves through the solid matrix at a velocity considerably less than the gas wave, as measured by the pressure transducers. This showed that correct modelling would have to include the separate behaviour of the two phases, taking account of the strong coupling between them, and in particular, the relative velocity with its associated pressure drop. This latter factor is seen to be the major reason for the spreading out of the transmitted wave. The Schlieren photographs also showed the emergence of a contact surface from the surface of the foam shortly before the position of maximum foam compaction. This feature was interpreted as being the front of the gas, which had been compressively heated within the foam, being driven out through the compacted material by the very strong pressure gradients developed due to the high pressure on the back wall and the short compacted length of foam. The presence of this feature also shows that the face of the foam should not be treated as a contact interface as assumed in the simple refraction model. The results were combined into the single consistent wave diagram shown in Fig.7.

There thus now appears to be a fairly general understanding of the physical processes occurring, although detailed effects for any particular fluid and foam type are not easy to predict. It is thus necessary to develop analytical and numerical models based on the controlling physical effects, so that the field may be more fully understood. There are fundamental difficulties in the measurement of some of the parameters and numerical and theoretical modelling offers the facility to explore individual effects in detail.

2.3. Recent theoretical and numerical studies

Probably the most comprehensive numerical simulation of shock wave impact on polyurethane foam is that of Baer (1992). The longitudinal pressure gradient is assumed to be dependent on the relative velocity V, as given by the Forcheimer equation, viz.

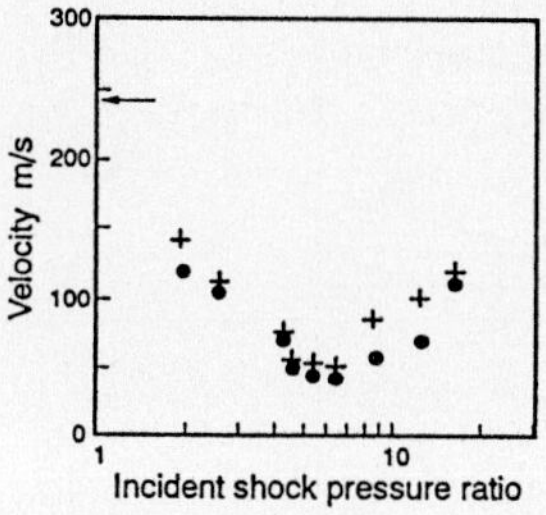

Fig. 5. Compression wave head velocity (Source: Henderson et al. 1989)

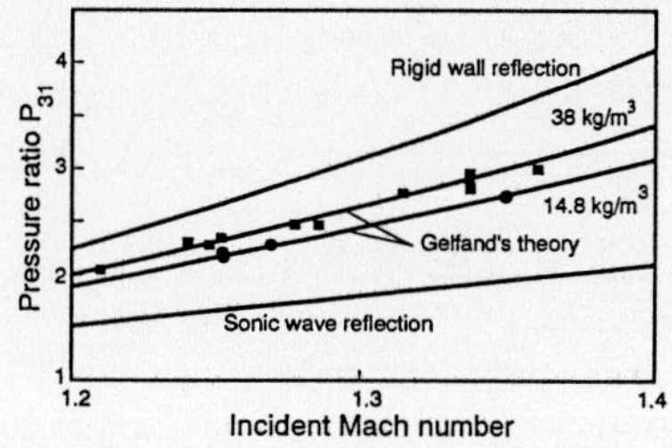

Fig. 6. First reflected wave behaviour (Source: Skews 1991)

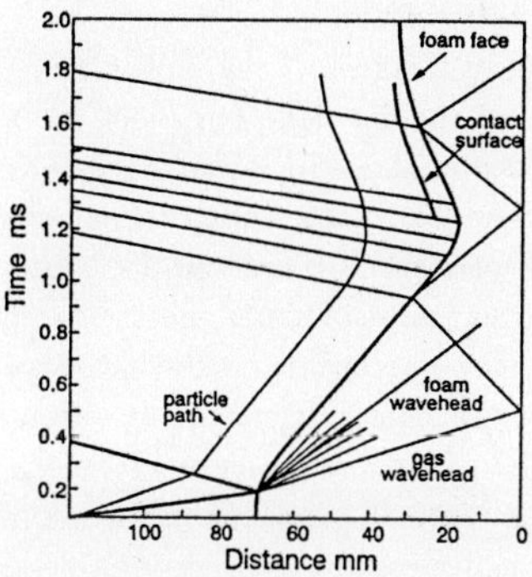

Fig. 7. Wave diagam from experiments (Source: Skews et al. 1991)

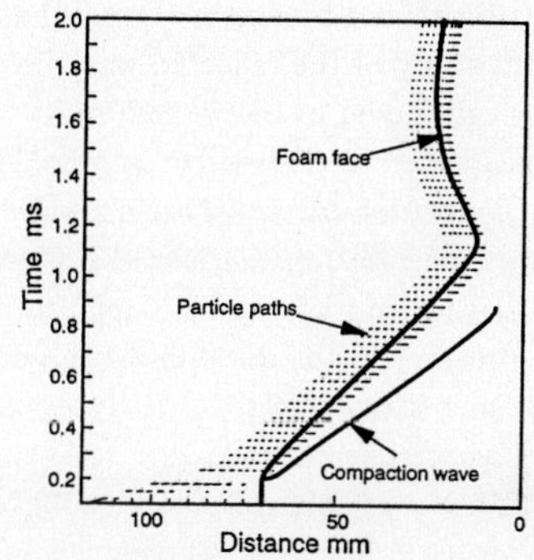

Fig. 8. Predicted flows (Source: Baer, unpublished)

$$\frac{\Delta p}{\rho} = \frac{\mu}{k}V + \frac{c\rho}{\sqrt{k}}V^2$$

where k, μ and ρ are the permeability, fluid viscosity, and fluid density, and c is the inertial coefficient. Of particular note is that the compaction wave in the material of the skeleton is correctly modelled. Such models now allow useful numerical experimentation to be conducted. As an example, this code has been run to ascertain the gas particle paths in the vicinity of the front face of the foam (Baer, private communication), as shown in Fig.8. This simulation shows a number of interesting features. Firstly, not only is there flow into the foam whilst it is accelerating, as inferred by Skews (1991), but there is a small percolation of gas into the material, and then out again followed by re-entry of gas into the foam during the recovery phase.

More recently Olim et al. (1993) have produced a simpler, yet very effective, numerical model. They assume the foam to be infinitely weak and to be relevant only when interactions between the phases are considered. This is reminiscent of the pseudo-gas model, whose success has been due to the minor effect the skeleton compressibility has. Furthermore it is assumed that the heat transfer coefficient is infinite. Good agreement is obtained with the experimental pressure traces (Fig.9).

Thus very good predictions are now available over the ranges for which these models have been validated ($M = 1.25$ to 1.4, and foam densities from 15 to 40 kg/m^3).

2.4. Stiff systems

One of the earlier studies of shock impact on materials with a stiff skeleton is that of Beavers and Matta (1972). They investigated the strength of the reflected wave for three different porous

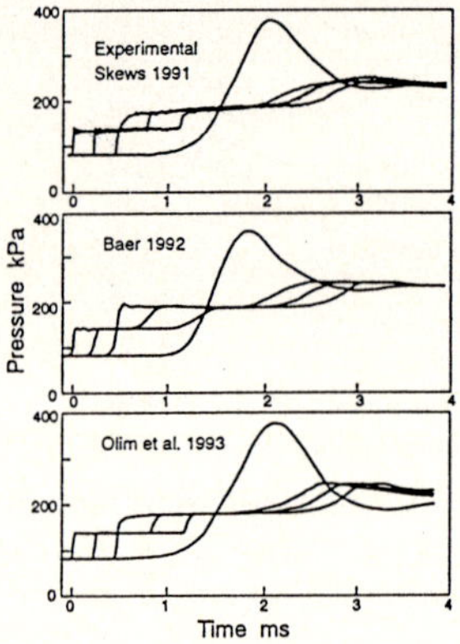

Fig. 9. Experimental and numerical results

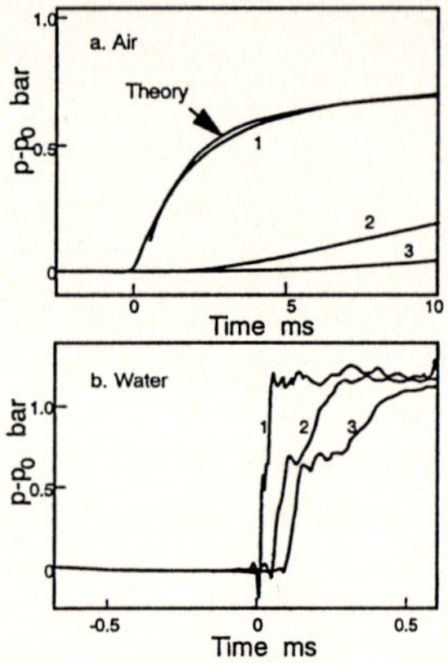

Fig. 10. Effect of interstitial fluid (Source: van der Grinten et al. 1985)

materials. The theoretical model assumes that there is no shock wave transmitted into the material and that the flow through the material is steady. Interestingly an allowance is made for a thin inlet and outlet region, a few pore diameters in extent, where the flow adjusts to the effects of area change. The flow through the bulk of the material is assumed to be according to the extended Darcy equation. Considering the fairly dramatic assumption of the neglect of transient effects, the predictions are generally good.

A potentially very interesting area of investigation is the use of shock tubes for the study of wave propagation in liquids saturating a solid skeleton. Limited experimental work is available in this field. This is somewhat surprising since it is nearly forty years since the existence of extra wave-mode solutions were discovered. Whilst the existence of the second wave had previously been shown to exist, the first study to do so using shock waves generated in a shock tube was by van der Grinten et al. (1985, 1987). The sample was placed in a vertical shock tube and both air and water were used as the interstitial fluid. With air, the compressibility of the skeleton is very small compared to that of the gas. Typical pressure records from three transducers positioned along the length of the specimen are shown in Fig.10a. The single transmitted wave is strongly damped. Agreement with theory was found to be good if the Forcheimer equation is used to describe the pressure gradient, with the non-linear term making a significant contribution. Experimentally determined steady state values of the coefficients in this equation were found to be adequate. A typical result with water is given in Fig.10b. The existence of the second wave is clearly evident, and is strongly damped compared to the first wave. The effect of added mass was found to be important, and the permeability had to be adjusted to about one third of the steady state value in order to get agreement with the theory.

3. Multi-dimensional studies

During the early seventies a number of two-dimensional studies were undertaken which have not received much attention although they are rather interesting from the point of view of the dramatic attenuation in shock wave strength that they illustrate. The first of these was by Cloutier et al. in 1971. What was needed was a method to attenuate the shock wave from a hypersonic model in a free-flight ballistic range. The attenuation material was attached to a solid table and a .303 bullet fired 100 mm above the surface. Measurements were made with Schlieren and shadowgraph photography and a pressure transducer. It was found that by correctly choosing the thickness and density of the material the reflected wave could be virtually eliminated. In 1973 Guy investigated shock wave attenuation in a sudden enlargement in a duct with an absorbent lining. Pressure measurements on the axis of the tube showed elimination of the high frequencies associated with the removal of the multiple transverse waves.

The first comprehensive theoretical analysis of oblique reflection of a shock wave from a porous slab was conducted by Clarke (1984). The interaction was assumed to comprise a reflected expansion wave terminated by a shock. The expansion wave was assumed to be necessary in order to account for the inflow into the porous material, although this does imply that the inflow velocity is greater than the vertical velocity component induced behind the incident wave. The material was theoretically modelled as an aggregate of thin tubes of different lengths, and extended previous work by the inclusion of inertial effects. The need to do this was prompted by some experimental evidence which indicated that the precursor expansion wave did not exist. These are possibly the first experimental results of wedge reflections using polyurethane foam. The analysis is restricted to very weak waves.

There are two linked effects resulting from the condition of a porous surface. Firstly it is essentially rough, and secondly there is a net inflow of gas through the surface. The effects of roughness have been identified by a number of workers for wedge interactions (see for example Reichenbach 1985, Ben-Dor et al. 1987, Adachi et al. 1992). Roughness reduces the transition angle between regular and Mach reflection as the roughness increases, and shows that the flow is not pseudo-steady. It appears that this effect is primarily due to the series of expansive and compressive wavelets arising from the roughness elements rather than boundary layer effects. Furthermore, the meticulous tests of Adachi et al., using two-dimensional square section roughness elements (gutters), have shown that a similarity parameter, viz. the ratio of the depth of the gutter to the distance the shock wave has moved up the wedge, is operating. It is also shown that the triple point trajectory angle approaches that of a smooth wall as the distance travelled by the shock increases. From these results it would then appear that pore size should be a factor in oblique reflection of shock waves from porous materials.

Significant insights into the mechanisms of oblique reflection off porous materials may be obtained from the study by Onodera and Takayama (1990) of shock reflection off an oblique wall with numerous transverse slits, and which thus not only includes the roughness effects but also a net flow of gas through the surface. Interestingly they found that the triple point trajectory was essentially straight and the process pseudo-stationary.

Tests on rigid porous wedges, with various mean pore sizes and a void fraction of about 80 per cent, were reported by Skews (1992). This study consisted of Schlieren photographs of the external flow field and pressure measurements underneath the wedge. Triple point trajectories were found to be straight lines within the accuracy of measurement. As expected, the Mach stem is considerably shorter than for a smooth wall at corresponding flow conditions, (in fact less than that for the guttered wedge tests of Adachi at similar scaled positions up the wedge), and the Mach number for transition is substantially different (Fig.11). This figure also shows the reduction in reflected wave angle which occurs under conditions of regular reflection. A typical Schlieren photograph is shown in Fig.12 which shows the complexity of the flow. It was found that at low Mach numbers and small wall angles (10°) the Mach stem is dragged backwards so that it lags behind the incident wave. As the wall angle increases so the Mach stem and the incident wave first become coplanar, and then the Mach stem curves forward in the more familiar pattern. An interesting feature of these flows is the degree to which the transmitted wave is slowed down as it passes through the wedge. Fig. 13 shows how the velocity of the disturbance on the lower wall of the shock tube is affected by wedge angle and pore size. It is significant to note that the wave front within the material may move slower than the sound speed in the gas, due presumably to the tortuous path that it has to follow.

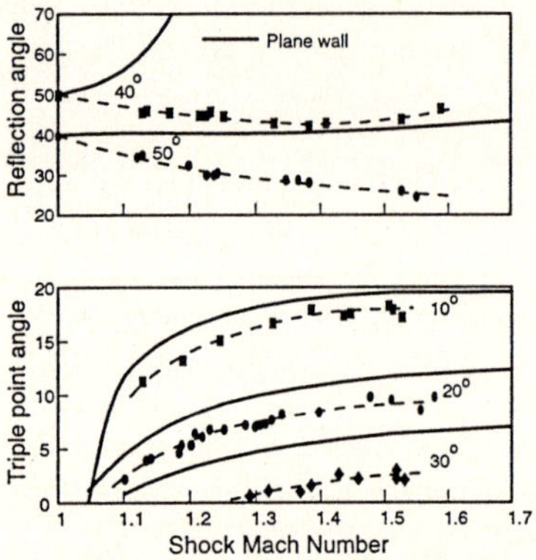

Fig. 11. Shock wave reflection characteristics

Fig. 12. Porous wedge reflection

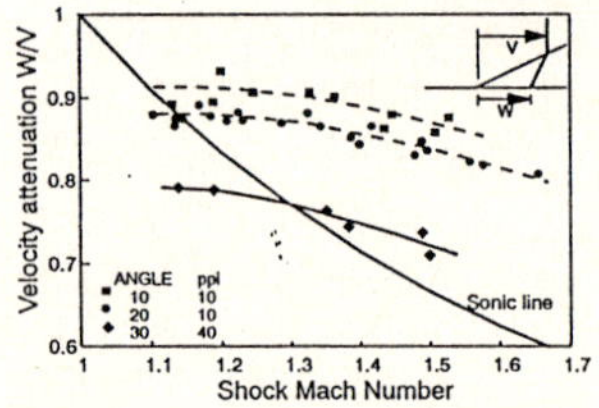

Fig. 13. Wave attenuation

Fig. 14. Flow over structured wedge

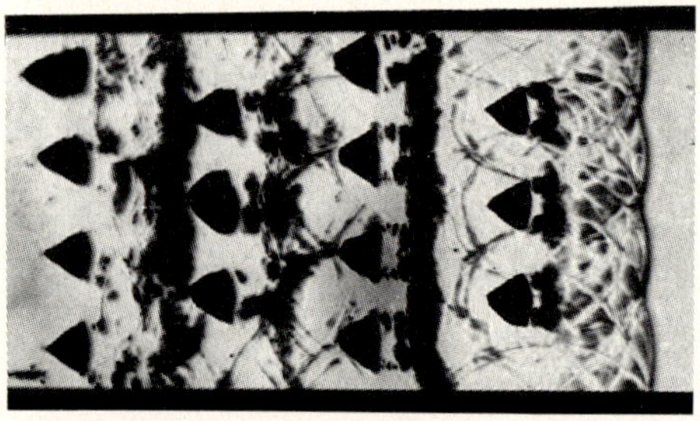

Fig. 15. Shock propagation through a matrix

In order to obtain greater insight into the flow features within such a wedge a special wedge model was constructed of square rods arranged in a staggered pattern transverse to the flow direction in a shock tube. Void fraction was 75 per cent and flow between successive layers is visible. A typical result is shown in Fig.14. The external flow field is similar in structure to that occurring with shock reflection from stepped and guttered wedges. The differences arise due to there being no compressive reflection at the bottom of the first layer and wavelets moving through to the second layer where they are again partially reflected. It is interesting to note that the leading front within the wedge is plane, and very strongly retarded.

Some insight into the complex flows within a matrix is evident in Fig.15, which shows a wave propagating through a matrix of triangular rods. Besides the expected reflection and diffraction phenomena, regions of vorticity are shed from the obstacles resulting in additional interaction sites. Examination of such flows makes it appear surprising that the numerical codes work as well as they do, particularly as they use empirical data determined from steady flow tests.

4. Status and outlook

Now that numerical schemes are starting to mature, more detailed comparisons with experiment are required. The exact nature of the functional dependency of the rates of heat and momentum transfer have still to be determined. The successful numerical schemes have shown that the extended Darcy equation adequately describes the pressure drop and that this is a primary controlling parameter. Therefore there is a need to examine in more detail the parameters that influence the coefficients of this equation. This is likely to require studies of the effects of foam morphology, and the influence of dynamic effects. It is significant to note from the work of van der Grinten et al. that whereas steady state coefficients were adequate when air was used as a working medium, they had to be reduced significantly when testing with water. This clearly raises the question of the behaviour of foams when the elasticity of the fluid and the skeleton are similar. Even for the case of polyurethane foams, this issue will be important for very weak incident waves ($M < 1.1$). The complexities of the flows shown in figures 14 and 15 make it clear that steady state results present a very poor description of the phenomena occurring.

Studies need to be extended to cover a wider range of materials. The mechanisms of collapse may differ, ranging from elastic buckling of the cell walls for elastomeric foams, through plastic yielding of the walls for elastic-plastic foams, to brittle crushing in the case of the more rigid materials such as ceramics. Furthermore the behaviour of the matrix will depend on the lateral constraints, as well as the possibility that some of these materials may have stress-strain relationships which are dependent on loading rate, or temperature.

These studies will bring additional challenges in measurement. No evidence has yet been presented of the reflection process of the compaction wave. All models to date have not accounted for wave propagation whilst the skeleton is within the densification region, and this is difficult to access experimentally. Even the measurement of the velocity of the head of the gas wave gives difficulties due to its relatively slow rise, which puts major demands on the sensitivity of a pressure transducer which still has to survive full shock reflection pressures.

Work on two-dimensional studies is only just starting and already show a number of intriguing issues which require resolution.

Acknowledgements

Appreciation is expressed to postgraduates Mark Atkins, Mike Seitz and Avi Levy (of the Ben-Gurion University of the Negev). I am also grateful to Dr Mel Baer for his willingness to supply unpublished results of his calculations.

References

Adachi T, Kobayashi S, Suzuki T. (1992) An experimental analysis of oblique shock reflection over a two-dimensional multi-guttered wedge. Fluid Dynamics Research 9: 119-132

Baer MR (1992) A numerical study of shock wave reflections on low density foam. Shock Waves 2: 121-124

Beavers GS, Matta RK (1972) Reflection of weak shock Waves from permeable materials. AIAA J. 10: 959-961

Ben-Dor G, Mazor G, Takayama K, Igra O (1987) Influence of surface roughness on the transition from regular to Mach reflection in pseudo-steady flows. J. Fluid Mech. 176: 33-356

Clarke JF (1984) The reflection of weak shock waves from absorbent surfaces. Proc. R. Soc. Lond. A396: 365-382

Cloutier M, Devereux F, Doyon P, Fitchett A, Heckman D, Moir L, Tardif L (1971) Reflections of weak shock waves from acoustic materials. Jnl. Acoust. Soc. Amer. 50: 1393-1396

Gelfand BE, Gubin SA, Kogarko SM, Popov OE (1975) Investigation of the special characteristics of propagation and reflection of pressure waves in a porous medium. Fh. Prikl. Mekh. Tekl. Fiz. 6: 74-77

Gelfand BE, Gubanov AV, Timofeev EI (1983) Interaction of shock waves in air with a porous barrier. Izv. Akad. Nauk SSSR, Mekh. Zhidk. Gaza 4: 79-84

Guy TB (1973) Attenuation of reflecting shock waves in a duct with absorbent lining. J. Sound Vib. 29: 501-503

Gvozdeva LG, Faresov YM (1985) Calculating the parameters of steady shock waves in porous compressible media. Zh. Tekh. Fiz. 55: 773-775

Gvozdeva LG, Faresov YM, Fokeev VP (1985b) Interaction of air shock waves with porous compressible materials. J. Appl. Mech. & Tech. Phys. 3: 401-405

Gvozdeva LG, Faresov YM, Brossard J, Charpentier N (1986) Normal shock wave reflection on porous compressible material. In: Bowen JR, Leyer JC, Soloukhin RI (eds.) Dynamics of explosions, Progress in Aeronautics and Astronautics 106, pp 155-165

Gvozdeva LG, Lyakhov VN, Raevskii DK, Faresov YM (1987) Shock-wave propagation in a gas and a porous medium. Fizika Goreniya i Vzryva 23: 125-129

Henderson LF, Virgona RJ, Di J, Gvozdeva LG (1989) Refraction of a normal shock wave from nitrogen into polyurethane foam. In: Kim YW (ed.) Proc. 17th Intl. Symp. on Shock Waves and Shock Tubes. American Institute of Physics Conference Proceedings 208, pp 814-818

Monti R (1970) Normal shock wave reflection on deformable solid walls. Meccanica, 5: 285-296

Olim M, van Dongen MEH, Kitamura T, Takayama K. (1993) On numerical simulation of propagation of shock waves in porous foams. Japan Symp. Shock Waves, Sendai

Onodera H, Takayama K (1990) Interaction of a plane shock wave with slitted wedges. Experiments in Fluids, 10: pp 511-516

Skews BW (1992) Shock reflection off porous wedges. 10th Mach Reflection Symposium, Denver

Skews BW (1991) The reflected pressure field in the interaction of weak shock waves with a compressible foam. Shock Waves 1: 205-211

Skews BW, Atkins MA, Seitz MW (1991) Gas dynamic and physical behaviour of compressible porous foams struck by a weak shock wave. In: Takayama K (ed.) Proc. 18th Intl. Symposium on Shock Waves, Sendai. Springer Verlag, pp 511-516

Reichenbach H (1985) Roughness and heated layer effects on shock wave propagation and reflection - experimental results. Report E24/85, Ernst Mach Institut, Freiburg

Van der Grinten JGM, Van Dongen MEH, Van der Kogel H (1985) A shock-tube technique for studying pore-pressure propagation in a dry and water-saturated porous medium. J. Appl. Phys. 58: 2937-2942

Van der Grinten JGM, Van Dongen MEH, Van der Kogel H (1987) Strain and pore pressure propagation in a water-saturated porous medium. J. Appl. Phys. 621: 4682-4687

Cellular and Tissular Effects of Shock Wave-Induced Cavitation: Potential Application to Digestive Cancers

F. Prat and A. Arefiev
INSERM U281; 151, Cours Albert Thomas, 69424 Lyon Cedex 03, France

Abstract. Cavitation is assumed to be the essential factor of toxic interactions between acoustic fields and tissues. This property is likely to offer therapeutic potentialities for some cancers. However, the natural behaviour of shock waves in tissues does not induce irreversible damage. The enhancement of cavitation, either through the saturation of the focal field with gas-containing micro-bubbles or by using a specific generator characterized by a negative half-cycle waveform, may overcome this difficulty. we present a series of works undertaken to validate this hypothesis and enhance the clinical potential on necrosis-inducing extracorporeal focused cavitation.

Key words: Cancer cells, Digestive cancer, Cavitation, Microbubbles

1. Introduction

Malignant tumors of the liver and hepatic metastases of digestive adenocarcinomas remain an important health problem throughout the world (Anon. 1983). The urgency of the problem is dictated by their high prevalence and their resistance to conventional methods of treatment such as surgery. Surgery requires resection of non-tumoral liver tissue and does not prevent recurrence. New methods, like percutaneous ethanol injection are probably useful only in very selected cases.

Subsequently, new therapeutic instruments are clearly needed to improve the long-term results of treatment of hepatic tumors. Despite disappointing results with radiotherapy and conventional hyperthermia, physical methods may represent an interesting alternative approach. Biological effects of acoustic waves, like shock waves (SW) have been studied for many years (Coleman and Saunders 1993). On our hypothesis, acoustic cavitation induced by SW can be used for local destruction of tissues (Prat et al. 1991 a). The expansion of cavitation bubble nuclei results from a decompression within the medium and their collapse may be induced by a sudden compression after expansion. It is generally assumed that when a bubble collapses near an interface with a cell membrane, it induces membrane rupture and cell death through mechanical, non-thermal effects (Flint and Suslik 1991, Morton et al. 1982). However, the therapeutic potentialities of cavitation have so far remained confined to experimental studies.

The decreased viability of cancer cells exposed to SW in suspension has been largely described (Gambihler et al. 1990, Russo et al. 1986, Smits et al. 1991). In the same conditions of exposition to SW the degree of cytotoxicity varies for different lines of tumor cells (Russo et al. 1986, Smits et al. 1991, van Dongen et al. 1989). The cellular damage was usually inferred from the presence of debris of membranes and organelles in a suspension, and in some studies from the inflation of mitochondria and distortion and inflation of plasmic membranes of still viable cells (Russo et al. 1987). Other cytotoxic effects were connected to the effects on cellular membrane (Randazzo et al. 1988). Most authors consider that the cellular damage induced by SW essentially results from cavitation (Morton et al. 1982, Zeman et al. 1990). This hypothesis is enforced by the fact that immobilisation of cells in a gel, which reduces the occurrence of cavitation, also inhibits cytotoxicity of SW (Gambihler et al. 1990).

The treatment by 1000 - 1500 SW with parameters such as those used for lithotripsy, with the help of piezoelectric and electrohydraulic lithotripters did not display a significant effect in the models of tumor implanted in rats and mice (Geldof et al. 1989, Laudone et al. 1989, Wilmer

Shock Waves @ Marseille III
Editors: R. Brun, L. Z. Dumitrescu

et al. 1989). In these studies different lithotripters were tested, e.g. Dornier HM3 (1000 shocks at 18 kW) or Piezolith 2300 Wolf (1500 shocks with the positive pressure 80 MPa).

Meanwhile, it was shown that different types of tumor cells exhibit differences in sensitivity. It was observed that fast- growing tumors (Dunning R3327 PAT 2) had their growth rate unchanged after treatment by SW *in vivo*, and that tumors with a slower growth exhibited a significant delay in their development (Oosterhof et al. 1990). Development was also delayed when suspensions of tumor cells were treated by SW before injection (Randazzo et al. 1988, Russo et al. 1986). The regression of tumors was observed after exposition of subcutaneous tumors to SW by placing the focal point very close to the surface of water, so reproducing the "air-water" effect observed *in vitro* (Weiss et al. 1990). The increase of SW doses administrated at the same point of the tumor did not seem to result in a significant reduction of tumor development. In contrast, application of SW to the different points of a tumor, and, especially to its periphery, decreased tumor growth rates (Weiss et al. 1990). Repetition of treatment sessions has equally improved the results from the point of view of local destruction (Hoshi et al. 1990). Finally, it was shown, that the initial tumor volume strongly influenced the effect of treatment (Oosterhof et al. 1990).

The potentiation of cytotoxic drugs by SW is very difficult to evaluate, because SW themselves are cytotoxic *in vitro*. Most of the studies concerning the association of chemotherapy and SW evaluated the cellular destruction by SW before administration of a chemotherapy, so obtaining a control group of cells affected only by SW (Durrant and Morris 1990, Holmes et al. 1990). After having treated by typically 2000 shocks, the time period separating these two methods varied from one hour to approximately 24 hours. Cellular growth, attachment to plates and viability were more decreased when SW were administered before treatment by vinblastine (8), cisplatyl or doxyrubicine (Berens et al. 1989, Wilmer et al. 1989). However most of these studies do not prove potentiation but a simple additive effect of SW to chemotherapy. It was shown in the same time, that tumor growth after injection of tumor cells NU1 (human kidney cancer) to the nude mice was reduced for a longer period if these cells had been treated with SW prior to adriamicin (Oosterhof et al. 1990).

When the following investigations were started by us in 1989 only part of the literature cited above was available. Concerning the biological effects of SW it was clear that:

1. SW possess a potential cytotoxicity for cells in suspension;
2. On the other hand, cytotoxic effects in vivo were insufficient and did not appear to find a clinical application in the field of destruction of tumors, particularly malignant ones;
3. Cytotoxicity was explained mostly, if not exclusively, by transitory acoustic cavitation produced by the SW.

Thus, it was suggested that focused extracorporeal SW present a potential interest for the local non-invasive treatment of certain types of tumors. It appeared necessary to stimulate the production of cavitation and to improve it, in order to achieve clinical potential. In order to test and to validate this procedure, different biological models could be proposed: transformed (cancer) cells in a continuous culture media, healthy and tumoral tissues *in vitro*, healthy and tumoral tissues treated extracorporeally and with exteriorization of the target organ *in vivo*. The works presented here were performed with the following objectives:

1. Investigation of two options designed to increase the occurrence of cavitation in the focal zone of a generator, i.e. a)to inject microbubbles of a gas into the focal area during exposition to SW induced by commercial spark-gap generator, and b)to increase the index of cavitation by creating a shock wave, whose waveform is characterised by a negative part of high amplitude (Fig.1). Such a shock wave was obtained by the mechanical arrangement of the different piezoelectric elements of the generator.
2. Investigation of the effect of cavitation on different models, evaluation of the feasibility of these methods, their cytotoxicity for cells, normal tissues and tumors and possible potentiation of

cytotoxic drugs. Considering that one of the principal future application of this procedure will be digestive oncology, most attention was paid to the treatment of the primary and secondary malignant liver tumors. We have used models of healthy liver and models of animal and human colon cancer.

2. Hepatic lesions in rabbits induced by acoustic cavitation (Prat et al. 1991 a)

The study population consisted of three groups of eight rabbits in each group. First group was treated only with SW. Second - with intra-arterial injection of a microbubble suspension produced by gelatine-air mixing. The last group was treated both with SW and microbubbles. The SW were produced by a commercial lithotripter Sonolith 3000 (Technomed International, Bron-France). Under general anaesthesia 500 shocks under a pressure of 500 bars at the focal point were administrated to the right liver lobe. The duration of treatment was 40 minutes.

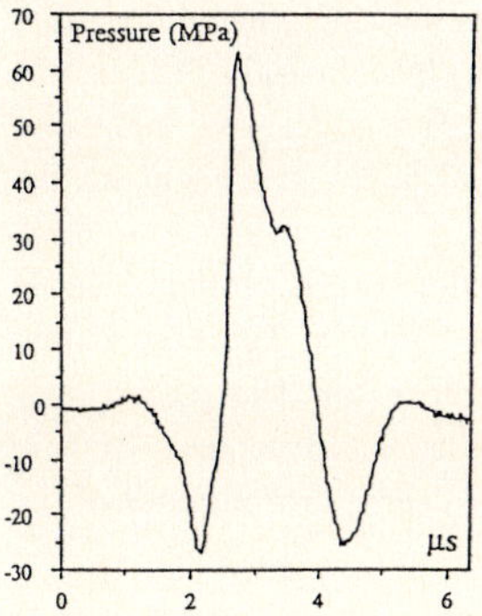

Fig. 1. Pressure tracing of the generator, with the initial half-cycle of depression (expansion wave) followed by a half-cycle of peak positive pressure (shockwave)

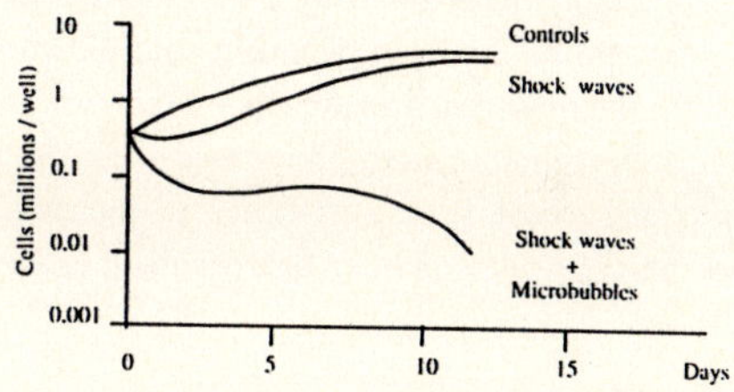

Fig. 2. Growth of HT-29 cells exposed to 1000 shock waves or 1000 shock waves with gas microbubbles and controls. Wells were studied with 0,3 million cells; cells were counted on Days 2,6,12

In the first group, two of eight animals had macroscopic signs of contusion of liver tissue. Microscopy showed that 5 of them had subcapsular and parenchymal micro-hematomas of at least 1 mm in diameter. In the second group nothing but a moderate diffuse congestion of the liver was observed in 3 rabbits. In the third group all 8 animals presented large hemorrhagic lesions (from 2 to 30 mm in diameter) localised exclusively or predominantly in the right lobe (50% of cases,) and spread in the depth of parenchyma in all cases. On light microscopy we observed intrahepatic periportal hematomas together with large lacunae of round shape, sometimes coalescing. Other digestive organs (stomach, duodenum, jejunum and colon) and gallbladder presented microscopically with severe hemorrhagic lesions of their anterior wall.

3. Cytotoxic effects of acoustic cavitation on HT-29 and Hep G_2 cells and a rat peritoneal carcinomatosis in vivo (Prat et al. 1991 b)

Treatment of the cell suspensions of a colon adenocarcinoma HT-29 and a hepatocellular carcinoma Hep-G_2 with the SW generated by the same generator as in the previous work, was performed separately and in the presence of gas bubbles. The microbubbles were infused into the test tube as an emulsion of a modified gelatin and air. Both cell lines were treated by 50 to 250 SW and immediate viability was investigated by the method of trypan blue exclusion. Cell survival in culture was evaluated by cell counts at the 2nd, 6th and 12th days of culture.

The viability of HT-29 cells was not affected by the presence of gas bubbles in suspension. The SW-treated samples had a significantly diminished viability. In comparison with the control group cellular growth was reduced and completely stopped during 48 hours in the group treated by SW in presence of gas bubbles. An additional group was treated by 1000 SW: in the presence of gas bubbles no recovery of growth was observed (Fig.2). On electron microscopy the cells treated in the presence of microbubbles presented ruptures of membranes, cytoplasmic vacuoles (Fig.3), and condensation of the chromatin such as observed in apoptosis. For the cells Hep-G_2 the effect on the immediate viability was more pronounced than for HT-29, and addition of microbubbles had also a significant effect in all cases.

4. *In vitro* effects of acoustic cavitation on the proliferation of peritoneal metastasis of colon cancer in rat (Prat et al. 1991 b)

To investigate the toxicity of SW for tumoral tissues, peritoneal carcinomatosis was induced by intraperitoneal injection of DHD K12ProB cells (syngenic adenocarcinoma of the colon). During the period between the 14th and 21st days after induction, metastases were sampled and treated by SW in a test tube, containing either a standard culture medium, or physiological solution with gas microbubbles. After treatment by 50, 100, 250, 500 or 1000 SW tumor nodes were incubated in the presence of 15mCi of ^{3}H thymidine. Free proteins were counted in the treatment media. Morphology was studied by light microscopy.

Thymidine incorporation in tumors reversally correlated with the dose of SW. It significantly further decreased in the presence of microbubbles and was absent after 1000 SW. Total concentration of proteins in media increased with the number of SW. This augmentation was significantly more important in the presence of microbubbles, resulting in larger zones of necrosis than in the control group, deeper hemorrhagic erosions, cellular pycnosis and decrease of number of the mitoses.

5. Increased chemotoxicity to colon cancer cells by shock-wave induced cavitation (Prat et al. 1993 b)

In order to determine if cavitation may potentate the toxicity of chemoterapeutic drugs HT-29 cells were exposed in suspension to SW+microbubbles and subsequently incubated with 5-Fluorocil (5FU). Exposing tumor cells to cavitation prior to 5FU reduced cellular growth in double, in certain conditions (in comparison with treatment with 5FU only). It was found that cavitation increased penetration of 5FU into the cells, and inhibited mitosis. Thus, submitting HT-29 cells to cavitation before treatment by 5FU increased significantly the effect of chemotherapy. The action of these two agents appeared to be synergetic, as they possess specific influence on the cell cycle.

6. *In vivo* effects of cavitation alone and in combination with chemotherapy (Prat et al. 1993 a)

The purpose of this work was to investigate the possibility to reproduce results of the previous study in vivo. The model of peritoneal carcinomatosis was the same as in the study of cavitation *in vitro*. The first group (macroscopic carcinomatosis) of rats was treated at the 21 -st day after injection of tumor cells. The metastases were treated with 250, 500 or 750 SW. The emulsion of gas microbubbles was injected into the peritoneal cavity during administration of SW. Before sacrifice animals received ^{3}H-thymidine. The animals of the second group were treated at the 4th and the 6th days (the stage of micrometastasis) with microbubbles in combination with 250 SW or with 5FU in the dose of 5 mg/kg/day or Cis Platinum (0,5-3 mg/kg at the day 4) or with

5FU together with cavitation and microbubbles. At the 30th day after implantation, a part of the animals was sacrificed. Survival was investigated in the animals left.

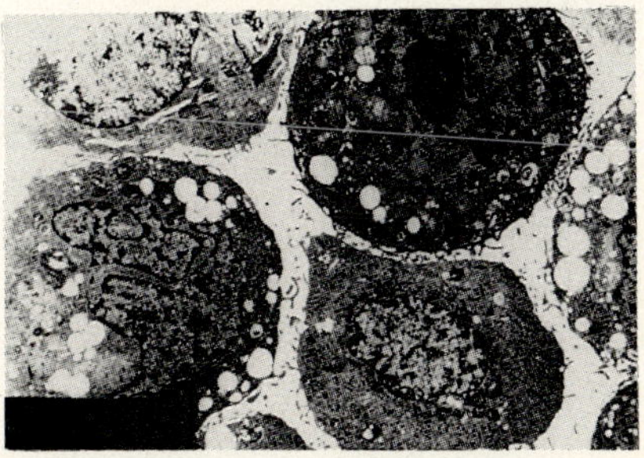

Fig. 3. Electron micrographs of HT-29 cells in suspension exposed to 250 shock waves with gas microbubbles. (Original magnification x 8000). Chromatin is distributed in compact lumps along perinuclear cisternae, the nuclear shape is irregular and numerous coalescent vacuoles are present on cytoplasma. The plasma membranes are irregular in shape with localized ruptures

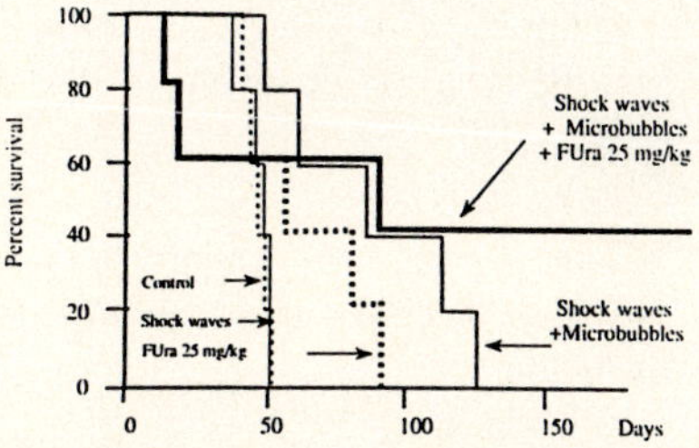

Fig. 4. Effect of shock waves on peritoneal carcinomatosis in rat. Survival of animals depending on the early-stage therapeutic protocols. —— controls; - - - FUra 5mg/kg day 4-8; ━━ FUra 25mg/kg day 4-8; · · · 250 SW+bubbles; — — — 250 SW+bubbles+FUra 5mg/kg day 4-8; — · — · — 250 SW+bubbles+FUra 25mg/kg day 4-8

The treatment of tumors by SW+microbubbles resulted in a dose-dependent decrease of the isotope incorporation. For 750 SW no incorporation was observed, so that the tumor could be considered completely destroyed. At day 30 100% of animals in the control group had a well-developed tumor. After treatment with 5FU or Cis Platinum only in some cases absence of ascitis and a more favourable clinical condition were observed. When one of the methods of chemotherapy was combined with SW+microbubbles, no macroscopic signs of peritoneal carcinomatosis was observed at the 30th day. The 3 month survival was 40% in the cavitation+chemotherapy group and 0% in the control group (Fig.4).

7. Extracorporeal focused liver ablation by cavitation in the rabbit (Prat et al. 1994)

Using the model of a healthy rabbit in vivo, we tested the possibility of liver tissue destruction with a prototype of a cavitation generator. Several groups of New Zealand rabbits were treated with the aid of the new generator consisting of 19 elementary piezoelectric transducers placed in a specific manner to produce a SW followed by a 1/2 cycle of negative pressure (Fig.1). A target was chosen by ultrasonography, and its volume was defined. Among the parameters studied were the influence of repetition frequency, the size of an elementary lesion, the depth of the target volume, the toxicity for the neighbouring organs, as well as biological side-effects and healing of the lesions.

The procedure succeeded in 100% of the animals. A lesion of a predetermined volume in the form of parallelepiped was obtained at depths between 20 and 30 mm from the abdominal wall (Fig.5). Elementary lesions occupied from 45 ± 15% to 92 ± 23% of this volume. the rate of tumor destruction was achieved from 52 ± 8% to 98 ± 17% depending upon parameters used. In 100% tissues surrounding the lesion were not found damaged both on macro- and microscopical examination. Elementary lesions consisted of foci of hemorrhagic lysis surrounded by necrosed hepatic cells. During the 24 hours period after the treatment insignificant variations of AST, ALT, hemoglobin and hematocrite were observed. These parameters normalised to the 21st day

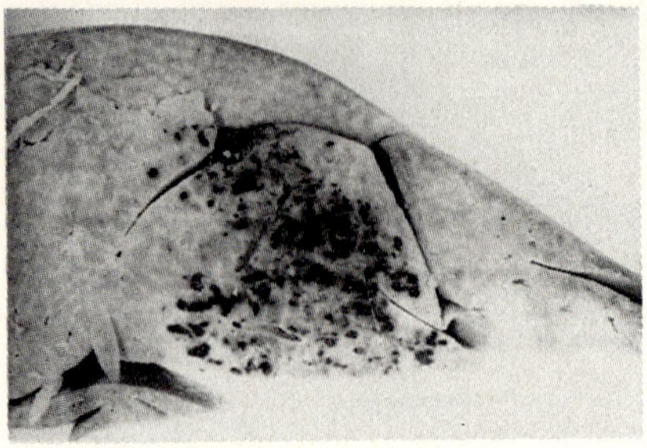

Fig. 5. Macroscopic view of a liver slice from the target zone (fixed in acetified formalin); the arrow indicates the SW direction and shows that the superficial parenchyma is not affected

after treatment and the zone of treatment was replaced by a scar having the volume of $15 \pm 12\%$ to initial lesion.

8. Discussion

In the first work we demonstrated the feasibility of destruction of a healthy hepatic tissue by extracorporeal SW, provoking cavitation by injection of microbubbles in the form of an emulsion and using a commercially available electrohydraulic generator. We established that a significant destruction cannot be achieved only by SW, but that pronounced effects were obtained when SW were combined with microbubbles. However these lesions were not selective, not localised, and mostly centered around portal spaces. This last observation indicates that the injected bubbles were too big to pass through capillaries and they were trapped in periportal spaces where cavitation occurred.

The consecutive investigations allowed us to demonstrate the effectiveness of this method using the concept of potentiation of cavitation induced by a commercial spark-gap lithotripter in combination with injection of gas-containing microbubbles. Two different lines of tumor cells (HT-29 and $HepG_2$) were tested for cytotoxicity of cavitation and its influence on cellular growth *in vitro*. We found that the $HepG_2$ cells were even more sensitive to cavitation *in vitro* than HT-29. We proved experimentally that this method allowed to destroy peritoneal metastases suspended *in vitro*. On the next step we proved the existence of synergism between cavitation and an antimetabolic inhibitor of synthesis of nucleic acids. It was found that in certain conditions cavitation leads to inhibition of mitosis and increases the number of cells in the S-phase of cell cycle, while it is known that 5FU exerts its specific activity in the same phase of cellular cycle.

In the next study a possibility of application of the same methodology was demonstrated *in vivo* for an animal cancer. We used the model of cavitation (SW+microbubbles) for extracorporeal treatment of peritoneal carcinomatosis of colonic origin in rats. The microbubbles were infused in the peritoneal cavity of an animal during treatment. Cavitation potentiated chemotherapy with 5FU or Cisplatinum. This allowed to use smaller doses of these drugs and, to obtain good long-term results and prolonged survival to the same extent as obtained with very high doses of these drugs. However, combination of high doses of drugs and cavitation induced very serious toxic effects.

At this stage of our work it seemed to be desirable to develop a procedure of cavitation, that would permit selective, focused, extracorporeal application of cavitation for models of cancer in animals bigger than rat. Using smaller microbubbles (less than 10 μm in diameter) that would be able to penetrate capillaries, and be distributed homogeneously in the liver parenchyma after injection into a peripheral vein was an option. Unfortunately at the present state of research this type of product is not sufficiently advanced. However it was possible to improve the index of cavitation of a piezoelectric generator by obtaining a SW including a deep negative half-cycle, following a positive wave with a pressure of about 50 mPa. In our last study we displayed the influence of different parameters of shot upon the quality of a lesion.

9. Conclusion

The work described above proves a potential interest of cavitation as a new method of extracorporeal surgery of liver tumors. It is proposed to apply the same method of focal tissue ablation induced by cavitation on different models of solitary and disseminated tumors. It will be necessary to prove its effectiveness not only for small metastasis, but also for tumor nodes bigger than 30 - 40 cm. in diameter. It is essential to examine the feasibility of this method for zones of liver deeper than 6 cm. This will pose a problem of evaluation of tissue attenuation and interference of costal ribs to the SW, that will be possibly solved by using pigs as an animal model because the size of a pig's liver is close to human's. Many things remain to be done to understand the role of physical interactions on long-term effects of the SW treatment of tumors.

References

Anon. (1983) World Health Organisation. Prevention of liver cancer: Report of WHO meeting In: World Health Organization Technical Report Series. Geneva: World Helth Organization 7-8

Berens ME, Welander CE, Griffin AS, McCullough DL (1989) Effect of acoustic shock waves on clonogenic growth and drug sensitivity of human tumour cells in vitro. J. Urol. 142: 1090-1094

Coleman AJ, Saunders JE (1993) A rewiew of the physical properties and biologcal effects of the high amplitude acoustic fields used in extracorporeal lithotripsy. Ultrasonics 31:75-89

Durrant LG, Morris DL (1990) Shock waves increase the sensitivity of human gastric and colorectal cancer cells to cytotoxic chemotherapy. Gut 31: A526 (Abtsr.)

Flint EB, Suslick KS (1991) The temperature of cavitation. Science 253: 1397-1399

Gambihler S, Delius M, Brendel W (1990) Biological effects of shock waves: cell disruption, viability, and proliferation of L1210 cells exposed to shock waves in vitro. Ultrasound Med. Biol. 16: 587-594

Geldof AA, De Voogt HJ, Rao BR (1989) High energy shock waves do not affect either primary tumor growth or metastasis of prostate carcinoma R3327 MatLyLu. Urol. Res 17: 9-12

Holmes RP, Yeaman Ll, Li W, Hart LJ, Wallen CA, Woodruff RD, McCullough DL (1990) The combined effects of shock waves and cisplatin therapy on rat prostate tumours. J. Urology 144: 159-163

Hoshi S, Orikasa S, Kuwahara MA, Yoshikawa K, Ohyama C, Satoh M, I Kawamura S, Nose M (1990) The effect of high energy shock waves on implanted urinary bladder cancer in rabbits. Jpn. J. Cancer Res. 81: 317-319

Laudone VP, Morgan TR, Huryk RF, Heston WD, Fair WR (1989) Cytotoxicity of high energy shock waves: methodologic considerations. J. Urology 141: 965-8

Morton Kl, Ter Haar GR, Stratford IJ, Hill CR (1982) The role of cavitation in the interaction of ultrasound with V79 chinese hamster cells in vitro. Br. J. Cancer 45: 147-150

Oosterhof GON, Smits GAHJ, De Ruyter AE, Schalken JA, De Bruyne FMJ (1991) Effects of high energy shock waves combined with biological response modifiers in different human kidney cancer xenografts. Ultrasound Med. Biol. 17: 391-399

Oosterhof GON, Smits GAHJ, De Ruyter AE, Schalken JA, De Bruyne FMJ (1990) In vivo effects of high energy shock waves on urological tumours. An evaluation of treatment modalities. J. Urology 144: 785-789

Prat F, Ponchon T, Berger F, Chapelon JY, Cathignol D (1991) Hepatic lesions in the rabbit induced by acoustic cavitation. Gastroenterology 100: 1345-1350

Prat F, Chapelon JY, Chauffert B, Ponchon T, Cathignol D (1991) Cytotoxic effects of acoustic cavitation on HT-29 cells and a rat colon carcinomatosis in vitro. Cancer Res. 51: 3024-3029

Prat F, Chapelon JY, Abou El Fadil F, Theilliere Y, Ponchon T, Cathignol D (1993 a) In vivo effects of cavitation alone or in combination with chemotherapy in a metastatic colon cancer in the rat. Br. J. Cancer 68:13-17.

Prat F, Sibille A, Luccioni C, Pansu D, Chapelon JY, Ponchon T, Cathignol D (1993 b) Increased chemotoxicity to colon cancer cells by shock-wave induced cavitation. Accepted to Gastroenterology

Prat F, Chapelon JY, Abou El Fadil F, Sibille A, Theilliere Y, Ponchon T, Cathignol D (1994) Focused liver ablation by cavitation in the rabbit: a potential new method of extra-corporeal therapy. Gut 35: 395-400

Randazzo RF, Chaussy CG, Fuchs GJ, Bhuta SM, Lovrekovitch H, De Kernion JB (1988) The in vitro and in vivo effects of extracorporeal shock waves on malignant cells. Urol. Res 16: 419-426

Russo P, Stephenson RA, Mies C, Huryk R, Heston WDW, Melamed MR, Fair WR (1986) High energy shock waves suppress tumor growth in vitro and in vivo. J. Urology 135: 626-628

Russo P, Myies C, Huryk R, Heston WDW, Fair WR (1987) Histopathologic and ultrastructural correlates of tumour growth suppression by high energy shock waves. J. Endourology 137: 338-341

Smits GHJ, Oosterhof GON, De Ruyter AE, Schalken JA, Debruyne FMJ (1991) Cytotoxic effects of high energy shock waves in different in vitro models: influence of the experimental set-up. J. Urology 145:171-175

van Dongen JW, van Steenbrugge GJ, Romitn JC, Schroder FH (1989) The cytocidal effect of high energy shock waves on human prostate tumour cell lines. Eur. J. Cancer Clin. Oncol. 25: 1173-1179

Weiss N, Delius M, Gambihler S, Dirschedl P, Goetz A, Brendel W (1990) Influence of the shock wave application mode on the growth of A-Mel 3 and SSK2 tumours in vivo. Ultrasound Med. Biol. 16: 595-605

Wilmer A, Gambihler S, Delius M, Brendel W (1989) In vitro cytotoxic activity of lithotripter shock waves combined with adriamycin or with cisplatin on L1210 mouse leukemia cells. J. Cancer Res. Clin; Oncol. 115: 229-234

Zeman RK, Davros WJ, Garra BS, Goldberg JA, Hayes WS, Catteu EL, Horii SC, Cooper CJ, Silverman PM (1990) Cavitation effects during lithotripsy. Part II. Clinical observations. Radiology 177:163-166

Part 1: Two-Phase Flows

Entrainment of a Spherical Particle by the Flow Behind a Shock Wave

M. Olim*, **H. Nagoya***, **K. Takayama*** **and F. Hiatt**†
*Shock Wave Research Center, Institute of Fluid Science, Tohoku University, Sendai, Japan
†The FSI International, Chaska Minnesota, USA

Abstract. The present study focuses on the interactions between shock waves and a solid particle which is at rest on a smooth solid surface. Shock wave loading on the solid particle is simulated numerically assuming a two-dimensional inviscid flow. For the simulation the flowfield is discretised using a boundary-fitting coordinate system in which the particle constitutes an impermeable boundary. The interaction of waves and vortices with the solid particle and the smooth surface generates unsteady drag and lift forces on the particle. The scaling of these forces with the radius of the particle is investigated using particles whose radius is 1, 10, and 100 μm. The calculated force components are also used to evaluate the unsteady lift and drag coefficients.

Key words: Two-phase flow, Dust particle entrainment

1. Introduction

In what has become a classical experiment, E. Mach (1878) used a shock wave undergoing a Mach reflection to remove carbon particles from a sooted glass plate. This was probably the first successful attempt at cleaning surfaces using shock waves. Various techniques of surface cleaning have been developed since then, and some recent reviews are given by Skydmore (1985) and Tam et al. (1992). The present study is focused on the effects of a shock wave on a single particle. The study is carried out in order to assess the feasibility of the use of shock waves for removal of dust from surfaces. Dust removal is often performed with the surface submerged in liquid, and the shock waves are usually generated by cavitation bubbles induced by an ultrasound source. The effectiveness of this process is greatly reduced when the particle diameter is less than 10 μm. For particles of this size the adhesive force, which consists of van der Waals, electrostatic, and capillary forces, may be six orders of magnitude larger than the gravity force (see Bowling 1985). Burgraff (1981) points out that in some studies the power of the sound source has been increased to a level which damaged the surface while having little effect upon micron-size particles.

However, shock waves are not usually generated with a single particle in mind, and it is conceivable that a shock wave generated at a specified distance and angle from a given dust particle is capable of removing that particle with no damage to the surface. The distance and angle will determine the reflection pattern of the shock waves when they reach the particle. The purpose of this study is to investigate the effects of various shock reflection patterns upon the particle.

It is postulated that two mechanisms contribute to the forces acting upon the particle as the shock wave interacts with it. The first mechanism is wave reflection between the particle and the solid surface on which the particle is located. The second mechanism is vortex activity behind the curved reflected shocks. The first mechanism exists if either a planar or a curved shock wave impinge upon the particle, while the second is possible only in the case of a curved shock wave.

Shock Waves @ Marseille III
Editors: R. Brun, L. Z. Dumitrescu

2. Theoretical background

Impingement of a shock wave upon a particle which is at rest on a smooth solid surface results in shock waves reflected and diffracted by that particle. As the shock reflected from the particle reaches the surface it reflects from it and interacts again with the particle. In addition to that, when the incident shock wave undergoes a regular or a Mach reflection the reflected shocks may be curved thus generating a vortex at the surface. Both the waves and the vortex are capable of exerting a force on the particle. The interaction between various waves, vortices, and solid surfaces generates a nonuniform flowfield around the particle. To quantify these effects the flowfield is simulated numerically. A preliminary experimental study is briefly presented.

3. Experimental study

An experiment was conducted in which shock waves were used to remove a particle from a solid surface. The experimental set-up consists of a glass laser and a lens focusing the laser beam. This set-up is capable of generating shock waves of intermediate strength. The pressures measured at a distance of several centimeters from the focal point of the lens correspond to shock Mach numbers in the range of 1.2 to 1.8. Fig.1 shows the pressure jumps across the shock measured at various distances from the focal points of two lenses with focal lengths, f, of 40 mm and 100 mm. As expected, the sharper focusing of the $f = 40$ mm lens generates stronger shocks.

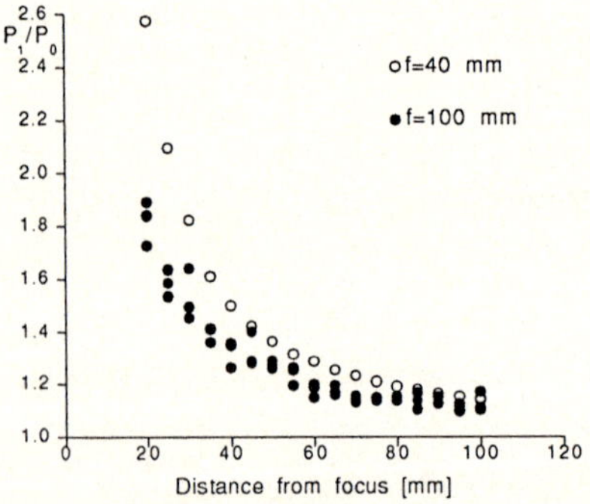

Fig. 1. Pressure jumps *vs.* distance from focus for lenses of $f = 40$ mm and $f = 100$ mm

4. Numerical solution

In the simulation, the heat conductivity and the viscosity of the gas are neglected, and the flow field is assumed to be two-dimensional. The interaction of a shock wave with a three-dimensional sphere will be studied in the near future. The conservation of mass, momentum, and energy in the field is described by the Euler equations. These equations, including the transformation thereof to a curvilinear coordinate system, are described in various texts and are not repeated herein. The ideal gas law is used to complement the system of conservation equations.

The simulation is performed using a second order TVD code described by Sasoh et al. (1992). The flowfield is discretised using a boundary-fitting coordinate system in which the surface of the particle constitutes an impermeable boundary.

5. Results

In all cases studies herein the flow field was initiated with the incident shock inclined at some angle θ to the solid surface and at some distance x_r from the particle. This situation is shown in Fig.2.

Depending on the incident shock Mach number, M_s, and the inclination angle, θ, the shock reaching the particle may be a planar wave or it may undergo a regular or a Mach reflection. At any given time step the pressure distribution is integrated around the particle to evaluate the components of the force vector acting upon the particle.

5.1. Impinging planar shock wave

A shock wave impinging upon the particle reflects and, later on, diffracts around the particle. A sequence of isopycnic plots obtained for $M_s = 1.5$ in air and $R = 10$ μm, where R is the particle radius, is shown in Figs.3a to 3d. At the early stages of the interaction between the shock wave and the particle (shown in Fig.3a) the flow around the particle is symmetric with respect to the direction of propagation of the incident shock wave. As the lower portion of the curved shock reflected off the particle reaches the solid surface, it moves upwards (Fig.3b). Since it is a shock, no waves from the surface reach the particle and therefore the flowfield in the immediate vicinity of the particle maintains its symmetry. In Fig.3c the shock wave reflected from the solid surface reaches the particle and reflects from it. This secondary shock wave, which had reflected from the surface and later on from the particle, travels towards and reflects from the surface, and qualitatively repeats the pattern of the first reflected shock. Fig.3d shows the flowfield after the secondary reflected shock wave had reached the particle surface.

These reflected shocks cause an increase of pressure at the lower left portion of the particle surface (assuming that the incident shock wave propagates from left to right) thus generating lift and drag forces. The lift forces acting upon particles of radius R=1, 10, and 100 μm are shown (normalised with respect to the gravitational force) in Fig.4a. The ordinate in this figure represents the distance traveled by the incident shock normalised with respect to the diameter of the particle, x/D. The figure indicates that the lift force exhibits sharp increases in two points. These instants correspond, respectively, to the arrival of the first and second shocks reflected from the solid surface at the lower surface of the particle. It is clear that the sudden increases in the lift force are due to the reflection of these shocks from the lower portion of the particle.

Table 1. $R \times F_{l_{max}}/F_g$ [μm] for a planar shock impinging on a particle of radius R.
$F_{l_{max}}$ is the largest lift force

R[μm]	M_S=1.2	M_s=1.5	M_s=1.8
1	7.26 E6	2.66 E7	6.03 E7
10	7.22 E6	2.65 E7	6.02 E7
100	7.23 E6	2.65 E7	6.02 E7

5.2. Impinging RR and MR shocks

To study the effects upon the particle of the flowfields behind shocks undergoing regular and Mach reflections the simulation was initiated with an oblique shock inclined to the solid surface at a distance $x_r = 2.5\ R$ ahead of the solid particle. The particles are similar to those used for the simulation of the flowfield generated by a planar impinging shock wave. The Mach number of the oblique incident shock is 1.5, and the reflection type depends upon the inclination angle. For example, a regular reflection takes place when the angle between the shock and the solid surface is 40°, and a Mach reflection occurs when the angle between the shock and the surface is 70°.

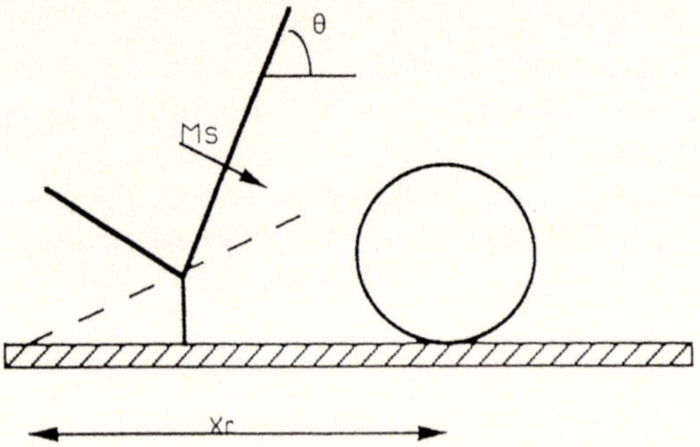

Fig. 2. Initial conditions

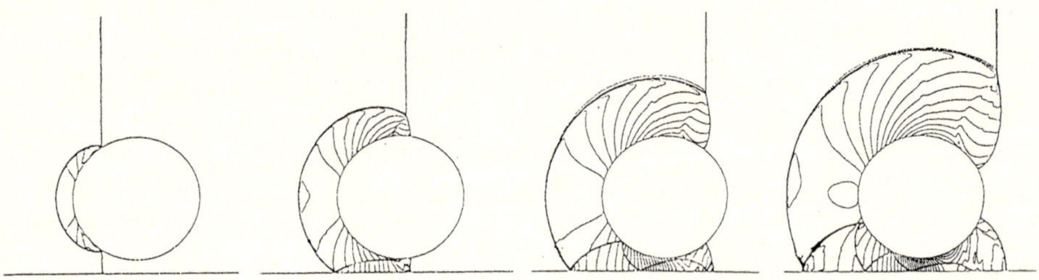

Fig. 3. Isopycnics generated by a planar impinging shock. $M_s = 1.5$ in air, $R = 10\ \mu m$.

The lift forces acting upon particles of radius $R = 1,\ 10$, and $100\ \mu m$ for these cases are shown in Figs.4b (for MR) and 4c (for RR).

Table 2. $R \times F_{l_{max}}/F_g$ for planar, MR, and RR shocks impinging on a particle of radius R.
The incident shock Mach number is 1.5

R [μm]	Planar	MR	RR
1	2.66 E7	3.35 E7	4.72 E7
10	2.65 E7	3.38 E7	4.72 E7
100	2.65 E7	3.38 E7	4.73 E7

In all cases the value of F_l/F_g scales as R, where F_l and F_g are the lift and gravity forces, respectively. This scaling is expected because the lift force is a function of the sphere surface, while the gravity force is a function of the sphere volume. The largest values of F_l/F_g, multiplied by R, are listed in Tables 1 and 2. In Table 1 the values were obtained from the simulation of interaction of a planar shock wave with various incident Mach numbers and particles of various diameters. The values in the table indicate that the R scaling is accurate to within 1%. In Table 2 the largest values of $R \times F_l/F_g$ for various particle radii and reflection configurations are presented for $M_s = 1.5$. Again, the R scaling is obvious, and it is also clear that the stronger lift force is generated by the RR configuration.

The distance between the particle and that point on the solid surface at which the reflection starts, x_r, affects the size of the reflection region as it reaches the particle, thus affecting the lift and drag forces. The flowfield was simulated with $M_s = 1.5$ and $R = 10\ \mu m$ for a range of inclination angles and $x_r/R = 2.5$ and 3.75. In both cases the lift force increases monotonically as the angle between the surface and the incident shock decreases, and it seems that the lift force approaches asymptotically the value (F_d) obtained by a planar shock descending perpendicularly to the solid surface, i.e. $\theta = 0$. These forces, normalised with respect to F_d, are shown in Fig.5.

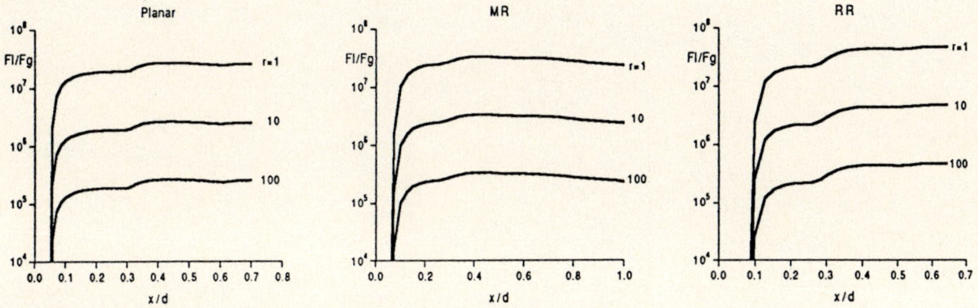

Fig. 4. Lift force exerted on the particles by the planar, MR, and RR impinging shock waves.

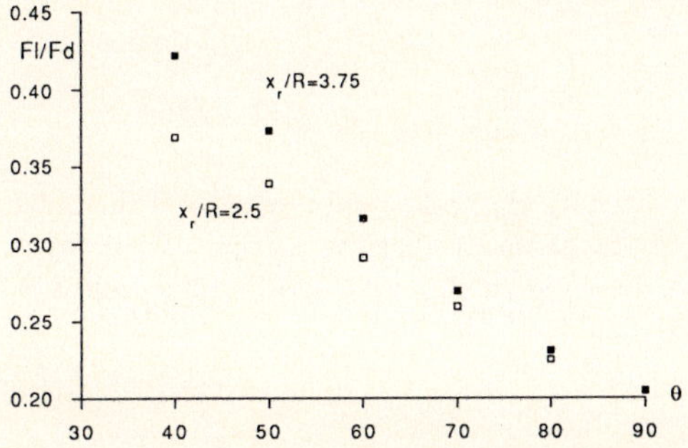

Fig. 5. Lift force for various inclination angles.

Clearly, for given initial conditions the pressure behind a shock reflecting from a solid surface is maximised when the reflection is head-on. The results obtained in the present study agree with that in the sense that they indicate that the lift force exerted upon the particle is maximised when the incident shock descends perpendicularly to the solid surface. The lift force, scaled with R, obtained in this case is given as a function of the incident Mach number in Table 3.

The lift and drag forces on the different particles were used to obtain the unsteady lift and drag coefficients, C_l and C_d. Their histories for the planar wave and the MR and RR reflections studied herein are shown in Fig.6. The plots show the coefficients for a given incident shock wave and all three radii, i.e. $R = 1,\ 10,\ 100\ \mu$m.

The figure indicates that the values of C_l and C_d are independent of the particle radius since the plots are indistinguishable (the values vary in the fourth significant digit). The figure also indicates that the higher lift is generated by the RR configuration.

Table 3. $R \times F_{l_{max}}/F_g$ for a planar shock reflecting head-on from the surface

M_s	1.1	1.2	1.3	1.4	1.5	1.6	1.7	1.8
$R \times F_{l_{max}}/F_g \times 10^{-7}$	1.49	3.65	6.38	9.56	13.01	16.58	20.25	24.01

The velocity of the particle after it is detached from the surface may be roughly estimated assuming that the impulse on the particle is approximately equal to particle momentum, i.e. $F\Delta t \approx mV_p$ where F, Δt, m, and V_p are, respectively, the force acting upon the particle, the time interval during which this force acts, the mass of the particle, and its velocity. The force is approximately equal to the overpressure, ΔP, multiplied by the normal area, πR^2, and the time interval is given by the particle diameter divided by the shock wave velocity, $2R/M_s a_0$. Thus one may write that

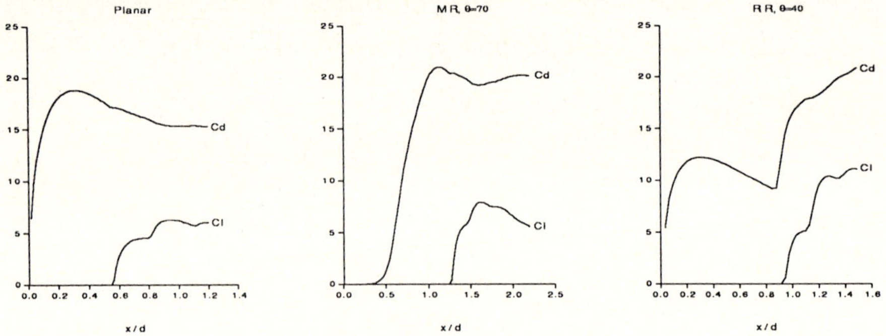

Fig. 6. Lift and drag coefficients, C_l and C_d, for planar, MR, and RR shock waves with $M_s = 1.5$.

$$V_p \approx \frac{\Delta P \pi R^2 2R/M_s a_0}{\rho_p 4/3\pi R^3} = \frac{6}{4}\frac{\Delta P}{\rho_p M_s a_0} \neq f(R).$$

For incident shock waves of Mach number in the vicinity of 1.5 the overpressure is of the order of the pressure ahead of the shock wave, i.e. 10^5 Pa. Assuming that the particle density, ρ_p is 10^3 kg/m^3 one obtains $V_p = O(0.1 \text{ m/s})$, and moreover, the particle velocity is independent of its size.

6. Conclusions

The flowfields generated by various incident shock waves impinging upon a spherical particle which is at rest on a smooth solid surface were simulated numerically. The results clearly indicate that the force exerted upon the particle by the flowfield scales with the radius of the particle. It is concluded that, for a given incident shock Mach number, the reflection angle has a crucial effect upon the lift and drag forces acting upon the particle. For example, the lift force provided by a $M_s = 1.5$ shock reflecting from the solid surface at an angle of 70° and thus resulting in a Mach reflection is larger than that provided by a planar shock. The lift force provided by a $M_s = 1.5$ shock reflecting from the solid surface at an angle of 40° and thus resulting in a regular reflection is larger than that provided by a Mach reflection. Since the RR force is larger than that obtained in the MR case it is concluded that, within the parameter range studied herein, a reflected shock generates a higher lift than a vortex. The largest lift force is generated by a shock wave descending vertically upon the particle after the shock reflected from the solid surface. The unsteady lift and drag coefficients were calculated from the forces and are found to be independent of the particle radius. Order of magnitude analysis indicates that the velocity of a moving particle is of the order of 0.1 m/s and independent of the particle radius.

References

Bowling RA (1985) An analysis of particle adhesion on semiconductor surfaces. Solid State Science and Technology: 2208-2214

Burgraff PS (1981) Wafer cleaning: sonic scrubbing. Semiconductor International: 97-99

Mach E (1878) Uber den verlauf der Funkenwellen in der Ebene und in Raume. Vienna Academy Sitzungsberichte 78: 819

Sasoh A, Takayama K, Saito T (1992) A weak shock reflection over wedges. 2: 277-281

Skydmore K (1985) Cleaning techniques for wafer surfaces. Semiconductor International: 81-85

Tam AC, Leung WP , Zapka W, Zielmlich W (1992) Laser-cleaning techniques for removal of surface particulates, J. Appl. Phys. 71: 3515-3523

On the Importance of the Basset History Term on the Particle Motion Induced by a Plane Shock Wave

M. Sommerfeld and S. Decker
Lehrstuhl für Strömungsmechanik, Universität Erlangen-Nürnberg, Cauerstr. 4, 91058 Erlangen, Germany

Abstract. The influence of the various forces in the particle momentum equation (i.e. drag force, added mass, Basset history force, and the force due to an external pressure gradient) on the shock wave propagation through a gas-particle mixture were evaluated by one-dimensional numerical analysis. The temporal evolution of the different forces at various locations in the low pressure section of the shock tube were evaluated for different particle response times and initial shock strengths. Furthermore, an effective drag coefficient which includes the influence of all forces affecting the particle motion was determined and compared to the standard drag curve and existing experimental data.

Key words: Shock wave propagation, Dusty gas, Forces on particles, Numerical computations

1. Introduction

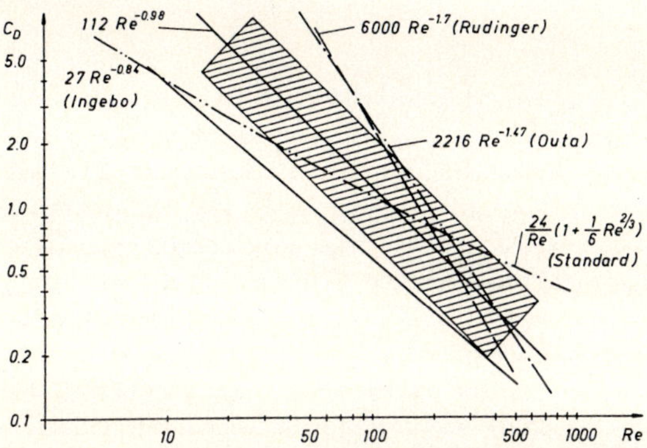

Fig. 1. Comparison of drag coefficients evaluated in shock tube experiments with the standard drag curve.

Shock wave propagation through particle-laden flows and particle-laden flows moving through shock waves were studied experimentally and numerically for many years due to their relevance for a number of practical problems, e.g. dust explosions and propulsion systems. Generally, shock tubes were used to experimentally study the propagation of shock waves through a mixture of gas and fine powder (Rudinger 1970, Outa et al. 1981, Sommerfeld 1985). In these studies, it was found that the dependence of the drag coefficient on the particle Reynolds number considerably deviates from the standard drag coefficient for spherical single particles valid for a steady flow. Different experimental data for the effective drag coefficient (Rudinger 1970, Outa et al. 1981, Sommerfeld 1985) are compared in Fig.1 with the standard drag correlation (Fuchs 1989).

For small particle Reynolds number (i.e. $Re_p < 200$), the experimentally determined drag coefficient is larger than the standard drag coefficient, while for larger values (i.e. $Re_p > 200$), the opposite is the case. This may have several reasons. At higher particle Reynolds numbers the aerodynamic interaction between the particles as a result of their wakes may cause a reduction

Shock Waves @ Marseille III
Editors: R. Brun, L. Z. Dumitrescu

of the drag for the individual particle. This effect can be estimated by calculating the average inter-particle spacing Δ as a function of the particle volume fraction α:

$$\frac{\Delta}{D_p} = \left(\frac{\pi}{6\alpha}\right)^{1/3} \tag{1}$$

The fact that the drag coefficient is higher than that of the standard drag correlation at lower particle Reynolds numbers is assumed to be the result of the increasing importance of other forces like the Basset history term which give rise to an additional acceleration. This is obvious from the equation describing the acceleration of single particles including all of the relevant forces (e.g. Maxey and Riley 1983)

$$F_p = \sigma\frac{D\boldsymbol{u}_p}{Dt} = \frac{3}{4}\frac{\rho\sigma}{\rho_p D_p}c_D(\boldsymbol{u}-\boldsymbol{u}_p)|\boldsymbol{u}-\boldsymbol{u}_p| - \frac{\sigma}{\rho_p}\frac{\partial p}{\partial x} + \frac{\rho\sigma}{\rho_p}c_A\frac{D}{Dt}(\boldsymbol{u}-\boldsymbol{u}_p) \tag{2}$$

$$+\frac{3}{2}\sqrt{\frac{\rho\mu}{\pi}}\frac{\sigma}{\rho_p D_p}c_H\int_{-\infty}^{t}\frac{\frac{D}{Dt}(\boldsymbol{u}-\boldsymbol{u}_p)}{(t-\tau)^{1/2}}d\tau$$

The terms on the right hand side of this equation are the drag force, the force due to a pressure gradient in the flow, the added mass force, and the Basset history force. External forces, e.g. gravity force, were assumed to be negligibly small and were not considered. For the coefficients c_A and c_H, the correlations experimentally obtained by Odar and Hamilton (1964) were used:

$$c_A = 1.05 - \frac{0.066}{A_c^2+0.12}; \quad c_H = 2.88 + \frac{3.12}{(A_c+1)^3}; \quad A_c = \frac{|u-u_p|^2}{D_p\left|\frac{D}{Dt}(u-u_p)\right|} \tag{3}$$

where A_c is the acceleration number.

Unfortunately, these correlations were only confirmed to be valid for small particle Reynolds numbers $\mathrm{Re}_p < 60$. In most cases, one can neglect all of the forces except the drag force in gas-solid flows where the density ratio ρ_p/ρ is about 2000 (Hjelmfelt and Mockros 1966). However, when the pressure and velocity gradients in the flow are very large, as for the flow through a shock wave, these additional forces should be retained in the equation given above.

The evaluation of the drag coefficient from shock tube experiments implies that all forces acting on the particle are included in an effective drag coefficient. When this effective drag coefficient becomes larger than the standard drag coefficient, the other forces seem to become relevant. In the present study, the effect of the different forces on the particle acceleration for the case of a shock wave propagating through a gas-particle mixture is evaluated by numerical calculations.

2. Basic equations and numerical method

For the calculation of the transient properties of the gas and particle phases, the one-dimensional conservation equations were solved. The gas phase was treated as a compressible, inviscid, and non-heat conducting flow. The vector form of the basic equations reads:

$$\frac{\partial}{\partial t}\boldsymbol{U} + \frac{\partial}{\partial x}\boldsymbol{F}(\boldsymbol{U}) = \boldsymbol{A}(\boldsymbol{U}) + \boldsymbol{B}(\boldsymbol{U}) \tag{4a}$$

The vector functions $\boldsymbol{U}$, $\boldsymbol{F}(\boldsymbol{U})$, $\boldsymbol{A}(\boldsymbol{U})$ and $\boldsymbol{B}(\boldsymbol{U})$ include the individual components for the continuity, the momentum, and the total energy equations for the gas and particle phase.

$$\boldsymbol{U} = \begin{pmatrix}\rho\\ \rho u\\ e\\ \sigma\\ \sigma u_p\\ e_p\end{pmatrix}; \quad \boldsymbol{F}(\boldsymbol{U}) = \begin{pmatrix}\rho u\\ \rho u^2+p\\ u(e+p)\\ \sigma u_p\\ \sigma u_p^2\\ u_p e_p\end{pmatrix}; \quad \boldsymbol{A}(\boldsymbol{U}) = \begin{pmatrix}0\\ -F_p\\ -u_p F_p\\ 0\\ F_p\\ u_p F_p\end{pmatrix}; \quad \boldsymbol{B}(\boldsymbol{U}) = \begin{pmatrix}0\\ 0\\ -Q_p\\ 0\\ 0\\ Q_p\end{pmatrix} \tag{4b}$$

with:

$$\sigma = \alpha \rho_{pi}, \qquad p = \rho R T \qquad \text{and} \qquad e = \frac{p}{\gamma - 1} + 0.5 \rho u^2 \tag{5}$$

The forces per unit volume on the particles, F_p, and the resulting influence of the particles on the gas phase are given by Eq. 2. The heat transfer term Q_p is given by:

$$Q_p = 6 \frac{\sigma \mu c_p}{\rho_p D_p^2} \frac{\mathrm{Nu}}{\mathrm{Pr}} (T - T_p) \tag{6}$$

The standard drag coefficient c_D and the Nusselt number Nu are calculated from:

$$\begin{aligned} c_D &= \frac{24}{Re_k}\left(l + \frac{1}{6} Re_k^{2/3}\right) \quad \text{if: } \quad \mathrm{Re} \leq 1000; \quad c_D = 0.44 \quad \text{if: } \quad \mathrm{Re} > 1000 \\ \mathrm{Nu} &= 2.0 + 0.6 \mathrm{Pr}^{1/3} \mathrm{Re}^{1/2}, \quad \mathrm{Pr} = 0.75; \quad \mathrm{Re} = \frac{\rho (u - u_p) D_p}{\mu} \end{aligned} \tag{7}$$

The gas phase equations were solved using the so-called Operator Splitting Technique in order to account for the inhomogeneous terms on the right side and the Random Choice Method to solve the homogeneous equations (Miura and Glass 1981, Sommerfeld 1984). The numerical method applied to solve the differential equations for the particle phase was based on a finite difference method by appropriately tracking the interface between particle-laden and particle-free regions (Miura and Glass 1981, Sommerfeld 1984).

The calculations were performed for the standard shock tube problem where initially a diaphragm separates the high-pressure section from the low-pressure section, which is completely filled with a dusty gas. After rupture of the diaphragm the shock wave propagates through the dusty gas and decays until the equilibrium shock strength is reached. The computations were carried out according to the experimental conditions of Sommerfeld (1984) where spherical glass beads (ρ_p = 2.500 kg/m^3) with a mean diameter of 27 μm were used.

In order to demonstrate the effect of the different forces appearing in the particle equation of motion on the shock wave propagation, calculations were also performed for various initial shock strengths, particle mass loadings, and particle sizes. All calculations were performed for a length of the low pressure section of 4 m using a grid size of Δx = 10 mm.

3. Importance of the different forces

In order to demonstrate the relative importance of the different forces (Eq. 2) on the particle motion in the considered basic shock tube flow (M_{S0} = 1.55, η = 0.66, D_p = 27 μm), the time-dependent evolution of these forces per unit volume was calculated at various locations, i.e. at x = 0.5, 1.5, 2.5 and 3.5 m from the diaphragm (Fig.2). It is obvious that the drag force constitutes the most dominant force. The force due to the pressure gradient across the shock wave is only of importance just across the initial frozen shock jump and is very small within the relaxation region. This is similarly observed for the added mass term. Hence, these forces do not considerably contribute to the acceleration of the particles. The Basset history force is the second dominant force besides the drag force. This force suddenly increases across the shock front to about 4% of the total force and then decreases in the initial relaxation region due to the larger gradients in the change of particle velocity compared to those in the gas velocity. Towards the end of the relaxation region, the Basset history force increases continuously due to steeper gradients in the change of gas velocity and reaches a considerable contribution to the total force as a result of its integral character at the end of the particle region. In this region, however, the total force on the particle has already considerably decayed, since the particles have almost attained the gas velocity.

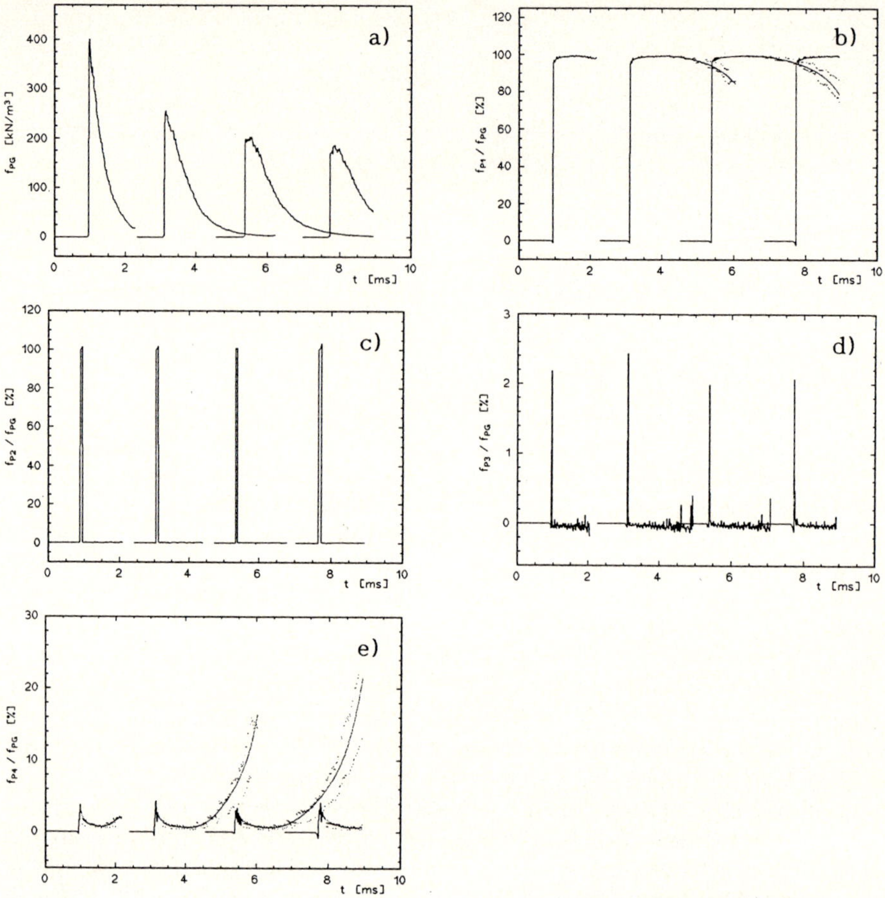

Fig. 2. Time-dependent evolution of the different forces acting on the particles at four locations downstream of the diaphragm (x = 0.5, 1.5, 2.5 and 3.5 m), (M_{S0} = 1.55, η = 0.66, D_p = 27 μm): a) total volumetric force, b) contribution of drag force, c) contribution of pressure force, d) contribution of added mass force, e) contribution of Basset history force

The comparison of the particle velocity within the relaxation region for calculations with all forces and those with only the drag force revealed that even the Basset history force did not considerably affect the particle acceleration for a shock wave propagating through a gas-particle mixture, due to the relatively short time available for particle acceleration. This is also the reason that the decay of the shock wave propagating through the dusty gas is not significantly influenced by including all forces affecting the particle motion. It was found only within a region of about one meter downstream of the diaphragm that the shock wave decayed at a slightly higher rate for the considered conditions.

4. Results for different flow conditions

In order to reveal the influence of the Basset history term for other flow conditions than those discussed above, the shock strength, the particle mass loading, and the particle diameter were systematically varied in the following calculations. In order to allow a more convenient comparison of the various conditions, the contribution of the Basset history force to the total force (i.e. all

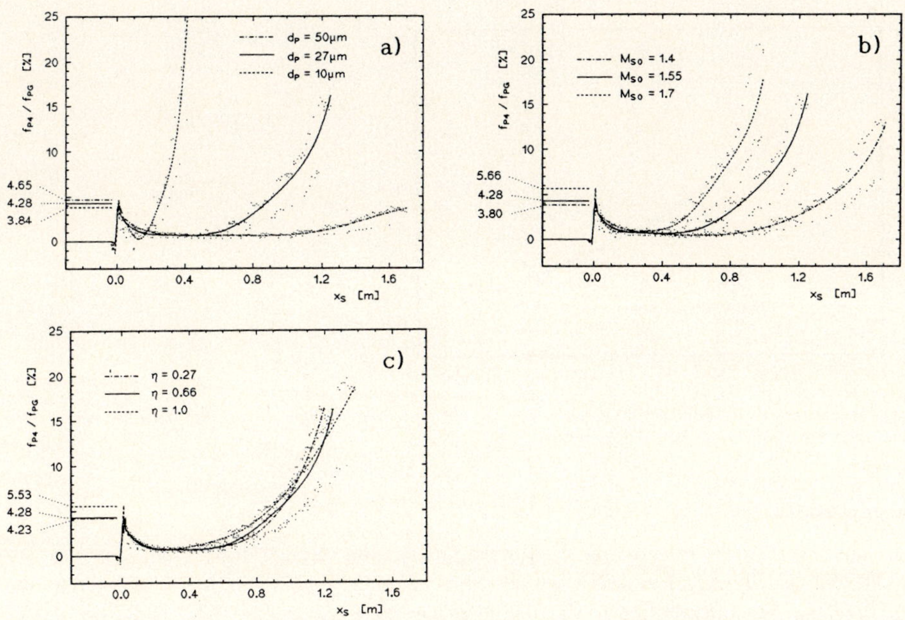

Fig. 3. Contribution of the Basset history force to the total force per unit volume for various flow conditions: a) influence of particle size ($M_{S0} = 1.55$, $\eta = 0.66$), b) influence of initial shock strength ($\eta = 0.66$, $D_p = 27$ μm), c) influence of particle mass loading ($M_{S0} = 1.55$, $D_p = 27$ μm).

other forces of Eq.2 included) was evaluated for the reference location $x = 2.5$ m and plotted versus distance from the shock front. These results are summarized in Fig.3.

The main conclusions from these results are: The particle mass loading has no considerable influence on the contribution of the Basset history term. With decreasing particle diameter and increasing shock strength, the contribution of the Basset history force increases but only towards the end of the relaxation region. The latter conclusion implies that the importance of the Basset history force increases with decreasing particle response time and hence decreasing length of the relaxation region behind the initial shock jump.

5. Effective drag coefficient

The final purpose of the present study was to evaluate an effective drag coefficient which incorporates the effects of all forces influencing the particle motion into a drag force with a modified drag coefficient. The comparison of the effective drag coefficient evaluated from the calculations with the standard drag coefficient reveals that both are almost identical for particle Reynolds numbers, larger than about 50 (Fig.4). For smaller Reynolds numbers the effective drag is considerably higher than the standard drag and some scatter is observed due to different particle sizes and shock strengths considered. This is similar to the trends observed from the evaluation of the drag coefficient from shock tube experiments (Fig.1), but the experimentally obtained drag coefficients are still much larger than those obtained from the present calculations. A conclusive explanation for this fact cannot yet be given. The experimentally observed reduction in the drag coefficient at larger Reynods numbers is assumed to be the result of aerodynamic particle-particle interactions. At higher Reynolds numbers, the length of the wakes of the particles increases to several particle diameters and can cause a dragging effect.

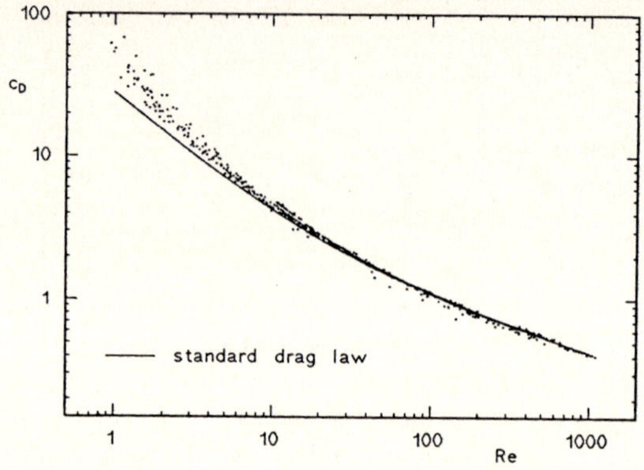

Fig. 4. Comparison of the standard drag curve with the effective drag coefficient evaluated from the calculations including all forces affecting the particle motion

6. Conclusions

The numerical studies presented on shock wave propagation through dusty gases demonstrated the effect of the different forces in the particle momentum equation. It was revealed that the other dominant force in addition to the drag force was the Basset history force. However, due to the short times available for particle acceleration, the Basset history force does not contribute considerably to the particle acceleration nor to the decay of the shock wave.

The determination of an effective drag coefficient involving all force contributions resulted in higher particle drag coefficients for particle Reynolds numbers below 50, similar to the experimental observations. The remaining question is the cause for the experimentally observed reduction in the particle drag at higher particle Reynolds numbers. It is supposed that this is the result of the aerodynamic and pressure wave interaction between particles. Detailed experiments under well-defined conditions are required to confirm this supposition.

References

Fuchs NA (1989) The Mechanics of Aerosols. Dover Publications, Inc., New York

Hjelmfelt AT, Mockros LF (1966) Motion of discrete particles in a turbulent fluid. Appl. Sci. Res. 16: 149-161

Maxey MR, Riley JJ (1983) Equation for a small rigid sphere in a nonuniform flow. Phys. Fluids 26: 883-889

Miura H, Glass II (1981) On a dusty-gas shock tube. UTIAS Rep. No. 250

Odar F, Hamilton WS (1964) Forces on a sphere accelerating in a viscous fluid. J. Fluid Mech. 18: 302-314

Outa E, Tajima K, Suzuki S (1981) Cross-sectional concentration of particles during shock process propagating through a gas-particle mixture in a shock tube. In: Treanor CE, Hall JG (eds) Proc. 13th Intl. Symp. on Shock Tubes and Waves, Albany, N.Y. pp 655-663

Rudinger G (1970) Effective drag coefficient for gas-particle flow in shock tubes. ASME J. Bas. Eng. 92:165-172

Sommerfeld M (1984) Instationäre Stoßwellenausbreitung in Gas-Teilchen Gemischen. Ph. D. Thesis, Stoßwellenlabor, RWTH Aachen

Sommerfeld M (1985) The unsteadyness of shock waves propagating through gas-particle mixtures. Exp. in Fluids 3: 197-206

Drag Coefficient Measurement of Spheres in a Vertical Shock Tube and Numerical Simulation

G. Rodriguez, P. Grandeboeuf, M. Khelifi and J.-F. Haas
Commissariat à l'Energie Atomique, Centre d'études de Vaujours-Moronvilliers BP7, 77181 Courtry, France

Abstract. The drag coefficient of spherical particles is measured in a vertical shock tube. The glass or nylon particles (diameter 1 or 2 mm) are initially in a free fall in air, SF_6 or helium. During the measurement time, they are subjected to an upward-propagating shock wave followed by a rarefaction wave or not, and anyway by a reflected shock wave. Particle trajectories recorded with a high-speed motion picture camera are compared to analytical and simulated trajectories. The drag coefficients (0.50-0.62) obtained for this unsteady situation in a high Reynolds number range ($5. \times 10^3 \cdots 1.2 \times 10^5$) are higher than the standard steady value (0.4).

Key words: Drag coefficient, Shock tube, Unsteady flow, Two-phase flow models

1. Introduction

There are many situations in engineering or nature leading to an unsteady two-phase flow of particles in a fluid: particle seeding for laser Doppler velocimetry in the presence of strong flow gradients, behaviour of the particles cloud ejected from the free surface of a plate subjected to a high-explosive generated shock wave, volcanic explosions. The use of the standard drag coefficient based on steady flow measurements is often questionable in unsteady situations. Horizontal shock tube experiments were recently performed in order to measure the drag coefficient of solid spheres subjected to an unsustained shock (Igra and Takayama 1991) in a higher Reynolds number range than previous experiments (see references herein). The particles, initially lying on the shock tube wall, were lifted from it and accelerated by the high-speed fluid behind the shock wave. Particle trajectories were constructed from measurements of particle positions recorded in a series of shock tube runs for different time delays after shock-particle interaction. Measured particle trajectory provided a series of points in a drag coefficient (C_d) vs. Reynolds number (Re) diagram for each particle and shock Mach number. The resulting averaged evolution of C_d over the Re range (6×10^3 to 10^5) is significantly higher (0.8 to 0.55) than the steady flow values (0.44, Henderson 1976, Bailey and Hiatt 1972). Complementary experiments are described here for different conditions: particles always outside the boundary layers and with non-zero initial velocities.

2. Shock tube experiments

The present experiments are carried out in a vertical shock tube (length 600 cm and cross section 13 × 13 cm) which generates upward propagating shock waves (Fig.1). In the region of measurement, the shock waves are sustained during a few milliseconds (constant gas velocity and density). Depending on the gas combination, this constant regime ends first with a rarefaction wave reflected from the lower plate or with the shock reflected from the top end-plate. A string of particles (glass or nylon spheres, density 2.6 or 1.05 g/cm^3 and diameter 1 or 2 mm) is dropped from the central area of the top end-plate. About 20 to 30 particles are dropped in order to insure that at least one of them is in the field of view at the time of shock interaction. As they are so far apart in the measurement zone, there is no interaction between them. The trajectory of the particles submitted to the incident shock wave and subsequent reflected waves is recorded with a high-speed motion picture camera. About 200 shadowgraph pictures with one or several particles in each frame are obtained over a period of 10 to 20 milliseconds. Wall-mounted pressure

Shock Waves @ Marseille III
Editors: R. Brun, L. Z. Dumitrescu

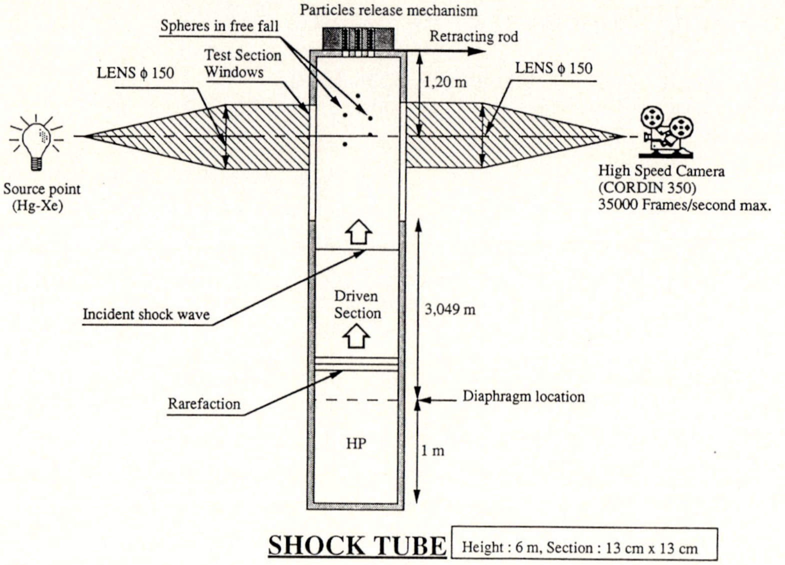

Fig. 1. Experimental set-up

gauges are used to measure shock velocity and shock pressure near the position of shock-particle interaction.

Table 1. Summary of the experimental results

driver/driven gases	M_s	part.dia mm	part.type	C_d	Re range
Air/air	1.291	2	glass	0.61	27000-30000
Air/air	1.298	2	nylon	0.57	25000-31000
Air/Air	1.29	1	glass	0.54	12000-14800
SF_6/SF_6	1.341	2	glass	0.62	80000-119000
SF_6/SF_6	1.341	2	nylon	0.62	79000-116000
SF_6/SF_6	1.354	1	glass	0.58	44000-62000
SF_6/Air	1.217	2	nylon	0.51	17000-21000
SF_6/Air*	1.218	1	glass	0.50	7200-10000
Air/Helium*	1.181	2	nylon	0.57	5000-6000

Driver gas is air or SF_6 (sulphur hexafluoride) and driven gas, initially at atmospheric pressure, is air, SF_6 or helium. When the same gas is used in both sections, we obtain an incident shock of Mach 1.3 to 1.34 followed by a rarefaction before the reflected shock. When the driving gas is heavier, the shock Mach number is 1.15 to 1.2 and the rarefaction is not seen during the observation of particle behaviour. The measured pressures for the two cases are shown in Fig.2. The variation of particle size and density as well as the choice of gases allow us to explore a range of Re from 5.0×10^3 to 1.2×10^5. Relative Mach number varies from 0.2 to 0.6.

In a typical experiment in air, 2 mm glass particles fall for 0.5 s and, about 1.2 m from the end-plate and with a downward velocity of about 5 m/s, encounter the incident shock wave creating an upward gas flow of 150 m/s. This constant flow lasts 4 milliseconds until the arrival of the rarefaction wave reflected from the bottom end-plate. After 2.4 ms, the shock reflected from the

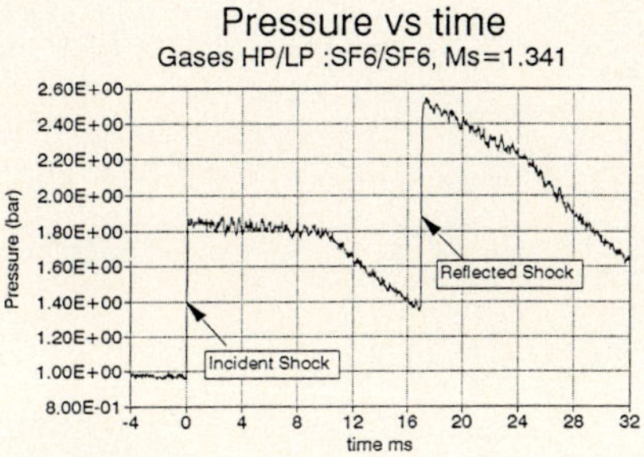

Fig. 2a. Pressure measurement in a case with rarefaction

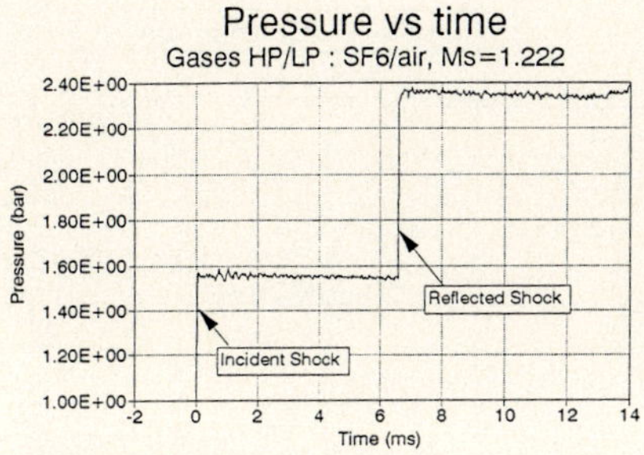

Fig. 2b. Pressure measurement in a case without rarefaction

top end plate slows down the gas flow to almost zero with the particle moving upward at about 15 m/s. Most particles fall vertically near the shock tube center. A few of them fall closer to the walls and are not taken into account since they penetrate the boundary layer and experience a slower gas flow.

3. Results

In the experiment, the trajectories of the particles are quasi-onedimensional and time scales are such that gravity plays no role after shock interaction. Because of the high density ratio between particle and gas, the added-mass and Basset forces can be neglected and only the drag force prevails. Therefore, the main equation describing the behaviour of a sphere, is the following:

$$\frac{du_p}{dt} = \frac{3\rho_g C_d(u_g - u_p)|u_g - u_p|}{8\rho_p r_p}$$

where u_g and ρ_g are gas velocity and density; u_p, ρ_p and r_p are particle velocity, density and radius.

The experimental and analytical trajectories of three spheres subjected to shocks at Mach about 1.3 in air appear In Fig.3. For instance, we compare the experimental trajectory of a 2 mm glass sphere with the one calculated using the above equation of particle motion subjected to the drag force with assumed constant drag coefficient. This yields a value of $C_d = 0.61$, higher than the value (0.44) published (Henderson 1976) for steady flows for the particle Mach (0.4) and Reynolds number (3×10^4 to 2.7×10^4). We note that the analytical trajectory diverges from the experimental points after 4 ms (time of arrival of the rarefaction) because the variations in u_g and ρ_g are not taken into account. The drag coefficients obtained by this method for the various cases are given in the table. We summarize our measurements in Fig.4 which includes our data points, the fit from Igra and Takayama (1991) as well as Henderson's law. Our data points fall between the two laws, and lie closer to Igra's fit for the higher Reynolds numbers. In fact our points are quite close to some of Igra's data points (not shown). Based on the error in reading the particle position, we estimate the error for C_d at ± 0.01. Another source of error (not evaluated) comes from the unmeasured gas velocity which is calculated using the Hugoniot relations from the measured shock velocity and pressure jump. This applies particularly to the last two experiments in the table (SF_6/air*, air/helium*) for which some driving gas was mixed to the test gas (SF_6 mass concentration in air 5.5% and air mass concentration in helium 6.75%).

We also follow the method (Igra and Takayama 1991) of fitting a polynomial (of order 4 to 6) to the experimental trajectory without the assumption of constant C_d. We differentiate twice this polynomial and, taking into account the evolution of inviscid gas flow (thus neglecting boundary

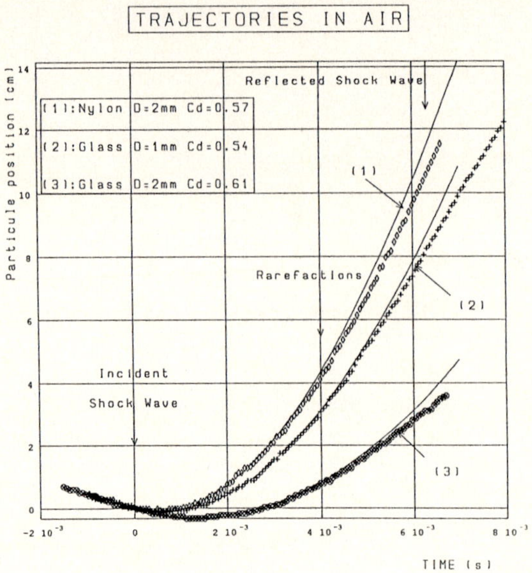

Fig. 3. Experimental and analytical trajectories for three particles in air, $M_s \approx 1.3$

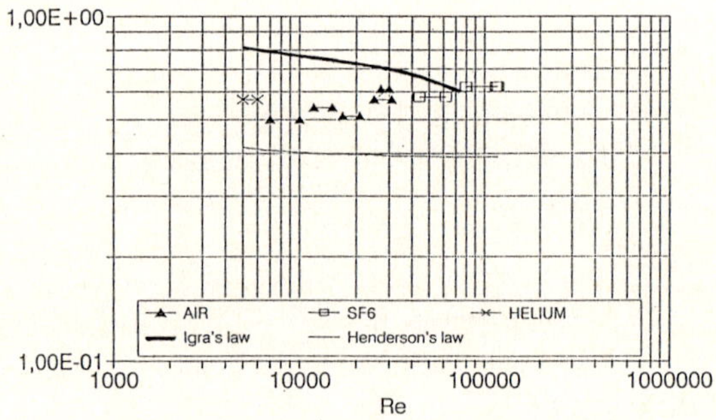

Fig. 4. Drag coefficient vs. Reynolds number: experimental results, Igra's and Henderson's laws

layer effects), use the drag formula to obtain a time evolution of C_d. For instance in the case of 2 mm glass sphere in air, using a fifth-order polynomial, we find that C_d remains at 0.6 during the period of constant gas flow, then increases with the reflected rarefaction which causes a Reynolds number decrease from 3×10^4 down to 1.5×10^4. As the value of C_d during the rarefaction depends on the order of the polynomial, this method has been discarded. Since we do not expect significant changes of C_d in each experiment anyway, we turn to comparisons with numerical simulations with assumed fixed C_d.

4. Numerical calculation

The behaviour of the particle dropped in the shock tube is simulated using two methods, one outlined here, the other to be presented in detail in another communication (Trophimova 1993). The two mathematical models are quite similar: they both assume that the interaction between particles is negligible and that the mass of the spherical particle remains constant (no coalescence, no fragmentation). The gas-particle interaction results in momentum and heat transfer between the fluid and the particles.

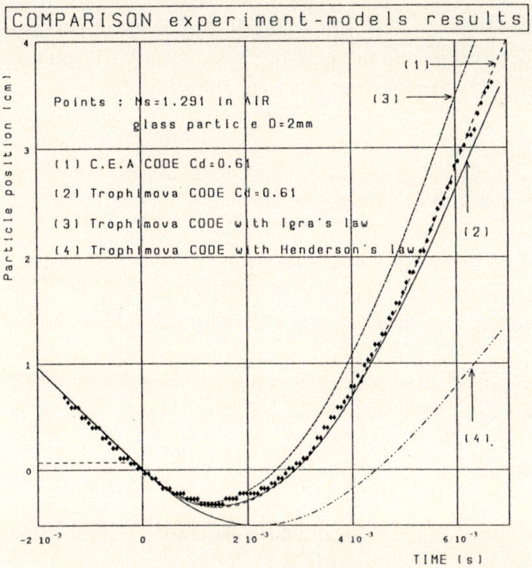

Fig. 5. Comparison of experimental and simulated trajectories in air for a glass particle

The first model is implemented in a 1D numerical Lagrangian code which uses a finite difference method with pseudo-viscosity. The method chosen is a generalization of the Richtmyer schemes traditionally used in Lagrangian hydrodynamic codes. A dispersed phase description is adopted for the particles: it means that each particle is considered as discrete. The gas is treated as continuous: the equations for the fluid are averaged on a control volume. This modelling is inspired by the ideas of Dukowicz (1980) and O'Rourke (1981).

The other model, developed by Trophimova (1993), also describes the behaviour of the particle distribution and the evolution of the gas characteristics. But, in this case, an Eulerian description of the two-component fluid flow is adopted and each component is treated as a continuous fluid. Heat transfer is switched off in the present calculations.

Both calculations give an accurate description of the inviscid gas flow in the shock tube. We test the validity of the various laws of drag coefficient by comparing the experimental and numerical trajectories. Fig.5 shows that each experiment is correctly simulated by both models (using the value of C_d given in the table) after interaction with incident shock, rarefaction and reflected shock. The slight difference between the two codes may be due to numerical factors: Eulerian vs. Lagrangian treatment, mesh size etc... In contrast, the trajectories calculated using C_d from the law of Henderson (1976) or the fit from Igra and Takayama (1991) disagree with experimental trajectories.

5. Conclusion

The experiments are performed in a vertical shock tube with particles initially in free fall and remaining outside the boundary layers at all time. The measured drag coefficients vary between 0.50 and 0.62 over a Reynolds number range of $5 \times 10^3 \cdots 1.2 \times 10^5$, which is higher (by 25% on the average) than the standard drag coefficient law in steady conditions. A simple fit of our points is: $C_d = 0.328\,\mathrm{Re}^{0.0539}$. This is somewhat lower than the tendency obtained from the horizontal shock tube experiments of Igra and Takayama (1991) in which the particles were initially at rest on the shock tube wall and the shocks were not sustained.

Acknowledgements

We thank M. H. Thorembey, G. Delouis, P. Fouquart, G. Fournier and P. Appourchaux for their contribution in the laboratory, R. Boria for assistance in using the first code, J. P. Choquin for several helpful discussions and M. G. Trophimova for allowing us to use the second code.

References

Bailey AB, Hiatt J (1972) Sphere drag coefficients for broad range of Mach and Reynolds numbers. AIAA Journal 10, 11:1436-1440

Dukowicz JK (1980) A particle-fluid numerical model for liquid sprays. Journal of Computational Physics 35, 2:229-253

Henderson CB (1976) Drag coefficient of sphere in continuum and rarefied flows. AIAA Journal 14, 6: 707

Igra O, Takayama K (1991) Shock tube study of the drag coefficient of a sphere in a nonstationary flow. In: Takayama K (ed) Shock Waves, Proc. 18th Intl. Symp. on Shock Waves, Sendai, pp 491-499

O'Rourke PJ (1981) Collective drop effects on vaporizing liquid sprays. Los Alamos National Laboratory Report LA-9069-T

Trophimova MG (1992) Mathematical and numerical modeling of multifield flows. Preprint, Keldish Institute of Applied Mathematics; see also these Proceedings

Experimental and Numerical Investigation of Shock Wave Attenuation in Dust-Gas Suspensions

O. Igra*, G. Ben-Dor*, F. Aizik* and B. Gelfand†

*Pearlstone Center for Aeronautical Engineering Studies, Department of Mechanical Engineering, Ben-Gurion University of the Negev, Beer Sheva, Israel
†Institute for Chemical Physics, Russian Academy of Sciences, Moscow, Russia

Abstract. The attenuation of planar shock waves while propagating through dust-gas suspensions was studied experimentally and numerically. The numerical solution was conducted twice, each time using a different correlation for the dust particle drag coefficient. It was shown that when the Reynolds number is relatively high (Re > 200) the solution based on the nonstationary drag coefficient provides better agreement with experimental findings. Also studied was the post shock pressure signature; good agreement was found between experimental and numerical results until the reflected rarefaction wave, from the driver's end wall, caught up with the transmitted shock wave in the suspension.

Key words: Shock attenuation, Dusty gas

1. Introduction

The propagation of shock waves through dust-gas suspensions has drawn a lot of attention in the past four decades due to its applicability to many engineering problems. For example, such events happen in explosive-generated shock waves over sandy terrains, in nozzle flows of solid propellant rockets and in accidental explosions in coal mines. A comprehensive description of the fundamentals of compressible, two-phase flows which develop behind shock waves propagating in suspensions can be found in textbooks (e.g., Clift, Grace and Weber 1978) and review articles (e.g., Marbel 1970; Igra and Ben-Dor 1988). Among the observed effects of the dust presence on the shock wave propagation and the post-shock suspension flow one should notice the shock wave attenuation and the change in the post-shock pressure signature, as compared with a similar dust-free case. These, and other effects were studied in the past both numerically and experimentally. Among the more recent studies one should note the works of Sommerfeld (1984, 1985) and Olim et al. (1990).

In numerical solutions of shock wave propagation in dusty gases one has to use correlations for the dust drag coefficient and its Nusselt number. Due to a lack of reliable data one is compelled to adopt correlations which were obtained for a single sphere immersed in a steady flow. However, the post-shock suspension flow, in general, involves a very large number of spheres (dust cloud) and it is truly nonstationary, at least in proximity behind the shock front.

Studying shock wave propagation through homogeneous dust-gas suspensions, experimentally, is not an easy task since it is very difficult to produce a uniformly distributed, single diameter sphere (dust) in a suspension which is in a state of equilibrium upstream of the shock wave.

It is the purpose of the present paper to study, numerically and experimentally, shock wave propagation in uniform suspensions.

Shock Waves @ Marseille III
Editors: R. Brun, L. Z. Dumitrescu

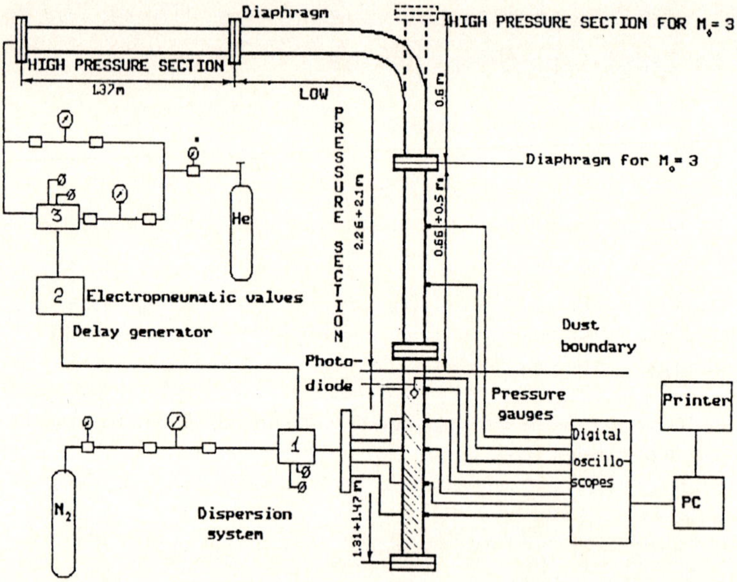

Fig. 1. Schematic description of the shock tube used in the experimental study

2. Theoretical and experimental background

The numerical solution of the conservation equations, shown in a vector form in Eqs.1 and 2, was done using the random choice method (RCM). Details regarding this numerical scheme can be found in Glimm (1965) and in Sod (1977).

For the gaseous phase:

$$\frac{\partial \boldsymbol{U}_g}{\partial t} + \frac{\partial \boldsymbol{F}_g}{\partial x} = -\boldsymbol{G} \tag{1}$$

Here

$$\boldsymbol{U}_g = \begin{bmatrix} \rho_g \\ \rho_g u \\ \rho_g \left(C_v T_g + \frac{u^2}{2}\right) \end{bmatrix}, \qquad \boldsymbol{F}_g = \begin{bmatrix} \rho_g u \\ \rho_g u^2 + P \\ \rho_g u \left(C_v T_g + \frac{u^2}{2} + \frac{P}{\rho_g}\right) \end{bmatrix}$$

where ρ_g, u, P, and T_g are the density, velocity, pressure and temperature, respectively and $e = C_v T_g$ is the specific internal energy of the gaseous phase. The total specific energy of the gaseous phase, E, is given by $E = e + u^2/2$. The source term $\boldsymbol{G}$ is due to the interaction between the gas and solid phases.

For the solid phase:

$$\frac{\partial \boldsymbol{U}_p}{\partial t} + \frac{\partial \boldsymbol{F}_p}{\partial x} = \boldsymbol{G}; \tag{2}$$

Here

$$\boldsymbol{U}_p = \begin{bmatrix} \rho_p \\ \rho_p v \\ \rho_p \left(C_m T_p + \frac{v^2}{2}\right) \end{bmatrix}, \qquad \boldsymbol{F}_p = \begin{bmatrix} \rho_p v \\ \rho_p v^2 \\ \rho_p v \left(C_m T_p + \frac{v^2}{2}\right) \end{bmatrix}$$

where the specific internal energy e_p, is given by $e_p = C_m T_p$, and the total specific energy of the particles, E_p, is given by: $E_p = e_p + v^2/2$. ρ_p and v are the density and velocity of the solid phase, respectively, and

$$G = \begin{bmatrix} 0 \\ F_D \\ Q_{HT} + F_D \dot{v} \end{bmatrix}$$

For solving Eqs. 1 and 2 the equations of state for a perfect gas (i.e., $P = \rho_g R T_g$) was used. The drag force, F_D, and the heat transfer, Q_{HT}, per unit volume, are given by

$$F_D = \tfrac{1}{2} n_p \rho_g C_D |u - v|(u - v)\frac{\pi D^2}{4} \tag{3}$$

$$Q_{HT} = n_p h (T_g - T_p) \pi D^2 \tag{4}$$

where n_p, D and C_D are the particles number density, diameter and drag coefficient, respectively; and h is the coefficient of heat transfer.

In the numerical solution, two different correlations for C_D were used: (i) the "universal drag coefficient" for a sphere in a steady flow, which is a curve fit to Bailey's (1974) figure showing C_D versus Re, and (ii) the correlation proposed by Igra and Takayama (1993) for a sphere in a nonstationary flow:

$$\log_{10} C_D = 7.8231 - 5.8137 \log_{10} \mathrm{Re} + 1.4129(\log_{10} \mathrm{Re})^2 - 0.1146(\log_{10} \mathrm{Re})^3 \tag{5}$$

The following Nusselt number was employed in the numerical computations:

$$\mathrm{Nu} = 2 + 0.459\, \mathrm{Pr}^{1/3}\, \mathrm{Re}^{0.55} \tag{6}$$

The experimental investigation was conducted at the Institute for Chemical Physics of the Russian Academy of Science, Moscow, using the shock tube facility shown schematically in Fig.1. The following parameter ranges were covered during the experimental investigation:

Initial shock wave Mach number	$1.3 \leq M_{s0} \leq 3.0$
Dust diameter	$20 \leq D \leq 200\,\mu$m
Dust loading ratio (ρ_p/ρ_g)	$0.22 \leq \eta \leq 27.3$

A sample of the obtained results is presented subsequently. The shock wave was produced in pure air and only after it became a well-formed shock wave it encountered the dust gas suspension. The suspension occupied the last part of the driven section; see Fig.1. Great care was taken to ensure as uniform dust distribution as possible and to narrow the range of variations in the particle diameter; details can be found in Igra et al. (1994)

3. Results and discussion

In all the experiments the shock tube section which contained the suspension was held in a vertical position. The shock wave attenuation was deduced from time records of its arrival at six positions along the tube, where pressure transducers were allocated; see Fig.1. Knowing the location of these gauges and the appropriate time of arrival, the shock wave average velocity can be deduced. This average velocity was assigned to the mid-distance between any given pair of pressure transducers. For each experiment the resulting flow was simulated numerically using the RCM with the appropriate initial conditions (i.e., M_{s0}, P_0, T_0, D and η). However, in the numerical simulations both the driver and the driven sections were assumed to be infinitely long; i.e., no wave reflections and/or interactions were included.

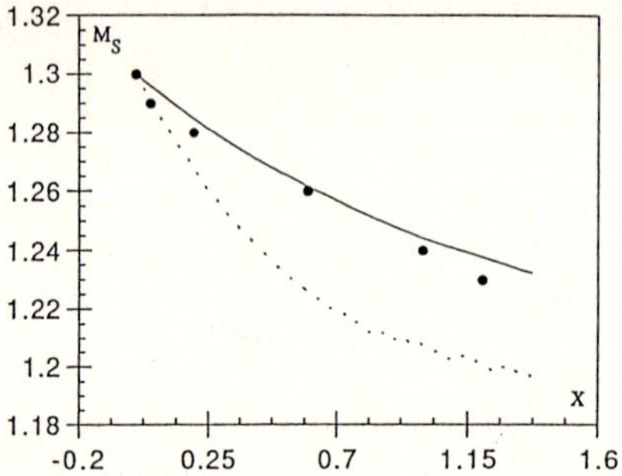

Fig. 2. Transmitted shock Mach number vs. distance for $M_{s0} = 1.3$, $D = 20\ \mu$m and $\eta = 0.22$.
— Numerical results based on steady-flow C_D,
- - Numerical results based on C_D of Eq.5.
• experimental findings

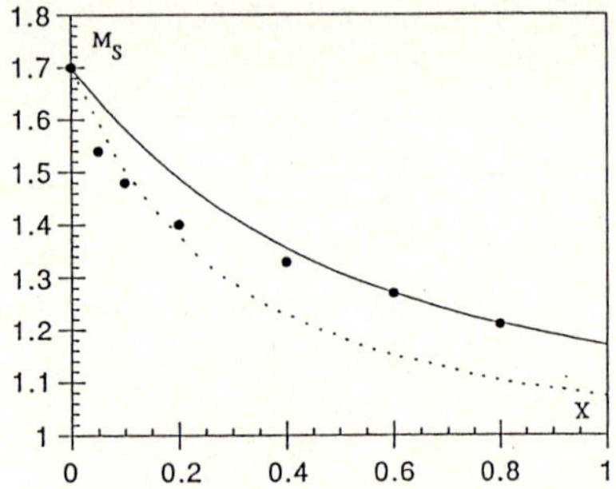

Fig. 3. Transmitted shock Mach number vs. distance for $M_{s0} = 1.7$, $D = 120\ \mu$m and $\eta = 12.9$
— numerical results based on steady-flow C_D,
- - Numerical results based on C_D of Eq.5.
• Experimental findings

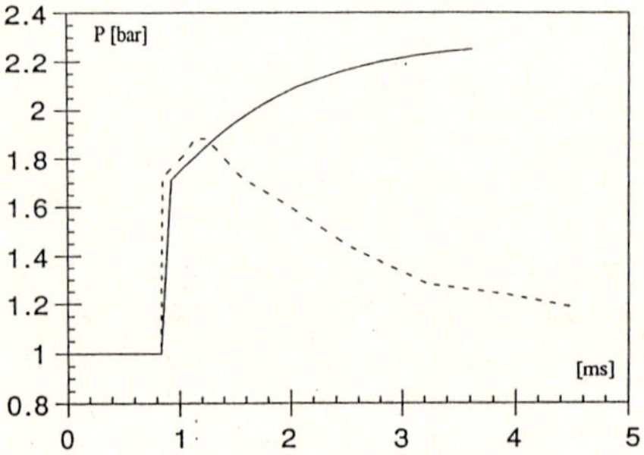

Fig. 4. Pressure variation behind the transmitted shock at $x = 50$ cm. x is measured from the beginning of the dust cloud.
— Numerical results
- - Experimental findings.
$M_{s0} = 1.6$, $D = 40\ \mu$m, $\eta = 3.1$

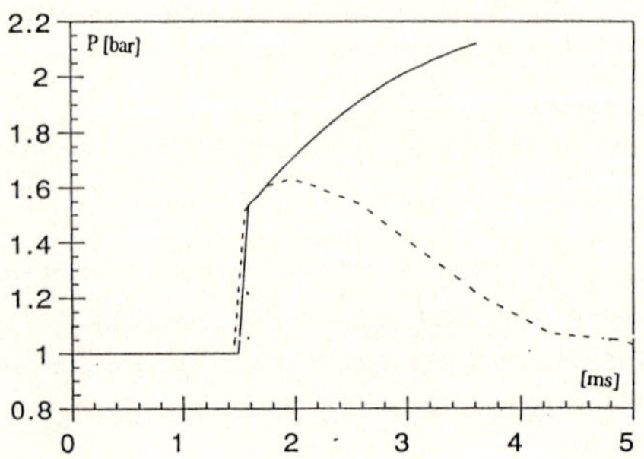

Fig. 5. Pressure variation behind the transmitted shock at $x = 75$ cm. x is measured from the beginning of the dust cloud.
— Numerical results
- - Experimental findings.
$M_{s0} = 1.6$, $D = 40\ \mu$m, $\eta = 3.1$

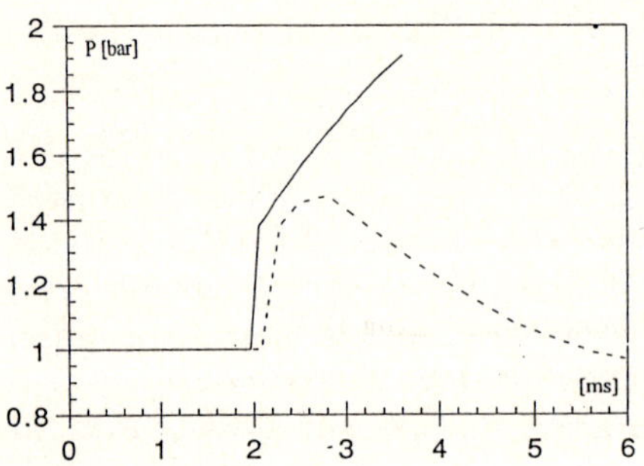

Fig. 6. Pressure variation behind the transmitted shock at $x = 105$ cm. x is measured from the beginning of the dust cloud.
— Numerical results
- - Experimental findings.
$M_{s0} = 1.6$, $D = 40\ \mu$m, $\eta = 3.1$

Typical results obtained for the incident shock wave Mach number in the dust gas suspension are shown in Figs.2 and 3. (M_s is defined as the shock wave velocity divided by the speed of sound ahead of the incident shock wave in pure air.)

For the case of a relatively low initial shock wave Mach number and small particles (M_{s0} = 1.3, $D = 20$ μm; Fig.2) the universal drag coefficient, obtained for a steady flow, reconstructs the experimental finding more accurately than when the nonstationary drag coefficient was used. This should not be surprising, since at such a low shock wave Mach number, the post-shock relative velocity between the dust and the gas is small. This yields a relatively small Reynolds number. Since the correlation given in Eq.5 is applicable only within the range of $200 \leq \mathrm{Re} \leq 101000$, a fixed value was assigned to C_D when Re < 200. As a result, a significant part of the dotted line shown in Fig. 2 was obtained for a fixed C_D ($C_D = 2$) while the solid line is based on a varying C_D obtained from a curve fit to the 'universal drag coefficient for a sphere'. When the initial Mach number of the shock wave in air was increased to $M_{s0} = 1.7$ and the dust particle diameter to $D = 120$ μm, the Reynolds number based upon the relative velocity increased as well. Now, at least during the initial part of the shock wave propagation in the dust-gas suspension, the Reynolds number exceeded 200 and, as a result, the numerical predictions based upon the nonstationary C_D (Eq. 5) gave better agreement with experimental findings. As M_s reduced due to attenuation of the transmitted wave, the relative velocity reduced and so did the Reynolds number. Once $M_s \leq 1.3$, using the 'universal drag coefficient for a sphere' provided better agreement with experimental findings due to the reason just given; see Fig.4.

It is of interest to check how the dust presence affects the pressure signature of the transmitted shock wave. In order to obtain an insight on this effect, static pressures were recorded at a few places along the tube wall, where the dust-gas suspension exists. Results from one experiment are shown in Figs.4-6. These results were obtained for $M_{s0} = 1.6$, $D = 40$ μm and $\eta = 3.1$. The dashed line connects the experimental findings while the solid line represents the numerical results. It is evident from Figs.4-6 that, at early times after the transmitted shock wave has past the position where the pressure was recorded (up to about 0.6 ms), good agreement exists between the experimental and the numerical results. In both cases, the peak suspension pressure is higher than the pressure jump obtained through a similar shock wave in a pure gas. However, at later times a great discrepancy exists between the experimental and numerical results. In the experimental findings (dashed line in Figs.4-6) a fraction of a millisecond after the transmitted shock wave passed the pressure recording station, a dramatic decrease in the pressure is recorded. This large pressure reduction is not observed in the numerical simulation. The reason for this is simple. In reality, the reflected rarefaction wave (from the driver's end wall) catches up with the transmitted shock wave in the suspension and weakens it significantly. This is shown clearly in the dashed line on Figs.4-6. In the present numerical solution both the driver and the driven sections of the shock tube were taken as infinitely long. Therefore no wave reflection takes place in the numerical solution and, as a result, the static pressure behind the transmitted shock wave in the suspension increases until it reaches a new equilibrium (not shown in Figs.4-6).

4. Conclusions

Using the RCM the shock wave attenuation in a suspension could be predicted accurately. For cases in which the incident shock wave initial Mach number is high and when the dust particle diameter is relatively large, better results are obtained when using the nonstationary drag coefficient (Eq. 5). This is so since in such cases the Reynolds number, based upon the relative velocity, exceeds 200 which is the lower limit of applicability of Eq.5. When the resulting Re is lower than about 200 the universal drag coefficient should be used due to lack of applicable information regarding nonstationary C_D. In addition to attenuating the transmitted shock wave, the dust presence causes an increase in the suspension pressure. Until the arrival of the reflected rarefaction wave good agreement is found between experimental and numerical results.

References

Bailey AB (1974) Sphere drag coefficient for subsonic speeds in continuum and free molecule flows. J. Fluid Mech. 65: 401

Clift R, Grace JR, Weber ME (1978) Bubble, Drops and Particles. Academic Press, New York

Glimm J (1965) Solution in the large for nonlinear hyperbolic system equations. Comm. Pure Appl. Math. 18: 95

Igra O, Ben-Dor G (1988) Dusty shock waves. Appl. Mech. Rev. 41:379

Igra O, Takayama K (1993) Shock tube study of the drag coefficient of a sphere in a nonstationary flow. Proc. R. Soc. Lond. A442: 231-247

Igra O, Gelfand B, Aizik F, Ben-Dor G (1994, to be published) Experimental and numerical investigations of dusty shock waves.

Marbel FE (1970) Dynamics of dusty gases. Ann. Rev. Fluid Mech. 2:397

Olim M, Ben-Dor G, Mond M, Igra O (1990) A general attenuation law of moderate planar shock waves propagating into dusty gases with relatively high loading ratios of solid particles. Fluid Dyn. Res. 6: 185

Sod GA (1977) A numerical study of a converging cylindrical shock. J. Fluid Mech. 83:785

Sommerfeld M (1984) Instationäre Stosswellenausbreitung in Gas-Teichengemischen. Ph.D. Thesis, Faculty of Mechanical Engineering, RWTH Aachen, Germany

Sommerfeld M (1985) The unsteadiness of shock waves propagating through gas-particle mixtures. Exp. Fluids 3: 197

Numerical Simulation of Shock Wave Reflection and Diffraction in a Dusty Gas

Shmuel Eidelman, Xiaolong Yang and Itzhak Lottati
Science Applications International Corporation, 1710 Goodridge dr, McLean, VA, USA

Abstract. The unsteady shock wave reflection and diffraction generated by a shock wave propagating over a semicircular cylinder in a dusty gas are studied numerically. The mathematical model is a multi-phase system based on a multi-fluid Eulerian approach. A Second Order Godunov scheme is used to solve the gas phase Euler equations and an upwind scheme is used to solve the particle phase conservation equations on an unstructured adaptive mesh. For the validation of the model, the numerically predicted one-dimensional shock wave attenuation is compared with experimental results. Shock wave reflection and diffraction over a semicircular cylinder in a pure gas flow is simulated first to show the excellent agreement between the present computation and the experimental results. For the shock wave reflection and diffraction in a dusty gas, the effects of particle size and particle loading on the flow field are investigated. Gas and particle density contour plots are presented. It has been shown that the shock wave configuration differs remarkably from pure gas flow depending on the particle parameters. The difference is explained as the result of momentum and heat exchange between the two phases.

Key words: Multi-phase flow, Numerical methods

1. Introduction

Shock wave propagation into a gas-particle suspension medium has attracted great attention in recent years due to its many engineering applications. Some of these applications include blast wave propagating over a dusty surface, exhaust from a solid propellant rocket, and coal or grain dust detonation. Many studies dealing with two-phase environment can be found in literature. See for example, Marble (1970), Rudinger (1964), Soo (1989).

In this paper, we study shock wave reflection and diffraction over a semicircular cylinder in a dusty gas. We numerically simulate the problem of a shock wave initiated in a pure gas section moving into a dusty region and impinging on a semicircular cylinder. We first formulate the compressible two-phase flow on the basis of a Eulerian multi-fluid formulation. We consider the two phases (i.e., gas and particle) to be interpenetrating continua. The dynamics of the flow are governed by conservation equations of each phase and the two phases are coupled by interactive drag force and heat transfer. We solve the system of conservation equations numerically on an unstructured adaptive grid. The objectives of the study are: (a) to solve the two-phase compressible flow field and compare the simulation with available experimental results; (b) to observe and investigate the reflection and diffraction wave patterns when a shock wave propagates over a semicircular cylinder in a dusty gas, with particle radius and loading as parameters.

2. Mathematical model and the numerical solution

2.1. Conservation equations

The mathematical model consists of conservation governing equations and constitutive laws that provide closure for the model. The basic formulation adopted here follows the gas and dilute particle flow dynamics model presented by Soo (1989). The following assumptions are used during the derivation of governing equations:

(1) The gas is air and is assumed to be ideal gas; (2) The particles do not undergo a phase change because for particles considered here (sand) the phase-transition temperature is much

Shock Waves @ Marseille III
Editors: R. Brun, L. Z. Dumitrescu

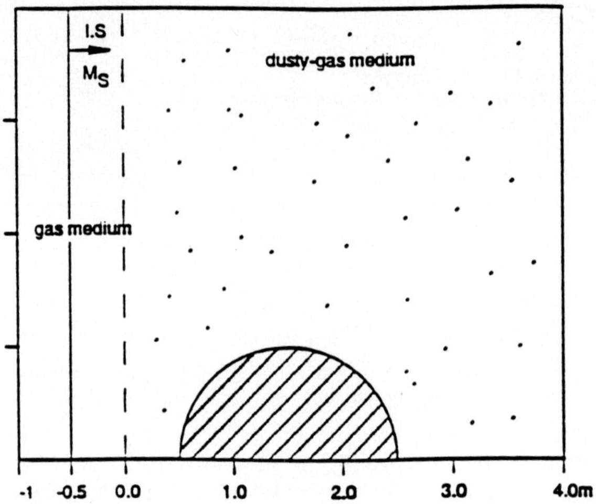

Fig. 1. An illustration of the considered flowfield

higher than the temperatures typical for the simulated cases; (3) The volume occupied by the particles is negligible; and (4) The interaction between particles can be ignored.

Under the above assumptions, distinct equations of continuity, momentum, and energy are written for each phase. The interaction effects between the two phases are listed as the source terms on the righthand side of the governing equation. The detailed mathematical formulation used in our study can be found in Yang et al. (1993).

2.2. Initial and boundary conditions

The geometry of the computational domain is shown in Fig.1. The initial conditions for the gas are ρ_0 = 1.2 kg/m^3 and p_0 = 101.3 kPa, with a coming shock at x = -0.5. There are no particles from -1.0 $\leq x \leq$ 0.0. From $x \geq$ 0.0, particles are initially in thermal and kinematic equilibrium with the surrounding gas. The particles, that are uniformly distributed in the dusty region, have the following parameters for different test problems:

Mass loading, ρ_p: 0.25 kg/m^3, 0.76 kg/m^3; Mass material density, ρ_s: 2500 kg/m^3; Particle radii, r_p: 10 μm, 25 μm, 50 μm; and Specific heat, c_s: 766 J/kg/K.

The lower boundary and cylinder surface are solid walls and assumed adiabatic and impermeable. A reflecting boundary condition is assumed for both the gas and particle phase. Particles are assumed to experience a perfect elastic collision with the wall and reflect from the wall. The right and upper boundaries are open boundaries where a nonreflection boundary condition is used for the gas phase and a zero normal gradient condition is used for the particle phase.

2.3. Numerical method of solution

The system of partial differential equations presented in Yang et al. (1993) is integrated numerically. In order to solve this equation numerically, an operator time-splitting technique is used. Assuming that all flow variables are known at a given time, we can calculate its advancement in time by splitting the integration into two stages.

In the first stage, the conservative part of equations is solved:

The Second Order Godunov method is used for the integration of the subsystem of equations describing the flow of the gaseous phase. The method is well documented in literature (Eidelman et al. 1984, Colella 1985). The subsystem of equations describing the particle phase flow is integrated using a simple first order finite difference upwind scheme (Peyret and Taylor 1983). This is done because there is no shock in the particle phase and the upwind scheme leads to a robust and accurate integration scheme.

In the second stage, the source term is added and the resulting system is integrated implicitly.

In order to produce a solution of the high spatial accuracy at minimal computational cost, an unstructured triangular grid with adaptive procedure is used. The adaptive procedure will automatically enrich the mesh by adding points in the high gradient (or high flow activity) region of the flow field and by removing points (coarsening the mesh) where they are not needed.

3. Results

3.1. Model validation for one-dimensional shock wave propagation in a dusty gas

To test the momentum and heat exchange mechanism for the current two-phase model, we first simulated a one-dimensional problem of a normal shock wave propagating into a dusty gas. We numerically simulated the experiments conducted by Sommerfeld (1985). In the experiments, small glass spherical particles of material density $\rho_s = 2500$ kg/m^3, specific heat capacity $c_s = 766$ J/kg/K, and average diameter of 27 μm were used as the suspension particle phase. The incoming shock Mach number M_S, and particle loading ratio $\eta = \rho_P/\rho$, are two varying parameters. The experimental results and our numerical simulation results of shock Mach number as a function of distance are in very good agreement. In fact the numerical results deviate less than 5% from the experiments over 3.5 m of shock wave atenuation.

3.2. Two-dimensional simulation results for two-phase flow

The basic setup for the two-phase simulation is shown in Fig.1. Here the planar shock with $M_S = 2.8$ propagates into an area of a dusty gas and impinges on a semicircular cylinder. The interface between pure air and dusty air is located at $x = 0.0$ of the computational domain. The area of the dusty air contains a semicylinder with a radius of 1 m. The size of the computational domain, initial parameters of the gas, parameters of the incoming shock, size of the semicylinder and its location in the computational domain, are the same as in the reflection and diffraction simulation presented in the previous section.

The main objective of this set of simulations is to study the effects of particle size and particle loading on the parameters of the reflected and diffracted shock waves. It is also valuable to study the dynamics of particle media, since it is extremely difficult to observe these interactions experimentally in an optically thick dusty gas.

The first set of simulation results is shown for the case with dust parameters $r_p = 10$ μm and $\rho_P = 0.25$ kg/m^3. The gas parameters and the parameters of the incoming shock wave are the same as in the pure gas case presented above. In Figs.2a and 2b, particle density contours and gas density contours are shown at the stage when the incident shock wave has reached the top of the semicylinder. At this stage, the largest difference of velocity and temperature between the two phases exists and the nonequilibrium between the two phases causes extensive heat and momentum exchange between particles and the gas. The presence of the particles causes a widening of the shock that is more noticeable for the incident shock. Also, an additional contour line is observed at the dusty gas/pure gas interface. Comparing gas density for the pure gas flowfield and the dusty gas density of Fig.2b, we see that the Mach stem and contact discontinuity resulting from Mach reflection are smeared in the dusty gas flow due to the presence of the particles. The particle density contours depict significant piling up of the dust particles at the leading edge stagnation point of the cylinder.

In Figs.2c and 2d, the particle density and gas density contours are shown at the stage where significant diffraction has taken place and the shock front is approaching the trailing edge of the cylinder. Further widening of the shock and some smearing of the slip line that originates at the triple point is evident. The particle density contours reveal that the particles were swept by the gas flow to the area of triple point and slip line for the gas flow, leaving a small amount of particles at the leading edge. We should note that this behavior is specific for our problem, where at $t = 0$, the dusty gas area was located at $x = 0$ and there is no influx of the dust from the left boundary. Also in Fig.2c, we note that the particles reach a distinct local maxima at the distance about 25 cm behind the incident shock front. At this maxima the particle density is 0.86 kg/m^3, which is more than three times the initial particle density. The particle density reaches a maximum value at the location of the gas slip line. We observe a significant accumulation of the particles that have been moved along the slip line by the shear flow. The larger concentration of particles in the vicinity of the triple point is, in fact, the remainder of the particles that were swept up with the flow. It is also interesting to observe that an essentially particle-free zone is formed due to the effects of particles slipping over the top of the cylinder and the rarefaction wave behind the cylinder.

To study the influence of particle loading on the dynamics of reflection and diffraction, we have simulated the case with a dust density of $\rho_p = 0.76$, and with $r_p = 10$ μm. The results for this simulation are shown in Figs.3a and 3b in the form of particle and gas density contour plots. In Fig.3a, the particle density contours are shown at the diffraction phase. Here we can observe two local maxima for particles accumulated in the regions along the slip line characteristic for the shock diffraction process. It should be noted here that in our problem the conditions behind the incident shock wave and its structure are in constant flux. At higher loading, dust will have a profound effect on the gasdynamics of reflection and diffraction. Fig.3b shows gas density contours for the reflection stage corresponding to the particle density contours shown in Fig.3a. We observe from Fig.3b that the incident shock wave is significantly smeared and the triple point cannot be clearly identified. Because of the widening of the incident shock, the area where the reflected and incident shock join is spread over 50 cm distance. From Fig.3a, we see that the high density particle region is spread wider than in the previous case, and the particle density reaches its maximum at about 25 cm behind the front. There is a visible maximum in gas density in the area where the reflected shock is interacting with the area of maximum particle density behind the incident shock. A part of the reflected shock front that is moving to the left side of the computational domain is not affected by the dust since it is propagating into an area with little dust concentration. The parameters and structure of this part of the front remain basically the same as in the case of pure gas flow.

To examine the effect of particle size on the reflection-diffraction process, we simulated a case where the particle loading and gas flow conditions are the same as in the previous case with particle density $\rho_p = 0.76$. However, the particle size is $r_p = 50$ μm. In Figs.4a and 4b, results for this simulation are illustrated by particle density and gas density contours correspondingly. The particle contour plots depict a significantly wider particle relaxation zone than in the previous case. The longer relaxation zone is caused by the larger inertia of larger particles. The maximum particle density of 2.64 kg/m^3 is reached 50 cm behind the incident shock front. This value is significantly lower than 4.01 kg /m^3 reached behind the shock in calculation with 10 μm particles. Larger particles skip above the apex of the cylinder creating a void where particle density is very small. Also, because of larger particle size, the maxima of particle concentration that has been created by a slip surface of the reflected Mach stem is indistinct. The main reason for this is that the particles do not follow the gas flow as closely as they did in the previous case due to the inertia of large particles. The maximum particle density is reached here at the slip line behind the Mach stem.

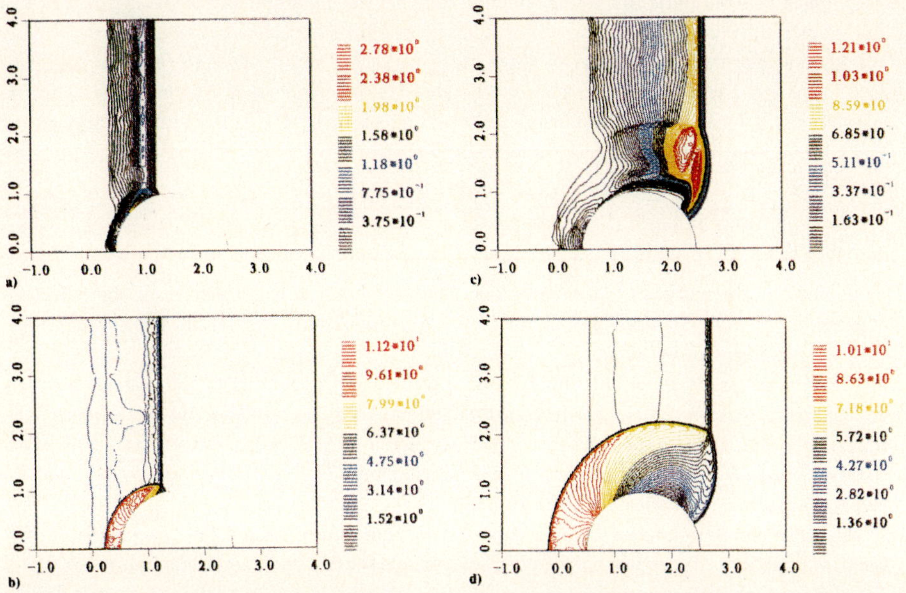

Fig. 2. Density contours for the case: $M_S = 2.8$, $\rho_p = 0.25$ kg/m^3, $r_p = 10$ μm at two different times, (a) particle density at t_1, (b) gas density at t_1, (c) particle density at t_2, and (d) gas density at t_2

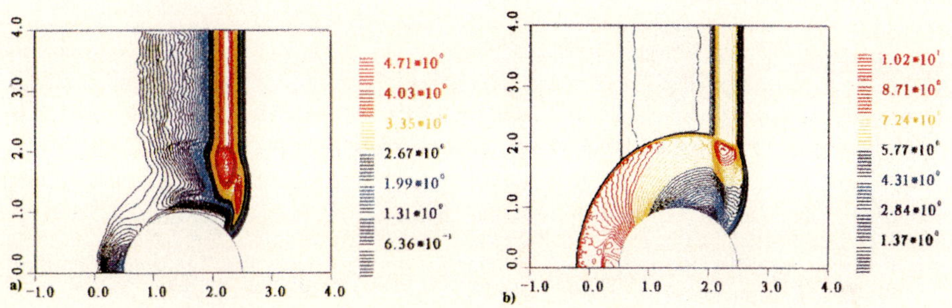

Fig. 3. Density contours for the case: $M_S = 2.8$, $\rho_p = 0.76$ kg/m^3 and $r_p = 10$ μm, (a) particle; (b) gas

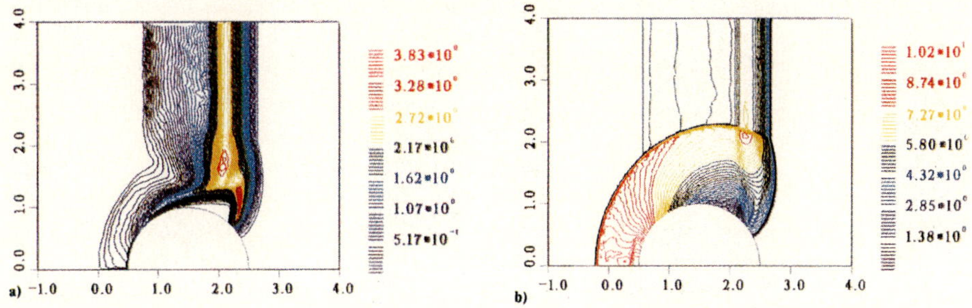

Fig. 4. Density contours for the case: $M_S = 2.8$, $\rho_p = 0.76$ and $r_p = 50$ μm, (a) particle density and (b) gas density

Comparing the gas density of Fig.4b to the previous case shown in Fig. 3b, we observe that the slip line behind the curved Mach stem becomes less distinguishable in Fig.3b. This result is expected, since at fixed particle loading, smaller particles have a larger surface/volume ratio and the larger surface/volume ratio increases momentum and heat exchange between the two phases.

One general comment regarding all three cases presented above: Due to the heat and momentum exchange between the two phases, the shock is decaying as it traverses the cylinder. Ultimately, it will reach a new equilibrium state as suggested by Fig.1. It should be noted that the shock considered in the previous three cases is still in the process of transition in the gas-particle mixture.

4. Conclusion

In this paper, a numerical study for a two-phase compressible flow is performed for the reflection and diffraction of a shock wave propagating over a semicircular cylinder in a dusty gas. The following conclusions can be made:

(1) The validation study for a one-dimensional shock wave propagating in a dusty gas shows a good agreement between the prediction of our model and the results of the experiment;
(2) Particles in the gas can have a profound effect on the shock wave reflection and diffraction pattern, which is a function of particle size and loading. The lesser the particle loading, the less the influence of particle on the flow field;
(3) In the three simulation cases, there is a particle accumulation behind the "back shoulder" of the semicircular cylinder due to the effect of particle inertia and gas rarefaction wave;
(4) For different particle sizes at fixed particle loading, the larger particle will have a longer relaxation zone and less accumulation at the "back shoulder" and behind the incident shock. The gas density contours show a less distinguishable slip line in the small particle case than in the large particle case.

Acknowledgments

The work reported here was partially supported by ARPA and AFOSR. The authors would like to thank Lt. Col. J. Crowley and Dr. A. Nachman for their interest in this project.

References

Colella P (1985) A direct Eulerian MUSCL scheme for gas dynamics. SIAM Journal Scientifical and Statistical Computation 6: 104-117

Eidelman S, Colella P, Shreeve RP (1984) Application of the Godunov method and its second order extension to cascade flow modelling. AIAA Journal 22, 11: 1609-1615

Marble F (1970) Dynamics of dusty gases. Annual Review of Fluid Mechanics Vol. 2, pp 369-377.

Peyret R, Taylor TD (1983) Computational Methods for Fluid Flow. Springer-Verlag, New York, pp. 18-31.

Rudinger G (1964) Some properties of shock relaxation in gas flows carrying small particles. Physics of Fluids 7: 658-663

Sommerfeld M (1985) The unsteadiness of shock waves propagating through gas-particle mixtures. Experiments in Fluids 3, 2: 197-206

Soo SL (1989) Particulates and Continuum. Hemisphere Publishing Corporation, New York, pp 266-324

Yang X, Eidelman S, Lottati I (1993) Computation of shock wave reflection and diffraction over a semicircular cylinder in a dusty gas. AIAA Paper 93-2940, Orlando, FL

Numerical Simulation of Dusty Shock Diffraction Around a 90 Degree Corner

B.Y. Wang*, **Q.S. Wu**†, **H. Zhu**† **and Y.H. Xu**†
*Institute of Mechanics, Chinese Academy of Sciences, Beijing 100080, China
†Department of Modern Mechanics, University of Science and Technology of China, Hefei 230026, China

Abstract. This paper analyses the diffraction problem of shock waves around a sharp 90 degree corner in a dusty gas. By using a GRP scheme and a MacCormack scheme respectively for the gas and particle phases, the governing equations are solved based on the operator-splitting technique and predictor-corrector method. The computational predictions for pure gas flows show good agreement with the experiments of Skews and the calculations of Carofano. The numerical simulations for dusty gas flows reveal the complicated structure of the post-shock flowfield and the effect of particles on the planar diffraction pattern.

Key words: Dusty gas flow, Shock wave diffraction, Sharp corner, GRP scheme

1. Introduction

The diffraction of a planar shock wave in a pure gas has been a subject of great interest and the computational modelling of such flows has still received considerable attention in recent years (Hillier 1991; Bazarov et al. 1992; Yang et al. 1992). Various difference schemes, especially those based on a high-order Godunov-type Euler solver, have been developed successfully to simulate the detailed diffraction pattern. However, the research work on shock diffraction phenomena in a gas-particle suspension (or dusty gas) is quite limited in either experimental or numerical aspects and the existing investigations are mainly concerned with the shock reflection problem (Sommerfeld et al. 1986; Miura and Glass 1986a,b; Kim and Chang 1991). In the present paper, we consider the plane shock wave propagation and diffraction around a 90 degree corner in a dilute dusty gas. By using a high-resolution shock-capturing GRP scheme for the gas phase and a second-order MacCormack scheme for the particle phase, the governing equations are numerically solved based on the operator-splitting technique and predictor-corrector method. The complicated flow patterns behind the dusty shock wave are obtained with the relaxation structure. The effect of particle presence and of their properties on the shock propagation and the post-shock flow are discussed.

2. Computational modelling

2.1. Basic assumptions

For this analysis, the two-continuum model is employed and the following assumptions are used: (1) The gas phase is a compressible perfect gas. Its viscosity and heat conductivity is involved only when particles interact with it; (2) The particle phase consists of uniform inert and rigid spheres without thermal and Brownian motions; (3) The particles have a constant heat capacity and uniform internal temperature; (4) The volume fraction of the particles is negligible and there is no mutual interaction among the particles; (5) There are momentum and energy exchanges but no mass transfer between the gas and particle phases; (6) The only force on the particles to be taken into account is viscous drag since the other mechanisms such as pressure gradient force, virtual mass force, Basset force, gravity force and lift force are much smaller for the present problem under consideration.

2.2. Governing equations

The conservation equations for two-dimensional unsteady two-phase flow in the Cartesian coordinate system (x, y) can be written as:

Shock Waves @ Marseille III
Editors: R. Brun, L. Z. Dumitrescu

$$\partial_t \boldsymbol{U} + \partial_x \boldsymbol{F}(\boldsymbol{U}) + \partial_y \boldsymbol{G}(\boldsymbol{U}) = \boldsymbol{H}(\boldsymbol{U}) \tag{1}$$

where $\boldsymbol{U}$ is the vector of conservative variables; $\boldsymbol{F}$ and $\boldsymbol{G}$ are corresponding vectors of fluxes respectively in $x-$ and $y-$ directions; $\boldsymbol{H}$ is the vector of source terms due to the gas-particle interaction. They are given by (in the following equations, subscript p refers to the particles):

$$\boldsymbol{U} = [\rho, \rho u, \rho v, \rho E, \rho_p, \rho_p u_p, \rho_p v_p, \rho_p E_p]^T \tag{2}$$

$$\boldsymbol{F} = [\rho u, \rho u^2 + p, \rho uv, u(\rho E + p), \rho_p u_p, \rho_p u_p^2, \rho_p u_p v_p, \rho_p u_p E_p]^T \tag{3}$$

$$\boldsymbol{G} = [\rho v, \rho uv, \rho v^2 + p, v(\rho E + p), \rho_p v_p, \rho_p u_p v_p, \rho_p v_p^2, \rho_p v_p E_p]^T \tag{4}$$

$$\begin{aligned}\boldsymbol{H} = [&0, -\rho_p D_x/m, -\rho_p D_y/m, -\rho_p(u_p D_x + v_p D_y + Q)/m,\\ &0, \rho_p D_x/m, \rho_p D_y/m, \rho_p(u_p D_x + v_p D_y + Q)/m]^T\end{aligned} \tag{5}$$

Here, ρ is the density; u and v are the velocity components in $x-$ and $y-$ directions; $E = CT + (u^2 + v^2)/2$ is the specific total energy where C stands respectively for the gas specific heat at constant volume C_V or the particle heat capacity C_m and T denotes the temperature. The pressure p is given by the gas state equation:

$$p = \rho RT \tag{6}$$

where R is the gas constant. In addition, the drag force D and the heat transfer Q to be exerted on a single particle of mass m and diameter d can be expressed as

$$D_x = \frac{\pi}{8} d^2 \rho (u - u_p) \sqrt{(u - u_p)^2 + (v - v_p)^2} (0.48 + 28\mathrm{Re}^{-0.85}) \tag{7}$$

$$D_y = \frac{\pi}{8} d^2 \rho (v - v_p) \sqrt{(u - u_p)^2 + (v - v_p)^2} (0.48 + 28\mathrm{Re}^{-0.85}) \tag{8}$$

$$Q = \frac{\pi}{\mathrm{Pr}} d\mu C_P (T - T_p)(2 + 0.6\mathrm{Pr}^{1/3}\mathrm{Re}^{1/2}) \tag{9}$$

where Re is the particle slip Reynolds number; Pr is the gas Prandtl number; μ is the gas viscosity and C_P is the gas specific heat at constant pressure.

2.3. Numerical procedure

The two-dimensional character and the interphase coupling of the dusty shock diffraction problem are dealt with in a fractional step approach and then the governing equations (1) are split into the following three equations:

$$\partial_t \boldsymbol{U} + \partial_x \boldsymbol{F}(\boldsymbol{U}) = 0 \tag{10}$$

$$\partial_t \boldsymbol{U} + \partial_y \boldsymbol{G}(\boldsymbol{U}) = 0 \tag{11}$$

$$\partial_t \boldsymbol{U} = \boldsymbol{H}(\boldsymbol{U}) \tag{12}$$

To obtain the solutions of Eqs.10 and 11, the GRP shock-capturing scheme (Ben-Artzi and Falcovitz 1984) and the MacCormack scheme (MacCormack 1969) are applied to the gas and particle equations, respectively. These two schemes have second-order accuracy both in time and space coordinates. Eq.12 is solved by the predictor-corrector method. The calculation of $\boldsymbol{U}_{ij}$, the grid point (i,j) value at the $(n+1)$ time step, is performed in the following five steps according to a Strang-type sequence of operator splitting (Strang 1968):

$$\boldsymbol{U}_{ij}^{n+1} = L_r(\Delta t/2) L_x(\Delta t/2) L_y(\Delta t) L_x(\Delta t/2) L_r(\Delta t/2) \boldsymbol{U}_{ij}^n \tag{13}$$

where L_x, L_y and L_r are the finite difference operators and Δt is the time interval.

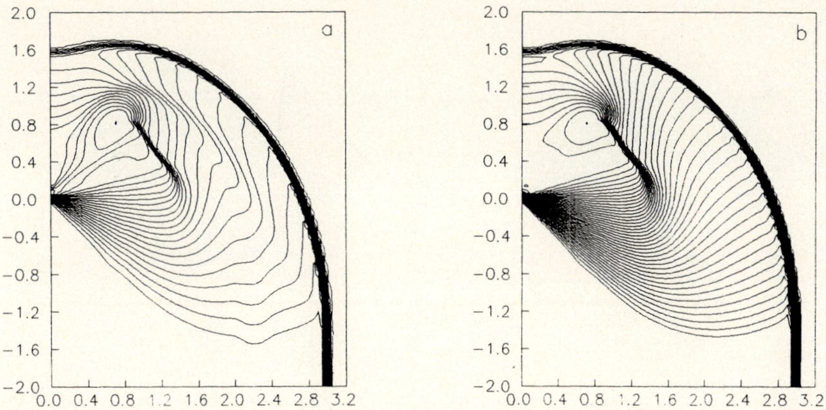

Fig. 1. Diffraction pattern of the pure gas flow in $(\overline{x}, \overline{y})$ plane ($M_s = 3.0$, $t = 78.3$ μs): a. density contours; b. pressure contours

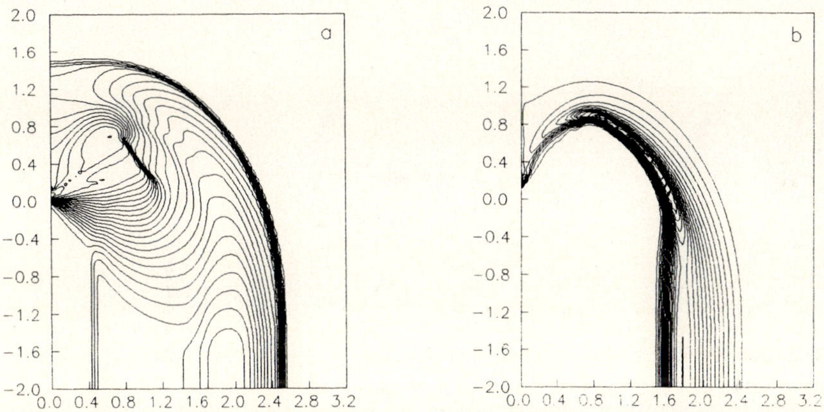

Fig. 2. Density contours of the dusty gas flow in $(\overline{x}, \overline{y})$ plane ($d = 10\mu m, \alpha = 1.0, \beta = 1.0$): a. gas phase; b. particle phase

3. Numerical results

In our computations, 90 degree diffraction patterns of a planar shock wave moving from a pure gas into a dusty gas are obtained with 130×120 grid points and a CFL number of 0.95. To test the validity of our numerical procedure, a pure gas flow was first simulated under the conditions: $M_S = 3.0$, Pr$= 0.75$, $\gamma = 1.4$, $p_1 = 1$ atm and $T_1 = 300$ K, where M_S is the Mach number of the shock wave in the pure gas and the subscript 1 refers to the pre-shock state. The density and pressure contours at the time $t = 78.3$ μs are plotted in Fig.1, where the similarity variables $\overline{x} = x/a_1 t$ and $\overline{y} = y/a_1 t$ are employed (here a is the sound speed). The predicted shock diffraction configuration agrees well with the experimental data by Skews (1967a,b) and the calculation results by Carofano (1984).

Then the numerical simulation of a dusty gas flow under the corresponding conditions as mentioned above was performed, assuming that the solid particles are uniformly distributed at the initial moment. Here the following particle parameters are used: particle material density

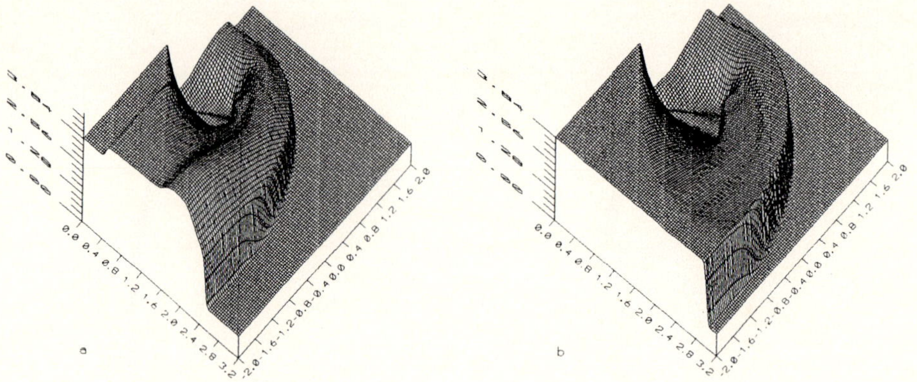

Fig. 3. Gas density distributions ($M_s = 3.0, t = 78.3\ \mu s$, $d = 10\ \mu m$, $\alpha = 1.0$, $\beta = 1.0$): a. dusty gas; b. pure gas

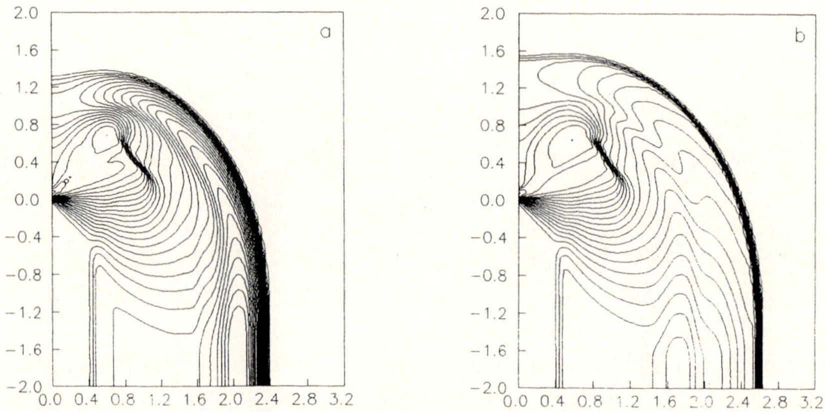

Fig. 4. Gas isopycnics of the dusty gas flow in $(\overline{x}, \overline{y})$ plane ($\alpha = 1.0, \beta = 1.0$): a. $d = 1\ \mu m$; b. $d = 50\ \mu m$

$\sigma_p = 2500\ \mathrm{kg/m^3}$, particle diameter $d = 10\ \mu m$, mass loading ratio $\alpha = \rho_{p1}/\rho_1 = 1.0$, heat capacity ratio $\beta = C_m/C_V = 1.0$. The resulting density contours for the gas and particle phases are given in Fig.2, which show a relaxation zone followed by a particle-free region.

For detailed comparison, the gas density profiles in the whole computation domain are shown in Fig.3a and 3b, respectively for the dusty and pure gas cases. The present numerical analysis leads to the following conclusions: (1) The dusty shock diffraction possesses similar features to those in the pure gas case: the transmitted, diffracted and secondary shock waves, the reflected expansion (sound) wave, the contact and slipstream surfaces as well as the vortex; (2) The particles have remarkable effects on the structure of the diffracting shock wave: a non-equilibrium relaxation zone and a reflected compression wave appear in the post-shock flowfield. Obviously, compared with the pure gas case, the shock wave diffraction in dusty gases is characterized by the relaxation. Due to this process, the strength and speed of the frozen transmitted and diffracted waves decrease and the imbedded contact surface is dispersed. In addition, the gas density jump across the frozen shock front is lower. However, after the first jump, the density continues to increase up to

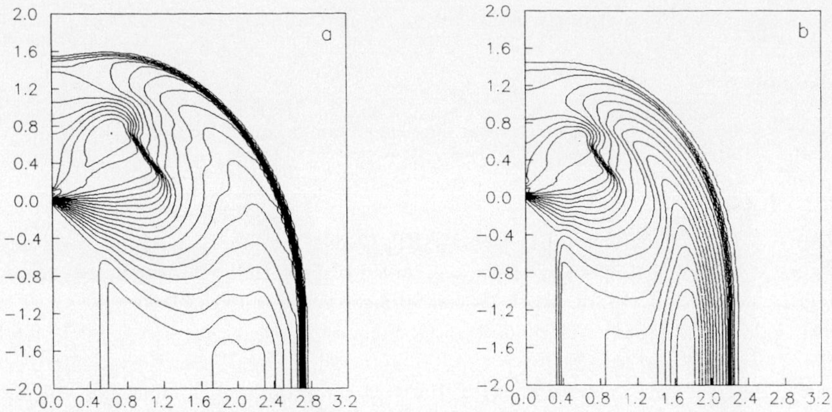

Fig. 5. Gas isopycnics of the dusty gas flow in $(\overline{x}, \overline{y})$ plane ($d = 10\ \mu$m, $\beta = 1.0$): a. $\alpha = 0.5$; b. $\alpha = 2.0$

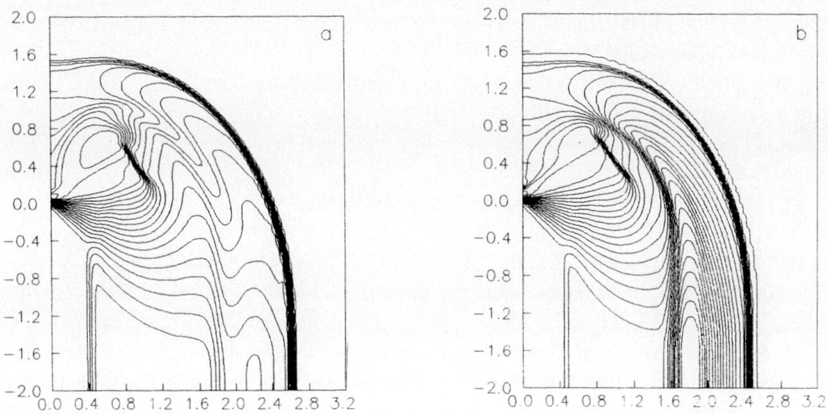

Fig. 6. Gas isopycnics of the dusty gas flow in $(\overline{x}, \overline{y})$ plane ($d = 10\ \mu$m, $\alpha = 1.0$): a. $\beta = 0.1$; b. $\beta = 10.0$

a maximum value and then gradually decrease. The maximum gas density behind the dispersed shock transmitted in the dusty gas case is higher than that in the pure gas case.

Finally, to illustrate the effects of particle parameters on the post-shock flow, the isopycnic curves of the gas phase are presented for $d = 1\ \mu$m and 50 μm (see Fig.4), $\alpha = 0.5$ and 2.0 (see Fig.5) and $\beta = 0.1$ and 10.0 (see Fig.6). It is found that, by decreasing the particle size or increasing the loading ratio and heat capacity ratio, the frozen shock front propagates at a lower speed but the peak density in the relaxation zone grows at a higher rate. These phenomena are attributed to the enhancement of the momentum and energy exchanges between the two phases, as a result of the variations in particle property.

4. Concluding remarks

The present computations show that the combined GRP and MacCormack scheme developed by the authors is effective to solve the dusty shock wave diffraction around a sharp corner with a high resolution. The numerical simulation indicates that the relaxation phenomenon due to the particle presence is a distinctive characteristic of diffracted shock waves in a dusty gas and that the post-shock flow structure is strongly influenced by the particle property parameters.

Ackowledgements

This work was supported by the National Natural Science Foundation of China.

References

Bazarov SB, Gvozdeva LG, Kryukov BP, Lagutov YP, Landin AA, Zharkov AP (1992) Study of shock wave diffraction in gases at sharp and curved corners. In: Takayama K (ed) Shock Waves, Proc. 18th Intl. Symp. on Shock Waves. Springer-Verlag, pp.1155-1160

Ben-Artzi M, Falcovitz J (1984) A second-order Godunov-type scheme for compressible fluid dynamics. J. Comp. Phys. 55:1-32

Carofano GC (1984) Blast computation using Harten's total variation diminishing scheme. ARLCB-TR-840029

Hillier R (1991) Computation of shock wave diffraction at a ninety degree convex edge. Shock Waves 1:89-98

Kim SW, Chang KS (1991) Reflection of shock wave from a compression corner in a particle-laden gas region. Shock Waves 1:65-73

MacCormack RW (1969) The effect of viscosity in hyperbolic impact cratering. AIAA Paper 69-354

Miura H, Glass II (1986a) Shock wave reflection from a rigid wall in a dusty gas. Proc. Roy. Soc. Lond. A404:55-67

Miura H, Glass II (1986b) Oblique shock waves in a dusty gas flow over a wedge. Proc. Roy. Soc. Lond. A408:61-78

Skews BW (1967a) The shape of a diffracting shock wave. J. Fluid Mech. 29:297-304

Skews BW (1967b) The perturbed region behind a diffracting shock wave. J. Fluid Mech. 29:705-720

Sommerfeld M, Selzer M, Grönig H (1986) Shock wave reflections in dusty-gas. In: Bershader D, Hanson RK (eds.) Proc. 15th Intl. Symp. on Shock Waves and Shock Tubes, Berkeley, pp 683-689

Strang G (1968) On the construction and comparison of finite difference schemes. SIAM J. Numer. Anal. 5:506-517

Yang JY, Hsu CA, Jiang CT, Müller B (1992) Computation of shock diffraction in external and internal flows. In: Takayama K (ed) Shock Waves, Proc. 18th Intl. Symp. on Shock Waves. Springer-Verlag, pp 1063-1068

Numerical Simulation of the Interaction of a Shock Wave with a Two-Phase Interface

Q. Yu, A. van de Ven, J.H. Geng, F. Zhang and H. Grönig
Stoßwellenlabor, RWTH Aachen, D-52056 Aachen, FRG

Abstract. The one-dimensional unsteady flow field induced by a planar incident shock wave propagating into a gas-particle mixture without chemical reactions has been calculated using the PLM-algorithm based on the high-resolution Riemann solver and the MacCormack scheme. The effects of Mach number ($M_S = 2.0 - 5.15$), inital loading ratio ($\eta = 0.1 - 4.0$) and particle diameter ($D_P = 10 - 100\,\mu$m) on the flow field behind the incident shock wave are investigated. The numerical results of the traces of the incident and reflected shock wave and the accelerated planar interface of the gas-particle mixture are in good agreement with the experimental data. It is also shown that the gasdynamic behaviour in the gas-particle mixture behind an incident shock wave of moderate strength is remarkably different from that behind a weak one due to the existence of a bow shock in front of each particle.

Key words: Particle acceleration, Particle-gas mixture, Numerical calculation

1. Introduction

For many years there has been an interest in the initial stage of interaction between an incident shock wave, which is mostly a weak one, and a planar interface of a gas-particle mixture because this is very important for the study of explosions in industrial dust-air mixtures (Rudinger 1980, Sommerfeld and Grönig 1983, Olim et al. 1989). According to the treatment of the transfer of momentum and energy between gas and particles, there is a "dissipation" of work due to the particles moving relative to the gas (Marble 1970, Wallis 1969). This dissipation appears as "heat" and is fully transferred into the gas. Once the incident shock wave propagates through the interface of a gas-particle mixture, besides the gas compression due to the interaction between gas and particles, the intense dissipation causes a significant increase in gas temperature behind the incident shock front. In the earlier analysis of this phenomenon of the unsteady gas-particle flow (Igra and Ben-Dor 1988), the numerical results have shown that the temperature rise in the gaseous phase resembles a double step shape. The recent experimental and numerical investigation has confirmed the occurrence of a reflected shock wave originating from a fan of compression waves during a delay time for the case of an incident shock wave of moderate strength and higher loading ratios (Geng and al. 1993).

The purpose of the present paper is to study numerically the effects of Mach number ($M_S = 2.0 - 5.15$), initial loading ratio ($\eta = 0.1 - 4.0$) and particle diameter ($D_P = 10 - 100\,\mu$m) on the one-dimensional unsteady flow field behind an incident shock wave. The variations of gas temperature are obtained for different conditions.

2. Physical model and numerical schemes

The flowfield considered is displayed schematically in Fig.1. A planar shock wave moving in the pure gas with a constant velocity propagates into a uniform gas-particle suspension which is in a state of thermal and kinematic equilibrium. The assumptions and simplifications of this physical phenomenon are similar to that described by Olim et al.(1989). Namely, the monodisperse particles are spherical, inert, occupy a negligible volume, have a uniform diameter, and do not interact with one another. The temperature distribution is uniform inside each particle. Based on

Shock Waves @ Marseille III
Editors: R. Brun, L. Z. Dumitrescu

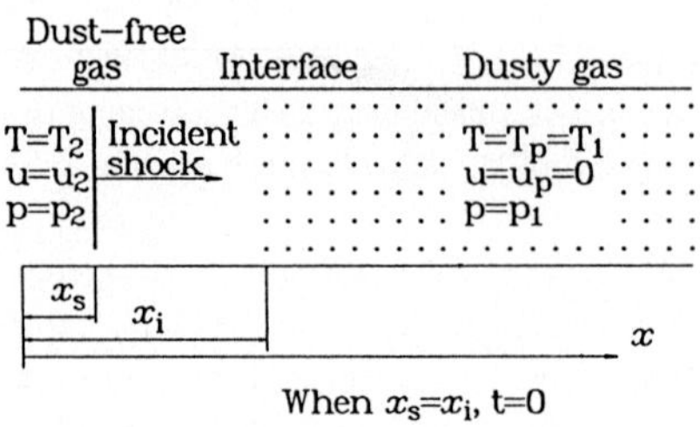

Fig. 1. Schematic description of the flowfield considered

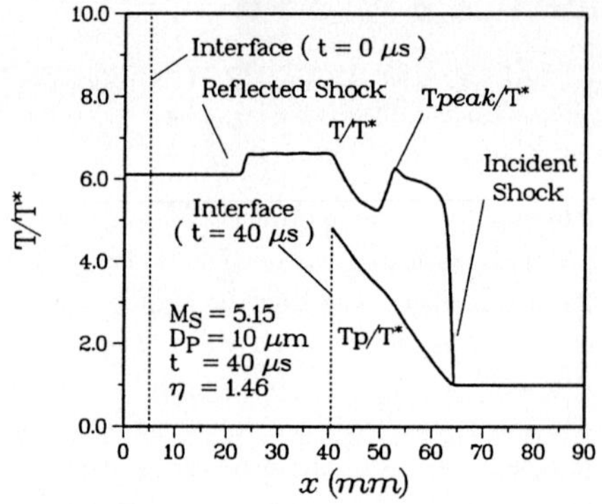

Fig. 2. Spatial distributions of the gas and particle temperature at $t = 40\mu s$

the above assumptions, the two-fluid model is employed since the particle parameters satisfy the dilute gas-particle flow proposed by Marble (1970) and Crowe (1982). The governing equations for the gas and particle phases can be obtained in the following form:

$$\frac{\partial \rho}{\partial t} + \frac{\partial \rho u}{\partial x} = 0 \tag{1}$$

$$\frac{\partial \rho u}{\partial t} + \frac{\partial(\rho u^2 + p)}{\partial x} = f_p \tag{2}$$

$$\frac{\partial e}{\partial t} + \frac{\partial[u(e+p)]}{\partial x} = u_p f_p + q_p \tag{3}$$

$$\frac{\partial \sigma_p}{\partial t} + \frac{\partial \sigma_p u_p}{\partial x} = 0 \tag{4}$$

$$\frac{\partial \sigma_p u_p}{\partial t} + \frac{\partial \sigma_p u_p^2}{\partial x} = -f_p \tag{5}$$

$$\frac{\partial e_p}{\partial t} + \frac{\partial u_p e_p}{\partial x} = -u_p f_p - q_p \tag{6}$$

where

$$e = \frac{p}{\gamma - 1} + 1/2\ \rho u^2 \tag{7}$$

$$e_p = \sigma_p C T_p + 1/2\ \sigma_p u_p^2 \tag{8}$$

$$p = R\rho T \tag{9}$$

The force exerted upon the particles is

$$f_p = 3/4\ C_D \frac{\rho \sigma_p}{D_P \rho_p}(u_p - u)\,|u_p - u| \tag{10}$$

The heat exchange q_p between the two phases is given by

$$q_p = 6 \frac{\sigma_p C_P \mu}{D_P^2 \rho_p}\left(\frac{\mathrm{Nu}}{\mathrm{Pr}}\right)(T_P - T) \tag{11}$$

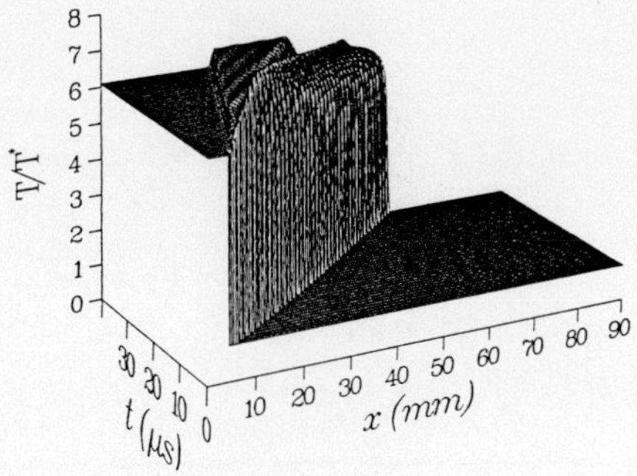

Fig. 3(a). Temperature

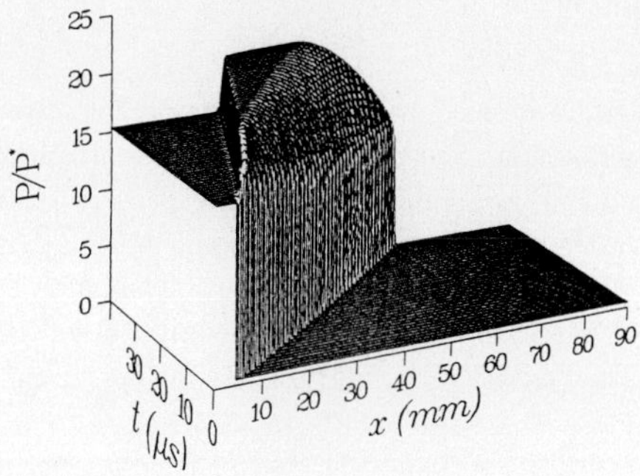

Fig. 3(b). Pressure

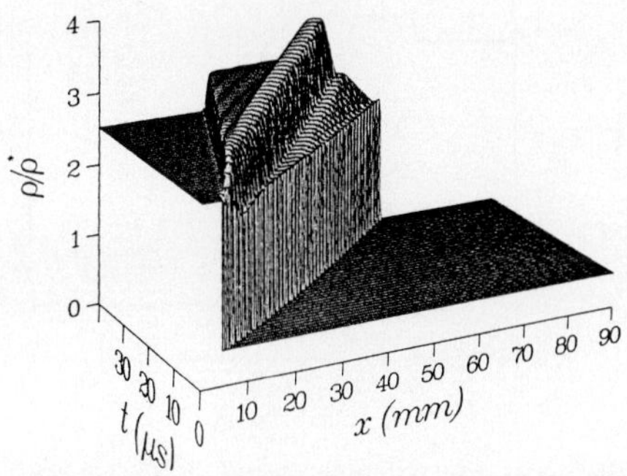

Fig. 3(c). Density

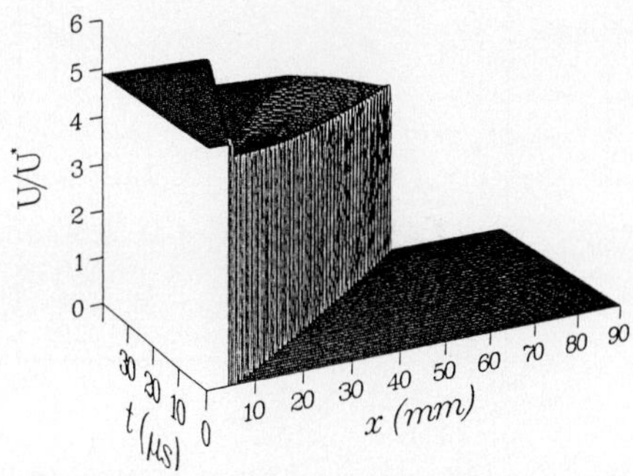

Fig. 3(d). Velocity

Distributions of gas temperature, pressure, density and velocity histories for M_S=5.15 and D_P=10 μm

For the condition of an incident shock wave of moderate strength, each particle behind the incident shock wave is imbedded immediately in a supersonic flow zone and accelerated due to the drag force caused by the gas flow. Thus the relative velocities of particles to the gas are changed from supersonic to subsonic values. In the supersonic flow zone, there is a bow shock in front of each particle, which plays an important role in the interaction between gas and particles. Therefore, more accurate values of the drag and heat-transfer coefficients, C_D and Nu, over a wide range of flow conditions are necessarily required for calculating the variations of phase viscosity and heat conductivity in the gas-particle interaction processes. In this paper, the drag coefficient, C_D, is given by a functional relation (Henderson 1976)

$$C_D = C_D(\mathrm{Re}_p, M_P, T_P/T, \gamma) \tag{12}$$

The Nusselt number, Nu, is given by Schaff and Chambre (1958)

$$\mathrm{Nu} = \mathrm{Nu}_0(1 + 3.42\ M_P \mathrm{Nu}_0/\mathrm{Re}_p\ \mathrm{Pr})^{-1} \tag{13}$$

where

$$\mathrm{Nu}_0 = 2 + 0.46\ \mathrm{Re}_p^{0.55}\mathrm{Pr}^{0.33} \tag{14}$$

$$M = |u - u_p|/(\gamma p/\rho)^{1/2} \tag{15}$$

$$\mathrm{Re}_p = |u - u_p|\rho D_P/\mu \tag{16}$$

$$\mu = \mu^*(T/T^*)^{0.65} \tag{17}$$

The above equations are numerically solved using the operator time-splitting technique. The numerical solution of the gas phase is obtained by using the PLM-scheme. The MacCormack scheme is applied to solve the particle-phase equations. The two explicit schemes are second-order accurate both in time and in space coordinates. In this paper, the gas phase is nitrogen, which behaves as a perfect gas. The particle phase is corn starch dust with ρ_p=1300 kg/m^3 and $C = 1.1313 \times 10^3$ J/kg K. The initial conditions and gas parameters are as follows: $p_1 = 0.51 \times 10^5$ Pa, T_1=293 K, u_1=0 m/s, ρ_1=0.6353 kg/m^3, Pr = 0.7146 and γ=1.4.

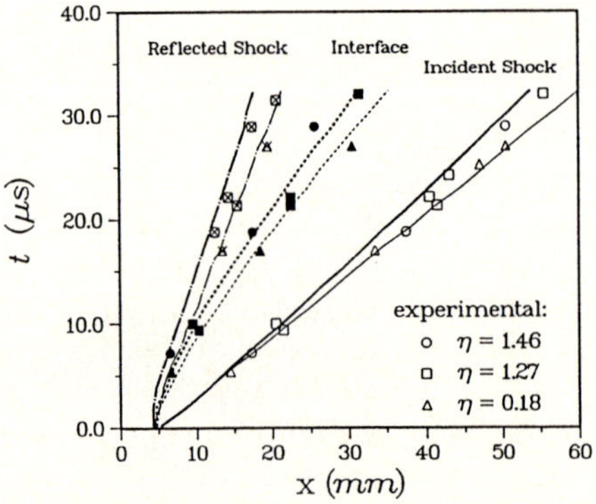

Fig. 4. Comparison between numerical results and experimental data for M_S=5.15 and D_P=10 μm

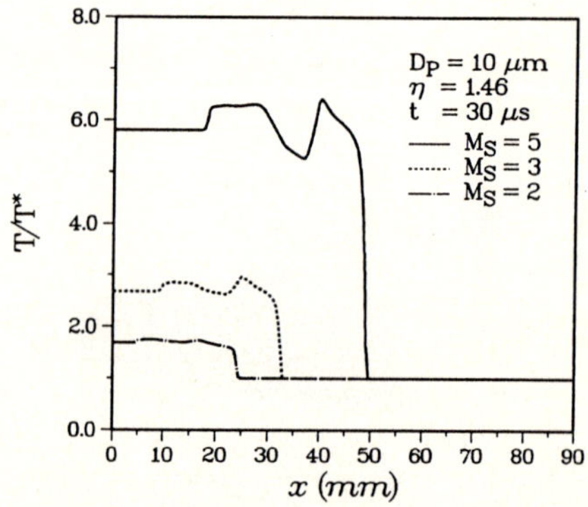

Fig. 5. Numerical results illustrating the effect of Mach number on gas temperature

3. Results and discussion

The spatial distributions of the gas and particle temperature are shown in Fig.2. Once an incident shock wave passes through the interface of the gas-particle mixture, the interaction between gas and particles leads to a compression of the gas, and the intense dissipation is fully transferred into the gas. It will be seen from this that the variations of the kinetic and internal energy of gas and particles causes the occurrence of a gas-temperature peak behind the incident shock front.

The distributions of the gas variables at different times shown in Fig.3a-d display more directly not only the formations and propagations of the reflected shock wave and the gas temperature peak, but also the attenuation process of the incident shock wave due to the interaction between gas and particles behind the shock front. The gas temperature peak follows the incident shock wave at a slower speed and its value varies with time. As distinguished from the incident shock wave, the reflected shock wave keeps a relatively stable strength and follows the incident shock wave as well.

Fig.4 gives the comparison between the numerical and experimental results (Geng et al. 1993) of the traces of the incident shock wave, reflected shock wave and accelerated planar interface of the gas-particle mixture for M_S=5.15 and D_P = 10 μm. As shown in Fig.3, the numerical results, which are represented by all kinds of lines, agree well with the experimental data.

The gasdynamic behaviour in the gas-particle mixtures behind an incident shock wave of moderate strength is remarkably different from that behind a weak one due to the existence of a

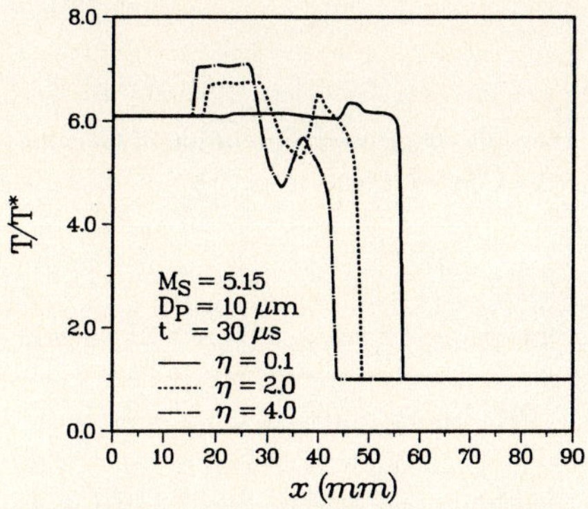

Fig. 6. Numerical results illustrating the effect of initial loading ratio on gas temperature

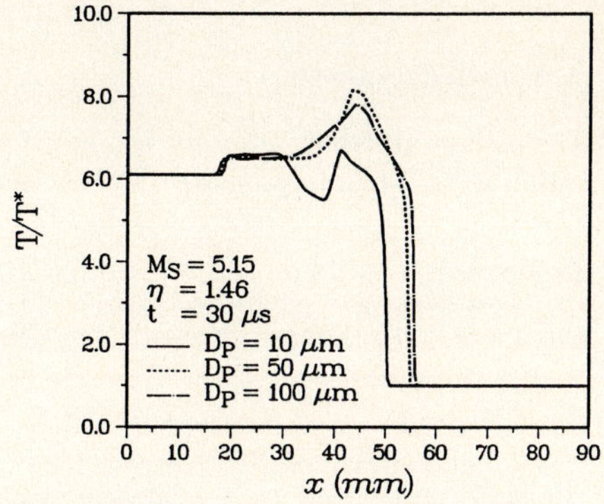

Fig. 7. Numerical results illustrating the effect of particle diameter on gas temperature

bow shock in front of each particle. Fig.5 shows the effect of Mach number on the gas temperature behind the incident shock wave for η=1.46 and D_P=10 μm. It is observed that a more intense transfer of momentum and energy between gas and particles caused by a stronger incident shock wave results in the formation of a stronger reflected shock wave and a more violent variation of gas temperature. In the case of $M_S = 2$, the strength of the reflected shock wave is weak and the value of the gas temperature peak is negligibly small.

The initial loading ratio η plays an important role in the interaction process between gas and particles. As shown in Figs.6 and 3, the propagation velocity of the incident shock wave decreases with the increase in initial loading ratio and so does its strength, but the propagation velocity of the reflected shock wave decreases with an increase in its strength.

The effect of the particle diameter on the flow field at the same initial loading ratio is displayed in Fig.7. For D_P >50 μm, the further increase of the particle diameter leads to a weaker attenuation of the incident shock wave. The variation of the particle sizes from D_P=10 μm to D_P=100 μm has only little influence upon the reflected shock wave. As shown in Fig.7, different spatial distributions of gas temperature would be expected since the transfer of momentum and energy between gas and particles is changed for the different particle sizes at the same initial loading ratio.

4. Conclusions

The one-dimensional unsteady flowfield induced by a planar incident shock wave propagating into a gas-particle mixture without chemical reaction has been investigated numerically. The numerical results of the traces of the incident and reflected shock wave and accelerated planar interface of the gas-particle mixtures are in good agreement with the experimental data.

The intense transfer of momentum and energy between the gas and particle phases results in a significant increase in gas temperature. A temperature peak is formed in the suspension and follows the incident shock wave at a slower speed. The reflected shock wave formed keeps a relatively stable strength and follows the incident shock wave as well. Changing the Mach number, initial loading ratio or particle diameter causes different variations of gas temperature peaks with time.

For the higher initial loading ratios, the incident and reflected shock waves propagate with lower speeds, and the strength of the reflected shock wave is increased. For the smaller particle sizes, the incident shock wave is attenuated, but the change of the particle sizes from D_P=10 μm to D_P=100 μm has little influence upon the reflected shock wave.

Acknowledgements

The authors gratefully acknowledge the support from the Volkswagenwerk Fundation of Germany within its program to promote Chinese-German Scientific Cooperation.

References

Crowe CT (1982) Review - Numerical models for dilute gas-particle flows. J. of Fluids Engineering 104: 297-303

Geng JH, van de Ven A, Yu Q, Zhang F, Grönig H (1993) Interaction of a shock wave with a two-phase interface. Shock Waves (to be published)

Henderson CB (1976) Drag coefficient of spheres in continuum and rarefied flows. AIAA J. 14: 707-708

Igra O, Ben-Dor G (1988) Dusty shock waves. Applied Mechanics Reviews 41, 11: 379-437

Marble FE (1970) Dynamics of dusty gases. Ann. Rev. Fluid Mech. 2: 397-446

Olim M, Igra O, Mond M, Ben-Dor G (1989) Numerical investigation of the flow behind a shock wave propagating into a carbon-oxygen suspension. In: Kim YW (ed) Proc. 17th Intl. Symp. on Shock Waves and Shock Tubes, Bethlehem, American Institute of Physics, pp 684-689

Rudinger G (1980) Fundamentals of gas-particle flow. Elsevier Scientific Publishing Co.

Schaff S, Chambre R (1958) Flow of Rarefied Gases. In: Fundamentals of Gas Dynamics, Princeton Series, Vol. III, Princeton University Press, Princeton, NJ Amsterdam Oxford New York

Sommerfeld M, Grönig H (1983) Decay of shock waves in a dusty-gas shock tube with different configurations. In: Archer, RD and Milton, BE (eds) Proc. 14th Intl. Symp. on Shock Waves and Shock Tubes, Sydney, pp 470-477

Wallis GB (1969) One-dimensional Two-phase Flow. McGraw-Hill Co.

Turbulent Boundary Layer Over a Dust Bed

K.-Y. Chien*, R.E. Ferguson*, J.P. Collins*, A.L. Kuhl† and R.G. Batt‡
*Naval Surface Warfare Center, Silver Spring, MD 20903, USA
†Lawrence Livermore National Laboratory, Livermore, CA 94550, USA
‡TRW, Redondo Beach, CA 90278, USA

Abstract. The dynamics of mixing and entrainment induced by a steady turbulent boundary layer flow over a loose dust bed is studied with our second-order Godunov code. The problem considered models the late-time flow of explosions over soil surfaces, and extends previous shock-tube studies that focused on the early-time lofting of dust behind the shock front. As in our previous work, an equilibrium model was used for the dust-air mixture. The calculated boundary layer growth and mean-flow profiles are in reasonable agreement with wind-tunnel measurements. This suggests that our Convective Mixing Model is capable of simulating a wide variety of turbulent dusty flows.

Key words: Dusty flows, Turbulent boundary layer, Explosions

1. Introduction

Turbulent dusty boundary layers are a common feature of flow interactions with soil or ground surfaces. Much of our fundamental understanding of the problem comes from laboratory experiments. For example, Ausherman (1973) studied the mechanisms of initial dust lofting induced by a normal shock in a shock tube, while Hartenbaum (1971) measured the stagnation pressure profiles and dust scouring rate for a turbulent boundary layer over a loose dust bed in a blowdown wind tunnel. Batt et al. (1988) measured the instantaneous velocity and density profiles at various distances behind a propagating shock in a shock tube; more recently, Batt et al. (1992) have measured mean velocity and density profiles in a dusty flow wind tunnel.

There are also theoretical studies of steady and nonsteady dusty boundary layer problems. Mirels (1984) published analytical solutions and estimated the dust scouring rate, but he assumed self-similar boundary layer profiles of a power-law form. Others used finite-difference methods with turbulence models and an empirical dust scouring rate function (Denison 1990, Frolov et al. 1991). The adequacy of this approach is uncertain since the dust scouring rate function is not well established, and turbulence models were developed for clean incompressible flow.

More recently, we have pursued a new approach -a Convective Mixing Model- that follows the dynamic evolution of turbulent eddies on the computational grid. Dust scouring occurs naturally in the calculation as the turbulent eddies entrain dust from the fluidized bed, thus alleviating the need for a dust injection model. This approach was used previously to simulate the turbulent boundary layer formed by a planar shock propagating along a dusty wall (Kuhl et al. 1990); the calculated mean velocity and density profiles were shown to agree well with the shock-tube measurements of Batt et al. (1988). This paper then extends our previous shock tube studies to the steady-flow regime of a turbulent boundary layer flow over a loose dust bed. The calculated profiles are compared with the dusty flow wind tunnel data measured by Batt et al. (1992).

2. Formulation

A schematic of the flow configuration is shown in Fig.1. The specific case considered was that of a uniform airflow over a dusty wall with freestream Mach number 0.3. Air percolates into the dust layer, forming a fluidized mixture. As in our previous work (Kuhl et al. 1990), the dust was assumed to be in velocity and temperature equilibrium with the air, and the entire mixture was

Shock Waves @ Marseille III
Editors: R. Brun, L. Z. Dumitrescu

modeled as a perfect gas with a constant ratio of specific heats, $\gamma_m = 1.4$. This "dense gas" model represents an equilibrium approximation that focuses on the inertia effects of the dust.

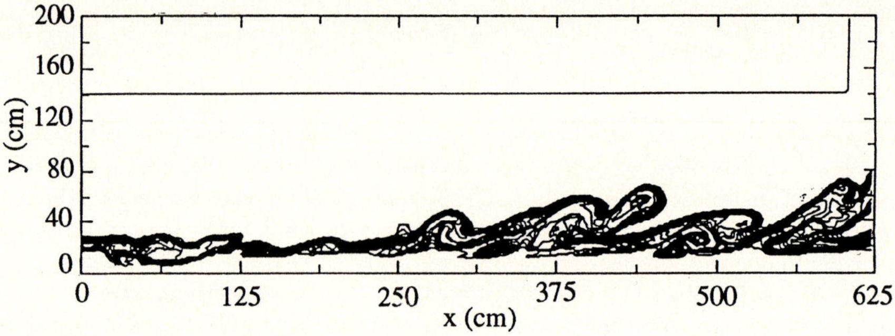

Fig. 1. Flow visualisation of dust entrainment

A turbulent, dusty boundary layer flow was fed into the left boundary of the computational grid. This grid consisted of 300 fine cells in the x direction and 71 fine cells in the y direction; a few coarse zones were used to the right of the fine mesh to remove downstream effects of the computational boundary. We chose $\Delta x = \Delta y = 2$ cm in the fine mesh region. To allow for the expansion of the fluidized bed, the top wall was placed at $y = 142$ cm. There is a gradual expansion of the wind-tunnel area in the experiment; this is approximated by a sudden expansion of the top wall to a value of $y = 612$ cm for $x > 602$ cm. This expansion is covered with 39 gradually expanding coarse cells in the y direction. The grid was initialized as follows: The fluidized bed was modeled as a four-cell-thick dense gas at a density of 50 mg/cc. To crudely model the soil underneath the fluidized bed, four additional cells of dense gas at a density of 1000 mg/cc were employed in the computation. The shear layer on the bottom wall was approximated by a $\tanh(y)$ profile with the proper asymptotes, and the density profile above the fluidized bed consisted of an eight-cell-thick transitional layer with freestream air above it. Note that the inflection point of the shear layer was located three-and-one-half cells above the fluidized bed. The left-hand boundary of the grid was then driven by these same profiles with sinusoidal perturbations on the velocity field only. Their frequencies corresponded to the frequency of maximum amplification rate from linear stability analysis and its first nine subharmonics. To reduce the total computational time for the calculation, the maximum perturbation amplitude used in the present study was five percent instead of the one percent value used in our previous study (Kuhl et al. 1990). A 5 cm thick boundary layer was placed at the top wall at $t = 0$. The dynamic evolution of the flowfield was calculated by the second-order Godunov scheme of Colella and Glaz (1985) that solves the two-dimensional nonsteady equations of gasdynamics. Note that the no-slip condition on the wall is imposed in the in-flow boundary condition of the calculation and this approach represents a large-Reynolds-number approximation to the dynamics of the flow. The calculation was run for 6000 cycles (about 210 minutes) on a Cray XMP computer to create a sufficient database for statistical analysis of the flow.

3. Results and discussion

The dust loading (comp-zone) depicted in Fig.1 provides a visualization of the mixing processes in the boundary layer region. The shear layer rolled-up into large rotational structures that entrained material from the fluidized bed. Thus, the dust scouring occurred naturally in the calculation without any modeling.

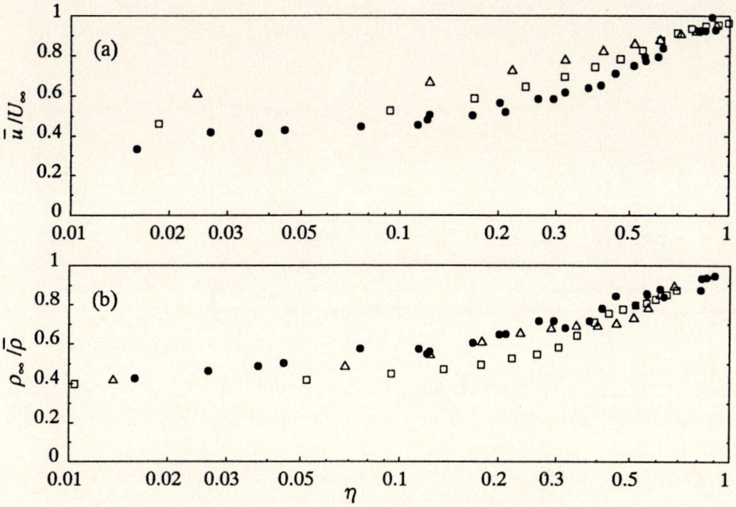

Fig. 2. Mean-flow profiles: (a) streamwise velocity; (b) inverse density. Present calcualations (△ : $x = 483$ cm , □ : $x = 577$ cm) vs. data (•)

Flowfield time histories were stored for all y-cells at stations $x = 215$ cm, 303 cm, 390 cm, 483 cm and 577 cm. These flowfield histories were then integrated over the last 2300 time steps to establish the mean-flow profiles. The boundary layer thicknesses of the first three stations are erratic (i.e., not well converged). Therefore, the mean velocity and specific volume profiles at only the last two stations are plotted in Fig.2. There, the dusty boundary layer scaling was employed:

$$\eta = (y - y_0)/\delta \tag{1}$$

where

$\delta =$ boundary layer thickness

$y_0 =$ fludized bed height (where $\bar{\rho}/\rho_\infty = 2.5$).

For comparison, we have also plotted in Fig.2 the wind-tunnel measurements of Batt et al. (1992) and the shock-tube calculations (Kuhl et al. 1990) and measurements (Batt et al. 1988). Note that the steady freestream flow present in the wind tunnel causes the density in the near wall region to be much lower than that of the corresponding transient shock tube case; this causes the velocity there to be higher than the corresponding shock tube value. Note also that the present calculated profiles are consistent with the wind-tunnel measurements of Batt et al.(1992). In addition, the calculated boundary layer thickness values of 24.8 and 36.4 cm at the last two stations are also in good agreement with the data (see Fig.3).

4. Concluding remarks

Dust lofting can be viewed as a two-step process: the formation of a fluidized bed, followed by entrainment of the dense material from the bed by the rotational structures (eddies) of the turbulent boundary layer. The present calculations focus on following the dynamic evolution of the large-scale turbulent eddies and the inertia effects of the dust. The calculated boundary layer growth and mean-flow profiles are seen to agree reasonably well with wind-tunnel measurements. This fact, together with the even better agreement between calculations and measurements observed in our previous shock-tube study, suggests that our Convective Mixing Model is capable of simulating a wide variety of turbulent dusty flows. The main limitation of the present study is

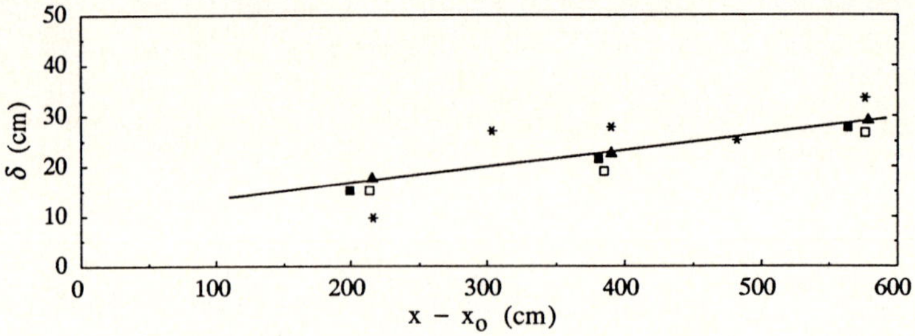

Fig. 3. Boundary layer thickness δ vs. distance
Data (□ : $u_\infty = 37$ m/s; ■ : $u_\infty = 67$ m/s; ▲ : $u_\infty = 98$ m/s). AMR calc. (∗)

the 2-D flow approximation; the 3-D effects are stronger for the present wind-tunnel configuration than the shock-tube problem previously studied, hence we are now starting some 3-D simulations.

Acknowledgment

Work performed under the auspices of the U.S. Dept. of Energy by the Lawrence Livermore National Laboratory under Contract number W-7405-ENG-48. Also sponsored by the Defense Nuclear Agency under contract DNA IACRO 93-818 and Work Unit 00003.

References

Ausherman DR (1973) Initial dust lofting: shock tube experiments. Defense Nuclear Agency Report 3162F

Batt RG, Kulkarny VA, Behrens HW, Rungaldier H (1988) Shock-induced boundary layer dust lofting. In: Gronig H (ed) Proc. 16th Intl. Symp. on Shock Tubes and Waves, VCH, Weinheim, Germany, pp 209-215

Batt RG, Petach MB, Peabody II SA, Batt RR (1992) Experimental investigation of dust and vegetation sweep-up. TRW Report (rough draft)

Colella P, Glaz HM (1985) Efficient solution algorithms for the Riemann Problem for real gases. J. Computational Physics 59, 2: 264-289

Denison MR (1990) NIAPE - Volume VI, a two-layer model of dust lofting. Defense Nuclear Agency Technical Report 89-48-V6

Frolov SM, Mack A, Roth P (1991) Diffusion model of dust lifting behind a shock. 13th International Colloquium on Explosions and Reactive Systems

Hartenbaum B (1971) Lofting of particles by a high-speed wind. Defense Nuclear Agency Report 2737

Kuhl AL, Chien K-Y, Ferguson RE, Collins JP, Glaz HM, Colella P (1990) Simulation of a turbulent, dusty boundary layer behind a shock. In: Kim YW (ed) Proc. 17th Intl. Symp. on Shock Waves and Shock Tubes, AIP Conference Proceedings 208, American Institute of Physics, New York, NY, pp 762-769

Mirels H (1984) Blowing model for turbulent boundary-layer dust ingestion. AIAA J. 22, 11:1582-1589

Optical Observation of Mach Reflection of a Partly Dispersed Shock Wave by Sub-Micron Water Droplets in Air

H. Hirahara and M. Kawahashi
Faculty of Engineering, Saitama University Urawa, Saitama 338, Japan

Abstract. Shock reflection process in fog was investigated experimentally. The fog was produced by non-equilibrium condensation of humid air in a test section. A double-diaphragm-type shock tube was used in order to generate sub-micron water droplets in fog, which was produced by the condensation generated when a mixture of water vapor and air expands adiabatically. After breaking-up of a diaphragm, the mixture expands to the vacuum tank so that sub-micron water droplets are generated. A shock wave, produced by breaking a second diaphragm, propagates into the shock tube 150 ms late after the first diaphragm breakup. The growth of the condensed water droplets was observed by measuring the light extinction rate. Shock speed was calculated by the recorded signals of pressure transducers. Shock reflection was observed by a high-speed image converter camera and still camera.

Key words: Shock waves in fog, Mach reflection, Shock tube, Optical observation, Shadowgraph

1. Introduction

Investigations of fully and partly dispersed shock waves have been carried out by many researchers with several interests (Cole and Dobbins, 1970; Jackson and Davidson 1983). A shock wave in a two-phase media is dispersed by the mass, momentum and energy transfer between the liquid drops and the ambient gas. These phenomena of shock dispersion in multi-phase media depend on the interphase mass, momentum and energy transfer processes behind the shock wave. Goossens et al. (1988) and Smolder et al. (1991) have studied the characteristics of a normal shock wave propagating through moist air experimentally. Young and Guha (1991), Guha (1991) investigated the normal shock structure in two-phase vapor-droplet flows theoretically. They have classified the dispersed shock wave structure into three types of fully dispersed shock waves and a partly dispersed one. They pointed out the criterion of shock strength between them. According to their analysis, each type is determined by the relation in magnitude of the wave velocity relative to equilibrium acoustic velocities. On the other hand, the reflected shock strength in Mach reflection varies according to the incident shock strength and wedge angle. Since the reflected shock wave is very weak relative to the incident shock wave, the reflected shock wave might be dispersed. Therefore, the flowfield around the triple point depends strongly upon the properties of the reflected shock.

In the present report, the shock reflection process in fog was investigated experimentally. Sub-micron droplets in mist were formed by non-equilibrium nucleation in a shock tube. Growth of the condensed water droplets was observed by measuring the light extinction rate. The reflection process has been observed by the shadowgraph method.

Shock Waves @ Marseille III
Editors: R. Brun, L. Z. Dumitrescu

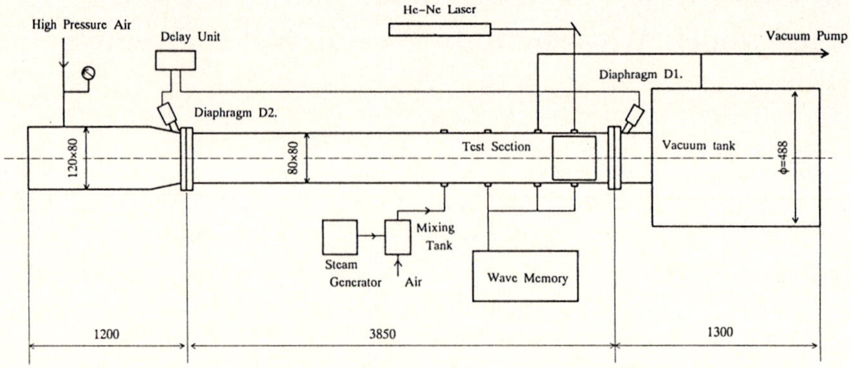

Fig. 1. Schematic diagram of the experimental apparatus

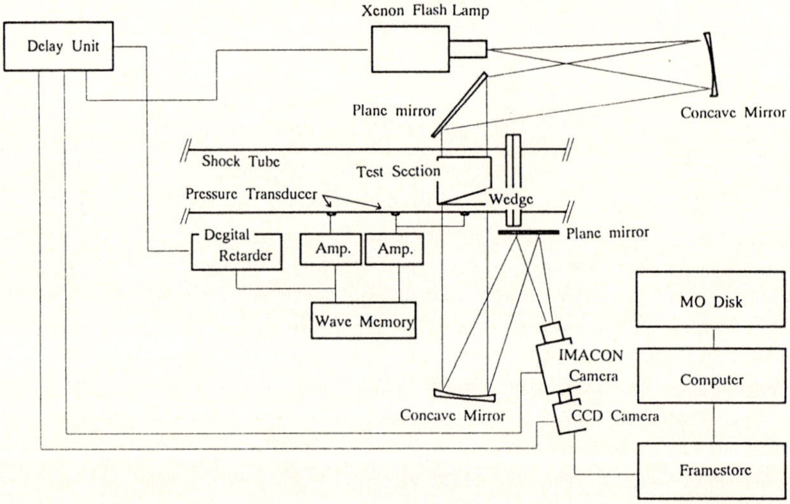

Fig. 2. Optical setup for visualization and measurements

2. Experiments

Fig.1 shows a schematic diagram of the experimental apparatus. A double diaphragm-type shock tube was used in order to generate sub-micron water droplets, which were produced by condensation when a gas mixture of water vapor and air expands adiabatically in the test section. The gas mixture was charged in the test section at atmospheric pressure and temperature. Pressure in the low-pressure chamber was set at 1.33 kPa (10 Torr). The high-pressure chamber was charged at 110 to 130 kPa with air. After the break-up of the right diaphragm, the mixture expands into the vacuum tank so that sub-micron water droplets are generated due to non-equilibrium condensation. Flow in the test section reached up to the steady state after 100 ms approximately. Pressure at the steady state was about 17.0 kPa (130 Torr). Temperature in this condition can be calculated numerically with the measured pressure. Calculated temperature was 180 K.

The shock wave which was produced by the break-up of the left diaphragm propagates into the shock tube with a 150 ms delay after the breakup of the first diaphragm. Delay time is controlled with one ms resolution by using a delay timer unit. Its accuracy depends on the initial relative humidity and uniformity of the fog. Therefore, the final accuracy in delay of shock arrival time resulted in 10 ms.

The optical setup is shown in Fig.2. Nucleation and growth of the condensed water droplets was observed with the measurement of light extinction rate at the center of the optical observation window. The intensity of transmitted light was measured by a PIN photo-diode (Hamamatsu S1722-01). The shock speed was calculated by the recorded signal of two pressure transducers (Kulite XCQ062). The reflection pattern of the shock wave was observed with a high-speed image converter camera (IMACON). The acquisition rate of the shadowgraph image was set at 10 μs.

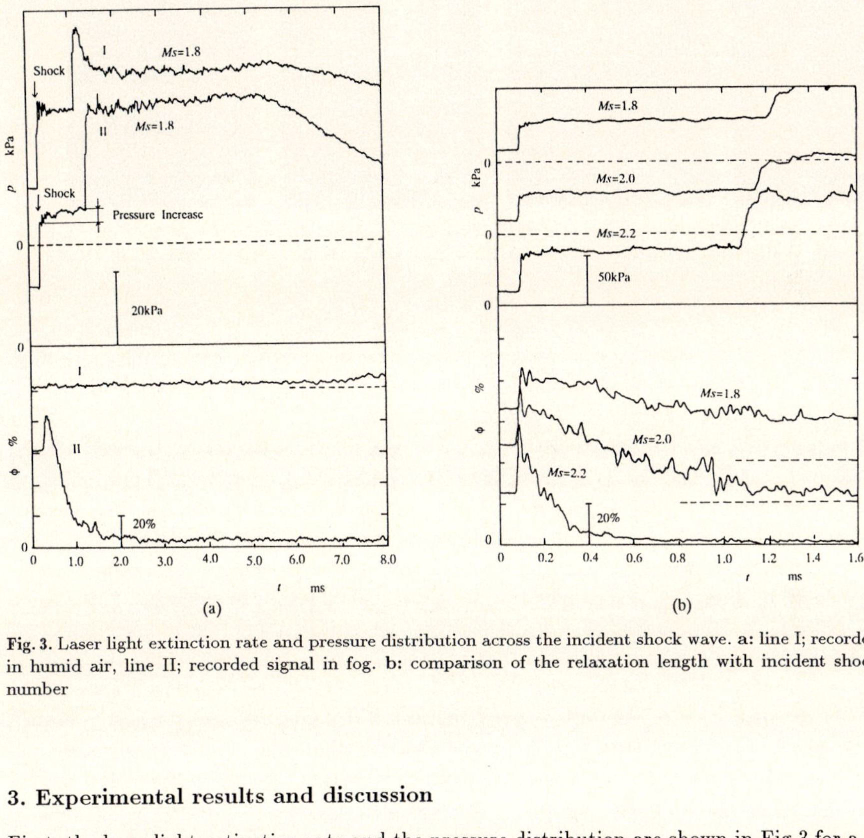

Fig. 3. Laser light extinction rate and pressure distribution across the incident shock wave. **a**: line I; recorded signal in humid air, line II; recorded signal in fog. **b**: comparison of the relaxation length with incident shock Mach number

3. Experimental results and discussion

First, the laser light extinction rate and the pressure distribution are shown in Fig.3 for a normal shock wave propagating into the test section. The light extinction rate was recorded at the center of the observation window as shown in Fig.1. Fig.3(a) shows a typical pressure signal behind the propagating normal shock. Line I is for humid air in the absence of sub-micron droplets and line II for two-phase flow with droplets. Immediately after the passage of the shock, the pressure is nearly constant in the first case (line I) as shown in Fig.3(a), while that of line II rises gradually as compared with line I. This pressure increase is due to the re-evaporation of the liquid droplets. The change of light extinction rate indicates the fluctuation of the mist presence ratio.

Fig.3(b) shows the change of extinction rate with incident shock Mach number. Experimental results show that the pressure increases discontinuously immediately after the shock so that the

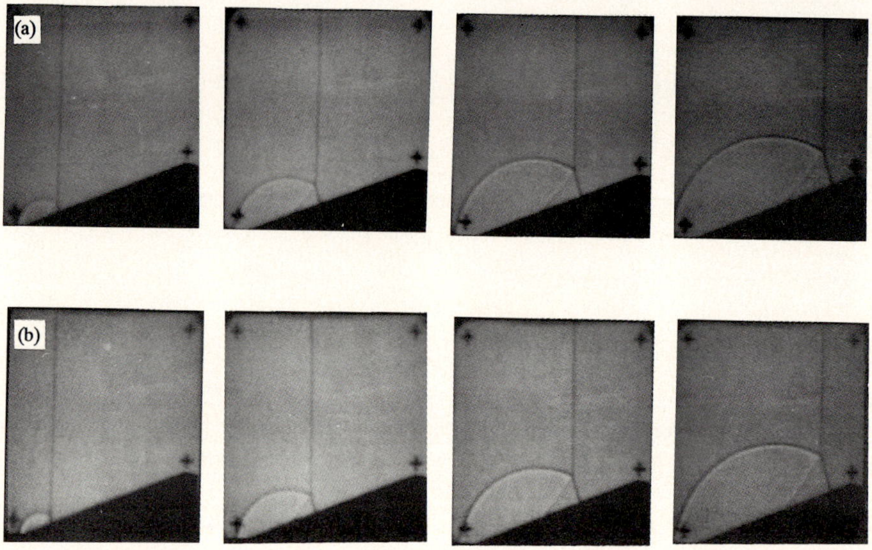

Fig. 4. Shadowgraphs of Mach reflection over a wedge in humid air, (absolute humidity $x = 4.1 \times 10^{-3}$), in the absence of water droplets. **a:** $M_s = 1.8$; **b:** $M_s = 1.9$. Photographs were taken every 20 μs after the shock passsage at the wedge apex

incident shock wave is partly dispersed and a relaxation region is formed behind the shock wave due to evaporation of the water droplets. For example, the length of the relaxation region was 100 mm approximately at the shock Mach number $M_s = 2.2$.

Shadowgraphs of Mach reflection in fog taken by the image converter camera (IMACON) are shown in Figs.4-6. The shock Mach number M_s is 1.8 to 2.4, and the absolute humidity $x = 5.5 \times 10^{-3}$ and 9.0×10^{-3} kg/kg, respectively. Photographs in Fig.4 were obtained in the absence of droplets, so the fog is not observed in this photograph. These photographs are shown in order to compare the shock pattern and flowfield over the wedge. The configuration of shock reflection shown in these photos corresponds to the experimental results in a real gas. Figs.5 and 6 show the reflection pattern in mist. The initial density of the mist in front of the propagating shock wave was increased by the shock wave, so that a dense cloud was observed behind the incident shock wave. The shock reflection process is regarded as a completely unsteady flow because of the existence of the relaxation region due to re-evaporation. Although a relaxation region is formed behind the shock wave, as described in the previous section, the triple point propagates along a nearly straight line within our experimental error, which intersects the apex of the wedge. So, the observed shock configuration near the triple point might be regarded as a pseudo-steady one.

The reflected shock wave was dispersed significantly, so that the triple point becomes indistinct. The strength of the reflected shock has an effect upon the formation of the triple point and Mach shock. A bent Mach shock was observed at the low shock Mach numbers, as a similar bending was observed for the lower incident Mach numbers which should be close to unity in a real gas. The dispersion of the reflected shock wave might be a very important and interesting problem.

Furthermore, we are able to recognize a linearly developed thermal boundary layer over the wedge surface when $M_s = 2.4$. It is suggested that the re-evaporation is more progressed than in the far region, e.g. an undisturbed region behind an incident shock wave. We could not notice such a rapid evaporation region in the boundary layer in the case of lower Mach numbers. Such

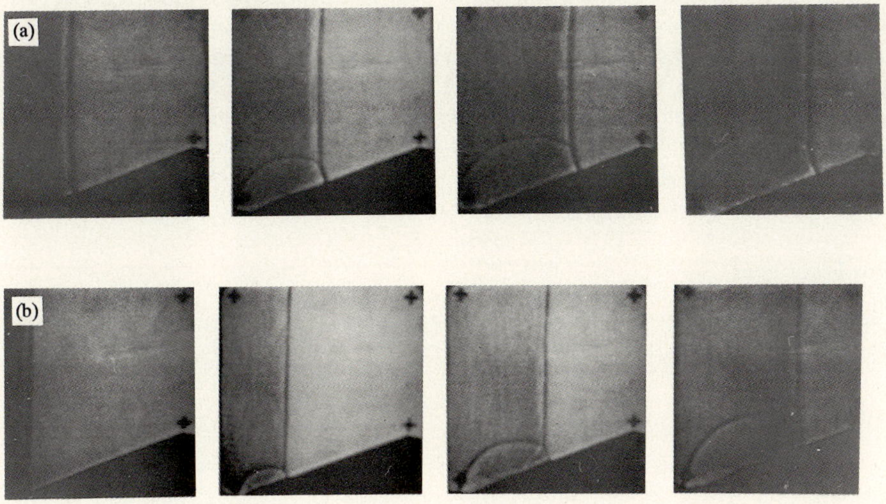

Fig. 5. Shadowgraphs of Mach reflection over a wedge in mist. a: $M_s = 1.9$, absolute humidity, $x = 5.5 \times 10^{-3}$. b: $M_s = 2.5$, absolute humidity, $x = 5.2 \times 10^{-3}$. Photographs were taken every 20 μs after the shock pasage at the wedge apex

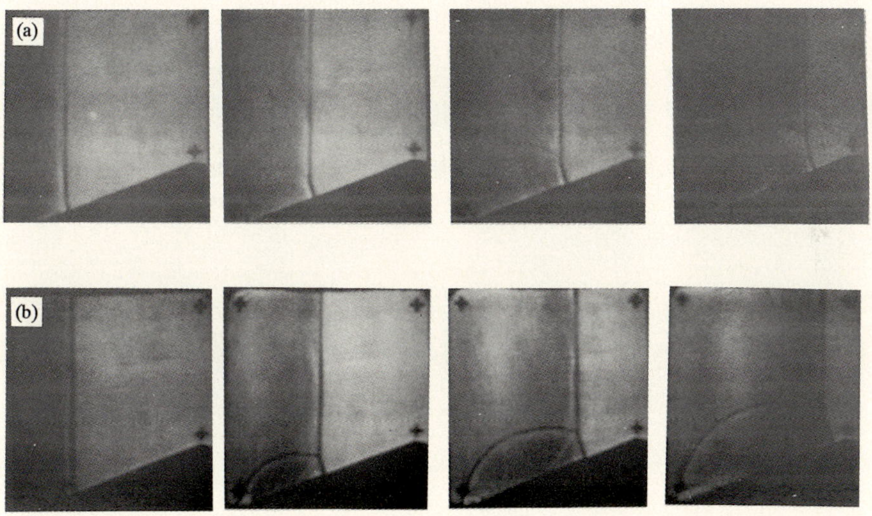

Fig. 6. Shadowgraph of Mach reflection over a wedge in moist air. a: $M_s = 1.8$; absolute humidity, $x = 9.0 \times 10^{-3}$. b: $M_s = 2.6$; absolute humidity, $x = 9.9 \times 10^{-3}$. The timing of each photograph is the same as in Fig.5

a rapid evaporation must be attributed to the turbulent flow following the slipstream, which is resulting from the interaction between the flow passing through the reflected shock and that passing through the Mach shock.

4. Concluding remarks

Reflection of a dispersed shock wave in fog was investigated experimentally. First, pressure measurements and optical measurements of the extinction rate of the transmitted light were carried out to measure the relaxation due to the evaporation behind an incident shock wave. Optical observation was performed to study the shock reflection in fog using a high-speed image converter camera. Experimental results show a partly dispersed shock structure, depending on the incident shock Mach number and absolute humidity. The reflected shock wave is dispersed so that a bent Mach shock was observed at a moderate shock Mach number. A rapidly evaporating region was observed in the boundary layer which is following the slipstream.

Acknowledgement

This work was supported by grants from the Ministry of Education, Science and Culture, Japan, (Grant-in-Aid for Scientific Research on Priority Areas, Shock Wave Research)

References

Cole JE, Dobbins RA (1970) Propagation of sound through atmospheric fog. J. Atmos. Soc. 27: 426- 434

Goossens HWJ, Cleijne JW, Smolder HJ, van Dongen MEH (1988) Shock wave induced evaporation of water droplets in a gas-droplet mixture. Exp. Fluids 6: 561-568

Guha A (1991) Jump condition across normal shock waves in pure vapour-droplet flows. J. Fluid Mech. 241: 349-369

Jackson R, Davidson BJ (1983) An equation set for nonequilibrium two-phase flow, and an analysis of some aspects of choking, acoustic propagation, and losses in low-pressure wet steam. Intl. J. Multiphase Flow 9: 491-510

Smolder HJ, de Kok CJGM, van Dongen MEH (1991) Condensation and evaporation relaxation induced by non-linear waves. In: Grönig H (ed) Proc. 16th Intl. Symp. on Shock Waves. Springer-Verlag, pp 475-480

Young JB, Guha A (1991) Normal shock-wave structure in two-phase vapour-droplet flows. J. Fluid Mech. 228: 243-274.

Effects of the Injection of Droplets on a Stationary Shock Wave in a Nozzle

F. Utheza, R. Saurel, E. Daniel and J.C. Loraud
I.U.S.T.I./ S.E.T.T.- Equipe Ecoulements Diphasiques et Réactifs - Case 321 - URA CNRS 1168 - Université de Provence - Centre de Saint Jérôme - 13397 Marseille Cedex 20 - France

Abstract. A numerical simulation of the flow of gases through a converging-diverging nozzle, where droplets are injected in a given section of the divergent, is presented. The two-dimensional equations are solved by a TVD scheme where fluxes are computed by using a new Riemann solver for the dispersed phase, and an exact Riemann solver for the gas phase equations. The behaviour of the initial shock wave as a function of the particles injection location is examined in this paper.

Key words: Nozzle flow, Droplet injection

1. Introduction

Shock wave behaviour in two-phase flows is an important engineering problem encountered in dust explosions, nuclear safety, rocket liquid motors etc. Numerous numerical studies have been carried out to understand the phenomena involved in a two-phase nozzle flow (Chang 1980; Sommerfield 1987; Daniel et al. 1992, 1993). The specific feature of the present work lies in the fact that the inlet flow is a one-phase flow and the dispersed phase is injected into this flowfield: the two-phase flow is the result of the injection of the water droplets. So there are two kinds of flows: a one-phase flow upstream the injection, then a two-phase one. The gas is assumed to be inviscid and the phenomenon is unsteady and two-dimensional.

In this paper, authors focus their attention on the influence of particle injection in a converging-diverging nozzle where a steady shock wave is settled. In particular, the behaviour of the initial shock wave as a function of the location of particle injection is examined (upstream or downstream the initial position of the shock wave). The two-phase model chosen is a two fluid-model where each phase is considered as a continuous medium. In order to emphasize the basic phenomena, only drag forces as momentum exchange mechanisms are considered. The numerical solution of this problem requires the use of a high-resolution scheme to handle and track discontinuities in the gas phase. In the present problem, particles are injected into a one-phase flow, then an accurate tracking of the droplet front is necessary. Several approaches are possible. The first consists in solving the gas phase equations by an Eulerian technique and the dispersed phase equations with a Lagrangian method. But this procedure is heavy in programming and is computer-time-intensive, when solving an injection problem. The second approach consists in solving the two-phase equations by the same Euler method based upon a centered scheme as generally used in gas dynamics. This technique requires the introduction of artificial viscosity terms to avoid nonlinear instabilities which could reduce the accuracy of the solution.

For all these reasons, the two-dimensional equations are solved by a TVD scheme in which the fluxes are computed by using a new Riemann solver (solution of a multivalued problem) for the dispersed phase, and an exact Riemann solver for the Euler equations.

Shock Waves @ Marseille III
Editors: R. Brun, L. Z. Dumitrescu

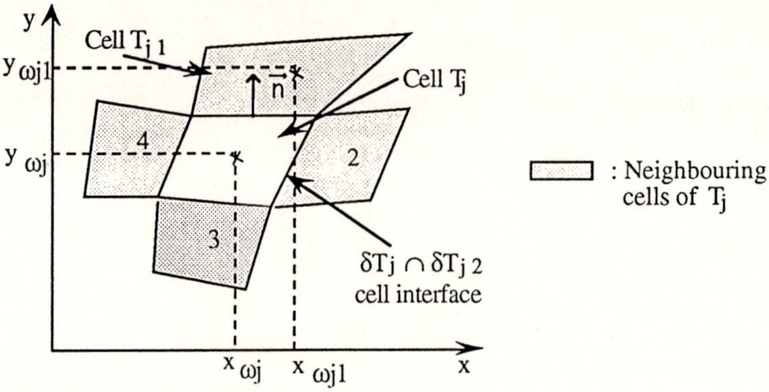

Fig. 1. Elements of a non-structured grid

2. Governing equations

In most numerical simulations of gas particle two-phase flows, available in the litterature, the "two-fluid" model is commonly retained. In this model, as said above, the gas and dispersed phases are treated as distinct continuous media, while momentum and energy exchanges take place across the particle surface, involving the viscosity and thermal conductivity. The two-phase flow considered here is a very dilute one, then the assumption $\alpha \approx 1$ can be used. Taking into account this assumption, the equation is written in the following conservative form, in two-dimensionnal Cartesian coordinates:

Gas phase:

$$\left\{\begin{array}{l}
\dfrac{\partial \rho_g}{\partial t} + \dfrac{\partial \rho_g u_g}{\partial x} + \dfrac{\partial \rho_g v_g}{\partial y} = 0 \\
\dfrac{\partial \rho_g u_g}{\partial t} + \dfrac{\partial (\rho_g u_g^2 + P)}{\partial x} + \dfrac{\partial \rho_g u_g v_g}{\partial y} = F_x \\
\dfrac{\partial \rho_g r_g}{\partial t} + \dfrac{\partial \rho_g u_g v_g}{\partial x} + \dfrac{\partial (\rho_g v_g^2 + P)}{\partial y} = F_y \\
\dfrac{\partial \rho_g e_g}{\partial t} + \dfrac{\partial u_g (\rho_g e_g + P)}{\partial x} + \dfrac{\partial v_g (\rho_g e_g + P)}{\partial y} = \\
\qquad = F_x u_p + F_y v_p
\end{array}\right. \qquad (1)$$

Dispersed phase:

$$\left\{\begin{array}{l}
\dfrac{\partial \rho_p}{\partial t} + \dfrac{\partial \rho_p u_p}{\partial x} + \dfrac{\partial \rho_p v_p}{\partial y} = 0 \\
\dfrac{\partial \rho_p u_p}{\partial t} + \dfrac{\partial \rho_p u_p^2}{\partial x} + \dfrac{\partial \rho_p u_p v_p}{\partial y} = -F_x \\
\dfrac{\partial \rho_p v_p}{\partial t} + \dfrac{\partial \rho_p u_p v_p}{\partial x} + \dfrac{\partial \rho_p v_p^2}{\partial y} = -F_y \\
\dfrac{\partial \rho_p e_p}{\partial t} + \dfrac{\partial \rho_p u_p e_p}{\partial x} + \dfrac{\partial \rho_p v_p e_p}{\partial y} = 0 \\
\dfrac{\partial N}{\partial t} + \dfrac{\partial N u_p}{\partial x} + \dfrac{\partial N u_p}{\partial y} = 0
\end{array}\right. \qquad (2)$$

The homogeneous part of system (1) is readily identified to the Euler equations. One should also remark that systems (1) and (2) are coupled only by the phase interaction terms, the void fraction being absent. The homogeneous left-hand parts of systems (1) and (2) being uncoupled, they may therefore be solved independently. To summarize, the following procedure results:

- Solve the homogeneous problem for the gas motion
- Solve the homogeneous problem for the particles
- Couple the two systems via the interaction terms.

One then recovers, for system (1), the three well-known real caracteristic directions: $u+a, u-a, u$ (a being the sound speed); this ensures the hyperbolic nature of the system and allows a relatively easy resolution, with the help e.g., of some high-resolution scheme (Van Leer 1979; Roe 1981). System (2) admits, on the other hand, only a single characteristic direction, u_s. It can be shown that this system is hyperbolic degenerated and may have multivaluated solutions.

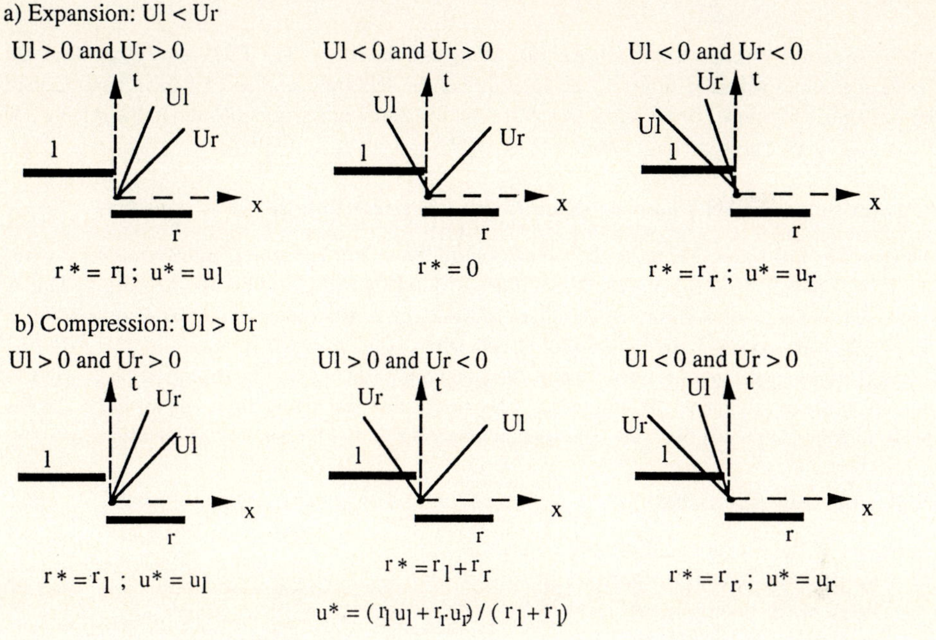

Fig. 2. Configurations for the Riemann Problem

3. Numerical procedure

The numerical algorithm is based upon the original ideas of Van Leer (1979) and extended to unstructured meshes following Forestier (1992). Systems (1) or (2) can be written in vector form:

$$\frac{\partial U}{\partial t} + \frac{\partial F_1}{\partial x} + \frac{\partial F_2}{\partial x} = 0$$

In the unstructured mesh, the following quantities are defined, as shown in Fig.1:

T_j, the current cell, where the solution is monotonic;

U_j, the average value of the solution and its slopes along x, d_j^n, and y, δ_j^n.

w_j is the gravity center, located at x_{wj} and y_{wj} with respect to the x and y-axes.

Let $I(j) = [T_j, \ldots T_{jk}]$ be the set of neighbouring cells, and δT_j the contour of cell T_j. The interface between T_j and another cell, T_{jk}, is denoted: $\delta T_j \cap \delta T_{jk}$.

The first step of the algorithm consists in evaluating the quantity U at each interface and at time $t = n + 1/2$. This is given by:

$$U^{n+1/2}_{\delta T_j \cap \delta T_{jk}} = U_j^n + \tfrac{1}{2} d_j^n \left(x_{wj} - x_{wjk}\right) + \tfrac{1}{2} \delta_j^n \left(y_{wj} - y_{wjk}\right) - \frac{\Delta t}{2} \frac{\partial F_1}{\partial U} \left(U_j^n\right) d_j^n - \frac{\Delta t}{2} \frac{\partial F_2}{\partial U} \left(U_j^n\right) \delta_j^n$$

At the second step, the fluxes of the various variables which cross the interface $\delta T_j \cap \delta T_{jk}$ along the normal are calculated. A one-dimensional Riemann problem is then solved along the outward pointing normal $\boldsymbol{n}$. For the gas-phase equations we use an exact Riemann solver following Gottlieb's (1988) algorithm. For the particle phase we use a new exact Riemann solver as detailed in the next paragraph.

Then, a conservation stage is applied:

$$U_j^{n+1} = U_j^n - \frac{\Delta t}{S_{T_j}} \sum_{k=1}^{I(j)} \left\{ F_1 \left(V_{\delta T_j \cap \delta T_{jk}}^{n+1/2} \right) n_{x\delta T_j \cap \delta T_{jk}} + F_2 \left(V_{\delta T_j \cap \delta T_{jk}}^{n+1/2} \right) n_{y\delta T_j \cap \delta T_{jk}} \right\} = 0$$

where we asume that the fluxes are constant along each interface. The TVD concept is then used by means of the Minmod limiter. The solution for the Riemann problem for ideal gases is well known since Godunov (1959) but it is not for the dispersed phase. Details about the scheme for two-phase flows can be found in Saurel et al. (1993).

4. Solution of the Riemann problem for the dispersed phase

This system has three real merged eigenvalues, u_p. So it is a hyperbolic multivaluated system. As a consequence, the trajectories of two different fluid elements could cross each other. This is expressed in Fig.2; subscripts r and l relate to the right and left-side states, while superscript $*$ marks the state retained at the interface of two cells.

With the knowledge of the solution of the Riemann problems for the dispersed phase, all the numerical fluxes are known. To summarize, the same numerical algorithm is employed for the gas phase and the dispersed one, only the Riemann solver changes for each set of equations.

5. Results and discussion

Geometry of the converging-diverging nozzle:

- lengh of the convergent part: 0.7 m
- lengh of the divergent part: 1.22 m
- area ratio (outlet / throat): 4.4

All the following results are established with the same tank conditions, $P_{res} = 5\dot{1}0^5$ Pa and T_{res}=1000 K. The initial radius of the injected particles is 1 μm. Particles are injected in the divergent at different locations: (Fig.3)

- Between the throat and the initial shock wave (abscissa: 1.02 to 1.11 m). At the location of droplet injection, the initial gas velocity is V_g=920 m/s and the temperature is T_g=542 K. Since the injected particles have an injection velocity lower than that of the gas, the gas phase is slowed down and the pressure increases. After the shock, in the subsonic part, the gas is accelerated, then the pressure reduces. So, the pressure discontinuity characterized by the shock wave disappears.
- Between the initial shock wave and the outlet section (abscissa: 1.42 to 1.48 m). The initial gas velocity is V_g=249 m/s and the temperature $T-g$=929 K, at the droplet injection station.

- If the particle speed is lower than the gas velocity, then the droplets brake the gas. Compressions waves are produced. They set the shock wave in motion and they move it toward the throat.

- If the droplet injection velocity is superior to the gas velocity, then the gas is accelerated. In the subsonic flow, when the velocity increases, the pressure decreases. Thus, expansions waves appear because of the gas acceleration. These expansion waves put the shock in motion toward the outlet section. This is equivalent to a decreasing pressure in the downstream tank. The physical behaviour previously discussed confirms the one-dimensional study of Daniel et al. (1993).

6. Conclusion

A numerical simulation of the flow in a nozzle, where an initial steady shock wave is settled and where droplets are injected, has been carried out. Results show a different behaviour of the flow and particularly of the shock wave, depending upon the particles injection location and the way in which they are injected.

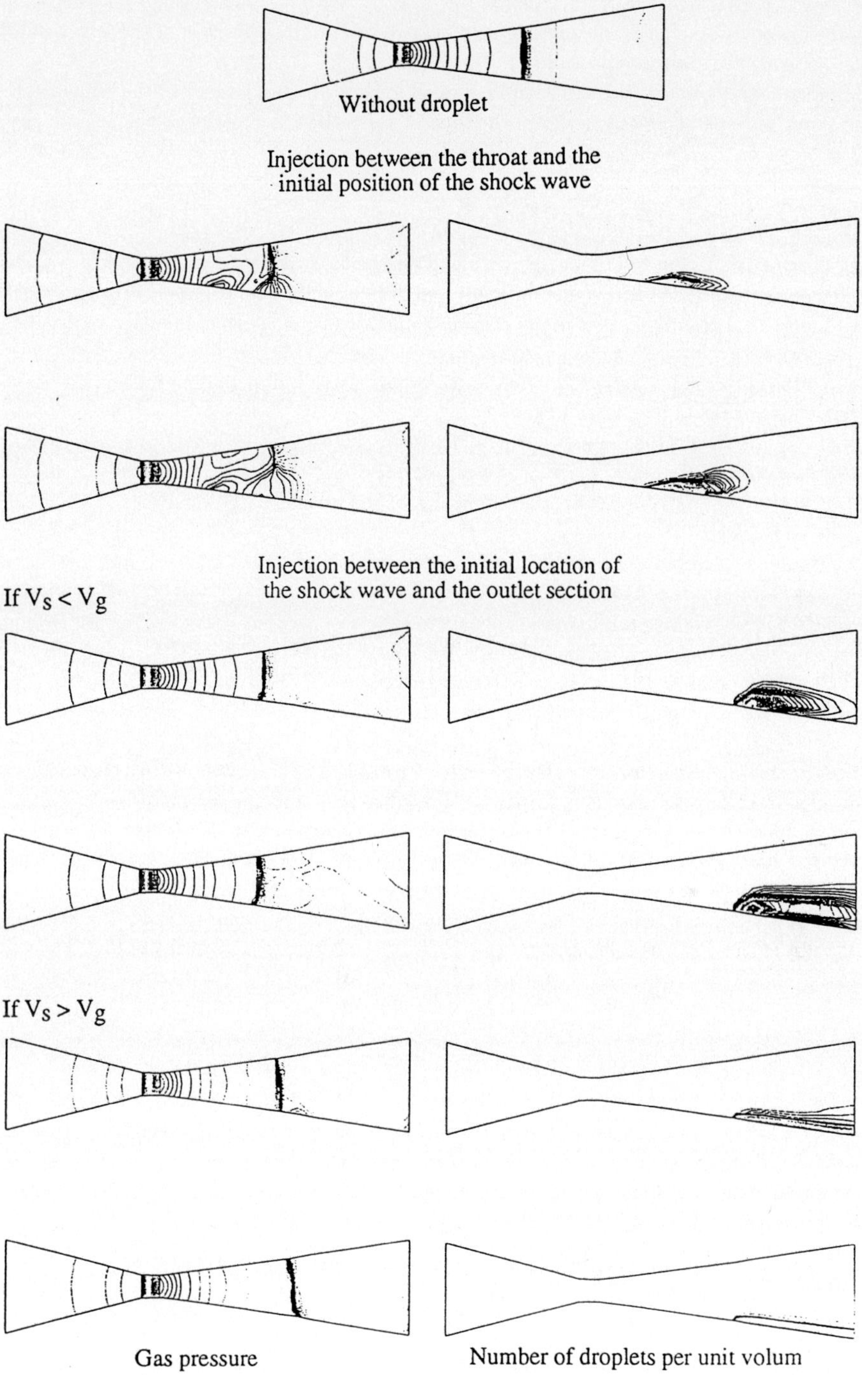
Without droplet
Injection between the throat and the initial position of the shock wave
Injection between the initial location of the shock wave and the outlet section
If $V_S < V_g$
If $V_S > V_g$
Gas pressure
Number of droplets per unit volum

Fig. 3.

In a first configuration, the injection is located between the throat and the initial location of the shock wave. For all velocities of injection of the droplets, the shock disappears or is weakened. On the other hand, when the droplets are injected between the initial position of shock wave and the outlet section, two cases are possible:

- the shock moves towards the throat, if the injection velocities are inferior to the gas velocity.
- the shock moves towards the outlet if the droplet velocities are superior to the gas velocity.

References

Chang IS (1980) One and two-phase nozzle flows. AIAA Journal 18: 1455-1461

Daniel E, Loraud JC, Larini M (1992) Influence de l'injection de gouttes d'eau dans de la vapeur d'eau en écoulement dans une tuyère. Intl. Journal of Heat and Mass Transfer 36, 6: 1619-1632

Daniel E, Larini M, Loraud JC, Porterie B (1992) A numerical simulation of injection of droplets in a compressible flow. AIAA Paper 92-2929, Nashville, TN

Forestier A (1992) Second order scheme for Euler equations in bidimensional unstructured geometries. Submitted to J. Comp. Phys.

Gottlieb JJ, Groth CPT (1988) Assessment of Riemann Solvers for unsteady one-dimensional inviscid flows of perfect gases. J. Comp. Phys. 78: 437-458

Roe PL (1981) Some contribution to the modelling of discontinuous flows. Proc. AMS-SIAM Seminar, San Diego

Saurel R, Daniel E, Loraud JC (1993) Two-phase flows: Second order schemes and boundary conditions. Submitted to AIAA Journal

Sommerfeld M (1987), Numerical simulation of supersonic two-phase gas-particles flow. In: Grönig H (ed) Proc. 16th Intl. Symposium of Shock Tubes and Waves, pp. 235-241

Van Leer B (1979) Toward the ultimate conservative scheme V. A second-order sequel to Godunov's method. J. Comp. Phys. 32: 101-136

Impact of Strong Shock Waves on Monodisperse Isopropanol Droplet Streams

N. Widdecke, W. Klenk and A. Frohn
ITLR Institut für Thermodynamik der Luft- und Raumfahrt, Universität Stuttgart, Pfaffenwaldring 31, 70550 Stuttgart, FRG

Abstract. In the present paper the impact of strong incident shock waves on monodisperse droplet streams is investigated in a shock tube. The droplets are produced with a vibrating-orifice droplet generator. With a new injection technique the monodisperse droplet streams are introduced into the test section of the shock tube. Droplet diameters range from 50 to 200 μm, Weber numbers from 100 to 15000 and Reynolds numbers from 50 to 15000. Droplet spacing is adjustable between one and three droplet diameters. Shock Mach numbers can be varied between $M_S = 2$ and $M_S = 6$.

Key words: Atomization, Droplet breakup, Monodisperse droplet streams

1. Introduction

Secondary atomization of droplets and sprays by shock waves has been studied in the past by many authors e.g. by Ranger and Nicolls (1969), by Wierzba and Takayama (1987) and by Eastes and Samuelsen (1992). These investigations have been performed mainly with drops in the millimeter-range or for Weber numbers below 200. A difficult problem for this work is to introduce liquid particles properly into the shock tube without disturbing the droplets and without changing the ambient conditions in the shock tube before the shock wave arrives. During the time between the introduction of the droplets into the shock tube and the arrival of the shock wave the droplets have to stay in the test gas. Droplet-droplet and droplet-wall interactions as well as droplet evaporation must be avoided. Especially droplets consisting of volatile fuels may cause a change of the initial conditions of the droplet parameters and of the thermodynamic properties of the driven gas. As a consequence the actual test conditions are not well defined. To avoid these problems a new injection facility has been developed.

2. Experimental technique

In the facility described here the droplet injection starts only few seconds before the shock wave arrives at the test section. Thus no substantial pre-evaporation of the droplet liquid can take place. By using monodisperse droplet streams droplet-droplet interactions are avoided. Accumulation of droplet liquid in the test section of the shock tube does not occur, as the droplets leave the test section through a small orifice opposite to the point of injection. With this injection system well-defined initial conditions at the beginning of an experiment could be established. Whereas previous investigations were carried out mainly with larger droplet diameters at lower Weber and Mach numbers, the experiments described here deal with Weber numbers up to We = 15000 and with shock Mach numbers up to $M_S = 6$, but with considerably smaller droplets. The diameters were in the range 50 μm $\leq d_T \leq$ 200 μm. The experiments have been performed in a shock tube having a circular cross-section with a diameter of 37 mm. The length of the driver section was 1.2 m and the length of the driven section 6.4 m. A mixture of helium, hydrogen and nitrogen at room temperature at pressures up to 100 bar has been used as driver gas, and synthetic air in the pressure range 1 mbar $\leq p_1 \leq$ 1 bar as driven gas. A schematic diagram of the experimental apparatus is shown in Fig.1. Monodisperse droplet streams are generated by means of a vibrating

Shock Waves @ Marseille III
Editors: R. Brun, L. Z. Dumitrescu

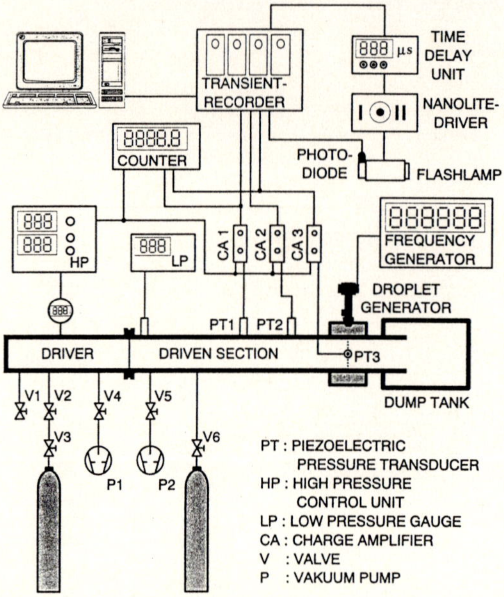

Fig. 1. Shock tube with associated instrumentation

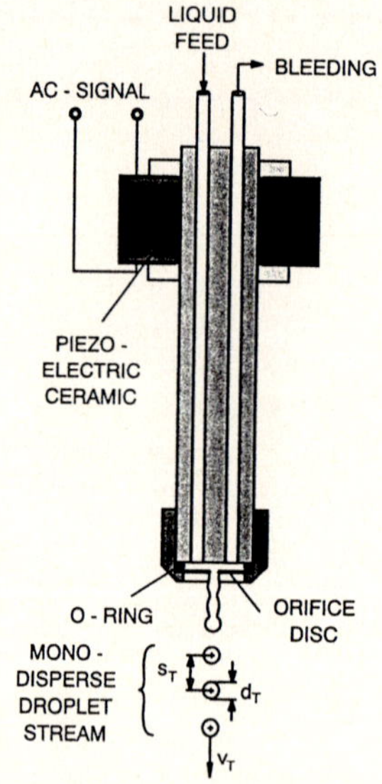

Fig. 2. Schematic of the droplet generator

orifice generator (Fig.2). The diameter of the generated droplets d_T may be obtained by measuring the liquid feed rate $\dot{V}$ of the generator using the equation

$$d_T = \left[\frac{6\,\dot{V}}{\pi\, f_G}\right]^{\frac{1}{3}} \tag{1}$$

where f_G is the excitation frequency of the droplet generator (Lord Rayleigh 1878, Berglund and Liu 1973). Assuming no loss of momentum, and negligible diameter change of the liquid jet behind the orifice, one obtains for the droplet velocity:

$$v_T = \frac{4\,\dot{V}}{\pi\, {d_G}^2} \tag{2}$$

where d_G is the diameter of the orifice. Initial droplet velocities may be varied in the range $10\ \mathrm{m/s} \le v_T \le 20\ \mathrm{m/s}$. The droplet spacing s_T is a function of the droplet diameter d_T and the orifice diameter d_G (Anders et al. 1992). One has

$$\frac{s_T}{d_T} = \frac{2}{3}\left(\frac{d_T}{d_G}\right)^2 \tag{3}$$

The droplets are introduced into the shock tube by means of a pulsed droplet injection facility which is shown in Fig.3. The injection orifice in the shock tube may be closed by a high-speed slide valve. The pressure in the small chamber between this valve and the droplet generator must be the same as in the test section of the shock tube in order to avoid disturbances of the droplet stream. The small droplets in the circular cross-section of the shock tube are visualized by a micro-projection equipment. Distortion-free images of the droplets are provided by a special arrangement of cylindrical lenses as shown in Fig.3. The droplets are illuminated by a Nanolite flashlamp, which is triggered by the approaching shock wave. With a microscope camera one

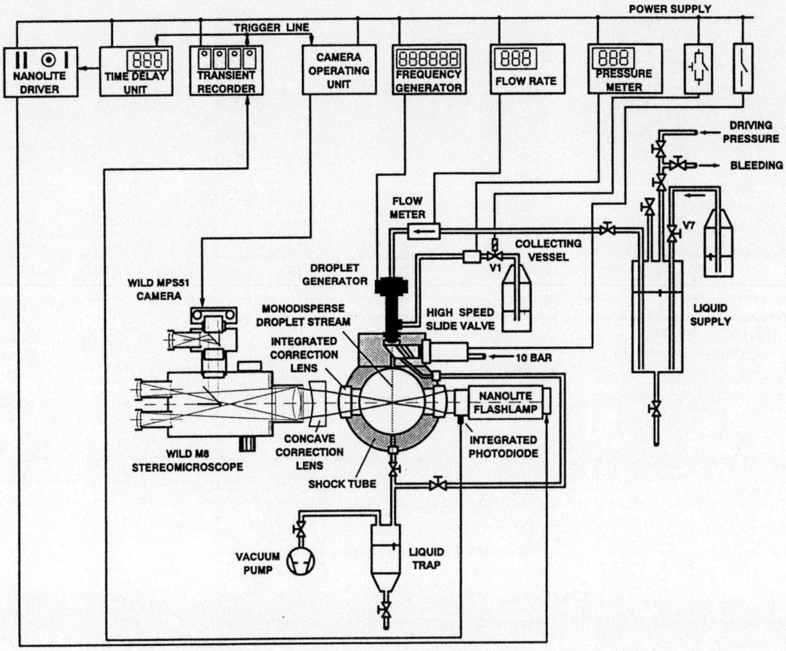

Fig. 3. Droplet injection and optical setup

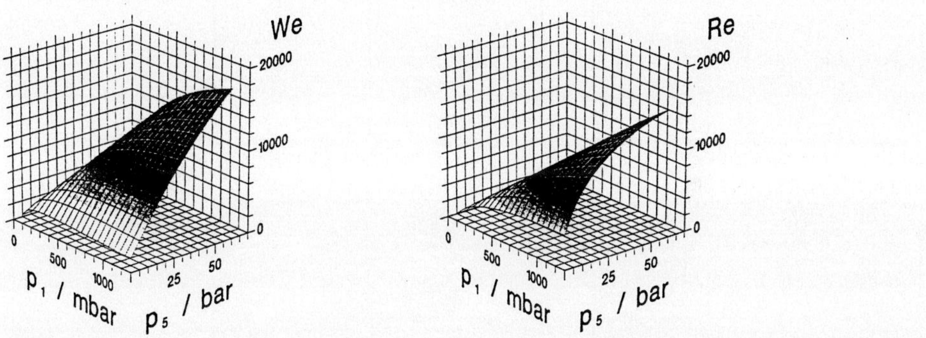

Fig. 4. Calculated regimes of Weber and Reynolds numbers for isopropanol droplets with a diameter of 120 μm, Helium as driver and synthetic air as driven gas

photograph can be taken in one experiment. Calculated regimes of Weber and Reynolds numbers which can be realized in the shock tube are shown in Fig.4. The Weber number

$$\mathrm{We} = \frac{\rho_2 {u_2}^2 d_T}{\sigma} \tag{4}$$

and the particle Reynolds number

$$\mathrm{Re} = \frac{\rho_2 u_2 d_T}{\mu_2}, \tag{5}$$

depend on the conditions behind the incident shock wave and the surface tension of the liquid σ.

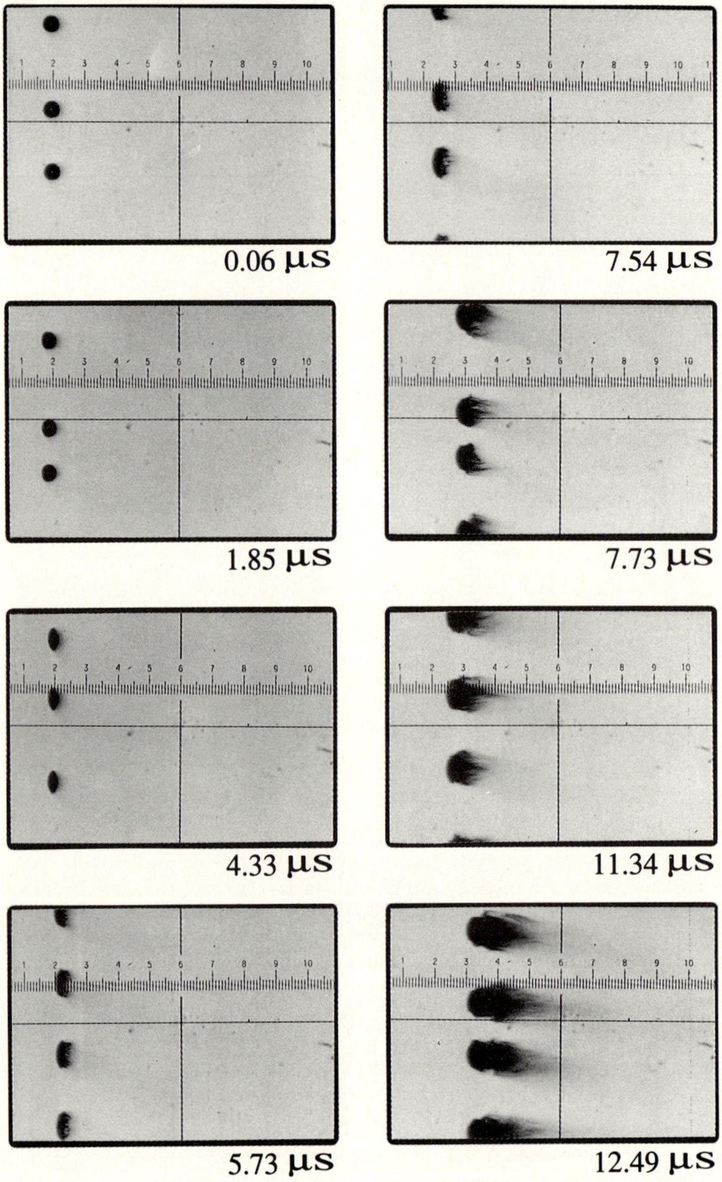

Fig. 5. Sequence of shadow photographs of droplet fragmentation; initial conditions: d_T = 120 μm, M_S = 2.5, M_2 = 1.2, We = 280, Re = 350

3. Results

Series of photographs have been taken for different Weber and Reynolds numbers. Two examples are presented in Figs. 5 and 6. The first series in Fig.5 shows the droplet stripping at different time steps for the Weber number We = 280 and for the Renolds number Re = 350. The first photograph in Fig.5 shows an undisturbed droplet stream 0.06 μs after the arrival of the shock;

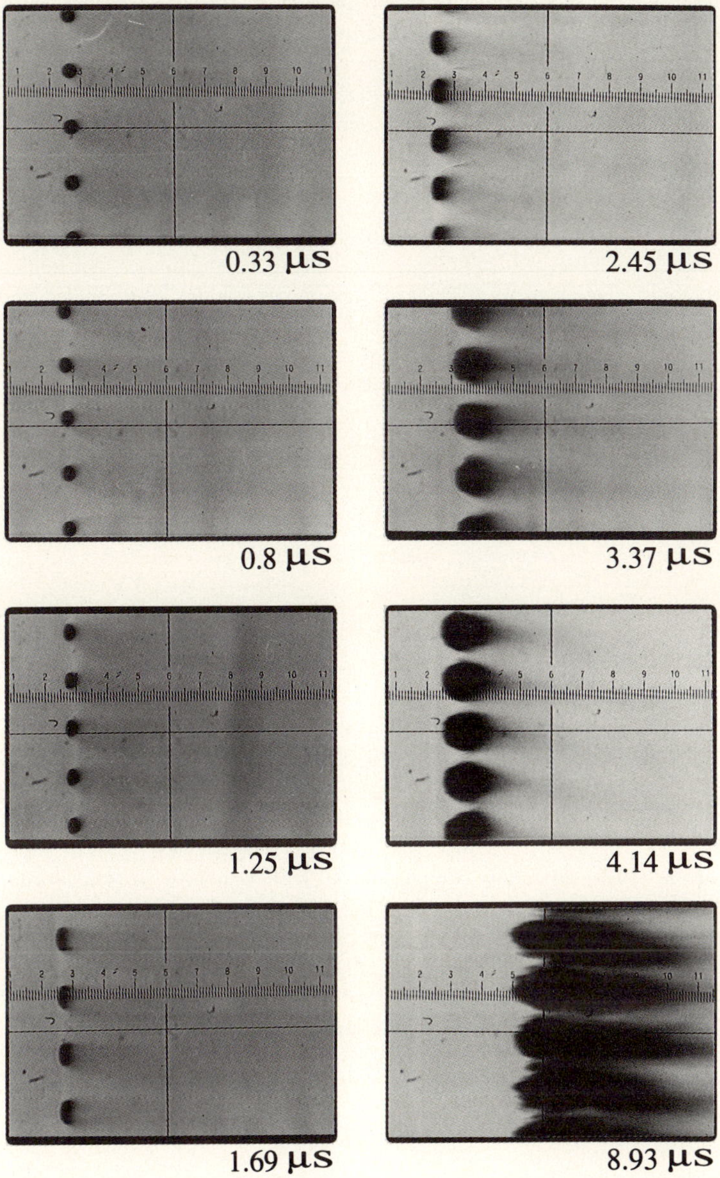

Fig. 6. Sequence of shadow photographs of droplet fragmentation; initial conditions: d_T = 120 μm, M_S = 2.7, M_2 = 1.3, We = 7500, Re = 8300

1.8 μs later, a deformation of the droplets may be observed, another timestep later the droplets are ellipsoids. Similar results have been obtained by Ranger and Nicolls (1969). Now the typical breakup phenomena of the stripping-type regime appears (Borisov et al. 1981; Deng and Jeng 1992). The gas flow tears off filaments of liquid from the droplets. Droplet breakup in the explosion-type regime is shown in Fig.6. The experiments showed, that at Weber numbers above

1000 and Reynolds numbers above 500 this type of droplet breakup prevails. In this regime the droplets will be flattened very fast due to the external pressure distribution. The droplets disintegrate and form clouds of micromist, which are expanded by the ambient flow. With the applied photographic technique, the core of the droplets cannot be observed, but it has been established that it takes only a few microseconds to destroy the core of the droplets. The last picture in Fig.6 shows the final breakup state where only the micromist remains. Inspection of the films reveals that the bow shocks of the individual droplets merge to a detached plane shock wave in front of the droplet stream.

4. Conclusions

The fragmentation process of monodisperse isopropanol droplet streams has been studied in a shock tube at high Weber and Mach numbers. The facility offers the possibility of pulsed injection of droplet streams into the shock tube. The stripping type and the explosion type breakup have been observed. The development of bow shocks can be shown. Future work will be devoted to flow studies around the droplets in order to investigate the influence of the bow shock on the breakup mechanism.

Acknowledgement

Financial support of the Deutsche Forschungsgemeinschaft is gratefully acknowledged.

References

Anders K, Roth N, Frohn A (1992) Operation characteristics of vibrating-orifice generators: The coherence length. Part. Part. Syst. Charact. 9: 40-43

Berglund RN, Liu BYH (1973) Generation of aerosol standards. Env. Sci. Tech. 7: 147-153

Borisov AA, Gelfand BE, Natanzon MS, Kossov OM (1981) Droplet breakup regimes and criteria for their existence. J. Eng. Phys. 40, 1

Deng ZT, Jeng SM (1992) Numerical simulation of droplet deformation in convective flows. AIAA J. 30: 1290-1297

Eastes TW, Samuelsen GS (1992) Secondary atomization by high amplitude pressure waves. AIAA Paper 92-3120

Lord Rayleigh (1878) On the instability of liquid jets. Proc. London Math. Soc. 10: 1047-1056

Ranger AA, Nicholls JA (1969) Aerodynamic shattering of liquid drops. AIAA J. 7: 285-290

Wierzba A, Takayama K (1987) An experimental investigation of the aerodynamic breakup of liquid drops. In: Grönig H (ed) Proc. 16th Intl. Symp. on Shock Tubes and Waves, Aachen. VCH Verlag, pp 335-341

Droplet Behaviour in Underexpanded Supersonic Jets

M. Sommerfeld* and J. Kurian†
* Lehrstuhl für Strömungsmechanik, Universität Erlangen-Nürnberg, Cauerstr. 4, 91058 Erlangen, Germany
† Department of Aerospace Engineering, Indian Institut of Technology, Madras 600036, India

Abstract. For improving the understanding of droplet break-up and behaviour in high-speed flows, experimental studies using an underexpanded free jet were performed. The outcome of the droplet break-up occuring under defined conditions in the converging nozzle, namely, the size distribution of the fragment droplets was measured using phase-Doppler anemometry. Furthermore, the interaction of the small fragment droplets with the strongly accelerating flow within the underexpanded jet and the embedded shock waves was of interest to the present study.

Key words: Underexpanded free jet, Droplet break-up, Droplet size measurements

1. Introduction

The study of the interaction of liquid droplets with high speed flows and shock waves is an interesting research problem of great practical significance. Airblast atomizers are, for example, found in numerous applications ranging from spray painting to fuel nozzles in gas turbine engines or furnaces. The air pressure normally used in airblast atomizers is high enough to give rise to sonic flow at the nozzle exit. Underexpanded sonic jets emanating from the nozzle result in a free jet structure characterised by a regular shock wave pattern (Ashkenas and Sherman, 1966). The break-up of the liquid sheet or jet occurs close to the nozzle as a result of the large velocity gradients between the injected liquid and the atomizing air. The droplets formed close to the nozzle then interact with a sequence of shock waves and strongly changing gas velocities which may result in further break-up or coalescence. For the efficient design of such atomizers, an understanding of the physical mechanisms of the interaction of droplets with the surrounding high-speed stream is needed.

Most investigations on droplet break-up have been carried out with very large drops falling in air at rest or in the airflow produced by a wind tunnel (Lane 1951, Krzeczkowski 1980). The associated subject of the deformation and disintegration of droplets in a high speed airflow induced by a shock wave passing over them has been investigated by many researchers (Hanson et al. 1963, Hirahara and Kawahashi 1992, Yoshida and Takayama 1991). In all these studies, investigations were carried out to determine the break-up mechanism of the droplets, the various factors which affect their break-up, and the time required for complete break-up. Most of the earlier researchers agree upon the point that the shattering of droplets is caused by relative velocity between them and the air and not by the shock wave itself. Also, it is understood that the break-up time is proportional to the drop diameter and inversely proportional to the relative velocity.

Additionally, for modelling droplet break-up, it is necessary to provide experimental information on the outcome of the break-up process, namely, the resulting size distribution of the fragment droplets. Available studies which provide this important result are, however, quite rare. An interesting result was obtained by Komabayasi et al. (1964) who found a bimodal size distribution of the fragment droplets for the bag-type break-up regime. Other data on the maximum size of stable fragments were reported by Lane (1951) and Hassler (1971). More recently, the resulting fragment size distribution for various break-up modes was measured by Hsiang and Faeth (1992) for various liquid properties in a shock tube. They found that the size distribution of the fragment droplets can be universally characterized by a root-normal distribution function, except for the shear type break-up which resulted in a bimodal size distribution.

In the present study, the droplet break-up in an underexpanded supersonic free jet issuing from a convergent nozzle was examined using phase-Doppler anemometry in order to determine

Shock Waves @ Marseille III
Editors: R. Brun, L. Z. Dumitrescu

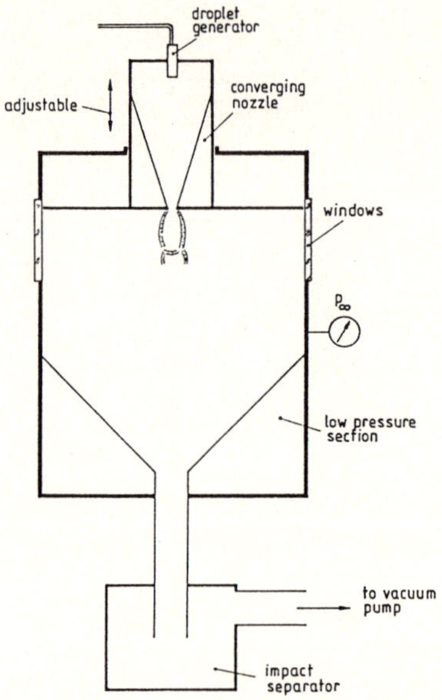

Fig. 1. Supersonic free jet test facility.

the fragment droplet size. This approach has the advantage that the outcome of the primary droplet break-up process can be analyzed as a steady-state process.

2. Experimental facility and procedure

The experiments were conducted in a supersonic free jet facility in the Lehrstuhl für Strömungsmechanik at the University of Erlangen-Nürnberg. The underexpanded free jet was produced by a converging nozzle mounted between a stagnation chamber and a downstream vacuum chamber (Fig.1). The exit diameter of the converging nozzle was 8 mm. The vacuum chamber (inner diameter 300 mm) was equipped with four glass windows of high optical quality, each having a diameter of 170 mm. The exit of the low-pressure chamber was connected to an impact separator and a large dump tank via a large ball-throttle valve. In all the experiments, the stagnation chamber was open to the ambient. Various levels of underexpansion were obtained by maintaining different pressures in the vacuum chamber which were achieved by setting different positions of the throttle valve.

Liquid droplets produced by a TSI droplet generator, which was positioned in the stagnation chamber, were introduced just upstream of the converging nozzle at a velocity of about 15 m/s. The initial droplet diameter could be varied in the range between 50 to 200 μm. For the present experiments, water and mixtures of water and glycerol were used. Due to large velocity differences between the droplets and the airflow developing within the nozzle, the initially large droplets break up into smaller ones. The change in droplet size distribution and the correlation between size and velocity along the center-line of the underexpanded free jet were measured using phase-Doppler anemometry (PDA). The measurements were performed at various downstream locations up to four times the nozzle exit diameter, starting from just outside of the nozzle.

For the transmitting optics of the PDA system, a front lens with a focal length of 1000 mm was used and a beam spacing of 50 mm was selected. In order to reduce the dimensions of the measurement volume, a beam expansion system was mounted between the laser and transmitting optics. The beam entering the transmitting optics was expanded to 3 mm. The receiving optics module of the PDA was positioned at 25° off-axis from the forward scattering direction, and a receiving lens with a focal length of 310 mm was used. More details about the receiving optics can be found in Qiu and Sommerfeld (1992). In order to allow the verification of the sphericity of the droplets, three photo-detectors at a separation of 20 mm were used in the receiving optics. This optical configuration can be employed for measurements of droplet sizes up to 120 μm. The signals from the photodiodes were band-pass filtered and stored in a LeCroy 9424, a 350 MHz digital storage oscilloscope, and then sequentially transferred to a 386/25 personal computer. The data processing used to evaluate signal frequency and phase was based on the cross spectral density method in conjunction with a Fast Fourier Transform (Qiu et al. 1991).

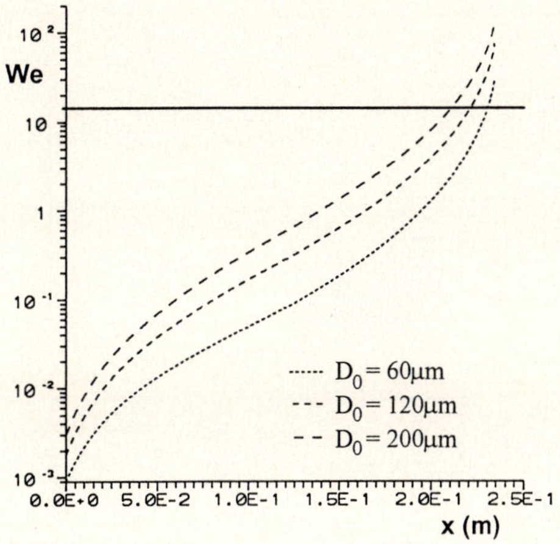

Fig. 2. Numerically calculated primary droplet We-number within the conveying nozzle.

3. Results for primary droplet-break-up

One objective of the present study was the determination of the droplet size distribution resulting from the break-up of the primary droplets in the conveying nozzle just downstream of the exit for different initial drop sizes and liquid properties. Due to the continuous increase of the relative velocity between gas and primary droplets in the conveying nozzle, the critical value of the Weber number We$\cong$ 13 (We $= \rho_g(u_g - u_d)^2 D_0/\sigma, \rho_g$ – local gas density, $|u_g - u_d|$ – relative velocity between gas and droplets, D_0 primary droplet diameter and σ – surface tension) is reached and break-up occurs in the bag mode (Krzeczkowski 1980). Since the droplet break-up in the converging nozzle could not be visualized, one-dimensional numerical calculations of the droplet acceleration process were performed. The results reveal that the location where the critical We-number is reached is located very close to the nozzle exit, depending on the size of the primary droplets (Fig.2). For the 120 μm droplets, the break-up occurs 15 mm upstream the nozzle exit. This implies that the break-up process is completed close to the nozzle exit, even with larger droplets, since here the break-up occurs earlier.

The droplet size distributions measured close to the nozzle exit (i.e. at $x = 2$ mm) for water droplets of various primary diameters revealed that the mean size of the fragment droplets was scattered around 12 μm with variations of ± 1 μm. This must be expected since at low viscosities or Ohnesorge numbers (Oh < 0.1, Oh $= \sqrt{\rho_l \nu_l^2/\sigma D_0}), \rho_l$ – liquid density, ν_l – liquid kinematic viscosity), the break-up occurs at about the same We-number and, hence, the process is independent of viscosity (Hinze 1955). This is also demonstrated in Fig.3 where the size distributions of the fragment droplets measured at $x = 2$ mm are shown for the various primary droplet diameters, expressed in terms of the Oh-number. The size distributions obtained for the water droplets have about the same shape with a mean number diameter of about 12 μm and a diameter RMS (root mean square) value of about 8 μm. For the water/glycerol mixture (40% water and 60% glycerol by mass), the break-up process is slightly different, since the size distribution becomes much wider, and the mean fragment diameter becomes larger, approximately 16 μm. This seems to be related to the larger viscosity and Oh-number (i.e. Oh = 0.115), since the break-up process occurs at higher We-number (i.e. at about We = 15, Krzeczkowski 1980). Unfortunately, it was not possible to use more viscous liquids with the droplet generator.

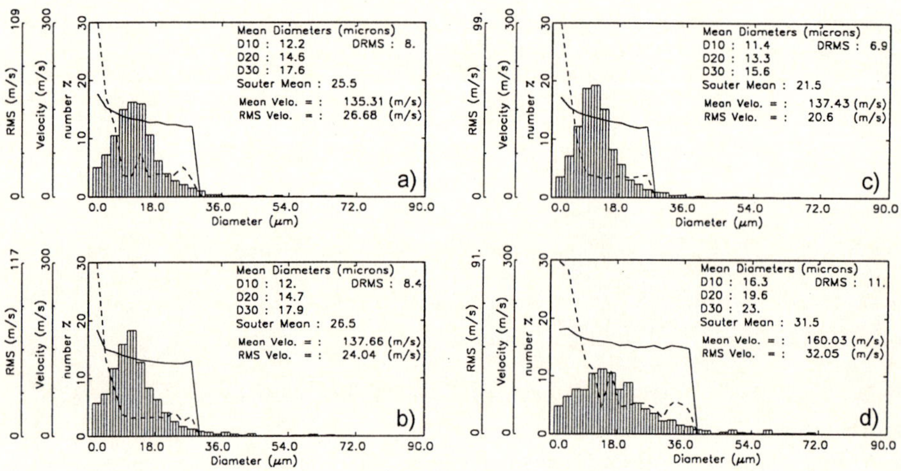

Fig. 3. Droplet size distributions measured at x = 2 mm for various primary droplet properties, a) Oh = 0.01 (p_0/p_∞ = 20, D_0 = 120 μm, water), b) Oh = 0.0135 (p_0/p_∞ = 20, D_0 = 80 μm, water), c) Oh = 0.01 (p_0/p_∞ = 5, D_0 = 104 μm, water), d) Oh = 0.115 (p_0/p_∞ = 5, D_0 = 110 μm, water/glycerol mixture, 40/60).

4. Droplet size and velocity characteristics within the underexpanded free jet

Within the underexpanded free jet, the fragment droplets resulting from break-up undergo further acceleration and spread in the radial direction. Furthermore, they interact with the Mach-disk or a sequence of normal shock waves at lower pressure ratios which is associated with strong changes in the relative velocity between gas and droplets. Both effects can result in a further break-up of the larger fragment droplets or cause an oscillation of the droplets. The development of the velocity and mean number diameter of water droplets along the jet center-line for three pressure ratios (i.e. p_0/p_∞ = 2.5, 5, and 20) is illustrated in Fig.4 for the same primary droplet diameter.

Close to the nozzle exit, the droplet mean diameter is about the same for all pressure ratios (Fig.4 a). This is expected, since the primary droplet break-up in the nozzle is not influenced by the pressure ratio, as the flow conditions within the nozzle are identical (see also Fig.3). Within the underexpanded free jet, the droplet mean diameter initially increases rapidly due to the radial spread of the droplet jet. Furthermore, there is no large difference in the development of the droplet mean diameter along the centerline for the different pressure ratios. It is also evident that no considerable changes exist in the droplet mean diameter across the Mach disks. This indicates

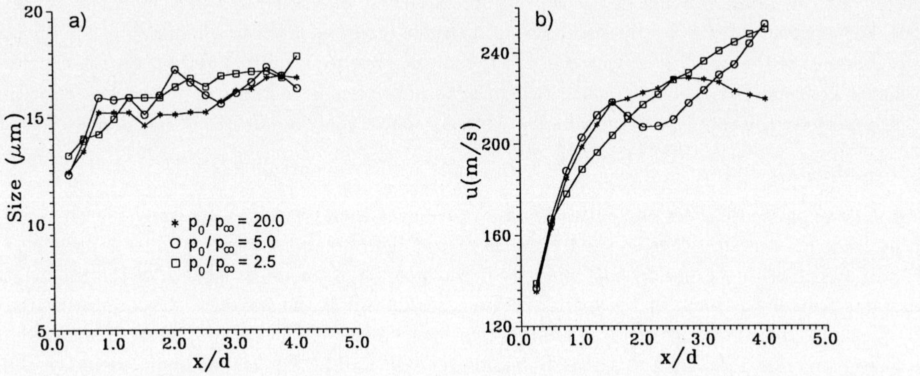

Fig. 4. Development of droplet mean diameter (a) and droplet velocity (b) along the jet center line for three different pressure ratios $p_0/p_\infty = 2.5$, 5, and 20 (water droplets with $D_0 = 120\ \mu m$).

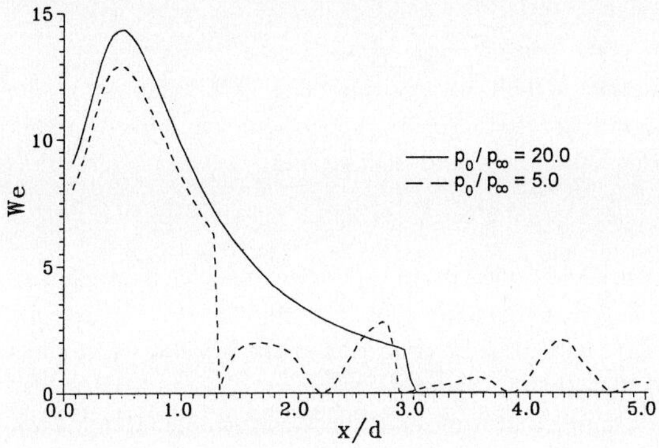

Fig. 5. Droplet We-number along the jet centerline for two pressure ratios (water droplets 30 μm).

that the droplet We-number remains below the critical value so that no oscillations or break-up occur.

The development of the droplet We-number along the jet centerline was therefore calculated based on the numerical method described by Sommerfeld (1993) for 30 μm droplets which are about the largest in the spectrum. Due to the low droplet concentration, the influence of the droplet phase on the gas flow could be neglected. The result shown in Fig.5 for two pressure ratios reveals that the We-number calculated for the 30 μm water droplets reaches maximum values of about 13 close to the nozzle exit for both pressure ratios. This suggests that further break-up of the larger droplets in the size spectrum might occur in this region. The We-number then continuously decreases up to the Mach disk. Although the relative velocity between gas and droplets continuously increases, the reduction in gas density causes this decrease of the We-number. For the high pressure ratio, the gas velocity across the Mach disk is reduced to a value close to the droplet velocity so that the We-number becomes very small. An increase in the We-number downstream of the first Mach disk is observed only for the lower pressure ratio due to the higher relative velocity. This reveals that across the Mach disk, no droplet break-up can be expected.

The droplet velocity along the centerline of the underexpanded free jet continuously increases up to the Mach disk location (indicated as vertical lines in Fig.4b) and then decreases again as a result of the sudden decrease in gas velocity for the higher pressure ratios (i.e. $p_0/p_\infty = 5$, and 20). For a pressure ratio of 5 the droplet velocity again increases downstream of $x/d = 2.2$ due to the presence of the multiple shock cells. For the lowest pressure ratio ($p_0/p_\infty = 2.5$), the droplet velocity continuously increases, since the droplets are not able to respond to the gas velocity changes across the shock patterns which are closely spaced, only about one nozzle diameter.

5. Conclusions

An underexpanded free jet emanating from a convergent nozzle was used to study the break-up process of primary droplets injected into the nozzle. The break-up occured under well-defined conditions close to the nozzle exit, whereby it was possible to measure the size of the fragment droplets just downstream of the exit. The size distribution of the fragment droplets measured by phase-Doppler anemometry was found to be independent of the initial droplet size and the imposed pressure ratio for low viscosity liquids (i.e. $Oh < 0.1$). For higher liquid viscosity, the mean size of the fragment droplets increased. Within the underexpanded free jet, further break-up might occur for droplets larger than about 30 μm at about half a nozzle diameter from the exit, since here the We-number reaches the largest values.

Acknowledgements

The present studies were initiated during the second authors stay at LSTM-Erlangen in the framework of the DAAD re-invitation programme. Furthermore, the authors would like to thank H.-H. Qiu, Ch. Mundo and G. Kohnen who supported the present studies.

References

Ashkenas H, Sherman FS (1966) The structure and utilisation of supersonic free jets in low density wind tunnels. 4th Symposium on Rarefied Gas Dynamics, Vol.2, pp 84-105

Hanson AR, Domich EG, Adams HS (1963) Shock tube investigation of the breakup of droplets by air blasts. The Physics of Fluids 6: 1070-1080

Hassler G (1971) Untersuchungen zur Verformung und Auflösung von Wassertropfen durch aerodynamische Kräfte im stationären Luft-und Querstrom bei Unterschallgeschwindigkeit. Ph.D. Thesis, Univerität Karlsruhe.

Hinze JO (1955) Fundamentals of the hydrodynamic mechanism of splitting in dispersion processes. A.I.Ch.E. J. 1: 289-295

Hsiang L -P, Faeth GM (1992) Near-limit drop deformation and secondary breakup. Intl. J. Multiphase Flow 18: 635-652

Kombayasi M, Gonda T, Isono K (1964) Life time of waterdrops before breaking and size distribution of fragment droplets. J. Met. Soc. Japan 42: 330-340

Krzeczkowski SA (1980) Measurement of liquid droplet disintegration mechanisms. Intl. J. Multiphase Flow 6: 227-239

Lane WR (1951) Shatter of drops in streams of air. Ind. and Engng. Chemistry 43: 1312-1317

Qiu H-H, Sommerfeld M, Durst F (1991) High-resolution data processing for phase-Doppler measurements in a complex two-phase flow. Meas. Sci. Technol. 2: 455-463

Qiu H-H, Sommerfeld M (1992) A reliable method for determining the measurement volume size and particle mass fluxes using phase-Doppler anemometry. Experiments in Fluids 13: 393-404

Sommerfeld M (1993) The structure of particle-laden, underexpanded free jets. Intl. J. Shock Waves (in print)

Yoshida I, Takayama K (1990) Interaction of liquid droplets with planar shock waves. Trans. ASME, J. Fluids Engng. 112: 481-486

Shock Tube Study of Droplet-Vapor Phase Transition of Hexane in Argon

F. Peters* **and B. Paikert**†
*Universität Essen, Schützenbahn 70,D-45117 Essen, Germany
†ASEA Brown Bovery CH 5405 Baden, Switzerland

Abstract. A shock tube technique in combination with Mie light scattering was employed to determine the state of hexane vapor at which homogeneous nucleation at a rate of 10^7 $cm^{-3}.s^{-1}$ takes place. The growth process of the nuclei into droplets was monitored as well as the reevaporation of droplets after shock heating. Comparisons with model calculations were carried out.

Key words: Homogenous nucleation, Droplets, Shock tube

1. Introduction

A liquid droplet suspended in a mixture of vapor and carrier gas grows by condensation or shrinks by evaporation according to the respective deviation from phase equilibrium. The rates of phase transition are controlled by two processes: the diffusion of vapor through the carrier and the countertransfer of heat. In case of condensation vapor diffuses to the droplet and the latent heat of condensation is conducted away from the droplet.

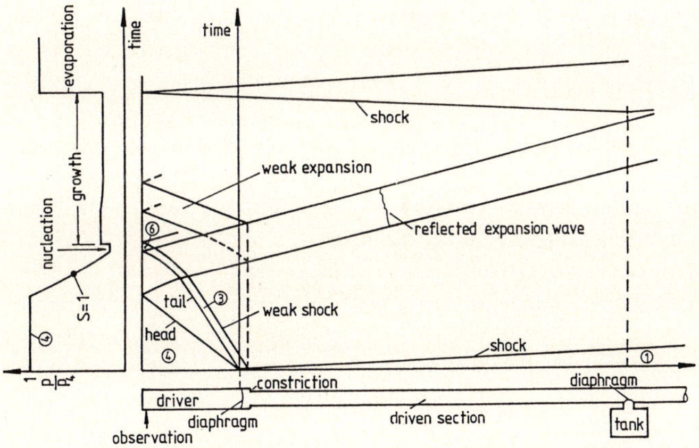

Fig. 1. Shock tube wave diagram with pressure history at the observation station

For the single droplet case, i.e. when neighbours are sufficiently far away to exclude interaction, theoretical models are available incorporating small droplets (high Knudsen number), big droplets (small Knudsen number) and the transition regime about Knudsen number one. We refer to a recent paper by Young (1993) which quotes most of the relevant previous work and provides updated rate equations applicable to our experimental results.

Shock tubes provide rapid cooling and heating capabilities in terms of the expansion and the shock wave, respectively. We utilize these to study homogeneous nucleation and subsequent growth and shrinkage. The droplet detection is based on optical Mie light scattering and photography.

Shock Waves @ Marseille III
Editors: R. Brun, L. Z. Dumitrescu

This work is linked to previous work focusing on nucleation rate measurements (Peters and Paikert 1989). Most of the experimental aspects have been dealt with therein. This paper outlines the experimental method, supplements a photographic technique and presents data on hexane in comparison with Young's theory. Hexane is a substance of low latent heat and thus subject to rather large growth and shrinkage rates. Its physical properties render it a representative fuel.

2. Experimental

The shock tube set-up comprises a driver section, a driven section featuring a constriction close to the diaphragm and a pressure tank attached to the side of the driven tube separated by a second diaphragm. The working principle may best be explained in terms of a wave diagram (Fig.1). The driver contains argon mixed with a small fraction of undersaturated hexane vapor which is to be nucleated and converted into droplets. The total pressure is close to atmospheric. The driven section holds pure carrier at a lower pressure. Everything is at room temperature. The experiment starts when the main diaphragm is broken. The carrier/vapor mixture is expanded and the gas is set into motion to the right. Along the head of the expansion wave the gas is in the undisturbed state "4". The expansion wave undergoes reflection at the end wall leaving the expanded state "6" behind the reflected tail of the centered wave. The expansion tail is closely followed by a small recompression originating from the shock/constriction interaction. The pressure vs. time plot to the left of Fig.1 depicts the situation at the observation station close to the end wall. The initial pressure starts to drop when the head reaches the wall. It drops to state "6" which persists only for a small period (≈ 0.1 ms) until the recompression elevates the pressure. Because of the wall the gas must be at rest at the observation station in this state.

Initial conditions are arranged such that the vapor passes saturation at some point in the expansion and reaches a certain supersaturation at state "6". According to homogeneous nucleation concepts a certain supersaturation is associated with a certain nucleation rate (Peters and Paikert 1989). Therefore a certain finite number of nuclei is generated during the nucleation period. In our experiments the rate is 10^7 $\text{cm}^{-3}.\text{s}^{-1}$. At nucleation the nucleus size is given by the Kelvin radius of a critical cluster (a few molecular radii). Since the recompression leaves some supersaturation the nuclei commence growing. They grow into the μm-range before a shock of variable strength passes the droplet suspension. This shock is produced by breaking the second diaphragm at a proper time opening the pressure tank to the tube. The shock transfers the vapor into the undersaturated state inducing evaporation, i.e. shrinkage.

For a quantitative evaluation of growth and shrinkage it is neccessary to know (i) the thermodynamic state of the droplet's surroundings, (ii) the droplet size and (iii) their concentration, all as function of time.

(i) The expansion wave expands the vapor/carrier mixture isentropically. Therefore, with the ratio of specific heats γ of the mixture, the nucleation temperature T_6 relates directly to the nucleation pressure p_6 and the initial values of state "4":

$$T_6/T_4 = (p_6/p_4)^{\gamma-1)/\gamma} \tag{1}$$

Since the initial mixing ratio remains constant throughout the expansion the ratio of vapor partial pressures p_{6v}/p_{4v} equals the ratio of total pressures p_6/p_4. The supersaturation at nucleation is found from

$$S = (p_6/p_4)\,(p_{4v}/p_\infty(T_6)) \tag{2}$$

where $p_\infty(T)$ is the equilibrium vapor pressure. p_4 and p_{4v} are known from mixing prior to the experiment. The mixing tank is shown in Fig.2. p_6 is measured by a pressure transducer also shown in Fig.2. During growth and shrinkage the total pressure is continuously monitored while

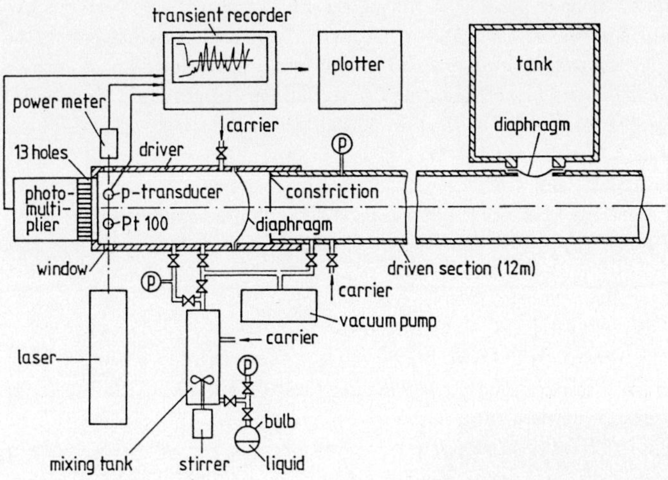

Fig. 2. Experimental set-up

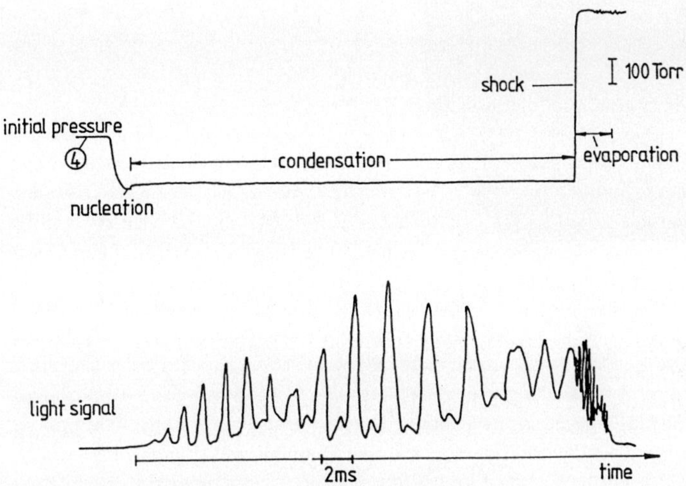

Fig. 3. Experimental recording of pressure and light signals

partial pressure and temperature are calculated from a mass and heat balance between vapor, carrier and condensed droplet mass.

(ii) The growing and shrinking droplets are observed by a Mie scattering technique. As drawn in Fig.2 an argon-ion laser (514.2 nm; 200 mW) is shone across the tube passing a stop of 1 mm diameter and two windows. Droplets present within the volume of the beam scatter in all directions. By means of 13 holes placed between glass wall and photomultiplier we select a scattering angle of $90° \pm 0.4°$ and 13 sensitive volumes along the beam of 8.67 mm^3 in total. The photomultiplier converts light power, i.e. intensity times sensitive area, proportionally into voltage. Mie theory provides scattered intensity normalized by incident intensity as function of droplet radius. This function oscillates exhibiting a sequence of peaks identifying certain radii. The first peak indicates a radius of 160 nm. Accordingly, by reading experimental peaks as function of time the radius is found as function of time.

(iii) Mie theory applies to a single droplet. When more than one droplet is present in the sensitive volume the scattered intensity is proportional to the number of scatterers supposing that the scatterers are independent and monodispersed. The second condition cannot be satisfied perfectly, however due to the shortness of the nucleation period the size spectrum of nuclei born at the beginning and at the end of the period is kept rather narrow. The combination of laser intensity, photomultiplier sensitivity, and scattering geometry brings about a minimum number of 50 droplets that need to be present at the first peak to produce a photomultiplier signal above noise level. In terms of nucleation rate (divide 50 by sensitive volume and nucleation period) this is the order of 10^7 $cm^{-3}.s^{-1}$ which is the nucleation rate the present experiments are based on.

Fig. 4. Photographic representation of growing droplets of about 2.5 μm radius illuminated by a light sheet crossing the tube

Fig.3 provides an experimental record showing the pressure signal as discussed in Fig.1 and the corresponding light signal. In this case we have rather moderate growth with peaks clearly apart followed by rapid reevaporation (peaks draw close) due to a strong shock. Mie scattering is supplemented by photographic detection of growing droplets. The method uses the set-up of Fig.2 with slight modifications. The laser beam is now expanded to form a sheet of parallel light passing the tube through a glass ring. The 13 holes are removed and the photomultiplier is replaced by a camera focused onto a plane within the sheet and facing the entire tube cross section. In Fig.4 we look at images of droplets of about 2.5 μm in radius 30 ms after nucleation (8 ms exposure time).

According to geometrical optics the droplets should hardly become visible on the film. The occuring images are magnifications due to diffraction at the camera aperture. Thus droplets are identified without permitting inference of their real size from the image size. The varying image sizes are due to depth of field effects. The droplets appear statistically distributed as expected from the statistical nucleation process. They also appear pointwise and not as streaks verifying the zero velocity boundary condition at the wall.

Finally the droplet concentration as measured by Mie scattering can be checked by droplet count. Actual comparisons have yielded very good agreement.

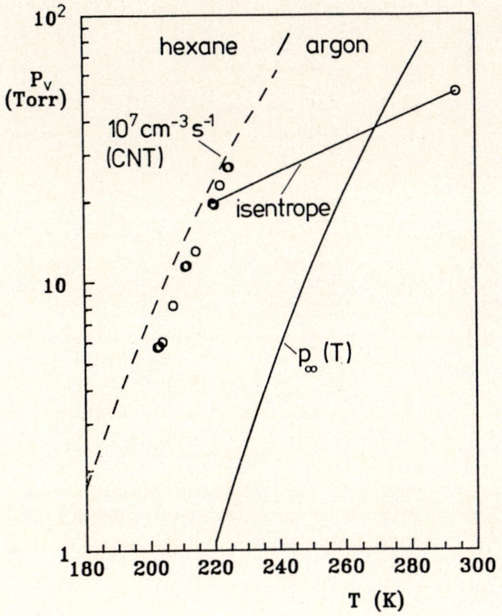

Fig. 5. $p - T$ states at which the nucleation rate of 10^7 $cm^{-3}s^{-1}$ was found. Dashed curve gives classical nucleation theory. $p_\infty(T)$ indicates phase equilibrium

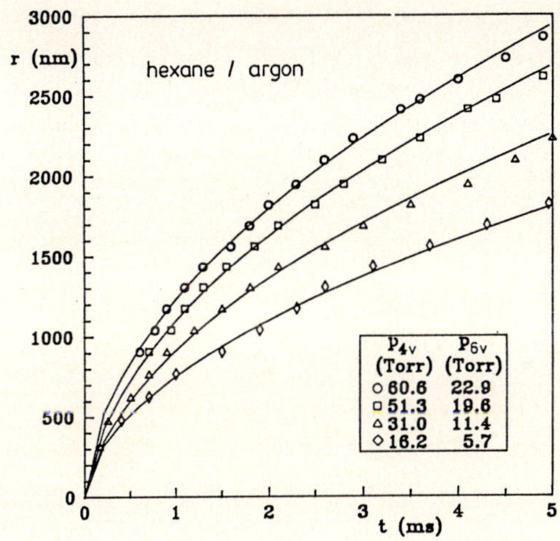

Fig. 6. Growth curves. Symbols represent experimental Mie peaks, curves Young's theory. p_{4v} gives initial, p_{6v} nucleation pressure

3. Results and conclusion

Fig.5 gives the states in terms of pressure and temperature (circles) at which a nucleation rate of 10^7 $cm^{-3}.s^{-1}$ was found. The isentrope exemplifies an experimental expansion between states "4" and "6". The dashed line represents the classical nucleation theory (as in Peters and Paikert 1993) predicting temperatures too low and thus supersaturations too high. Fig.6 shows results on growth in comparison with Young's theory. Nucleation states (as in Fig.5) form the beginning of these growth curves. The top curve represents the highest vapor pressure the bottom the lowest. Fig.7 gives droplet growth and susequent shock evaporation. All growth curves start at the same initial conditions. The shock strength is varied according to the after-shock conditions (box). Growth and evaporation data are extremely well represented by the theory.

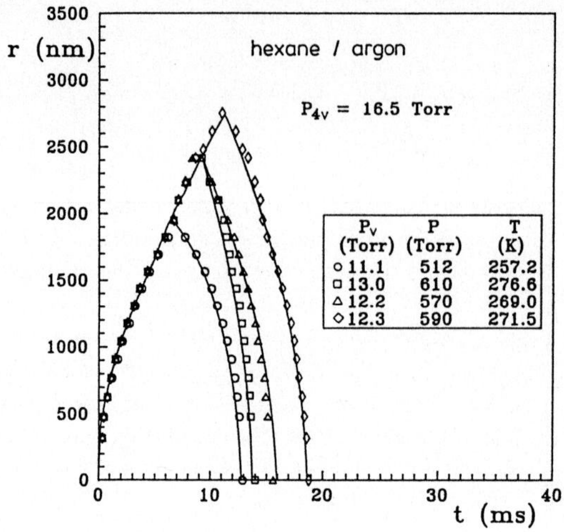

Fig. 7. One repeated growth curve with various evaporation curves. Box gives states at beginning of evaporation, i.e. behind shock

References

Peters F, Paikert B (1989) Nucleation and growth rates of homogeneously condensing water vapor in argon from shock tube experiments. Exp. Fluids 7: 521-530

Peters F, Paikert B (1989) Experimental results on the rate of nucleation in supersaturated n-propanol, ethanol and methanol vapors. J. Chem. Phys. 91, 9: 5672-5678

Young JB (1993) The condensation and evaporation of liquid droplets at arbitrary Knudsen number in the presence of an inert gas. Intl. J. Heat Mass Transfer 36, 11:2941-2956

Thickness Measurements of Partial Liquefaction Shock Waves

M. Herrmann and F. Obermeier
Max-Planck-Institut für Strömungsforschung, Göttingen, FRG

Abstract. The nucleation and condensation behind a partial liquefaction shock wave could influence the shock structure if the characteristic times for nucleation and change of state in the shock are of the same order of magnitude. To investigate this effect, thicknesses of partial liquefaction shock waves and shocks without condensation in perfluoro-n-hexane, obtained with the light reflectivity method, are compared. It could be shown that inside our parameter range, nucleation has no measurable influence on the thickness of the shock, in accordance with calculated nucleation rates. As a function of Mach number, shock thickness scales with the mean free path ahead of the shock also in case of condensation behind the shock. Droplet condensation behind the shock is monitored using light scattering and extinction. Calculated nucleation rates and droplet growth are in good agreement with our measurements.

Key words: Shock structure, Retrograde fluids, Nucleation, Condensation

1. Introduction

Retrograde fluids find technical applications, e.g. as fossil fuels, working fluids for organic Rankine cycles or solvents. Research on the gas dynamics of retrograde substances was reviewed by Meier (1987). Retrograde fluids can be condensed isentropically from the vapor phase by compression because of their high molar specific heat, which limits the temperature increase due to compression and release of latent heat. If the molar specific heat is large enough, even partial (Thompson et al. 1987) or complete liquefaction (Dettleff et al. 1979) initiated by a shock is possible. In the present paper, the condensation process behind a partial liquefaction shock is investigated.

Bratos and Meier (1979) tried to treat this phenomenon by solving simultaneously the equations for conservation of mass, momentum and energy of a one-dimensional two-phase flow, the equations of state and the equations for nucleation rate and droplet growth. For the surface tension, which is a crucial quantity for nucleation theory, a Tolman correction was adopted, which is questionable if used for nuclei consisting of several hundred molecules only. Applying the correction to conditions similar to experiments reported by them, significant changes of thermodynamic variables are caused by condensation over a distance of 10^{-7}m. This length is comparable to the shock thickness, which was neglected in the calculations. This result would suggest a broadening of the shock. On the other hand, without the Tolman correction a condensation zone longer by two orders of magnitude was calculated.

The structure of shocks in monatomic gases is well understood (Fiszdon et al. 1974). As retrograde fluids consist of complex polyatomic molecules with low excitation temperatures for the vibrational degrees of freedom, rotational and vibrational relaxation occur in our experiments. To investigate whether experimental evidence of an influence of nucleation on maximum slope shock thickness can be found, comparative measurements of shock thicknesses with and without condensation were performed for the test fluid perfluoro-n-hexane (supplier ISC, Bristol). For this purpose we used the light reflectivity method developed by Cowan and Hornig (1950), because the shock thickness to be measured is of the order of a micrometer, which rules out all other methods like the electron beam method. For further details, also concerning the shock tube construction and thermodynamic properties, see Herrmann (1993).

Shock Waves @ Marseille III
Editors: R. Brun, L. Z. Dumitrescu

2. Experimental methods and set-up

The shock tube has an inner diameter of 80 mm and consists of driver and driven sections of 2.33 m length each and a test section of 0.3 m. Cellulose tri-acetate shatter-type diaphrams are used. Kistler 603b piezoelectric transducers are used to record transient pressure and to determine shock velocity. Initial temperature and pressure are measured with a thermocouple and a MKS Baratron pressure gauge, type 221AA, respectively. The tube can be heated up to 100°C by a thermal syphon system. The heating room is made up by the space between the tube itself and a concentric tube of larger diameter. The tube temperature can be held uniform over the whole tube length within ±0.5°C. Fig. 1 shows a sketch of the optical system fixed to the tube.

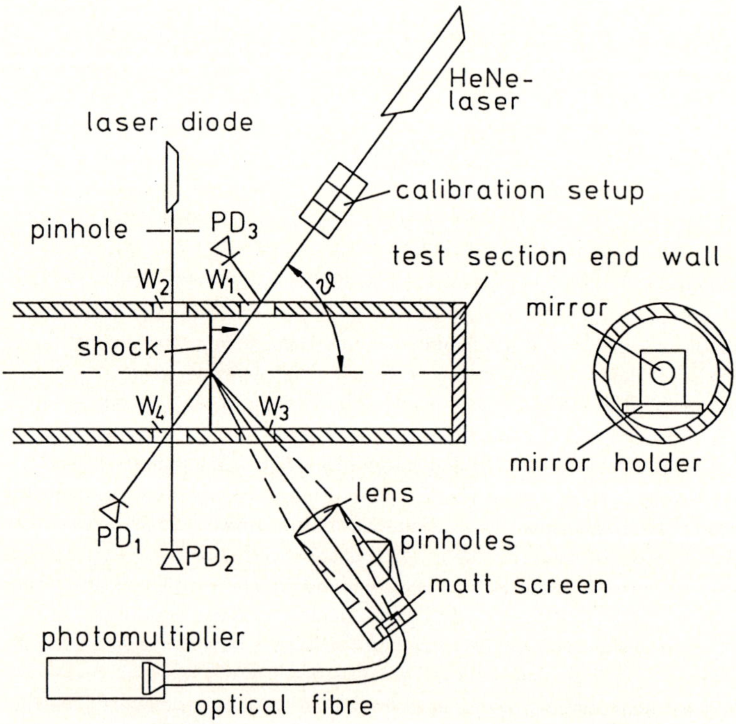

Fig. 1. Shock tube test section with optical system; PD photodiodes, W windows

The test section is equiped with synthetic quartz glass windows. All sketched beam paths lie in the horizontal plane through the tube axis. A light barrier is set up using a laser diode and photodiode PD_2. For shock thickness measurement, the reflection coefficient R has to be determined. The beam of a He-Ne laser enters the tube under an angle ϑ to the tube axis. When the shock comes into the measuring volume defined by the lens at the intersection of laser beam and tube axis, a reflection signal is registered by the photomultiplier. The light intensity is averaged over the reflection signal. The set-up is calibrated after about every ten runs using a mirror as model shock front and filters of known attenuation to obtain the reflection coefficient. Variations in light intensity are monitored by photodiodes PD_1 and PD_3.

For small reflection coefficients (here $10^{-7} < R < 2 \cdot 10^{-5}$) caused by the density gradient of the shock, integration of the Fresnel formulas for infinitesimal layers over the shock leads to

$$R_l = \frac{1 + \tan^4 \vartheta}{4} |F(y)|^2 \quad ; \quad F(y) = \int_{-\infty}^{\infty} \frac{dn}{dx} e^{-2\pi i y x}\, dx \tag{1}$$

with $y = 2\cos\vartheta/\lambda_l$, where λ_l is the light wave length. The profile of the index of reflection n is approximately determined from the Gladstone-Dale relation, the total jump Δn over the shock, which enters the resulting equation, from the more accurate Lorentz-Lorenz equation. Assuming a hyperbolic tangent profile of density yields

$$F(y) = \Delta n \, \frac{\pi^2 y L/2}{\sinh(\pi^2 y L/2)} \tag{2}$$

From Eqs.1 and 2 the shock thickness is determined via a bisection method.

The hyperbolic tangent is the exact solution of the Mott-Smith theory and also results from the Navier-Stokes equations in the limit $M \rightarrow 1$. For perfluoro-n-hexane both rotational and vibrational relaxation are so fast, that they can be accounted for by an additional bulk viscosity, which increases the shock thickness, but does not change measurably the shape of the density profile. This could be shown by Walenta (Chaves et al. 1989) using electron beam attenuation and laser differential interferometry. If nucleation would lead to an increase of shock thickness, including a significant change in shape of the density profile, one parameter – its thickness – would not be sufficient to describe the profile. Nevertheless the broadening of the shock would be detected. Measurements with a similar set-up using argon and nitrogen and reported earlier (Chaves et al. 1989) lead to satisfactory results, whereas shock thicknesses for perfluoro-n-hexane shown there exhibit large scatter. The reason for that has been identified, i.e. not all reflected light reaches the photomultiplier and the method is more sensitive to errors in R for thinner shocks. In the present work, scatter could be reduced by using a thicker fiber optics, improved calibration and changing ϑ from 80° to 60°. A discussion of the range of application of the method and error estimation is given in Herrmann (1993).

3. Results

Initial temperatures T_1, pressures p_1 and saturation pressures p_{sat} at T_1 for all experiments are summarized in Table 1. In this paper, only the results for 30°C are presented, the other experiments give a similar picture.

Fig.2a shows measured shock thicknesses scaled with the mean free path λ ahead of the shock as a function of Mach number M. Also included are results obtained by Walenta using electron beam densitometry for initial conditions $p_1 = 5.33$ μbar and at room temperature. To make comparison easier, an estimation of shock thickness developed by Thompson et al. (1983) is plotted, which contains an unknown bulk viscosity. The resulting value of total viscosity – 2.036 times the shear viscosity – was taken from Chaves et al. (1989). Within their scatter, measured shock thicknesses lie on one curve, which may be described by the estimation of Thompson et al. (1983). For comparison, Fig.2b shows the supersaturation behind the shock. No influence of supersaturation on shock thickness is perceptible, only the results for $p_1 = 300$ mbar and $M > 1.8$ show a noticeably greater scatter. This indicates the limit of the method to small shock thicknesses, which lies at about 1 μm in the present set-up.

Table 1. Initial conditions

T_1 [°C]	p_1 [mbar]	p_{sat} [mbar]
30	100, 150, 200, 300	364
50	300, 500, 700	789
80	100, 150, 200, 300	2077

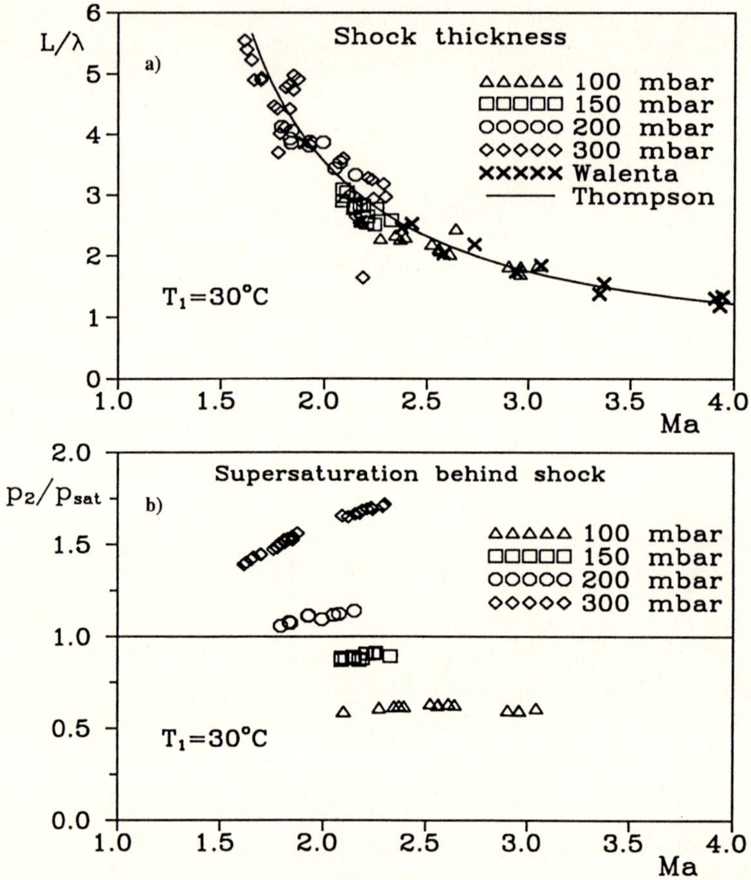

Fig. 2. a: Shock thickness L scaled with mean free path λ ahead of the shock as a function of Mach number M. b: Supersaturation p_2/p_{sat} behind the shock

On the contrary Chaves et al. (1989) found a systematic growth of L/λ with initial pressure in experiments without condensation. This is due to the sources of error discussed in Section 2. They discussed an effect of molecule axis alignment caused by the velocity gradient in the shock. Its magnitude was estimated in the present work and it seems to be negligible.

In experiments with condensation, after the reflection signals the photomultiplier also records scattered light from droplets in the measuring volume. Fig.3a,b shows traces of the runs with lowest and highest Mach number for the initial condition leading to strongest condensation. In Fig.3a the scattered light increases, but it shows a superimposed modulation which can be explained by Mie scattering (Kerker 1969). During the observation time, the droplets grow to a diameter of approximately 1 μm. The droplet growth theory of Gyarmathy (1963) assuming constant thermodynamic state yields diameters about two times larger, which is to be expected, as vapor depletion is not accounted for. In Fig.3b condensation is so strong that scattered light suffers strong attenuation on its path to the outlet window and that the recorded light decreases

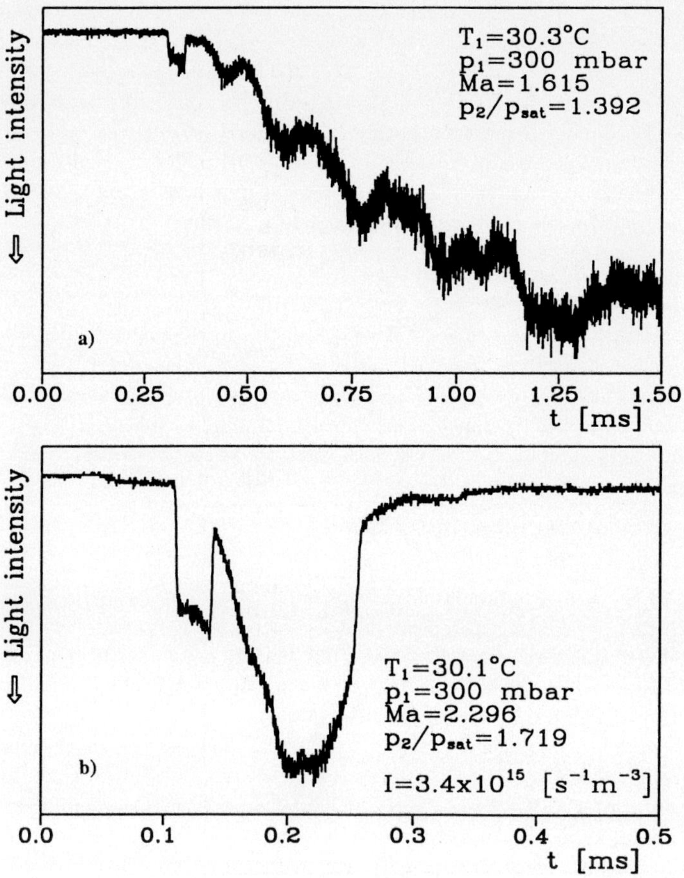

Fig. 3a-b. Photomultiplier signals showing light scattering from droplets in the measuring volume.

rapidly. This figure also points out the limitation of the method due to strong light scattering, which adds to the reflection signal; here only half of the signal was used for evaluation. Homogeneous nucleation rates calculated from Dillmann and Meier (1991) theory suggest that in Fig.3b nucleation is predominantly homogeneous, whereas in Fig.3a heterogenous nucleation prevails. In no case nucleation rates can be expected to influence the shock structure.

The light barrier signals show the passing of the shock and droplet condensation in the mean flow. Quantitative evaluation, which could yield droplet sizes and concentrations, is hindered by boundary layer condensation.

4. Conclusion and outlook

To investigate possible effects of nucleation and condensation on the structure of liquefaction shock waves, shock thicknesses were measured using the light reflection method. Within our measurement range, no influence of nucleation on shock thickness was found, in accordance with nucleation rate calculations. Measured thicknesses are in good agreement with results obtained by Walenta using electron beam densitometry at low densities.

As in our range of parameters nucleation appears to be much slower than changes of the flow variables in the shock, observation of the relaxation zone behind the shock is feasible. Corresponding experiments are under way in a purpose-built apparatus, where the shock leaves the tube into a chamber of larger diameter to avoid boundary layer and wall condensation.

A detailed investigation of nucleation and droplet growth would be possible by measuring scattered light intensity in one or more directions or extinction at several wavelengths.

Only in connection with the experimental investigations mentioned above a simultaneous calculation of nucleation, condensation and the changes of thermodynamic state would be useful, as otherwise the uncertainty in the nucleation rate would not justify the effort.

Acknowledgement

The authors would like to express their gratitude to Mr. M. Kalisch, the designing engineer in charge of the construction of the shock tube, who contributed many ideas of his own.

Mr. T. Teske, in the course of his own work, developed many computer programs for creating graphs and for further evaluation of the experimental data, which could be used here.

We thank Mr. Z. Walenta for helpful discussions on shock structure and for advice concerning construction details during his stay at our Institute.

References

Bratos M, Meier GEA (1979) Shock wave induced condensation in retrograde vapour. Arch. Mech. 31:679-693

Chaves H, Herrmann M, Meier GEA, Kowalczyk P, Walenta ZA (1989) Measurements of shock-wave thickness in PP1. In: Kim YW (ed) Proc. 17th Intl. Symp. on Shock Waves and Shock Tubes, American Institute of Physics, New York, pp 108-113

Cowan GR, Hornig DF (1950) The experimental determination of the thickness of a shock front in a gas. J. Chem. Phys. 18:1008-1018

Dettleff G, Thompson PA, Meier GEA, Speckmann H–D (1979) An experimental study of liquefaction shock waves. J. Fluid Mech. 95:279-304

Dillmann A, Meier GEA (1991) A refined droplet approach to the problem of homogeneous nucleation from the vapor phase. J. Chem. Phys. 94:3872-3884

Fiszdon W, Herczynski R, Walenta ZA (1974) The structure of a plane shock wave of a monatomic gas: Theory and experiment. In: Proc. 9th Intl. Symp. on Rarefied Gas Dyn. Appendix B.23, DFVLR–Press, Köln

Gyarmathy G (1963) Zur Wachstumsgeschwindigkeit kleiner Flüssigkeitstropfen in einer übersättigten Atmosphäre. ZAMP 14:280-293

Herrmann M (1993) Stoßstruktur und Relaxation bei stoßinduzierter Kondensation in Substanzen hoher molarer spezifischer Wärme. Mitteilungen aus dem Max-Planck-Institut für Strömungsforschung Nr. 106, Göttingen, Germany

Kerker M (1969) Scattering of Light and Other Electromagnetic Radiation. Academic Press, New York

Meier, GEA (1987) Zur Realgasdynamik der Fluide hoher spezifischer Wärme. Mitteilungen aus dem Max-Planck-Institut für Strömungsforschung Nr. 89, Göttingen, Germany

Thompson PA, Strock TW, Lim DS (1983) Estimate of shock thickness based on entropy production. Phys. Fluids 26:48-49

Thompson PA, Chaves H, Meier GEA, Kim Y, Speckmann H–D (1987) Wave splitting in a fluid of large heat capacity. J. Fluid Mech. 185:385-414

Interferometric Measurements and Numerical Simulation of Relaxation in an Argon-Steam Mixture

F. Maršík, J. Blaha and P. Sopuch
Institute of Thermomechanics, Academy of Sciences of the Czech Republic, Prague, Czech Republic

Abstract. The onset of condensation of an argon–steam mixture in the expansion fan in a shock tube driver is experimentally measured by a Mach–Zehnder interferometer. The effect of a high expansion rate is determined and compared to the classical Volmer theory and the new Dillmann–Meier theory. The expansion of steam from states above of the saturation curve to states under this curve is simulated by a finite-volume method. An adaptive computational mesh is used to suppress the dissipative properties of the numerical procedure. The numerical results show a strong influence of the two-temperature approach on the magnitude of the condensation rate. The influence of the value of the sticking and accommodation coefficients on heat and mass transfer on the droplets can also be studied in this way.

Key words: Phase transition, Two-phase flow, Optical measurement

1. Introduction

The influence of the phase kinetics on the flow pattern is usually investigated numerically (Hill 1966, Guha and Young 1991) and experimentally (Schnerr and Dohrmann 1989). The theory of water drop nucleation in steam is based on the capillary approach (Dillmann and Meier 1991). The influence of the condensation model (i.e. the sticking and accommodation coefficients) and as well as the influence of the dissipative propreties of the numerical procedure are intensively studied (Maršík et al. 1993). The numerical solutions are carried out for both the expansion in the shock tube driver and for a duct flow. A finite-volume method accurate to the second order is usually used.

Shock tube experiments allow to investigate the kinetics of condensation and evaporation in a wide range of temperature gradients and temperature rates. The homogeneous nucleation of water vapor in a carrier gas was studied in the expansion part of a shock tube using light scatering by Barschdorff (1975). The nucleation and growth rate of condensing vapor in argon was measured by Peters and Paikert (1989) using the Mie–scattering method. The newest optical measurements of the shock–induced evaporation in a gas–droplet mixture were performed by Goossens et al. (1988). The influence of a carrier gas was experimentally investigated by Barschdorff (1975) and shows that supersaturation p_v/p_s of steam in argon at the Wilson point is lower than in pure steam.

To improve the physical models of steam–water transition it is necessary to concentrate on:

i) Clusters formation and the influence of their internal degree of freedom (Dillmann and Meier 1991);
ii) Mass an energy transfer between drops and their surroundings (Gyarmaty 1982, Hill 1966);
iii) Dispersive and dissipative properties of the used numerical procedures (Maršík et al. 1993)

The experimental data are somewhat affected by the presence of the carrier gas, therefore its influence on the onset of the condensation must also be taken into account. For further developement of the condensation theory it is also very important to gain the most precise data on the tested substance. For water see e. g. Saul and Wagner (1987).

Shock Waves @ Marseille III
Editors: R. Brun, L. Z. Dumitrescu

2. Balance equations for the two–velocity, two–temperature model

Transfer of mass, momentum and energy in a mixture of carrier–argon and water for one-dimensional flow in a shock tube of cross section area A are considered in the following simplified conservative form:

$$\frac{\partial \boldsymbol{U}}{\partial t} + \frac{\partial \boldsymbol{F}(\boldsymbol{U})}{\partial x} = \boldsymbol{H}(\boldsymbol{U}) \tag{2.1}$$

where

$$\boldsymbol{U} = \begin{bmatrix} A\rho_c \\ A\rho_v \\ A\rho_l \\ A(\rho_c v_c) \\ A(\rho_l v_l) \\ A(\rho_c e_c) \\ A(\rho_l e_l) \\ A\rho Q_0 \\ A\rho Q_1 \\ A\rho Q_2 \end{bmatrix}, \quad \boldsymbol{F}(\boldsymbol{U}) = \begin{bmatrix} U_1 v \\ U_2 v \\ U_3 v \\ U_4 v \\ U_5 v \\ U_6 v \\ U_7 v \\ U_8 v \\ U_9 v \\ U_{10} v \end{bmatrix} + \begin{bmatrix} Aj_{Dc} \\ \frac{U_2}{U_1} Aj_{Dc} \\ Aj_{Dl} \\ A(\frac{U_4}{U_1} j_{Dc} + p_c) \\ A(\frac{U_5}{U_3} j_{Dl} + p_l) \\ A(\frac{U_6}{U_1} j_{Dc} + \frac{U_4}{U_1} p_c) \\ A(\frac{U_7}{U_3} j_{Dl} + \frac{U_5}{U_3} p_l) \\ 0 \\ 0 \\ 0 \end{bmatrix}, \tag{2.2a}$$

$$\boldsymbol{H}(\boldsymbol{U}) = \begin{bmatrix} 0 \\ -Aw_l \\ Aw_l \\ \frac{dA}{dx} p_c + Ak_c \\ \frac{dA}{dx} p_l + Ak_l \\ A\sigma(e_c) \\ A\sigma(e_l) \\ AJ \\ A(Jr_0 + \rho Q_0 \frac{d\bar{r}}{dt}) \\ A(Jr_0^2 + 2\rho Q_1 \frac{d\bar{r}}{dt}) \end{bmatrix} \equiv \begin{bmatrix} \text{Balance of} \\ \text{-carrier mass} \\ \text{-mass of vapor} \\ \text{-mass of liquid fraction} \\ \text{-carrier momentum} \\ \text{-momentum of liquid fraction} \\ \text{-carrier total energy} \\ \text{-total energy of liquid} \\ \text{-number of drops} \\ \text{-sum of radius of drops} \\ \text{-sum of square radii of drops} \end{bmatrix}. \tag{2.2b}$$

The mean velocity of the gases is the same, i.e.: $v_g = v_v = v_c$, but the velocity of the droplets v_l is different. The temperature of the droplets T_l is different from the temperature of the gases $T_g = T_v = T_c \neq T_l$. The density (with respect to the total volume) of the entire mixture is

$$\rho = \rho_c + \rho_v + \rho_l \quad (c: \text{ carrier}, \ v: \text{ vapour}, \ l: \text{ liquid}), \tag{2.3}$$

where $c_l = \rho_l/\rho$ is the wetness and $c_w = (\rho_l + \rho_v)/\rho$ is the water concentration. During the experiment the total amount of water in the carrier is constant (c_w =const).

The barycentric velocity is defined as

$$v = \frac{(\rho_c v_c + \rho_l v_l + \rho_v v_v)}{\rho} = \frac{U_4 + U_5 + (U_4 U_2/U_1)}{U_1 + U_2 + U_3} \tag{2.4}$$

and the diffusion flux densities are

$$j_{Dc} = \rho_c(v_c - v) = (U_4 - U_1 v)/A, \quad j_{Dl} = \rho_l(v_l - v) = (U_5 - U_3 v)/A. \tag{2.5}$$

The gaseous part of this mixture is considered as a mixture of gases with the same temperatures $T_c = T_v = T_g$. The partial pressure of the carrier p_c and the steam pressure p_v are given by

$$p_c = \frac{\rho_c \bar{R}_c}{c_{pc} - \bar{R}_c} \left(e_c - v_c^2/2\right) = \frac{\bar{R}_c}{c_{pc} - \bar{R}_c} \left(U_6 - \frac{U_4^2}{2U_1}\right)/A,$$

$$p_v = \frac{\bar{R}_v U_2}{\bar{R}_c U_1} p_c, \qquad p_l = 0. \tag{2.6}$$

In the vicinity of the saturation curve the water is taken as a caloric perfect liquid with the different temperature $T_l \neq T_g$. The partial pressure of the droplets is zero. Its internal energy is

$$u_l(T_l) = u_{l_{ref}} + c_{vl}(T_l - T_{ref}) \tag{2.7}$$

where $u_{l_{ref}}$, T_{ref} are standard values for the liquid (e.g. at the triple point). The evaporation heat is $h_{vs} - h_{ls} = h_{vl} = -h_{lv}$.

3. Model of the phase transition

3.1. Nucleation rate

To improve the usual capillary approximation we take into account more contributions to the cluster free energy (Dillmann and Meier 1991) than is usually used in the Gibbs capillary approach. The nucleation rate developed by Dillmann and Meier (1991) depends on the following macroscopic quantities:

1) mass of molecule $m_v = \frac{M}{N_A}$
2) the values of the thermodynamic variables T_c, p_c, ρ_c at the critical point
3) surface tension $\sigma(T)$
4) density of saturated liquid ρ_l
5) second virial coefficient $B(T)$.

The final form of the nucleation rate $J\left[\frac{\text{nuclei}}{\text{m}^3.\text{s}}\right]$ is (Dillmann and Meier 1991):

$$J = \frac{1}{3}\left[\frac{\sigma s_1}{\pi kT}\left(1 + \alpha_1 i_*^{-1/3}\right) + \frac{9\tau}{2\pi} i_*^{-2/3}\right]^{1/2} \times$$

$$\times \frac{s_1 p q_0}{m_v\sqrt{2\pi m_v kT}} \exp\left[-K_{i_*}\frac{\sigma s_1}{kT} i_*^{-2/3} - \tau \ln i_* + i_* \ln\frac{p_v}{p_s}\right], \tag{3.1}$$

s_1 is the average surface of a single molecule and K_i, α_i are the cluster shape constants. The critical cluster size i_* is given by the real root of the cubic equation (Dillmann and Meier 1991).

The value of the coefficients τ, q_0 describing the contributions from translational, rotational and vibrational degrees of freedom and their values are determined from the equation of state at the critical point: $\tau = 2.166$ and $q_0[\text{kg/m}^3] = 0,151\rho_c$ for water. For $\tau = 0$, $K_i = 1$ (i.e., $\alpha_1 = 0$) the nucleation rate (Eq.3.1) yields the classical Becker–Döring–Zeldovich theory.

3.2. Condensation rate

The amount of liquid (in kg) created in m^3 per one second is called the condensation rate w_l. This rate, for a given material point in the mixture depends on the nucleation rate J and on the critical size of droplets r_* not only at the actual time t but during the whole history of this material point. One of the good models is:

$$w_l = \frac{4\pi}{3}\bar{\rho}_l\left\{r_*^3 J + 3\dot{\bar{r}}\int_{-\infty}^{t}\left[r_*(t_1) + \int_{t_1}^{t}\dot{r}dt_2\right]^2 J|_{t_1} dt_1\right\}. \tag{3.2}$$

The time integral describes the condensation rate due to the growth of all droplets created sometime before (Hill 1966). We consider the Hill approximation of the formula (3.2) in the form:

$$w_l = \frac{4\pi\bar{\rho}_l}{3}\left[r_0^3 J + 3\rho\dot{\bar{r}}Q_2\right], \tag{3.3}$$

where $\bar{r}$ is a "surface-averaged" droplets radius defined by $\bar{r} = \left[Q_2/Q_0\right]^{1/2}$ for $r_0 \doteq 1.3r_*$ (Hill 1966).

3.3. Mass and heat transfer on the drops

The interaction between the liquid and gaseous components is accounted for by the right-side vector $\boldsymbol{H}$ (Eq.2.2) and is divided into three parts:

i) mass transfer - the component H_2, which accounts for the growth of the drops (clusters) in the case of molecular free flow (Gyarmaty 1982, Hill 1966) ($r_* \approx 10^{-8}[\mathrm{m}] \leq l_{\mathrm{mol}} \approx 10^{-6}[\mathrm{m}]$) is successfully modeled by the formula:

$$\frac{d\bar{r}}{dt} = \frac{\xi}{\bar{\rho}_l} m_v (\beta_v - \beta_l), \; \beta_v = \frac{p_v}{\sqrt{2\pi m_v k T_v}}, \; \beta_l = \frac{p_l}{\sqrt{2\pi m_v k T_l}}, \; p_s = p_{s\infty} \exp\left[\frac{2\sigma}{\bar{\rho}_l \hat{R}_v T_g \bar{r}}\right] \tag{3.4}$$

The impingement rate $\beta = \frac{p}{\sqrt{2\pi m k T}}$ depends on the actual pressure p, temperature T and on the mass of molecule m.

ii) momentum transfer - the quantities k_v, k_c (where $k_c + k_v + k_l = 0$) in the components H_4, H_5, are:

$$k_l = 6\pi\rho Q_0 \bar{r} \mu_g (v_g - v_l) \left[\left(1 + 0.15\mathrm{Re}^{0.687}\right)^{-1} + 4.5\mathrm{Kn}\right]^{-1} + w_e v_l,$$

$$\text{for} \quad \mathrm{Re} = \frac{2\rho_g |v_g - v_l| \bar{r}}{\mu_g}, \quad \mathrm{Kn} = \frac{\mu_g \sqrt{4/3 R_g T_g}}{4\bar{r} p_g}, \quad k_c = -\frac{\rho_c}{\rho_v + \rho_c} k_l \tag{3.5}$$

iii) energy transfer - the densities of production of the total energy $\sigma(e_c)$, $\sigma(e_v)$, $\sigma(e_l)$, introduced in the components H_6, H_7, have to fulfil the following identity:

$$\sigma(e_c) + \sigma(e_v) + \sigma(e_l) = \sigma(u_c) + \sigma(u_v) + \sigma(u_l) + k_l(v_l - v_g) + w_e \frac{v_g^2 - v_l^2}{2} = 0 \tag{3.6}$$

The density of energy transfer between drops and the gaseous part (Hill 1966, Gyarmaty 1982), which is given by:

$$\begin{aligned} \sigma(e_l) = \sigma(u_l) + \sigma\left(\frac{v_l^2}{2}\right) =& 4\pi Q_2 \rho \left\{\xi c_{vv} m_v (\beta_v T_g - \beta_l T_l) - (1 - \xi)\, \alpha_v (T_l - T_g) c_{vv} m_v \beta_v\right\} \\ &- 4\pi Q_2 \rho \left\{\alpha_c (T_l - T_g) c_{vc} m_c \beta_c\right\} + k_l v_l - w_l \frac{v_l^2}{2} \end{aligned} \tag{3.7}$$

is proportional to the number of molecules of water going through the surface Q_2 both from the vapour and the liquid sides. Because all of the impinged molecules are not going through, the stick coefficient (Hill 1966, Gyarmaty 1982) $\xi = 0,3 \div 1$ is introduced. The transfers of energy caused by reflected molecules of the vapour and by the carrier gas are taken into account using the accommodation coefficients $\alpha_v, \alpha_c \doteq 0.8$ (Gyarmaty 1982). The internal energy production of vapor is supposed to be described in the following way:

$$\sigma(u_v) = \frac{c_{vv} \rho_v}{c_{vc} \rho_c} \sigma(u_c) - c_{vv} T_g w_l. \tag{3.8}$$

The desired density of the production of the carrier total energy is:

$$\sigma(e_c) = \frac{c_{vv} T_g w_l + k_l v_g - \sigma(e_l)}{1 + \frac{c_{vv}\rho_v}{c_{vc}\rho_c}} + k_c v_g. \tag{3.9}$$

4. Numerical procedure

For the numerical solution of Eq.2.1 the McCormack scheme was adopted, with artificial viscosity incorporated. The numerical scheme is accurate to second order in time and in space. The oscillations at the shock are damped by the artificial aditional flux $F_{(AV\,k)} = -\Delta x D_k \left(\Delta_x, \lambda, \frac{\partial U_k}{\partial x}, \frac{\partial^2 U_k}{\partial x^2}\right) \frac{\partial U_k}{\partial x}$, for $k = 1, \cdots 10$ or by introducing limiters which prevent the occurence of parasitic waves (TVD - schemes). All flux vector components $F_k, k = 1, 2, \cdots 10$

of Eq.2.2 have then the form: $\hat{F}_k = F_k + F_{(AV\,k)}$ where the function $\Delta x D_k\left(\Delta_x, \lambda, \frac{\partial U_k}{\partial x}, \frac{\partial^2 U_k}{\partial x^2}\right)$ plays the role of numerical damping and for the balance of momentum is equivalent to physical viscosity, $\lambda = v + a$ is the maximum eigenvalue of the linearized Eq.2.1 and Δx is the mesh size.

The additional flux introduces always a nonphysical contribution to our mathematical model. To minimize this defect we have introduced a flexible computational mesh with $\Delta x^{(n)}$ going to more than an usually physically acceptable value, i.e. $\Delta x^{(n)} \sim 10^2 \cdot l_{\text{mol}}$ free molecular path. The physical kinematic viscosity is approximately equal to $a \cdot l_{\text{mol}}$ (a is the speed of sound). We have used the function D_k in the form $D_k = \min\left(\frac{\lambda}{2}\Delta x \frac{\left|\frac{\partial^2 U_k}{\partial x^2}\right|}{\left|\frac{\partial U_k}{\partial x}\right|}, \frac{\lambda}{2}\right)$. The maximum value of the numerical viscosity $\Delta x D_k$ is in the vicinity of sharp changes, e.g. shock waves and is equal to $\frac{\lambda}{2}\Delta x$.

The time step Δt for the explicit numerical scheme for Euler flow equations is given by the CFL= $\frac{(a+v)\Delta t}{\Delta x}$ criterion. However the relaxation time of droplet growth is different. For our computational mesh $\Delta x_{\text{min}} = 5 \cdot 10^{-4}$ [m] it follows that $\Delta t^{(n)} = 0.2\,[\mu\text{s}]$. The shortest time having a physical meaning in this condensation model is however the time in which the system reaches the steady state of nucleation rate. For $T = 273.3$ [K], $p_s = 6 \times 10^2$ [Pa], $p_v/p_s = 5$, it is 0.2 [μs]. The time step for the two-temperature model is about 10^{-7} [s] and the mesh size for CFL ≤ 1 is $\Delta x \approx 5 \cdot 10^{-5}$ [m].

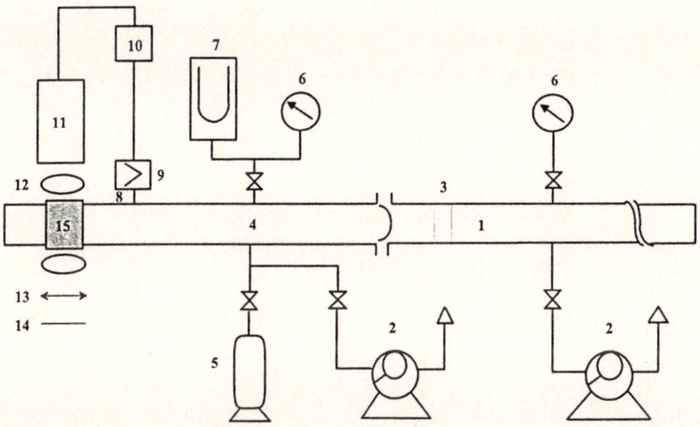

Fig. 1. Experimental set-up: 1–driven section of the shock tube ID 50 [mm], 2–vacuum pump, 3–adjustable shock wave generator, 4–driver section, 5–filling station, 6–pressure gauge, 7–vacuummeter, 8–pressure transducer, 9–amplifier, 10–delay, 11–light source, 12–Mach-Zehnder interferometer, 13–lens, 14–camera, 15–test section, 16–water vapor filling system

5. Experimental method

The experiment has been carried out with a water-argon mixture at room temperature, in the experimental facility schematically drawn in Fig.1. One of the aims of this experiment is to carry out simultaneously and independently measurements of the pressure and density of the gaseous part during condensation and evaporation.

6. Results and discussion

Interaction of the reflected shock wave with the argon-wet steam mixture produced by foregoing expansion is shown in Fig.2 together with an interferogram which was taken under the same experimental conditions.

The influence of the phase transition on the density change is apparent. The open questions in these models are the values of the second virial coefficient for steam and the magnitude of

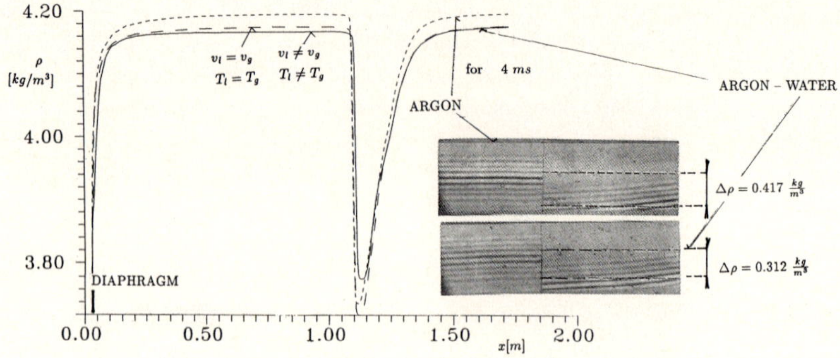

Fig. 2. Density distribution of the shock wave–expansion fan interaction. Comparison of the optical measurement with the numerical calculation (two–temperature, two–velocity model) for:
- - - - pure argon $p_1 = 256$ kPa, $p_4 = 100$ kPa, $T_1 = T_4 = 293$ K; ——— argon–water mixture $p_1 = p_{1Ar} + p_{1w} =$ 256 kPa, $p_{1w} = 2.2$ kPa,$p_4 = p_{2Ar} = 100$ kPa, $p_{4w} = 0$, $T_1 = T_4 = 293$ K, $\xi = 0.7$, $\alpha_c = \alpha_v = 0.8$.

the sticking and accommodation coefficients ξ, α_c, α_v, respectively. The density relaxation caused by evaporation is also possible to measure in this way directly, but in the given range of the parameters it is not visible.

References

Barschdorff D (1975) Carrier gas effects on homogeneous nucleation of vapor in shock tube. Phys. Fluids 18, 5:529–535

Dillmann A, Meier GEA (1991) A refined droplet approach to the problem of homogeneous nucleation from vapor phase. J. Chem. Phys. 94, 5: 3872–3884

Goossens HWJ, Cleijne JW, Smolders HJ, van Dongen MEH (1988) Shock wave-induced evaporation of water droplets in a gas–droplet mixture. Experiments in Fluids 6: 561–568

Guha A, Young JB (1991) Time–marching prediction of unsteady condensational phenomena due to supercritical heat addition. In: Proc. Conf. Turbomachinery: Latest Developements in a Changing Scene, London, IMechE, pp. 167–177.

Gyarmaty G (1982) The spherical droplet in gaseous carrier stream: Review and synthesis. in: Hewitt GF et al. (eds) Multiphase Science and Technology, Vol. 1, Hemishere Publishing Corp., Wasington, pp 99–279

Hill PG (1966) Condensation of water vapour during supersonic expansion in nozzles. Journal of Fluid Mechanics 25:593–620.

Maršík F, Blaha J, Sopuch P (1994) Adiabatic expansion in superheated steam with condensation. Transport Theory and Statistical Physics 23, 1-3:365-384

Peters F, Paikert B (1989) Nucleation and growth rate of homogeneously condensing water vapor in argon from shock tube experiments. Experiments in Fluids 7:521–530

Saul A, Wagner W (1987) International equations for the saturation properties of ordinary water substance. J. Phys. Chem. Ref. Data 16, 4: 893–901

Schnerr G, Dohrmann U (1989) Transonic flow around airfoils with relaxation and energy supply by homogeneous condensation. AIAA Paper 89–1834, AIAA 20th Fluid Dynamics, Plasma Dynamics and Lasers Conference, Buffalo, New York

Propagation of Shock Waves Through a Gas Mixture Due to Strong Evaporation from a Plane Condensed Phase

Y. Onishi* **and** **H. Miura†**
*Department of Applied Mathematics and Physics, Faculty of Engineering, Tottori University, Tottori 680, Japan
†Department of Mechanical Engineering, Faculty of Science and Technology, Kinki University, Higashi-Osaka 577, Japan

Abstract. Flowfields associated with strong evaporation from a plane condensed phase in a binary mixture of condensable and noncondensable gases occupying the half space have been investigated by solving numerically the BGK model of the Boltzmann equation subject to the boundary condition of diffuse-reflection at the interface. One can see how the disturbances in the form of shock and expansion waves caused by the evaporation process propagate through the mixture. The effects of the presence of the noncondensable gas on the propagation of the disturbances and on the establishment of the final state of the flowfields are clarified. In the strong evaporation case, the noncondensable gas, when it is present in small amount, is driven almost completely into the contact region owing to the vapor flow and there appears a region almost free from the noncondensable gas between the condensed phase and the contact region.

Key words: Kinetic theory, CFD, Evaporation, Shock wave, Expansion wave

1. Introduction

Flowfields associated with the evaporation process from the condensed phase are of fundamental importance in gas dynamics, in the sense that one may be able to understand how the disturbances in the form of shock and expansion waves caused by the sudden change of the state of the system propagate through the flowfield and affect it during the course of their propagation. Especially, when a certain amount of noncondensable gas exists, the effects of its presence on the propagation of the disturbances and also on the establishment of the final steady state, if it exists, are worth being clarified. Recently, an evaporation problem of this kind has been studied numerically by Sone and Sugimoto (1990) for a single-component condensable gas (pure vapor case) and the transient behavior as well as its final steady state has been clarified. Here we investigate the corresponding problem for a mixture of a condensable gas and a noncondensable one by solving numerically the BGK model of the Boltzmann equation (Hamel 1965) subject to a diffusive boundary condition at the interface.

2. Fundamental kinetic equations, initial and boundary conditions

Now, let the plane condensed phase be located at $X = 0$ and the half space ($X \geq 0$) be occupied by a binary mixture of a condensable gas (A-gas) and a noncondensable gas (B-gas). Let the temperature of the condensed phase be raised by a certain amount from the original temperature at which the condensed phase and the mixture gas phase have been in complete equilibrium. Then the evaporation begins to take place at the interface and the mixture is set to motion. The Boltzmann equation of BGK type (Hamel 1965) describing this behavior of the mixture can be written as

$$\frac{\partial f^s}{\partial t} + \xi_x \frac{\partial f^s}{\partial X} = N^s \kappa_{ss}(F_e^s - f^s) + N^t \kappa_{st}(F_e^{st} - f^s) \tag{1}$$

$$\begin{bmatrix} N^s \\ N^s U^s \\ \frac{3}{2} N^s \mathrm{k} T^s \end{bmatrix} = \iiint \begin{bmatrix} 1 \\ \xi_x \\ \frac{1}{2} m_s [(\xi_x - U^s)^2 + \xi_y^2 + \xi_z^2] \end{bmatrix} f^s \, d\xi_x d\xi_y d\xi_z \tag{2}$$

Shock Waves @ Marseille III
Editors: R. Brun, L. Z. Dumitrescu

$$F_e^s = \frac{N^s}{(2\pi R_s T^s)^{3/2}} \exp\left\{-\frac{(\xi_x - U^s)^2 + \xi_y^2 + \xi_z^2}{2R_s T^s}\right\} \tag{3}$$

$$F_e^{st} = \frac{N^s}{(2\pi R_s T^{st})^{3/2}} \exp\left\{-\frac{(\xi_x - U^{st})^2 + \xi_y^2 + \xi_z^2}{2R_s T^{st}}\right\} \tag{4}$$

$$U^{st} = U^{ts} = \mu_s U^s + \mu_t U^t, \qquad T^{st} = T^s + 2\mu_s\mu_t(T^t - T^s) + \frac{m_s}{3\mathrm{k}}\mu_t^2 (U^t - U^s)^2 \tag{5}$$

where $\mu_s = m_s/(m_s+m_t)$. s appearing as superscript and subscript refers to A or B and t refers to B or A. t is the time; X is the coordinate; (ξ_x, ξ_y, ξ_z) the molecular velocity vector; f^s, N^s, U^s, T^s and P^s $(= N^s \mathrm{k} T^s)$ are, respectively, the distribution function, the number density, the mean flow velocity, the temperature and the partial pressure of s-gas. m_s and R_s are the molecular mass of s-gas and its gas constant per unit mass, respectively; k is the Boltzmann constant. F_e^s and F_e^{st} are the local Maxwellian distributions for molecules of s-gas characterized by the macroscopic quantities (N^s, U^s, T^s) and (N^s, U^{st}, T^{st}), respectively. $N^s\kappa_{ss}$ and $N^t\kappa_{st}$ represent the number of collisions per unit time of a molecule of s-gas due to the self and cross collisions, respectively, thus $N^s\kappa_{ss} + N^t\kappa_{st}$ being the average collision frequency for a molecule of s-gas irrespective of its collision partners. The collision frequencies are related to the transport coefficients of the mixture and its component gases and, hence, κ_{ss} and κ_{st} $(= \kappa_{ts})$, which are assumed to be constant and of the same order of magnitude, can be evaluated from these coefficients at some equilibrium reference state (see e.g., Hamel 1965 or Onishi 1992), say at infinity, as

$$\left.\begin{aligned} \eta^M &= \frac{P_\infty^A}{N_\infty^A \kappa_{AA} + N_\infty^B \kappa_{AB}} + \frac{P_\infty^B}{N_\infty^A \kappa_{AB} + N_\infty^B \kappa_{BB}}, \\ D_{AB} &= \frac{(m_A + m_B)\mathrm{k} T_\infty}{m_A m_B N_\infty \kappa_{AB}}, \qquad \eta^s = \frac{P_\infty^s}{N_\infty^s \kappa_{ss}}, \qquad \lambda^s = \frac{5}{2} R_s \eta^s \end{aligned}\right\} \tag{6}$$

where the fluid dynamic quantities with subscript ∞ such as N_∞^s, P_∞^s and T_∞ refer to those at the equilibrium reference state at infinity, and $N_\infty = N_\infty^A + N_\infty^B$ is the reference number density of the mixture. η^M and D_{AB} are the viscosity and diffusion coefficients of the mixture, η^s and λ^s being, respectively, the viscosity and thermal conductivity of s-gas.

Now, initially, the condensed phase and the mixture gas phase are assumed to be in complete equilibrium with temperature T_∞, the number density and pressure of the component gas in the mixture then being N_∞^s and P_∞^s, respectively. The distribution function of s-gas at this state is given by

$$f^s = \frac{N_\infty^s}{(2\pi R_s T_\infty)^{3/2}} \exp\left\{-\frac{\xi_x^2 + \xi_y^2 + \xi_z^2}{2R_s T_\infty}\right\} \quad \text{for} \quad 0 \le X < \infty \tag{7}$$

and for all possible values of the molecular velocity. At time $t = 0$, the temperature of the condensed phase is suddenly raised to T_W $(T_W > T_\infty)$, then the distribution function f^s becomes

$$f^s = \frac{N_W^s}{(2\pi R_s T_W)^{3/2}} \exp\left\{-\frac{\xi_x^2 + \xi_y^2 + \xi_z^2}{2R_s T_W}\right\} \quad \text{for} \quad \xi_x > 0 \quad \text{at} \quad X = 0 \tag{8}$$

for $t \ge 0$, whereas f^s at infinity $(X \to \infty)$ for $\xi_x < 0$ remains to be the same as that given by Eq.7 for $t \ge 0$. Here N_W^s is the number density of the molecules of s-gas leaving the interface after the interaction with it. N_W^A is a unique function of T_W and is given as the saturated vapor number density corresponding to T_W given by the Clapeyron-Clausius relation as

$$N_W^A = \frac{P_W^A}{\mathrm{k} T_W}, \qquad \frac{P_W^A}{P_\infty^A} = \exp\left\{-\frac{h_L}{R_A T_\infty}\left(\frac{T_\infty}{T_W} - 1\right)\right\} \tag{9}$$

where P_W^A is the saturated vapor pressure corresponding to T_W just as P_∞^A is to T_∞. h_L is the latent heat of vaporization per unit mass of A-gas. N_W^B, on the other hand, is to be determined, as part of the solution, by the condition of no net mass flow of B-gas across the interface, i.e., $N^B U^B = 0$ at $X = 0$. Hence, using Eq.8, we have for N_W^B

$$N_W^B = -\frac{2\pi}{(2\pi R_B T_W)^{1/2}} \int_{-\infty}^{\infty}\int_{-\infty}^{\infty}\int_{\xi_x<0} \xi_x f^B(t,\, X=0,\, \xi_x,\, \xi_y,\, \xi_z)\, d\xi_x d\xi_y d\xi_z \tag{10}$$

Now, for the numerical analysis, let us introduce the following functions (see e.g. Chu 1965)

$$\begin{bmatrix} g^s \\ G^s \\ G^{st} \end{bmatrix} \equiv \iint \begin{bmatrix} f^s \\ F_e^s \\ F_e^{st} \end{bmatrix} d\xi_y d\xi_z, \qquad \begin{bmatrix} h^s \\ H^s \\ H^{st} \end{bmatrix} \equiv \iint (\xi_y^2 + \xi_z^2) \begin{bmatrix} f^s \\ F_e^s \\ F_e^{st} \end{bmatrix} d\xi_y d\xi_z \tag{11}$$

then

$$\begin{bmatrix} N^s \\ N^s U^s \\ \frac{3}{2} N^s \mathrm{k} T^s \end{bmatrix} = \int \begin{bmatrix} g^s \\ \xi_x g^s \\ \frac{1}{2} m_s [(\xi_x - U^s)^2 g^s + h^s] \end{bmatrix} d\xi_x \tag{12}$$

Rewriting the original equations with these variables, we then get the simultaneous equations for the four unknown functions (g^A, h^A, g^B, h^B) of three independent variables (t, X, ξ_x) instead of having two unknown functions (f^A, f^B) of five independent variables $(t, X, \xi_x, \xi_y, \xi_z)$. This new form renders the numerical calculation far easier than the original one. Appropriate nondimensionalization of the set of the governing equations, initial and boundary conditions gives us the following parameters which characterize the motions of the mixture and its component gases:

$$\frac{T_W}{T_\infty}, \quad \frac{N_\infty^B}{N_\infty^A}, \quad \Gamma = \frac{h_L}{R_A T_\infty} \ \left(\text{or } \frac{P_W^A}{P_\infty^A}\right), \quad \frac{m_B}{m_A}, \quad \frac{\kappa_{AB}}{\kappa_{AA}}, \quad \frac{\kappa_{BB}}{\kappa_{AA}} \tag{13}$$

where the last two parameters can be calculated from the transport coefficients using Eq.6.

3. Results and discussion

Actual calculations have been carried out for more than 20 cases, setting $\Gamma = 11$, $m_B/m_A = 2$ and $\kappa_{AA} = \kappa_{AB} = \kappa_{BB}$. The transient behaviours for a few of these cases are shown in Figs. 1-3 where U, T, P and N are the fluid dynamic quantities of the mixture. c^s and c are the sound speeds of s-gas and of the mixture at the local state, respectively, and c_∞^s and c_∞ the corresponding ones at the initial state. l_∞^A and τ_∞^A are the molecular mean free path and the mean collision time of A-gas (condensable gas) at the initial state, which are defined by

$$l_\infty^A = \frac{(8R_A T_\infty/\pi)^{1/2}}{(N_\infty^A \kappa_{AA} + N_\infty^B \kappa_{AB})}, \qquad \tau_\infty^A = \frac{l_\infty^A}{(8R_A T_\infty/\pi)^{1/2}}. \tag{14}$$

Fig.1 shows the propagation of a shock wave accompanied behind it by the formation of the uniform region, the contact region (corresponding to the contact surface in inviscid terms), the second uniform region, the expansion region due to the swept expansion waves and the Knudsen layer (see the graphs for the mixture and A-gas). This layer, the thickness of which is of the order of the molecular mean free path, always manifests itself near the boundary. The noncondensable B-gas is now being pushed into the contact region by the strong flow and is also compressed by the shock wave, having a peculiar behavior behind it. The contact region has a complex structure because of almost all the B-gas having been driven into it, whereas the Knudsen layer and the swept expansion region are quite simple in the sense that the behavior of A-gas (and, of course, of the mixture) in those regions is virtually identical to that for a single vapor case (Sone and Sugimoto 1990). The expansion waves, which are propagating in the opposite direction to the shock wave relative to the flow, are being swept backward by the supersonic flow existing outside of the Knudsen layer, except a part of the waves which may remain indefinitely at sonic point near the edge of the Knudsen layer (see U^s/c^s graphs). The flowfield in this case seems to remain unsteady and never to attain a steady state (cf. Sone and Sugimoto (1990) for a single vapor case).

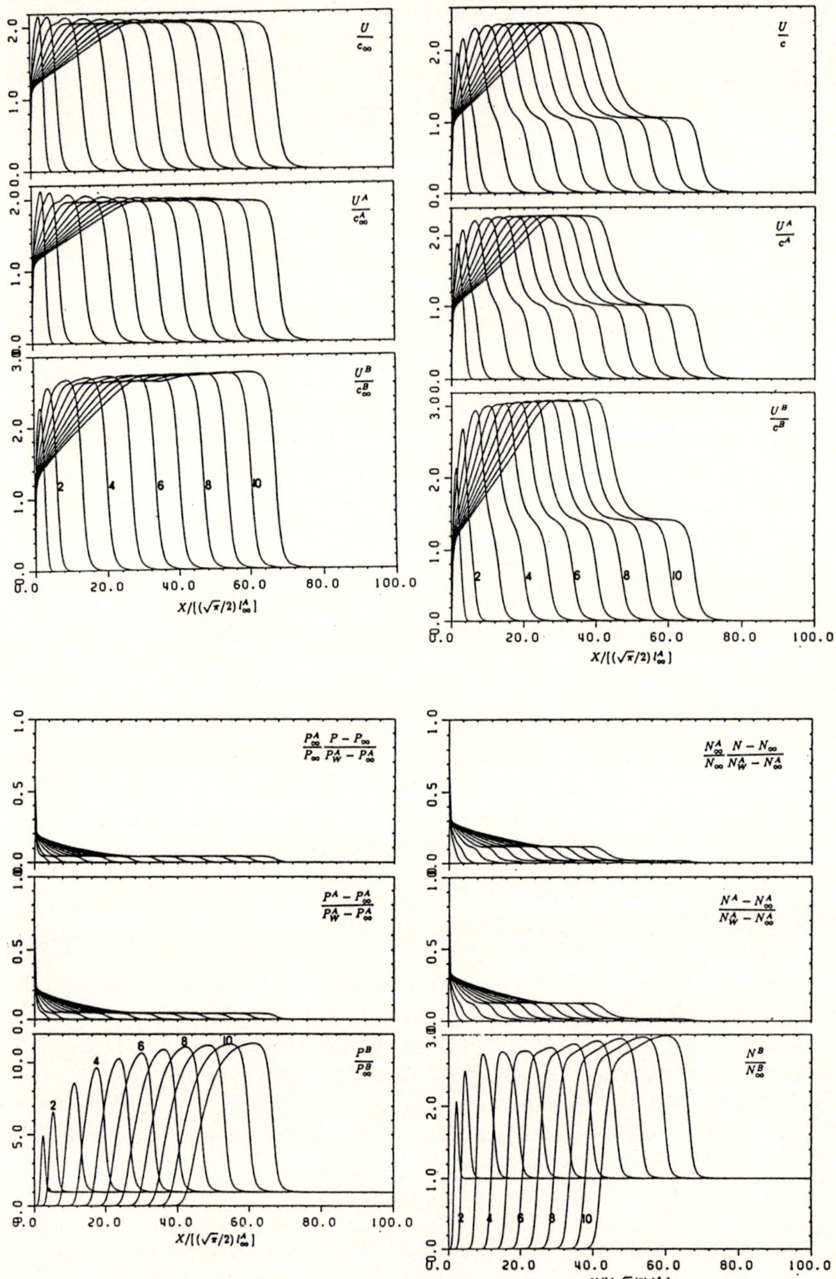

Fig. 1. Transient behaviour of various fluid dynamic quantities. $T_W/T_\infty = 2.0$, $N^B_\infty/N^A_\infty = 0.1$, $\Gamma = 11$ (or $P^A_W/P^A_\infty = 244.692$), $m_B/m_A = 2$, $\kappa_{AA} = \kappa_{AB} = \kappa_{BB}$.
1 : $t/\tau^A_\infty = 1.25$, 2 : $t/\tau^A_\infty = 2.5$, 3 : $t/\tau^A_\infty = 5.0$, 4 : $t/\tau^A_\infty = 7.5$, 5 : $t/\tau^A_\infty = 10.0$, 6 : $t/\tau^A_\infty = 12.5$, 7 : $t/\tau^A_\infty = 15.0$, 8 : $t/\tau^A_\infty = 17.5$, 9 : $t/\tau^A_\infty = 20.0$, 10 : $t/\tau^A_\infty = 22.5$, 11 : $t/\tau^A_\infty = 25.0$

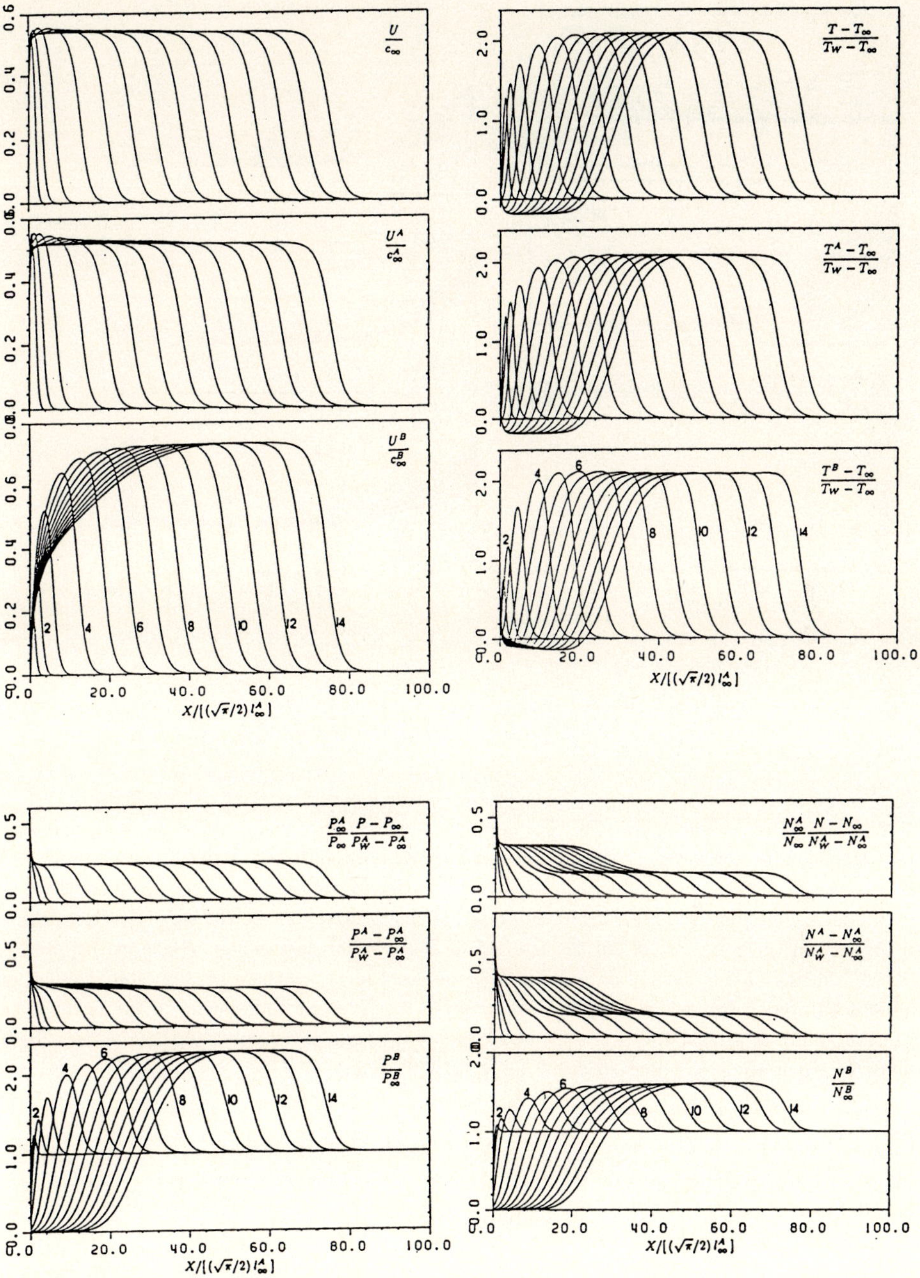

Fig. 2. Transient behaviour of various fluid dynamic quantities. $T_W/T_\infty = 1.2$, $N^B_\infty/N^A_\infty = 0.1$, $\Gamma = 11$ (or $P^A_W/P^A_\infty = 6.2547$), $m_B/m_A = 2$, $\kappa_{AA} = \kappa_{AB} = \kappa_{BB}$.
1 : $t/\tau^A_\infty = 1.25$, 2 : $t/\tau^A_\infty = 2.5$, 3 : $t/\tau^A_\infty = 5.0$, 4 : $t/\tau^A_\infty = 10.0$, 5 : $t/\tau^A_\infty = 15.0$, 6 : $t/\tau^A_\infty = 20.0$, 7 : $t/\tau^A_\infty = 25.0$, 8 : $t/\tau^A_\infty = 30.0$, 9 : $t/\tau^A_\infty = 35.0$, 10 : $t/\tau^A_\infty = 40.0$, 11 : $t/\tau^A_\infty = 45.0$, 12 : $t/\tau^A_\infty = 50.0$, 13 : $t/\tau^A_\infty = 55.0$, 14 : $t/\tau^A_\infty = 60.0$

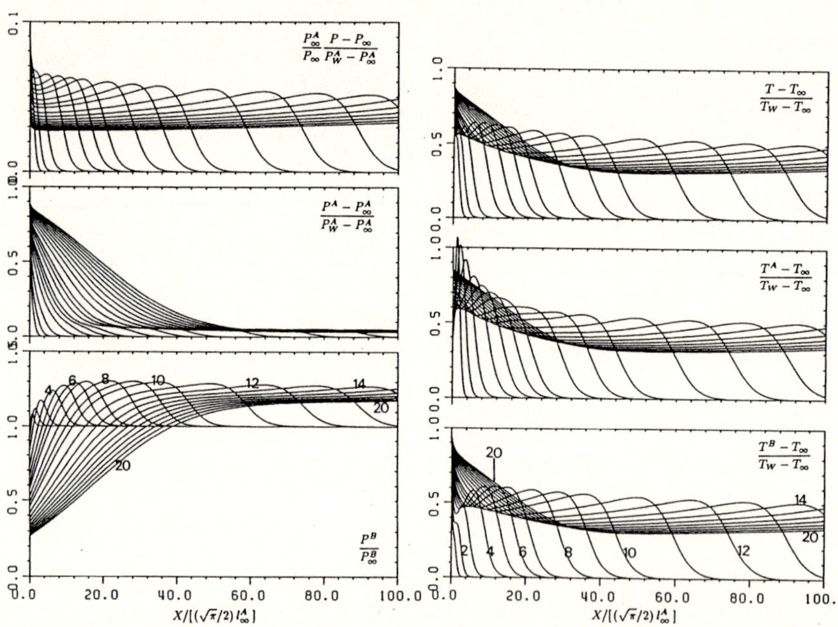

Fig. 3. Transient behaviour of various fluid dynamic quantities. $T_W/T_\infty = 1.2$, $N_\infty^B/N_\infty^A = 5.0$, $\Gamma = 11$ (or $P_W^A/P_\infty^A = 6.2547$), $m_B/m_A = 2$, $\kappa_{AA} = \kappa_{AB} = \kappa_{BB}$.
1 : $t/\tau_\infty^A = 1.25$, 2 : $t/\tau_\infty^A = 2.5$, 3 : $t/\tau_\infty^A = 5.0$, 4 : $t/\tau_\infty^A = 10.0$, 5 : $t/\tau_\infty^A = 15.0$, 6 : $t/\tau_\infty^A = 20.0$, 7 : $t/\tau_\infty^A = 25.0$, 8 : $t/\tau_\infty^A = 35.0$, 9 : $t/\tau_\infty^A = 45.0$, 10 : $t/\tau_\infty^A = 55.0$, 11 : $t/\tau_\infty^A = 75.0$, 12 : $t/\tau_\infty^A = 95.0$, 13 : $t/\tau_\infty^A = 115.0$, 14 : $t/\tau_\infty^A = 135.0$, 15 : $t/\tau_\infty^A = 155.0$, 16 : $t/\tau_\infty^A = 175.0$, 17 : $t/\tau_\infty^A = 195.0$, 18 : $t/\tau_\infty^A = 215.0$, 19 : $t/\tau_\infty^A = 235.0$, 20 : $t/\tau_\infty^A = 255.0$

When the temperature ratio T_W/T_∞ becomes small, the expansion waves no longer appear and, instead, a uniform region almost free from B-gas prevails between the Knudsen layer and the rear part of the contact region, although the behavior of B-gas scarcely present there is continuously changing (see Fig.2). This uniform flow may correspond to the steady state solution for A-gas and the mixture. When the amount of B-gas is large, although B-gas in this case is also continuously being pushed away from the condensed phase by the flow, the establishment of the uniform region for A-gas and the mixture takes much longer time owing to the fairly large diffusion ability of B-gas, and it is difficult to pursue the time development further on (see Fig.3 where the calculation was done up to $X/[(\sqrt{\pi}/2)\, l_\infty^A] = 255$).

References

Chu CK (1965) Kinetic-theoretic description of the formation of a shock wave. Phys. Fluids 8:12-22

Hamel BB (1965) Kinetic model for binary gas mixtures. Phys Fluids 8:418-425

Onishi Y (1992) Kinetic theory-based nonlinear treatment for motions of a binary gas mixture involving slightly strong evaporation and condensation processes – A system of macroscopic equations, the boundary conditions and the Knudsen-layer corrections. ZAMP 43:875-910

Sone Y, Sugimoto H (1990) Strong evaporation from a plane condensed phase. In: Meier GEA, Thompson PA (eds) Adiabatic Waves in Liquid-Vapor Systems. Springer-Verlag, Berlin, pp 293-304

Measurements on Film Condensation Behind Incident and Reflected Shock Waves

T. Teske and F. Obermeier
Max-Planck-Institut für Strömungsforschung, 37073 Göttingen, Germany

Abstract. Measurements on film condensation of n-heptane and mixtures of n-heptane with nitrogen have been performed in the low-pressure section of shock tubes. Previous work has been continued with emphasis on a phenomenological description of the growth coefficients of films. The growth of the liquid film at the end-wall and the side-walls has been observed from the very beginning up to the millisecond range. Experimental results corresponding to different flow situations are presented and compared.

Key words: Film condensation, Phase change, Stefan problem

1. Introduction

When a gas is supersaturated compared to the temperature of a surface, it starts to condense forming a thin liquid layer. The known literature is concerned with the initial stage, a period of a few μs (Maerefat 1990, Mizutani 1990). During the very early stage condensation is governed by molecular gasdynamic effects. Here the effect on vapour condensation due to the presence of a second gas has been studied by Pong and Moses (1986) on the basis of kinetic theory.

The growth rate at a later stage is a function of the thermal conductivities and heat capacities of the wall, the fluid and the gas, and of the state of the gas (T_g, P_g). Several transport coefficients are needed to formulate the problem. For a detailed analysis we refer to the textbook by Carslaw and Jaeger (1959).

In our case the wall temperature T_W has been constant for series of runs. At the end-wall the condensation process can approximately be described by the classical Stefan problem, which originally explains the movement of a phase boundary of a freezing liquid. For a condensing gas one has additionally to account for the different densities of liquid and gas, whose ratio is nearly 1000. The predicted asymptotic time dependence of the film thickness $d(t) \propto \sqrt{t}$ has been confirmed in all our measurements at the end-wall for time scales much larger than the μs range. At the side-wall the situation is somewhat more complicated, as will be shown in sections 3.2.1. and 3.2.3.

2. Setup and measurement technique

2.1. Shock tubes and instrumentation

The experiments described in the present work were performed using two different shock tubes. These tubes were described in detail by Teske (1992), series A, and Herrmann (1993), series B and C. Both have an inner diameter of 80 mm and they are made of brass (A) and stainless steel (B,C). The length of driver and driven sections is 2.33 m each. The test sections of both tubes mounted to the driven sections are 0.3 m in length. A diffusion pump has been mounted to tube B to allow runs with higher purity and/or lower initial pressure.

Kistler 603b piezoelectric transducers (PET) are used to record transient pressure and to determine the shock velocity. In order to protect them against the large heat flux we have insulated them thermally with a thin layer of silicon rubber. Initial temperature and pressure are measured by thermocouples placed on the window and 15 mm inside the tube, respectively, and by a heated MKS Baratron absolute pressure gauge. The tube temperature can be held uniform over the

Shock Waves @ Marseille III
Editors: R. Brun, L. Z. Dumitrescu

whole tube length within $\pm 0.5^{\circ}$C. Cellulose-tri-acetate and aluminum shatter-type diaphragms have been used. All transient signals are recorded by digital storage oscilloscopes (DSO).

2.2. Optical measurement technique

The principle of the optical method has already been described (Teske 1991). The condensed film is used as part of an interferometer whose reflectivity changes during deposition. Optical access is provided by synthetic quartz glass windows. Our light sources are linear polarised He-Ne lasers with different wavelengths (red 632.8 nm, green 543.5 nm) tuned for stable light intensity. Variations in light intensity are monitored by photodiodes.

For test reasons we performed test runs with light barriers to check if the mean flow is uniform, e.g. if droplets appear or if particles move downstream. In general the signals of light barriers sum up all the effects at the walls and inside the flow. An extraction of wall effects is done by simultaneous measurement in reflection and transmission (light barrier). The actual setup consists of several beams, partly sharing the same geometric path, partly illuminating different areas on the same window.

3. Experimental results

3.1. Pressure traces

According to Maerefat (1990) the pressure behind the reflected shock is smaller than compared to the Hugoniot values. The ratio of the measured and calculated jump is a function of the shock Mach number M_{is}, the inverse Stefan-number Ph (defined in Eq.4a) and other parameters. The measured jumps range between 75 to 100 % of the expected ones.

It is undecided to which extent the temperature jumps differ from the expected ones. The absence of a general explanation of this difference indicates the need for further work on shock wave reflection even at large heat and mass transfer rates.

In practice the pressure behind the reflected shock seems to be time-dependent: The pressure decreases for $t < 25$ μs by $\approx 5\%$ compared to stationary values maintained for hundreds of μs. Since one does not observe this effect without having condensation, we attribute it to be a consequence of the large heat flux when the film starts to condense, since the signals from a PET are similar for a decrease in pressure and a large heat flux towards the PET.

At the instant of shock reflection, one observes an overshot which is characteristic for our metal-on-metal mounting and the measurement chain.

3.2. Monitoring of the liquid film thickness

In all our experiments we found the liquid films to be homogeneous. This has been checked by direct comparison of signals recorded simultaneously in three modes: different wavelengths of the light source, different angles of incidence over a wide range from almost perpendicular incidence to the limit of total reflection between film and gas ($n_{\text{liquid}} = 1.385, \alpha_T = 46.2°$) and different sizes of illuminated areas ($\approx$1.7 mm^2 to 50 mm^2). The refractive index of the layer and the gas change in time and spatial coordinates in general.

In our case the temperature jump across the films, with thicknesses in the μm range, is a few degrees. This results in very small absolute changes in refractive index across the film. The changes within the gas are even smaller compared to those in the film. Both the observation time and the vertical resolution of the DSOs are not sufficient for measurements with the required high precision to detect very small changes in refractive indexes with thickness or time. Up to now there is no experimental evidence for non-uniformity. The experimental uncertainty due to digitizing and further processing steps can be neglected.

3.2.1. Condensation behind the incident shock

Behind the incident shock the gas condenses from a moving gas at the side wall, whereas it condenses behind the reflected shock from a gas at rest. When the reflected shock returns and reaches the observation area at the side-wall, condensation is again enforced. A discussion will be given in Section 3.2.3.

Here a simple function for the film thickness $d(t)$ does not exist. In addition to the dependencies described by Fujikawa et al. (1987) for end-wall measurements, we expect the growth to be influenced by the center flow and the boundary layer behind the incoming shock.

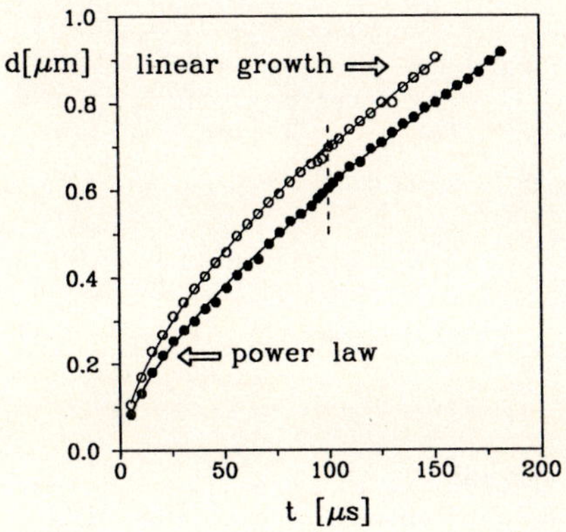

Fig. 1a. Samples showing the two $d(t)$ laws given in Eqs.1,2 behind the incident shock

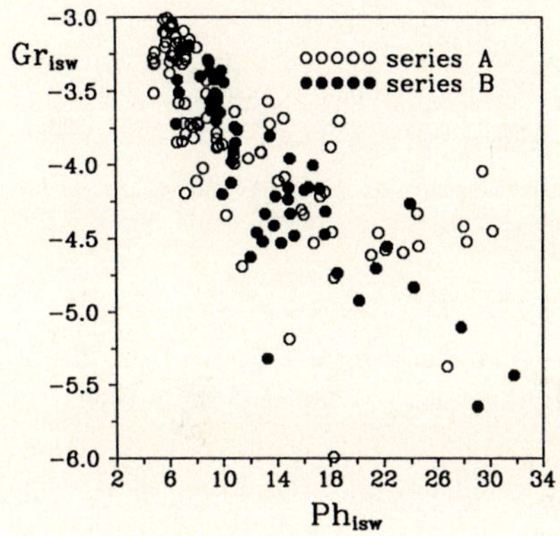

Fig. 1b. Growth coefficient as defined in Eq. 1 as function of Ph$_{isw}$

Since our range of flow velocities is quite narrow we found experimentally two simple functions describing $d(t)$ in two different intervals:

$$\ln(d(t)) = Gr_{isw} + \text{pow} \times \ln(t) \quad \text{for} \quad t \in 5 \ldots 100\mu s \quad \text{and} \tag{1}$$

$$\Delta d(t) \propto \Delta t \quad \text{for} \quad t > 100\mu s. \tag{2}$$

Note that Gr_{isw} given in Fig.1b corresponds to times t and thicknesses $d(t)$ scaled to μm resp. μs in Eq.1. For the exponent pow in Eq.1 we found pow $\in 0.5 \cdots 0.8$.

By transformation of Gr_{isw} into rates using pow $= 0.5$ for the fit, as done in Section 3.2.2, the curve obtained is similar to the one for the end-wall measurements shown in Fig.3a. The ratio of latent to sensible heat behind the incoming shock at the side-wall, Ph$_{isw}$, reaches large values for small Gr_{isw}, since the rise in temperature is fairly small.

$$\text{Ph}_{isw} = h_{vap}(T_{isw})/(c_p \times (T_{isw} - T_W)) \tag{3a}$$

$$T_{isw} \quad \text{is calculated from} \quad p_2 = p_{sat}(T_{isw}) \tag{3b}$$

Here T_W is the wall temperature, T_{isw} is the temperature at the interface between gas and liquid, h_{vap} is the enthalpy of vaporisation and c_p the specific heat at constant pressure. Fig.1b describes the dependence of Gr_{isw} (Eq.1) on Ph$_{isw}$ from side-wall measurements. Left with the uncertainty of the physically adopted temperature we estimate that the scatter of Gr_{isw} increases with Ph$_{isw}$.

Fig.2a,b shows an example of transmitted and reflected light measurements. The modulation of the reflected light is much larger than for transmitted light. In contrast to transmitted light

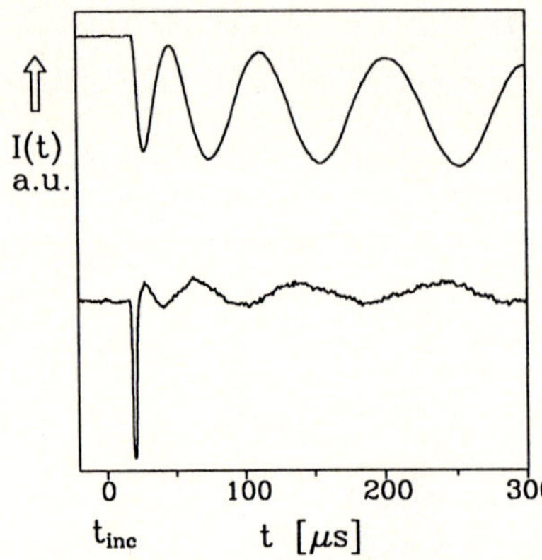

Fig. 2a. Light signals from sidewall measurements (upper signal: reflected light, lower signal: light barrier)

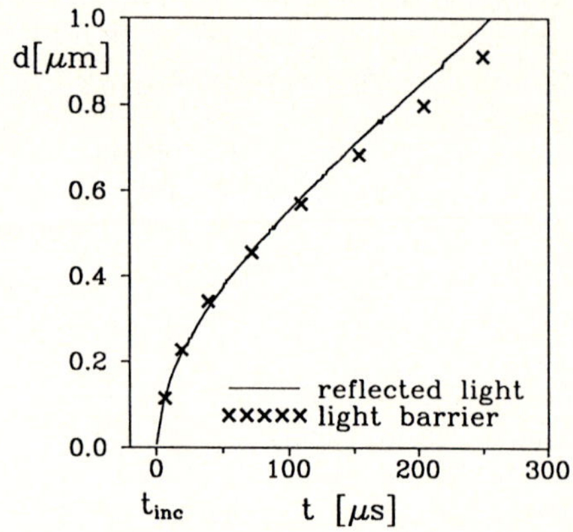

Fig. 2b. Evaluated thickness from Fig.2a

there is no black-out in reflected light when the shock passes the observation area, thus it is easier to fix the onset of condensation in time.

3.2.2. Condensation behind the reflected shock at the end-wall

The growth of the liquid film at the end-wall is compared to Neumann's solution of the problem of Stefan. Similar to Eqs.3a,b we obtain the temperature T_{pb} at the phase boundary and Ph from

$$\mathrm{Ph} = h_{vap}(T_{pb})/(c_p \times (T_{pb} - T_W)) \quad \text{and} \tag{4a}$$

$$T_{pb} \quad \text{from} \quad p_3 = p_{sat}(T_{pb}). \tag{4b}$$

For the gas at rest this yields the square root law for the thickness $d(t) \propto Gr\sqrt{t}$.

Fig.3a shows the asymptotic growth coefficient Gr as a function of Ph. The solid line has been calculated by means of Neumann's solution to the Stefan problem, within the following assumptions: all thermal properties remain constant, the heat flux is not limited by transport coefficients, the transient process in the beginning is negligible. Values predicted in this way for end-wall measurements are overestimated by 20 to 30 %. We attribute this to the fact, that the heat flux is limited. In practice, contrary to the assumptions made to solve the Stefan problem, there is always a resistance for heat and mass fluxes.

Fig.3b shows samples from experiments with initial pressures of 4.0 kPa n-heptane and additional nitrogen ranging from 0.025 to 1.0 kPa at initial temperature $T_1 = 303$ K. Here the experimental results show, in agreement with Maerefat's calculations (1987) on mixtures of methanol and air, that small impurities of non-condensable gas decrease the growth rate very efficiently.

This happens at least in two ways: Firstly, for a given initial state and strength of the initial shock, the mixture is heated to a higher temperature than the pure gas – thus leading to smaller supersaturation. Secondly, non-condensable gas accumulates in front of the liquid layer forming a concentration layer stacked onto the usual boundary layer. A quantitative study on the influence of impurities is in progress.

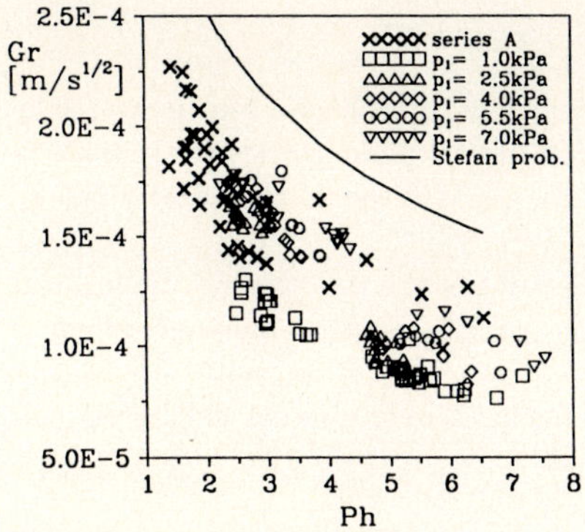

Fig. 3a. Asymptotic growth coefficient *Gr* for pure n-heptane at the end-wall

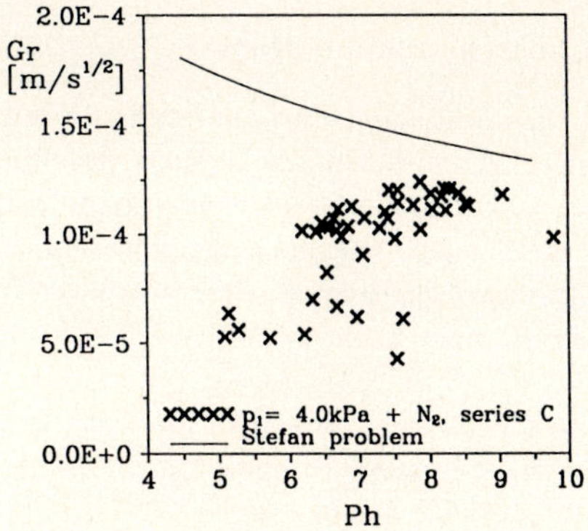

Fig. 3b. Asymptotic growth coefficient *Gr* for mixtures of n-heptane and nitrogen at the end-wall

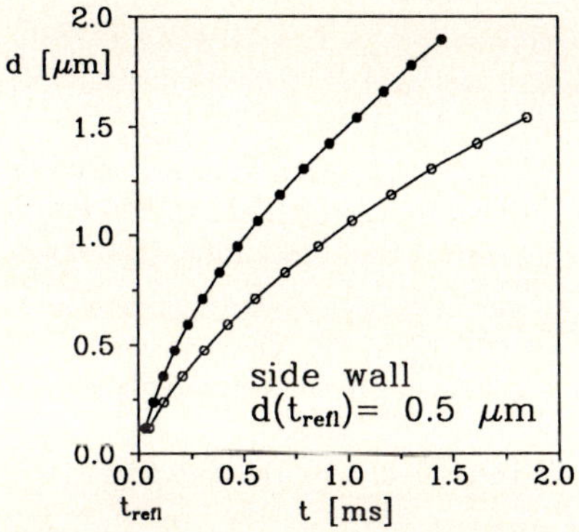

Fig. 4a. Change in thickness at the side-wall

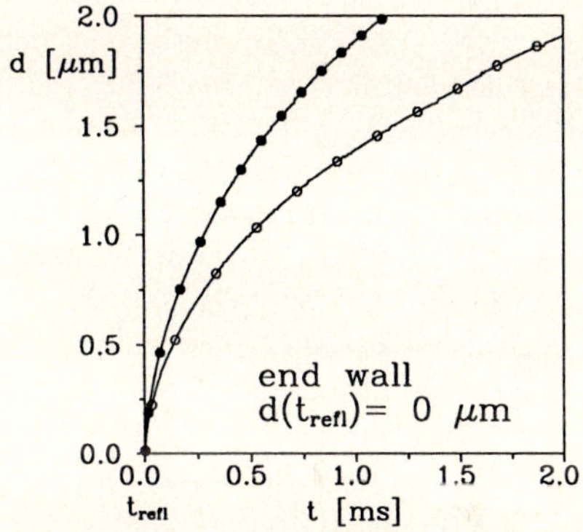

Fig. 4b. Change in thickness at the end-wall

3.2.3. Condensation behind the reflected shock at the side-wall

Measurements have shown significant difficulties arising from the fact that the reflected shock bifurcates. A recent work by Kleine et al. (1991) illustrates the complicated flow pattern when the shock moves into the boundary layer caused by the the incident shock.

Due to its high molecular specific heat, typical experiments with n-heptane exhibit already film condensation behind the incident shock. A comparison between films condensed on the side-walls behind the reflected shock and films condensed on the end-wall is shown in Fig.4a,b. The growth is governed by the amount of heat which is transferred to the wall through the liquid film deposited behind the initial shock. The layer formed behind the incident shock is a heat resistor whose influence decreases as the total film thickness increases. Thus their total thicknesses reach similar values for long observation times.

The set of parameters needed to predict the growth rate behind the reflected shock at the side-wall is not fixed up to now, it may e.g. depend on the distance to the end-wall and the development of the reflected shock. Comparing experimental results with a 2D or 3D flow model seems to be necessary to fix the parameter set.

4. Conclusion and outlook

Thicknesses of liquid films condensing behind shock waves in a shock tube have been measured by means of an interferometric method. Resulting growth coefficients are compared with the solution of the simple Stefan problem adapted to condensation of a gas. The square root growth fails to describe the very beginning of condensation.

The time dependence of films condensing at the side-walls of the tube behind the incident shock has been found to be fitted best by using a power law followed by a linear growth with time.

Quantitative measurements for mixtures are subjected to further experiments and descriptions including the main flow as well as the diffusion of species.

References

Carslaw HS, Jaeger JC (1959) Conduction of Heat in Solids, Ch. XI Change of State. 2nd Edition, Clarendon Press, Oxford

Fujikawa S, Okuda M, Akamatsu T, Goto T (1987) Non-equilibrium vapour condensation on a shock-tube end-wall behind a reflected shock wave. J. Fluid Mech. 183:293-324

Herrmann M (1993) Stoßstruktur und Relaxation bei stoßinduzierter Kondensation in Substanzen hoher molarer spezifischer Wärme. Mitteilungen aus dem Max-Planck-Institut für Strömungsfoschung Nr. 106, Göttingen, Germany

Kleine H, Lyakhov VN, Gvozdeva LG., Grönig H (1991) Bifurcation of a reflected shock wave in a shock tube. In: Takayama K (ed) Proc. 18th ISSW, Sendai. Springer Verlag, pp 261-266

Maerefat M (1990) Studies on non-equilibrium vapour condensation by a shock tube. Diss., Dep. of Mech. Eng. , Kyoto Univ., Japan

Maerefat M, Fujikawa S, Goto T, Akamatsu T (1987) Unsteady nonequilibrium condensation of a vapour-gas mixture induced by a shock wave on a shock tube end wall. In: Grönig H (ed) Proc. 17th ISSW, Aachen. VCH Verlagsgesellschaft, Weinheim, pp 327-333

Mizutani T (1990) A study of non-equilibrium vapour condensation in a shock tube. Diss., Dep. of Mech. Eng. , Kyoto Univ., Japan

Outa E, Matsuda E, Tajima K (1988) Shock tube study on non-equilibrium wall condensation of a super-heated freon vapour flow. In: Grönig H (ed) Proc. 17th ISSW, Aachen. VCH Verlagsgesellschaft, Weinheim, pp 289-295

Pong L, Moses GA (1986) Vapor condensation in the presence of a noncondensable gas. Physics of Fluids 29:1796-1804

Teske T, Obermeier F (1991) Comparative measurements of film condensation on the side and the end-wall of a shock tube. In: Takayama K (ed) Proc. 18th ISSW, Sendai. Springer Verlag, pp 485-490

Teske T (1992) Filmkondensation im Stoßrohr. Bericht 22/92, MPI für Strömungsforschung, Göttingen, Germany

On the Onset of Condensation in Rarefaction Waves in Shock Tubes

Can F. Delale* and Günter H. Schnerr

Institut für Strömungslehre und Strömungsmaschinen, University of Karlsruhe, Germany

Abstract. The expansion of a mixture of a condensible vapor and a carrier gas in the rarefaction wave generated in a shock tube is considered. Using centered expansion wave theory the condensation zones and the onset of condensation are exhibited along pathlines from the asymptotic solution of the condensation rate equation. Consequently the condensation wave front is located accurately in the rarefaction wave of a given shock tube flow with homogeneous condensation. Results obtained for the expansion of moist air and of water vapor in nitrogen are compared with measurements and possible sources for the differences are discussed.

Key words: Condensation onset, Asymptotic solution, Shock tube

1. Introduction

The shock tube offers an alternative technique with desirable features in addition to the well-known cloud chamber and nozzle experiments in understanding condensation dynamics. The experimental conditions in this case can be better controlled. The mixture of a carrier gas and a condensible vapor, initially at rest in the driver section, is expanded after the sudden rupture of the diaphragm until droplets formed essentially by homogeneous condensation can be detected by light scattering (onset of condensation). In later times the sudden release of latent heat sets in giving rise to a pressure "bump". In what follows we discuss a method based on the asymptotic solution of the condensation rate equation along pathlines to accurately locate the onset of condensation. Results obtained for the expansion of moist air and of water vapor in nitrogen are also presented and discussed.

2. Basic equations of condensing flows in the rarefaction wave

We consider a mixture of a condensible vapor and a carrier gas which for time $t' < 0$ is at rest in the driver section (region 4 in Fig.1) of a shock tube with initial relative humidity ϕ_4, initial specific humidity ω_4 and initial temperature T'_4. We examine the expansion of the mixture in the expansion fan produced in the driver section after the removal of the diaphragm. We define the normalized pressure p, the normalized density ρ, the normalized temperature T and the normalized velocity u by

$$p \equiv \frac{p'}{p'_4} \quad , \quad \rho \equiv \frac{\rho'}{\rho'_4} \quad , \quad T \equiv \frac{T'}{T'_4} \quad , \quad u \equiv \frac{u'}{a'_4}$$

where we have distinguished the actual variables by a prime and where subscript 4 is reserved for the initial state of the mixture in the driver section with a'_4 denoting the initial speed of sound in the driver section. Similarly the normalized time and space coordinates t and x are defined by

$$t \equiv \frac{t'}{\Theta'} \quad , \quad x \equiv \frac{x'}{\Theta' \, a'_4}$$

* Alexander von Humboldt Fellow. On leave from the Department of Mathematics, Bilkent University, Ankara, Turkey.

Shock Waves @ Marseille III
Editors: R. Brun, L. Z. Dumitrescu

where Θ' is some characteristic flow time. The inviscid flow equations of the mixture, assumed to consist of perfect gases, can then be cast into the characteristic form (Sislian and Glass 1976, Delale et al. 1993) as

$$\left(1-\frac{\mu_m}{\mu_v}g\right)\,dLn\,p \pm \frac{\gamma\, a_f}{T}\,du + \frac{[(c_{pm}-L_1\,g\,\mu_m/\mu_v)-(1-g\,\mu_m/\mu_v)\;L/T]}{[c_{pm}+(1-L_1)\;g\,\mu_m/\mu_v-1]}\frac{\mu_m}{\mu_v}\chi(t)dt=0 \tag{1}$$

on the characteristics $dx/dt = u \pm a_f$ and

$$dg = \chi(t)\,dt \tag{2}$$

$$\left(1-\frac{\mu_m}{\mu_v}g\right)\,dLn\,p - \left(c_{pm}-\frac{\mu_m}{\mu_v}L_1\,g\right)\,dLn\,T + \frac{\mu_m}{\mu_v}\frac{L}{T}\chi(t)\,dt = 0 \tag{3}$$

on the pathlines $dx/dt = u$. In Eqs.1-3, μ_m, μ_v are respectively the mixture molecular weight and the vapor molecular weight, $L \equiv L'\,\mu_v/(\Re\,T_4')$ with L' denoting the latent heat of condensation and $\Re$ denoting the universal gas constant, $L_1 \equiv dL/dT$, g is the condensate mass fraction, $c_{pm} \equiv c'_{pm}\,\mu_m/\Re$ with c'_{pm} denoting the mixture specific heat, γ is the adiabatic exponent defined by $\gamma \equiv c_{pm}/(c_{pm}-1)$ and $a_f \equiv a'_f/a'_4$ with a'_f denoting the actual frozen speed of sound is given by

$$a_f^2 = \frac{[c_{pm}-L_1\,g\,\mu_m/\mu_v]\;[1-g\,\mu_m/\mu_v]}{\gamma\;[c_{pm}+(1-L_1)\;g\,\mu_m/\mu_v-1]}\,T$$

Eqs.1-3 together with the thermal equation of state of the mixture, namely

$$p = \rho T\left(1-\frac{\mu_m}{\mu_v}g\right) \tag{4}$$

form the basic equations for the complete solution of the nonstationary flow field provided that the function $\chi(t)$ of Eq.2 is known. The identification of the function $\chi(t)$ follows from the integral condensation rate equation constructed from a conveniently normalized nucleation rate equation (Delale et al. 1993)

$$J \equiv \frac{J'}{\zeta'} = \Sigma\,(p,T,g)\;exp\left[-K^{-1}\,B(p,T,g)\right] \tag{5}$$

and a normalized droplet growth law (Delale et al. 1993), assumed to be independent of droplet radius

$$\frac{dr}{dt} = \lambda\,\Omega(p,T,g) \tag{6}$$

where J' is the actual nucleation rate, ζ' is the normalization constant for nucleation, B is the normalized activation function for nucleation, Σ is the pre-exponential function of nucleation theories, Ω is the normalized growth function of the droplets and the parameters K ($\ll 1$) and λ ($\gg 1$) are respectively the nucleation parameter signifying a relatively slow process and the droplet growth parameter signifying rapid droplet growth. The integral condensation rate equation on pathlines then assumes the form (Delale et al. 1993):

$$g(t) = \int_{t_s}^{t}\left[r^*(\tau)+\lambda\int_{\tau}^{t}\Omega(\eta)\,d\eta\right]^3\frac{\Sigma(\tau)\exp\left[-K^{-1}\,B(\tau)\right]}{\rho(\tau)}\,d\tau \tag{7}$$

where t_s is the normalized time when the state of the condensible vapor crosses the saturation line and where r^* is the normalized critical radius beyond which condensation nuclei grow into droplets. Now $\chi(t)$ of Eq.2 can be obtained from Eq.7 by differentiation under the integral sign.

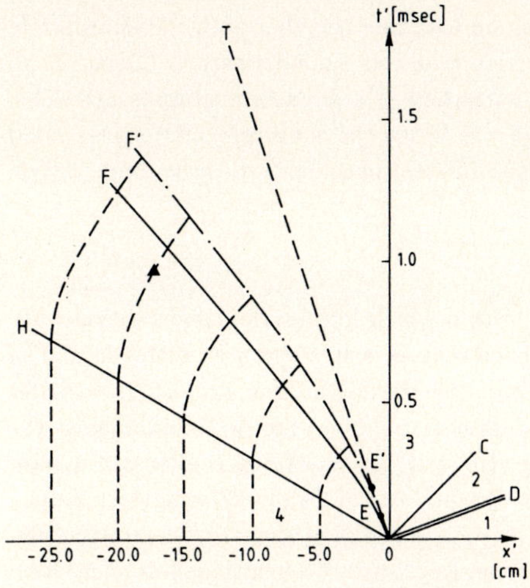

Fig. 1. Condensation wave front for the expansion of water vapor in nitrogen in the rarefaction wave of a shock tube with initial driver section mixture pressure $p_4' = 680$ Torr, mixture temperature $T_4' = 295.3$ K, water vapor pressure $(p_v')_4 = 18.6$ Torr, and driven section air pressure $p_1' = 100$ Torr and air temperature $T_1' = 295.3$ K (OH is the rarefaction wave head, OT is the virtual isentropic rarefaction wave tail, OC is the contact surface, OD is the shock wave, E′F′ is the condensation wave front computed by Sislian and Glass (1976) by the method of characteristics, EF is the condensation wave front computed by the present asymptotic method and the symbol ▲ shows the experimental point of Kalra (1975))

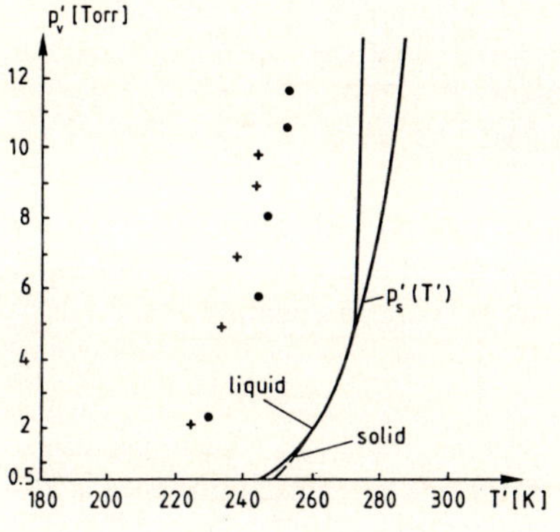

Fig. 2. The onset conditions for the expansion of moist air in a shock tube, where $p_s'(T')$ denotes the saturation pressure.

- • the experimental data of Barschdorff (1975);
- + the present asymptotic predictions

3. The condensation wave front

The gasdynamic analysis of condensing flows in the rarefaction wave described by the equations of the preceding section is substantially simplified when the flow is taken to be isentropic on each pathline until the actual onset of condensation. This assumption has already been confirmed in many experiments and was used by different investigators (Barschdorff 1975, Wu 1977). Thus, in locating the condensation wave front, Eqs.1 and 3 can as a first approximation be described by the frozen isentropic solution obtained by the Riemann invariants on characteristics. The equations of pathlines and the classical solution of isentropic flow variables then follow immediately. The rate equation (Eq.7) on pathlines is now decoupled from the rest of the equations and its solution plays the dominant role in locating the condensation wave front. The rate equation (Eq.7) on pathlines is solved by Laplace's method in the double limit as $K \to 0$ and $\lambda \to \infty$ similar to the procedure given in Clarke and Delale (1986). The condensation zones, namely initial growth zone (IGZ), further growth zone (FGZ), rapid growth zone (RGZ) and onset zone (OZ) of Clarke and Delale (1986) can then be exhibited along pathlines together with the appropriate asymptotic expressions for the condensate mass fraction g and its time derivatives (Delale et al. 1993). The onset of condensation on each pathline is reached at a point of maximum accessible supersaturation, called the onset point, where $dB/dt = 0$. When the above procedure is repeated along different pathlines, the locus of the onset points of different pathlines yields the condensation wave front.

4. Results and discussion

Following the method outlined in Section 3, an algorithm which locates the condensation wave front in the rarefaction zone is developed for the expansion of water vapor in nitrogen and of moist air utilizing the classical nucleation theory and the Hertz-Knudsen droplet growth law (details are given in Delale et al. 1993, Schnerr and Dohrmann 1989) and is tested against the experimental data of Barschdorff (1975) and Kalra (1975). Fig.1 shows the condensation wave front (EF) of the present method for a particular experimental run of Kalra (1975) against the one (E′F′) obtained by Sislian and Glass (1976) using the same nucleation and droplet growth laws, but different values for some thermodynamic functions (e.g. surface tension and condensation coefficient) entering the rate laws. The experimental point at a fixed observation station by Kalra is seen to lie closer to the present condensation wave front in spite of the fact that there are still differences between the experimental onset point and its asymptotic prediction. The onset conditions obtained by the present method for the experiments of Barschdorff (1975) with moist air are plotted in Fig.2. As can clearly be seen there are differences between measurements and predictions. Possible sources for these differences may be listed as

(a) the neglect of the heterogeneous nucleation mechanism,
(b) the use of the classical nucleation theory and the Hertz-Knudsen droplet growth law,
(c) the choice of surface tension and accommodation coefficients that enter the nucleation and droplet growth theories,
(d) the use of the planar centered-expansion wave theory, and
(e) the neglect of viscous effects.

Differences between theory and experiments are expected to diminish when improvements with respect to the above list are made in the theoretical model and also when the experimental techniques in locating the onset of condensation in shock tubes are further refined.

Acknowledgement

One of us (C.F.D.) acknowledges the support by the Alexander von Humboldt Foundation during the performance of this work in the Institut für Strömungslehre und Strömungsmaschinen at the University of Karlsruhe.

References

Barschdorff D (1975) Phys. Fluids 18,5:529-535

Clarke JH, Delale CF (1986) Phys. Fluids 29, 5:1398-1413

Delale CF, Schnerr GH, Zierep J (1993) Asymptotic solution of shock tube flows with homogeneous condensation. Submitted to J. Fluid Mechs.

Kalra SP (1975) Institute of Aerospace Studies, University of Toronto, Rep. No. 195

Schnerr GH, Dohrmann U (1989) In: Zierep J, Oertel H (eds) Proc. IUTAM Symp. Transsonicum III. Springer, Berlin Heidelberg, p 131

Sislian JP, Glass II (1976) AIAA J. 14,12:1731-1737

Wu BJC (1977) In: Pouring AA (ed) Condensation in high speed flows. Proc. Symp. ASME-UEC, Yale University p 73

Influence of Shock Waves on the Condensation Process in a Nozzle

A.L. Itkin*, A.V. Rodionov* and E.G. Kolesnichenko†
*International Institute for Problems of the Asteroid Hazard, 191187, St.-Petersburg, Nab. Kutuzova, 10, Russia
†Moscow State University, Institute of Mechanics, Moscow, Michurinsky pr. 1,Russia

Abstract. Results of numerical computations of a condensable gas flow in a nozzle with a break of the contour are presented. The unimolecular condensation theory of Itkin and Kolesnichenko is used to describe the condensation kinetics. The main advantage of the method is that it allows to solve the hydrodynamic equations in conjunction with a single eq. for the supersaturation S. Other cluster concentrations of the energetic levels are analytically expressed via S and T. It turns out that behind a shock wave in each cross-section x =const. the concentrations of a j-cluster are uniformly distributed in y for each x. This effect may be of great importance, e.g. for applying a cluster coating by precipitation.

Key words: Condensation, Nozzle, Analytical solutions, Numerical simulation

1. Introduction

For modern technologies it is very important to produce clusters of various substances and flows with the cluster size distribution function (CSDF) determined. At the present time there are a few manners of influencing the CSDF: ultraviolet radiation, charged particles, a carrier gas and some others. In the present report we consider a new approach for forming CSDF of a predetermined type by the action of a shock wave on a condensable flow. In order to study this problem the numerical simulation of a two-dimensional flow with condensation in a supersonic nozzle with a wall bend was carried out.

For the calculation of the flow of the gas mixture where one of the gas components is able to condense it is necessary to close the respective gasdynamic equations by the kinetic ones. A short review of the familiar kinetic methods is given by Itkin and Kolesnichenko (1991). The theories described in Itkin and Kolesnichenko (1991) do not take into account nonequilibrium of the cluster internal energy distribution, although this effect influences dramatically the type of kinetic equations for small clusters (Itkin and Kolesnichenko 1990a,b, Itkin et al. 1991a,b,c). Taking this into account in the framework of the unimolecular condensation theory (see above references) allows to interpret a number of experimental data which cannot be explained by available theories, and to reach a good agreement between theoretical and experimental data in a wide range of the initial parameters of the flow for various substances. In addition to that, a convenient and economic algorithm for the calculation of such a one-dimensional flow with condensation was proposed on the basis of the analytical solution of the kinetic equations. The main advantage of this algorithm is that in order to calculate the parameters of the condensable flow one should update the gasdynamic equation system for uncondensable gas only by the kinetic equation for the supersaturation S. The other parameters such as the cluster concentrations or the populations of the cluster energetic levels can be determined from the analytic formulae given in Itkin and Kolesnichenko (1990a,b), Itkin et al. (1991a,b,c). A generalization of this approach for multi-dimensional flows is given in Itkin and Kolesnichenko (1991).

In the present work the above-mentioned approach is used for the numerical simulation of condensable vapor flow in plane and axisymmetric nozzles and special attention is paid to the cluster size distribution function.

Shock Waves @ Marseille III
Editors: R. Brun, L. Z. Dumitrescu

2. Modelling of the nozzle flow

Let us consider a stationary supersonic flow of condensable vapor in a plane or axisymmetric nozzle. We assume as usual that the vapor is a mixture of ideal gases of molecules and clusters and that processes of translational and rotational relaxation of the clusters are quick enough so that it is possible to introduce the rotational - translational temperature for the whole system T. The basic hydrodynamic equations for such a mixture can be written in conservative form

$$\frac{\partial \rho u}{\partial x} + \frac{\partial \rho v}{\partial y} + \nu \frac{\rho v}{y} = 0 \tag{1}$$

$$\frac{\partial (\rho u^2 + P)}{\partial x} + \frac{\partial \rho u v}{\partial y} + \nu \frac{\rho u v}{y} = 0 \tag{2}$$

$$\frac{\partial \rho u v}{\partial x} + \frac{\partial (\rho v^2 + P)}{\partial y} + \nu \frac{\rho v^2}{y} = 0 \tag{3}$$

$$\frac{\partial \rho u h_0}{\partial x} + \frac{\partial \rho v h_0}{\partial y} + \nu \frac{\rho v h_0}{y} = 0 \tag{4}$$

$$\frac{\partial \rho u \alpha_j(k)}{\partial x} + \frac{\partial \rho v \alpha_j(k)}{\partial y} + \nu \frac{\rho v \alpha_j(k)}{y} = \rho \Phi_j(k) \qquad j=1 \ldots N, \quad k=0 \ldots M_j \tag{5}$$

Here x and y are the Cartesian coordinates, u and v are the velocities of the mixture in the direction x and y, respectively, ρ, T and P are the density, temperature and pressure of the mixture, $h_0 = (u^2 + v^2)/2 + h$ is the specific stagnation enthalpy, $\alpha_j(k)$ are the mass fractions of the clusters consisting of j molecules with the total internal energy $E_j(k)$, $\Phi_j(k)$ are the terms describing the nonequilibrium condensation, $\nu = 0$ or 1 for plane or axisymmetric configuration, respectively.

To complete Eqs.1-5 it is necessary to introduce the thermodynamic and caloric equations of state of the mixture and to determine explicitly the kinetic terms $\Phi_j(k)$ describing the evolution of $\alpha_j(k)$ along a streamline due to condensation and relaxation of the vibrational energy of the clusters. In accordance with Itkin and Kolesnichenko (1990b) one should get

$$P = \rho R T \sum_{j=1}^{N} \frac{\alpha_j}{\mu_j}, \qquad h = \sum_{j=1}^{N} \sum_{k=0}^{M} \alpha_j(k) h_j(k) \tag{6}$$

$$\alpha_j(k) = \frac{j m x_j(k)}{\rho}, \qquad \alpha_j = \sum_{k=0}^{M} \alpha_j(k)$$

$$\Phi_j(k) = \sum_{i=0}^{M_1} \left[j \sum_{l=0}^{M_{j-1}} I_{j-1}(l, i \mid k) - j \sum_{l=0}^{M_{j+1}} I_j(k, i \mid l)(1 - \delta_{jN}) + \sum_{l=0}^{M_j} \sum_{\nu=0}^{M_1} J_j(l, i \mid k, \nu) \right],$$

$$j = 2 \ldots N, \quad k = 0 \ldots M_j$$

$$\Phi_1(k) = - \sum_{j=1}^{N-1} \sum_{i=0}^{M_{j+1}} \sum_{l=0}^{M_j} I_j(l, k \mid i)(1 + \delta_{j1} \delta_{lk}) - \sum_{j=1}^{N} \sum_{i,l=0}^{M_j} \sum_{\nu=0}^{M_1} J_j(l, k \mid i, \nu)(1 + \delta_{j1} \delta_{lk}),$$

$$k = 0 \ldots M_1$$

$$I_{j-1}(l, i | k) = \frac{C_{j-1}(l, i | k) \alpha^*_{j-1}(l) \alpha^*_1(i) \rho}{m(j-1)} - \frac{E_j(k | i, l) \alpha^*_j(k)}{j}$$

$$J_j(l, i | k, \nu) = \left[R_j(l, i | k, \nu) \alpha^*_j(l) \alpha^*_1(i) - R_j(k, \nu | l, i) \alpha^*_j(k) \alpha^*_1(\nu) \right] \frac{m \rho}{j}$$

$$\alpha^*_j(k) = \alpha_j(k) / s_j(k)$$

where $\mu_j = j\mu_1$ is the molecular mass of the cluster consisting of j molecules, m is the monomer mass, $h_j(k)$ are the specific enthalpies of the corresponding components of the mixture, α_j are the mass fractions of clusters of size j, $x_j(k)$ are the populations of the energetic state k in the cluster of size j, R is the gas constant, $C_{j-1}(l,i \mid k)$ is the rate constant of the j-cluster formation with the energy $E_j(k)$ from the $(j-1)$-cluster with the energy $E_{j-1}(l)$ by absorbing a monomer with the energy $E_1(i)$; $E_j(k \mid l,i)$ is the rate constant of dissociation of the j-cluster with the energy $E_j(k)$ to the $(j-1)$ cluster with the energy $E_{j-1}(l)$ and a monomer with the energy $E_1(i)$; $R_j(l,i \mid k,\nu)$ is the relaxation rate constant for the j-cluster from the energy state $E_j(l)$ to the state $E_j(k)$ simultaneously with a monomer transition from the energy state $E_1(i)$ to the state $E_1(\nu)$; M_j is the number of the top energy state of the cluster of size j, $s_j(k)$ is the statistic weight of the k-th state of the j-cluster.

In deriving Eq.6 the processes of formation (dissociation) of a cluster by absorbing (emitting) a dimer, trimer etc. as well as three-particle collisions were neglected. But the law of conservation of the number of molecules in the system has been already taken into account, because the following relationship follows from Eq.6:

$$\sum_{j=1}^{N}\sum_{k=0}^{M} \alpha_j(k) = 1 \tag{7}$$

The system (Eqs.1-6) is extremely difficult to solve because of the high dimension and the lack of reliable data for the rate constants. It can be simplified on the basis of the idea of "contracted description" (Itkin and Kolesnichenko 1990ab, Itkin et al. 1991c). This idea consists in the separation, in Eqs.1-6, of the slow and fast variables and searching for the steady state solutions in which the fast variables depend upon the time only indirectly through the slow variables. The result of applying this method is that instead of Eqs.5 and 6 the system (Eqs.1-4) is coupled with the single equation

$$\frac{\partial \rho u \alpha_1}{\partial x} + \frac{\partial \rho v \alpha_1}{\partial y} + \nu\frac{\rho v \alpha_1}{y} = -\frac{S^{m+1} n_{m+1,e} K_{m+1} b}{9m^{4/3}}\alpha_{1e}, \qquad m = \max(r, j_{*,cl}) \tag{8}$$

$$h_0 \approx \frac{u^2+v^2}{2} + h_1 - L(1-\alpha_1), \quad P \approx \rho R T\frac{\alpha_1}{\mu_1}$$

Here r is the new physical parameter of the unimolecular theory (the particular expressions of r are given in Itkin and Kolesnichenko (1991)), K_j is the equilibrium constant of the j-cluster formation, n_{je} is the equilibrium cluster distribution function over the size (Bogdanov et al. 1992), $bj^{2/3}$ is the nondimensional surface energy of the j-cluster, $S = \alpha_1/\alpha_{1e}$ is the supersaturation, $h_1 \approx C_p T$ is the vapor enthalpy.

The other variables like

$$\alpha_j(k), \quad (j = 2\ldots N, \quad k = 0\ldots M_j)$$

and

$$\alpha_j = \sum_k \alpha_j(k), \quad (j = 2\ldots N)$$

have to be determined by the expressions given in Itkin et al. (1991a,b) using values of T and α_1 obtained from Eqs.1-4. Thus the algorithm proposed in Itkin et al. (1991a,b) allows to manage a radical simplification of the initial system and therefore gives the possibility to simulate multi-dimensional flows with condensation using low-powered computers, even PCs.

3. Numerical method

A finite-difference method proposed in Rodionov (1987a,b) is used for the numerical prediction of steady supersonic flows of the nonequilibrium gas. This method is a modification of the well-known Godunov's one. The method is based on a piece-wise linear distribution of parameters in each section of the nozzle x = const. (x is the marching direction of the calculation) and has second order of accuracy on the retention of monotonicity. One of the main features of the procedure used is the application of a predictor-corrector recomputation. This fact makes the algorithm comparatively simple and universal.

The main features of the method's implementation for the calculation of the nonequilibrium flow were described in Rodionov (1987a). These include: 1) the freezing of the nonequilibrium processes during the Riemann problem calculation; 2) the semi-implicit approximation for the source terms of the kinetic equations. Such an approximation of the source terms does not change the structure of the finite-difference scheme and provides a reliable calculation of the flow parameters using the unified algorithm even for the substantially different relaxation times of the nonequilibrium processes. In such a case it is possible to perform the calculation with the step defined by the condition of stability of the gasdynamic equations. The implicit parts of these equations for each elementary finite-difference cell are resolved iteratively by a Newton procedure.

4. Computed results

Computed results of H_2O vapor flow in the supersonic part of a plane nozzle under condensation are presented. The nozzle shape is specified with a break of the contour that leads to the generation of a shock wave coming from the break point. The initial data of the problem are the following: critical section radius $r_k = 0,005$ m; pressure and temperature at the critical section $P_k = 10^5$ Pa, $T_k = 380$ K; Mach number at the critical section $M_k = 1.05$.

The isolines of the pressure [Pa] (Fig.1), temperature [K] (Fig.2) and the dimer concentration (Fig.3) are given in the figures. It is seen that condensation begins in the central part of the flow at $x \approx 3r_k$. In spite of the further stream expansion the temperature in the condensation region remains almost constant because of the latent heat of condensation.

As the result of the shock wave passage the gas temperature increases and the condensation processes slows down sharply. A new flow expansion behind the shock wave induces resumption of vapor condensation. It turns out that behind the shock wave, in each section x =const. concentrations of a j-cluster are uniformly distributed in y. This effect may be of great importance, e.g. for applying a cluster coating by precipitation.

5. Conclusions

A numerical method for the coupled solution of gasdynamic and kinetic equations is developed. It consists in a modification of Godunov's method of second order accuracy. To simulate condensation processes the unimolecular condensation theory based upon the analytical solutions of the condensation kinetic equations is used. The main advantage of the method is that it allows to solve the hydrodynamic equations in conjunction with a single kinetic equation for S. The other cluster concentrations are analytically expressed via S and T.

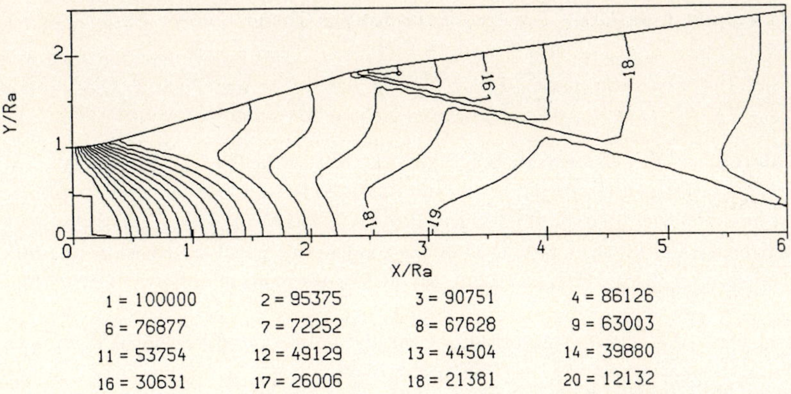

Fig. 1.

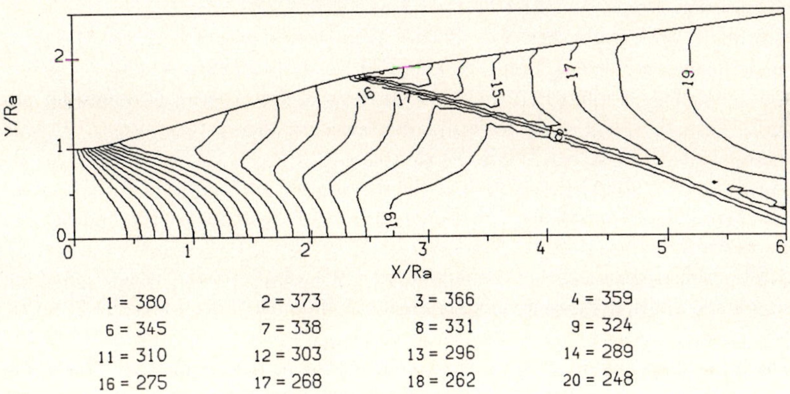

Fig. 2.

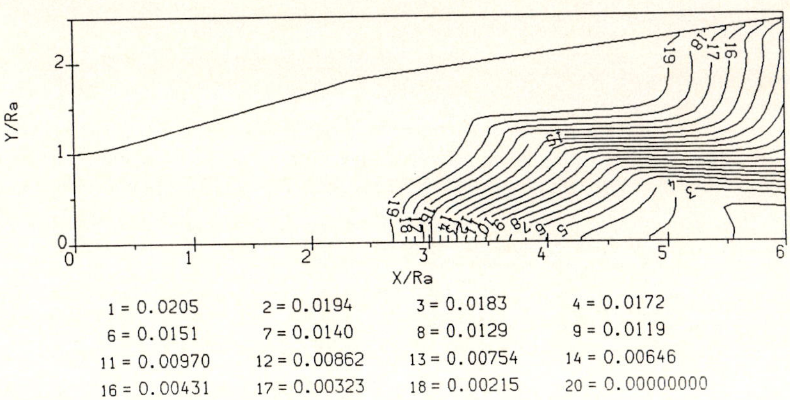

Fig. 3.

Acknowledgement

We acknowledge support from Grant No. 286 of the Russian Foundation for Basic Research program.

References

Bogdanov AV, Gorbachev YE, Dubrovskii GV, Itkin AL, Kolesnichenko EG, Strelchenya VM (1992) Analytical investigation of equilibrium and quasisteady solutions for a quasichemical condensation model. Soviet Journal of Chemical Physics 9, 5:1113-1129

Itkin AL, Kolesnichenko EG (1990) Calculation of the condensable gas flows based on the unimolecular condensation theory. Izvestiya Acad. Sci. USSR. Mechanika Zhidkosti i Gaza 5:138-148

Itkin AL, Kolesnichenko EG (1991a) Algorithm of calculation for two-dimensional flows with condensation. Mathematical simulation (USSR) 3, 2:76-82

Itkin AL, Kolesnichenko EG (1991b) Role of nonequilibrium in condensation kinetics. Doklady Acad. Sci. USSR 311, 3:569-571

Itkin AL, Kolesnichenko EG, Gorbachev YE (1991a) Quasy - stationary state method in the theory of condensation. Prepr. A. Ioffe FTI Acad. Sci. USSR N1488

Itkin AL, Kolesnichenko EG, Gorbachev YE (1991b) Asymptotical methods in the theory of condensation (unimolecular model). Prepr. A. Ioffe FTI Acad. Sci. USSR. N1490

Itkin AL, Kolesnichenko EG, Gorbachev YE (1991c) Analysis of the experimental data for nucleation kinetics from the point of unimolecular condensation theory view. Prepr. A. Ioffe Physics-Technical Inst. of Acad. Sci. USSR. N1523

Itkin AL, Kolesnichenko EG (1991) Calculation of the dissociation rate constant for vibration-excitational clusters and an estimation of the main parameter of the unimolecular condensation theory. Chimicheskaya Fyzika. 1991 11:1545-1553

Rodionov AV (1987a) Monotonic scheme of the second order of accuracy for the calculation of the nonequilibrium flows. Zhurnal Vychislitelnoi Matematiki i Matematicheskoi Fiziki (USSR) 27, 2:175-180

Rodionov AV (1987b) Increase of the order of the Godunov scheme approximation. Zurnal Vychislitelnoi Matematiki i Matematicheskoi Fiziki (USSR) 27, 6:164-169

Propagation of Shock Waves in Dilute Bubbly Liquids

M. Kameda, Y. Matsumoto and Y. Haneda
Department of Mechanical Engineering, The University of Tokyo 7–3–1, Hongo, Bunkyo-ku, Tokyo 113, Japan

Abstract. Transient shock wave phenomena in a liquid containing noncondensable gas bubbles are investigated experimentally and numerically. In the experiment, using a vertical shock tube, the time evolution of the propagation velocity and the waveform are successfully measured. In the numerical analysis, the thermal condition of the bubble interior is directly simulated by using full equations for mass, momentum and energy conservations, and the results are combined to simulate shock waves in bubbly liquids. Relative translational motion between the bubbles and the liquid is also taken into account. Comparison between the numerical results and the experimental ones shows that there is a discrepancy of the period of pressure oscillation behind the shock front, even though the rough feature of both pressure profiles agrees well.

Key words: Multiphase flow, Shock wave, Bubble, Experiment, Numerical analysis

1. Introduction

A bubbly mixture has some unique features with wave phenomena. For example, even though the concentration of bubbles is quite small, the compressibility of the mixture increases so greatly that the speed of sound in the mixture becomes much lower than the speed of sound in either the gas or the liquid separately. For pressure waves of finite amplitude, the effects of bubble oscillations tend to make the waves dispersive and dissipative (van Wijngaarden 1972). On the other hand, the relaxation effects caused by various kinds of phenomena, such as the heat transfer through the bubble wall and the translational relative motion between the bubbles and the liquid, induce a change of the waveform as the wave propagates. Noordzij and van Wijngaarden (1974) state that the relative motion plays a major role in the relaxation process. However, as pointed out already by Nigmatulin (1982), Beylich and Gülhan (1990), Prosperetti (1991) and Matsumoto and Kameda (1993a), it is the heat transfer, rather than the relative motion, that mainly affects the relaxation process. Experimentally, the propagation velocity of shock waves in bubbly mixtures was first measured by Campbell and Pitcher (1958), and the time evolution of the waveform was first observed by Noordzij and van Wijngaarden (1974). Owing to their effort, the basic features of shock waves in bubbly mixture have been shown qualitatively. However, there is no experimental result good enough to compare quantitatively with the numerical result, except for the results by Beylich and Gülhan (1990). They additionally pointed out an interesting fact, namely that the effect of bubble/bubble interaction is not negligible even though the void fraction is very small.

In this paper, the propagation of shock waves in liquids containing noncondensable gas bubbles is investigated experimentally and numerically. In the experiment, the time evolutions of the propagation velocity and the waveform are successfully measured using a vertical shock tube. In the numerical analysis, the conservation equations for mass, momentum and energy of each bubble interiors are solved directly in order to precisely estimate the effect of internal phenomena of the bubbles. And the results are combined to simulate shock waves in bubbly liquids. Size distribution of the bubble radius and translational relative motion between the bubbles and the liquid are also taken into account. Finally, the experimental results are compared with the numerical results.

Shock Waves @ Marseille III
Editors: R. Brun, L. Z. Dumitrescu

2. Numerical calculation

Let us consider one-dimensional flow. The equations for the conservation of mass and momentum of bubbly liquids are written as

$$\frac{\partial}{\partial t}[(1-\alpha)\rho_l] + \frac{\partial}{\partial x}[(1-\alpha)\rho_l U_l] = 0 \tag{1}$$

$$\frac{\partial}{\partial t}[(1-\alpha)\rho_l U_l] + \frac{\partial}{\partial x}[(1-\alpha)\rho_l U_l^2] = -\frac{\partial P}{\partial x} - (1-\alpha)\rho_l g \tag{2}$$

where the liquid density is denoted by ρ_l, the liquid velocity by U_l, the average mixture pressure by P, the gravitational acceleration by g and the void fraction by α. In these equations, the total mixture density and momentum have been simplified by neglecting the gas density ρ_g against that of the liquid ρ_l. Also the mixture viscosity is ignored because it has little influence on wave phenomena (Matsumoto and Kameda 1993b).

In this numerical calculation, bubbles are dealt with individually, and they are assumed to retain their spherical shape. The equation of translational relative motion between each bubble and the liquid is written as

$$\frac{d}{dt}\left[\beta\rho_l\frac{4}{3}\pi R^3 U_g\right] - \frac{D_l}{Dt}\left[\beta\rho_l\frac{4}{3}\pi R^3 U_l\right] + \frac{4}{3}\pi R^3\frac{\partial P}{\partial x} + C_D\pi R^2\rho_l\frac{(U_g-U_l)|U_g-U_l|}{2} = 0 \tag{3}$$

where the bubble radius is denoted by R, the translational velocity of bubble by U_g, and the liquid viscosity by μ_l. The substantial derivative D_l/Dt, the virtual mass coefficient β, and the drag coefficient C_D, are defined as

$$\frac{D_l}{Dt} = \frac{\partial}{\partial t} + U_l\frac{\partial}{\partial x}, \qquad \beta = \frac{1}{2}, \qquad C_D = \frac{24}{Re} + \frac{4}{\sqrt{Re}} + 0.4, \qquad Re = \frac{2\rho_l R|U_g-U_l|}{\mu_l}$$

The equation of the motion of a single bubble in an infinite incompressible liquid, called the Rayleigh-Plesset equation (Plesset and Prosperetti 1977), is used as the equation for the radial motion of a bubble in the bubbly liquid. It is written as

$$R\ddot{R} + \frac{3}{2}\dot{R}^2 = \frac{1}{\rho_l}\left(p_{gw} - \frac{2\sigma}{R} - 4\mu_l\frac{\dot{R}}{R} + \frac{\rho_l(U_g-U_l)^2}{4} - P\right) \tag{4}$$

where the gas pressure at the bubble wall is denoted by p_{gw} and the surface tension by σ. The gas pressure p_g is estimated by solving the conservation equations for mass, momentum and energy of bubble interior. On solving these conservation equations, the following assumptions are employed: (1) The gas inside the bubble, consisting of noncondensable gas, obeys the perfect gas law; (2) The temperature at the bubble wall is equal to the constant liquid temperature; (3) The amount of gas inside the bubble is constant; (4) The viscosity of the gas is ignored.

The wave propagation phenomena in bubbly liquids are calculated by solving the equations mentioned above. The details of the numerical procedure are discussed in a recent paper (Matsumoto and Kameda 1993a).

3. Experimental apparatus

Fig.1 shows the schematic diagram of the experimental apparatus. A vertical glass shock tube of 51.6 mm ID and 5.4 m total length is used for this experiment. The tube is separated into three parts: 1.5 m-long high pressure gas section, 1.0 m-long low (atmospheric) pressure gas section, 2.9 m-long liquid column. The two gas sections are separated by a thin plastic membrane. A shock wave is generated by rupturing the membrane. Six piezoelectric pressure transducers (Kistler 603B) are flush mounted on the wall of the tube to measure the shock waves. The signals from

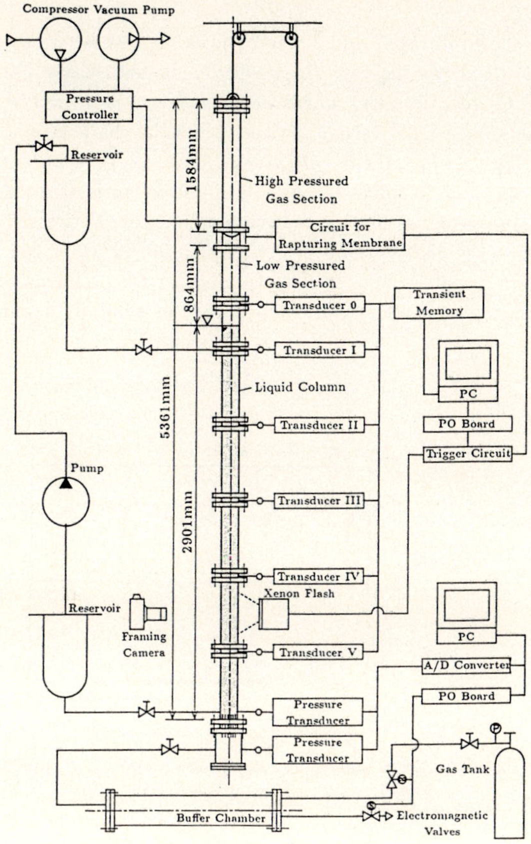

Fig. 1. Experimental apparatus

the transducers are recorded in a digital transient memory through charge amplifiers. A number of glass needles of 0.1 mm ID and 80 mm length are mounted at the bottom of the tube to serve as the bubble generator. The void fraction is almost proportional to the number of glass needles. To make the variance in size distribution of the generated bubbles small, the pressure difference between the liquid at the bottom of the shock tube and the gas chamber is controlled to be constant. The initial bubble radius is measured from a photograph taken by a framing camera just before rupturing the membrane. The mean void fraction before the passage of the shock wave is determined by measuring the difference in the levels of the free surface of the liquid column.

In the present experiments, the initial mean void fractions were in the range from 0.17% to 0.99%. The averaged initial bubble radii were from 0.52 to 0.57 mm at 2.1 m below the free surface, and the standard deviation was 7%. The liquid used is silicone oil (dimethyl siloxane polymer; kinetic viscosity 50×10^{-6} m^2/sec). The gas inside the bubbles is nitrogen.

4. Results and discussions

4.1. Time evolution of shock waves in bubbly liquids: Summary of the numerical calculation

The propagation of shock waves in dilute bubbly liquids is calculated numerically as a kind of piston problem, where the pressure at one end of the tube is raised from the initial pressure P_{L0}=100 kPa to the high pressure P_{H0}=150 kPa at time 0 and is then maintained. Initially, all the bubbles, whose radii are equal to R_{L0}, are assumed to distribute uniformly throughout the calculation region. The liquid is water, and the gas inside the bubbles is nitrogen. The initial temperature is 20°C.

The effects of initial bubble radius are shown in Figs.2 and 3. Fig.2 shows the case where the initial bubble radius R_{L0} is 50 μm and Fig.3 shows the case R_{L0}=100 μm. In both cases, the initial void fraction α_{L0} is 1.0%. In these figures, time and length are normalized as $t^* = t/\tau$, $x^* = x/L$, where the asterisk denotes a nondimensional variable. τ and L are the characteristic time and length, and are defined as $\tau = R_{L0}\sqrt{\rho_l/P_{L0}}$, $C_0 = \sqrt{P_{L0}/(\rho_l\alpha_{L0}(1-\alpha_{L0}))}$, $L = C_0\tau$ $= R_{L0}/\sqrt{\alpha_{L0}(1-\alpha_{L0})}$, where C_0 is the speed of sound in the bubbly liquid in an isothermal equilibrium condition. Fig.2(a) shows the pressure distributions at the nondimensional times t^* of 4, 10, 20, 50, 100, 150, and Fig.3(a) shows the distributions at t^* = 10, 20, 50, 100, 200, 300. Fig.2(b) and 3(b) show the changes of the propagation velocity C as the shock wave propagates. In these figures, the solid line shows the propagation velocity of a shock wave in a bubbly liquid in isothermal equilibrium conditions C_i.

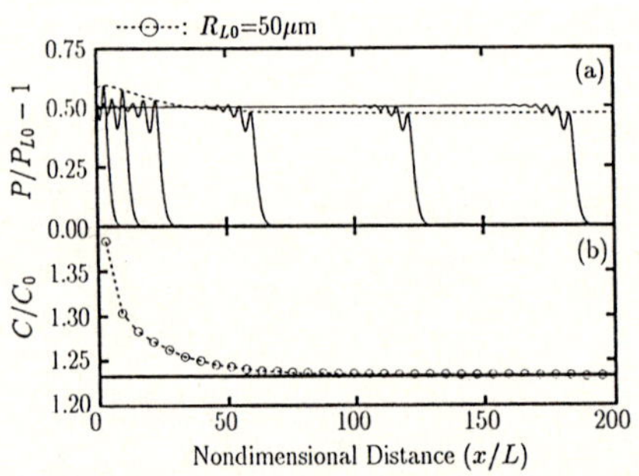

Fig. 2. Time evolution of a shock wave (1) R_{L0}=50 μm

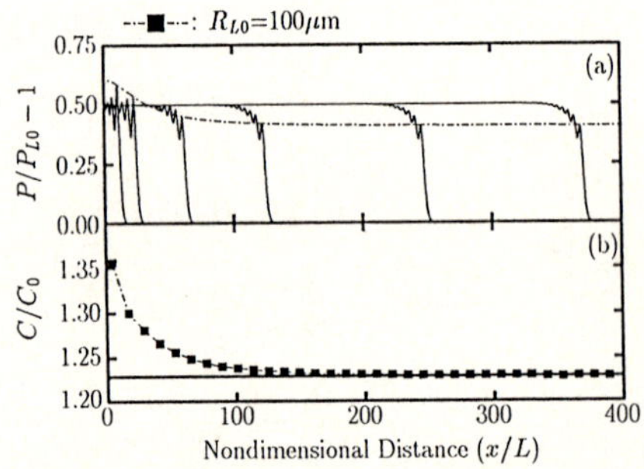

Fig. 3. Time evolution of a shock wave (2) R_{L0}=100 μm

In both cases of the initial bubble radius, the shock wave propagates faster than the reference propagation velocity C_i at the initial stage. Then the shock wave gradually develops to steady conditions, where the propagation velocity is equal to C_i. As the initial bubble radius R_{L0} becomes larger, the nondimensional length for developing a shock wave to steady conditions increases, and the first maximum of the pressure profile in steady conditions becomes lower.

The physical meaning of these phenomena can be explained with the consideration of thermal condition of the bubble interior. As the size of bubble becomes larger, the apparent thermal diffusivity of the gas phase becomes lower. When the thermal diffusivity is low, the relaxation time of the bubble oscillation increases, because the thermal dissipation process is prevented. As the time required for the relaxation process becomes longer, the length required to develop the shock wave to steady conditions becomes longer. On the other hand, the relation between the averaged pressure P and the void fraction α is almost linear in steady conditions. When the thermal diffusivity is low, the internal conditions of the bubble become so adiabatic that the rate of the internal pressure change becomes large. This is the reason why the first maximum of the pressure profiles becomes lower as the initial bubble radius becomes larger.

4.2. Comparison with the experimental results

Figs.4 and 5 show the comparison between the experimental results and the numerical ones. Fig.4 shows the case where the mean void fraction α_{mean} is 0.17% and Fig. 5 shows the case of 0.36%. The initial pressure is atmospheric. The averaged bubble radius at 2.1 m below the free surface is 0.57 mm, the pressure step is 48.1 kPa, and the liquid temperature is 18.2°C. The conditions mentioned above are the same in both figures.

Figs.4(a) and 5(a) show the development of the propagation velocity as the shock wave propagates. The filled circles represent the velocity obtained from the experiment, which is determined from the travelling time of the shock wave between two pressure transducers. The solid curves shows the velocity obtained from the present numerical calculation. The solid line shows the propagation velocity of a shock wave in isothermal equilibrium conditions, C_i. The void fraction at the free surface is larger than that at the bottom of the shock tube due to the difference of the bubble radius caused by the gravitational force. Therefore C_i at the bottom of the tube is higher than C_i at the free surface.

Figs.4(b) and 5(b) show the pressure profile at 0.7 m below the surface and Figs.4(c) and 5(c) show the profile at 2.35 m below the surface. The solid line shows the experimental results, and the thin line shows the numerical results. All profiles of the experimental data were averaged by superimposing 21 shots.

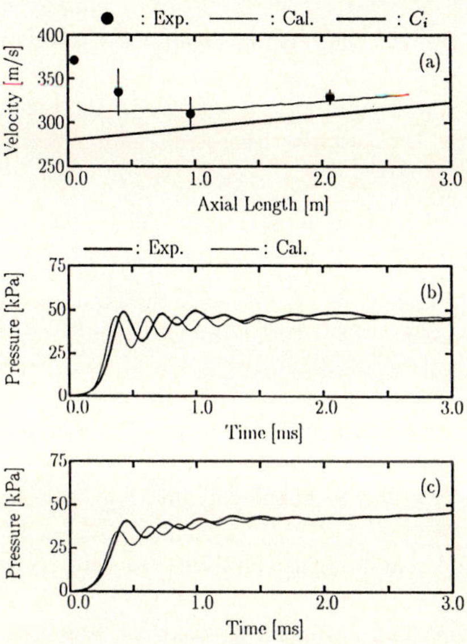

Fig. 4. Comparison between the experimental result and the numerical one (1) α_{mean}=0.17%

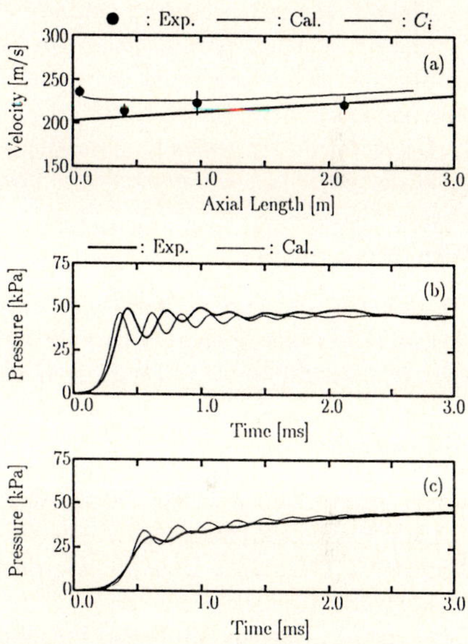

Fig. 5. Comparison between the experimental result and the numerical one (2) α_{mean}=0.36%

All propagation velocities measured from the experimental results are higher than C_i. This means that the shock waves do not develop sufficiently even though the shock wave has reached 2.35 m below the free surface. The experimental work by Beylich and Gülhan (1990), was carried out under the conditions that the mean void fraction is 0.28%, which is the same as our experiment, and the mean initial bubble radius is 1.15 mm, which is twice as large as our experiment. Therefore, the waves observed by Beylich and Gülhan were far from the steady conditions so that the comparison between their experimental results and their theoretical results for the steady conditions may not be appropriate.

The numerical results agree well with the experimental results in terms of the time evolution of propagation velocity. The rough feature of both pressure profiles agrees rather well. However, the periods of oscillation behind the shock front are obviously different from each other. We do not explain clearly the cause of this discrepancy. One of the prospective reason is that the Rayleigh-Plesset equation (Eq.4) is applied to estimate the bubble radius. As already pointed out by Beylich and Gülhan, bubble/bubble interaction forces in the near field may be necessary to be considered in the shock wave propagation process of bubbly liquids. Furthermore, these experimental results cannot be regarded as the result under the conditions of one-dimensional flow. Three-dimensional structure of the bubbly liquids may affect the wave propagation phenomena.

5. Concluding remarks

Propagation of shock waves in bubbly liquids is investigated numerically and experimentally. In the numerical analysis, various kinds of phenomena, such as the heat transfer at the gas-liquid interface, the relative motion between the bubbles and the liquid, are taken into account. A vertical glass shock tube is available for the experiment, and the time evolution of the propagation velocity and the waveform are successfully measured.

Finally, our conclusions are summarized as follows:

(1) The thermal behavior of the bubble interior has a significant influence on the wave propagation. For instance, as the initial bubble radius becomes larger, the length required for developing shock waves to steady conditions increases and the first maximum of the pressure profile in steady conditions becomes lower.

(2) Comparison between the numerical results and the experimental ones shows that the rough features of both pressure profiles agree well. However, it also shows that there is a discrepancy of the period of pressure oscillation behind the shock front.

Acknowledgements

The authors wish to express their appreciation to Y. Kurashiki for his help with the experiment. This study has been supported partially by the Grant-in-Aid for Scientific Research on Priority Areas (No. 04222103) of the Japanese Ministry of Education, Science and Culture.

References

Beylich AE, Gülhan A (1990) On the structure of nonlinear waves in liquid with gas bubbles. Phys. Fluids A 2:1412-1428

Campbell IJ, Pitcher AS (1958) Shock waves in a liquid containing gas bubbles. Proc. Roy. Soc. A243:534-545

Matsumoto Y, Kameda M (1993a) Pressure wave phenomena in bubbly liquid. Nucl. Eng. Des. 141:123-134

Matsumoto Y, Kameda M (1993b) Propagation of shock waves in dilute bubbly liquids. Trans. Jpn. Soc. Mech. Eng. B 59: 2386-2394 (in Japanese)

Nigmatulin RI (1982) Mathematical modeling of bubbly liquid motion and hydrodynamical effects in wave propagation phenomenon. Appl. Sci. Res. 38:267-289

Noordzij L, van Wijngaarden L (1974) Relaxation effects, caused by relative motion, on shock waves in gas-bubble/liquid mixtures. J. Fluid Mech. 66:115-143

Plesset MS, Prosperetti A (1977) Bubble dynamics and cavitation. Ann. Rev. Fluid Mech. 9:145-185

Prosperetti A (1991) The thermal behaviour of oscillating gas bubbles. J. Fluid Mech. 222:587-616

van Wijngaarden L (1972) One-dimensional flow of liquids containing small gas bubbles. Ann. Rev. Fluid Mech. 4:369-396

"Resonance Solitons" in a Bubbly Liquid

N.V. Malykh
Institute of Thermophysics of the Siberian Branch of the Russian Academy of Sciences, Novosibirsk, Russia

Abstract. Steady solitary waves were revealed experimentally in a region of strong nonlinearity and dispersion in a liquid containing gas bubbles. They form at some fixed point of the medium from short initial high-amplitude disturbances. Upon interaction between solitons and the bubble medium, the bubbles change their size by an order of magnitude. Such solitons may be described only within the framework of the system of equations taking into account all the nonlinearities of the gas-liquid medium.

Key words: Bubbly liquids, Solitons, Strong nonlinearities, Dispersion

1. Introduction

A bubbly liquid is a model of a medium with properties of nonlinearity, dispersion and dissipation, the quantity and development of which depends as well on parameters of the medium as on the parameters of the perturbations applied. The structure of a wave in a bubbly liquid is formed as a result of the interaction between disperse- and -dissipative and nonlinear effects. Depending on the ratio of the perturbation duration to the bubble collapse time, these effects manifest themselves to a various extent, which results in various types of structures: wave packets $t \gg \tau$, solitary $t \equiv \tau$ or oscillating shock waves $t \ll \tau$.

Korteveg-de-Vries solitons (KdV) propagating in a liquid containing gas bubbles over the regions being far from the resonance ones, and possessing slight nonlinearity and dispersion effects, have been studied rather well. Bubbles subjected to the action of KdV solitons deviate weakly from initial sizes, and the soliton velocity exceeds insignificantly the equilibrium speed of sound in the medium (Nakoryakov 1983). In the case of the propagation of strong and short pressure pulse perturbations in a bubbly medium the dissipation can be small, and dispersion and non-linearity are very large throughout, but can compensate each other. In this situation the generation of stable solitary waves ("solitons") in the region of strong nonlinearity and dispersion of the "resonance" region are possible. Typically, the duration of the perturbation (soliton) is equal to the time of "collective collapse" of the bubbles (Kedrinsky 1968).

2. Experimental procedure

In a hydroacoustic basin and shock tube, gas bubble layers were formed by electrolysis and by gas injection through micropore inserts which were specially made. The void fraction distribution along the X-axis across the bubble layer $\varphi_0(x)$ and the bubble size $R_0(x)$ were determined with the help of a hot-wire sensor. The dynamics of the gas bubbles was studied using high-speed recording. The incident shock wave was generated by the explosion of an electro-detonator at a certain distance in front of the bubble layers. By changing the distance between the detonator and the bubble region, one could vary the intensity of the incident pulse because its amplitude decreases with the increase of the distance from the explosion location, due to spherical divergence. This distance was varied from 0.5 m to 6.0 m in the experiment. Thereby the amplitude varied from 0.6 MPa to 10 MPa, respectively.

3. Wave models 1 and 2 for the calculation of wave structures

The equations of a homogeneous model in the one-velocity approximation, using a simplified variant of the uniform wave Lighthill equation for a liquid flow containing gas bubbles, and the Rayleigh equation were adopted for the numerical study of structure of the solitons (Ogorodnikov 1979).

Shock Waves @ Marseille III
Editors: R. Brun, L. Z. Dumitrescu

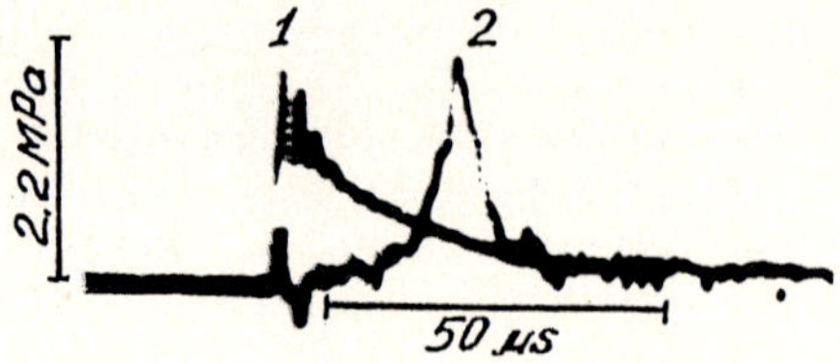

Fig. 1. Soliton formation

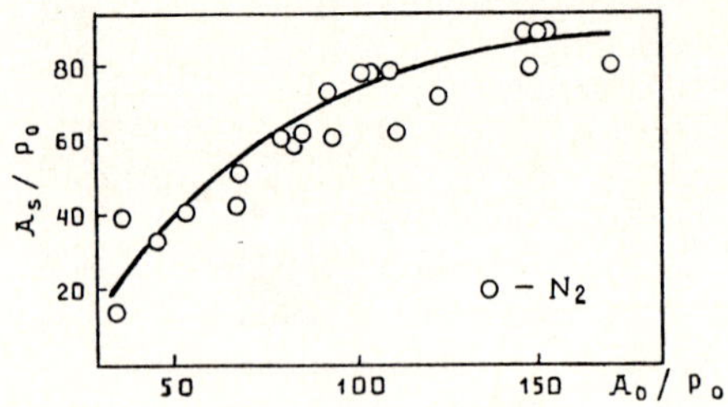

Fig. 2. Amplitude of solitons

$$\frac{1}{C^2}\cdot\frac{\partial^2 P}{\partial t^2}-\frac{\partial^2 P}{\partial x^2}=-\frac{\partial}{\partial t}\left(\rho\cdot\frac{\partial\ln(1-\varphi)}{\partial t}\right);\qquad P=(\rho-\rho_0)\cdot C^2$$
$$R\ddot{R}+\frac{3}{2}\dot{R}^2=\frac{P_0}{P}\cdot\left(\frac{R_0}{R}\right)^{3\gamma}-\frac{P}{\rho_0};\qquad \varphi=\frac{4}{3}\pi R^3 N \tag{1}$$

where P is the pressure in the wave; R the bubble radius; φ the void fraction varying in the wave. This system was obtained within the framework of the following assumptions: a) $R \ll l \ll L$, i.e., the bubble sizes are smaller that the distances between the bubbles l, which, in turn, are smaller than the characteristic scale L of the waves considered (Iordansky 1960). b) In model 1 the motion of the centre of mass of the bubble with respect to the liquid may be neglected. This is why, in the Lighthill equation, dipole and quadrupole terms connected with these motions were omitted (Crighton 1969). c) Nonsphericity of the bubble form in the wave is neglected. Waveforms of the soliton type are described most exactly within the framework of a steady-state solution of the complete nondissipative system of equations, first suggested by Iordansky (1960). The only difference between them is that the equation of state for the liquid is in the form of Tait's nonlinear equation, as in Plaksin (1983):

$$\frac{\partial\rho}{\partial t}+\frac{\partial(\bar\rho U)}{\partial x}=0,\qquad \frac{\partial(\bar\rho U)}{\partial t}+\frac{\partial(P+\bar\rho U^2)}{\partial x}=0$$
$$\frac{\partial N}{\partial t}+\frac{\partial(NU)}{\partial x}=0,\qquad \bar\rho=(1-\varphi)\cdot\rho,\qquad \varphi=(4/3)\pi R^3 N \tag{2}$$
$$R\ddot{R}+(3/2)\dot{R}^2=\left(P_0\cdot(R_0/R)^{3\gamma}-P\right)/\rho_0$$
$$(P-P_0)/\rho_0 C^2=((\rho/\rho_0)^n-1)/n,\qquad n=7.15$$

U, $\bar\rho$, P are the velocity, density and pressure in the medium, ρ the density of water, n, R the number and radius of bubbles, φ the volumetric gas content, C the sound velocity in water, γ the adiabatic exponent. Subscript "0" means the initial, equilibrium state.

4. Experimental data and discussion

A powerful shock wave can be characterized by the maximum manifestation of nonlinear effects, the negligible influence of heat transfer and the small time to produce stationary or quasistationary waveforms. In a medium with resonant $t \equiv \tau$ bubbles solitary waves have been experimentally found out to form at the distance of $x = 0.3$ m, which amounts to approximately five lengths of the initial perturbation λ_0.

In Fig. 1 the oscillograms of signals are given, which coincide in time at the distance of 0.3 m $\equiv 5.0\lambda_0$ from the beginning of the bubbly medium. The initial signal 1 with a steep leading front was recorded in a liquid without bubbles. Signal 2 is a solitary wave which has been formed at the

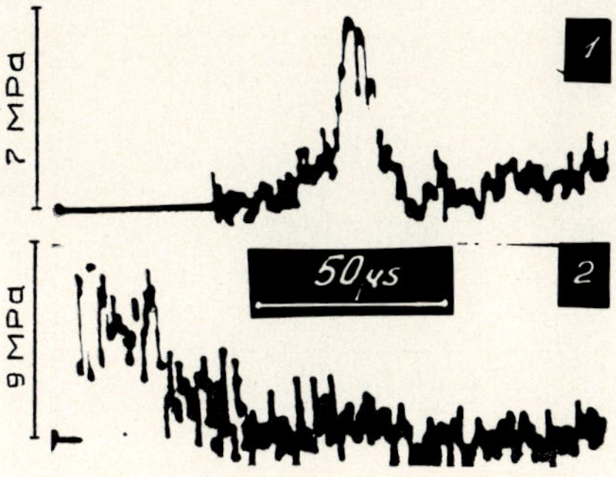

Fig. 3. Transformation of the wave structure $\varphi_0 = 7 \times 10^{-4}$, $R_0 = 4 \times 10^{-4}$ m

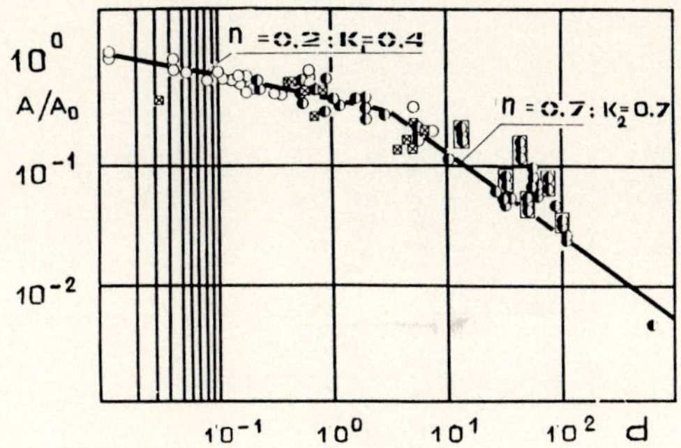

Fig. 4. Decrease in wave amplitude;
◐ - He, ○ - N_2, ⊗ - air
$d = A_0 \cdot \varphi \cdot x / P_0 \lambda_0$, $k_1 = 0.4$, $k_2 = 0.7$

same point of the medium with gas volume concentration $\varphi_0 = 10^{-3}$, with He bubbles and the mean radius $R_0 = 4 \times 10^{-4}$ m. The formed solitary wave begins with the front of a disperse-nature forerunner which has a small amplitude at this distance and arrives at the point of observation at the sound velocity in the liquid without bubbles C. Then the symmetric, acute-angled "main part" of the wave is formed, which is referred to as a soliton. Its amplitude A_s is higher than that of the initial signal A_0, $A_s \equiv 1.2 \cdot A_0$, its duration τ_s is less than the duration of the initial pulse. The energy of the initial pulse remains practically unaffected and during the formation of the solitary wave it dissipates only slightly.

Let the parameters of the medium be fixed (e.g., $\varphi_0 = 7 \times 10^{-4}$, $R_0 = 4 \times 10^{-4}$ m). We will follow the shape and the amplitude of a soliton formed at the point of observation with the increase in the amplitude of the wave falling onto a bubbly medium. The soliton amplitude will grow nonlinearly, up to a certain limit. The dependence of the soliton amplitude on the increase in the initial wave amplitude is shown in Fig.2. For the given parameters the maximum amplitude of the soliton was equal to 7 MPa, as the amplitude of the initial wave increased up to 17 Mpa. The amplitude of the wave in the bubble medium having achieved 9 MPa, nonlinear turning-over occurs and an oscillating shock wave is formed. Model 2 describes the effect of soliton reversal with the generation of a shock wave.

In Fig.3 the oscillograms elucidating the structural transformation that occurs in this case are shown. In this case the wave damping increases markedly. The experimental law for the decrease of the soliton amplitude can be written as $A = k_1 \cdot A_0 \cdot d^{-0.2}$, while the oscillating shock-wave amplitude can be described by $A = k_2 \cdot A_0 \cdot d^{-0.7}$ (see Fig.4). The experimentally observed decrease in the amplitude depending on distance seems to be due to the fact that the medium is not ideal and to the impossibility to maintain the spatial constancy of the parameters, as well as to the existence of a bubble distribution according to size, which cannot be ignored in the experiment. A comparison of soliton forms, obtained experimentally, with calculations by model 1 and 2 is given in Fig.5. The bubbles are distributed in such a way that the "resonant" bubbles constitute the main part of the gas volume. Comparison shows that even for small ($P \leq 2$ MPa) amplitudes of the perturbations, keeping into account both the nonlinearity of the liquid state equation and the hydrodynamic nonlinearity leads to a more exact description of the "resonance" soliton form (model 2). In contrast with model 1, which does not consider either the nonlinearity of the liquid-state equation or the nonlinearity of the hydrodynamic equations, this approach considers all the possible nonlinearities to a maximum extent. As a result of these modification we obtained that the estimate for the maximum possible velocity of solitons and their amplitude changed. As for the linearized models, there the maximum estimate for the velocity of solitons u

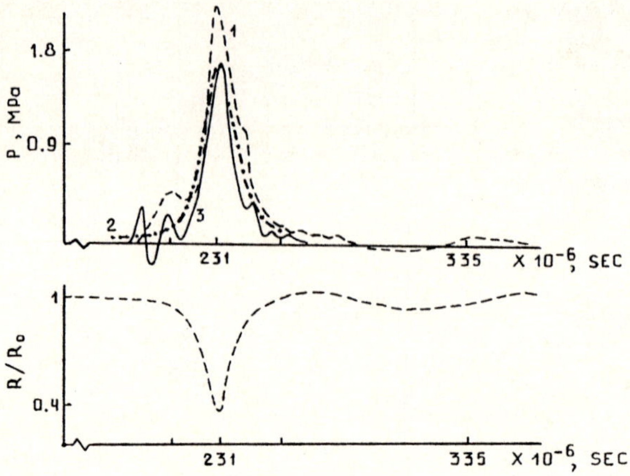

Fig. 5. Comparison of experimental and calculated soliton structures. 1 - by model 1; 2 - by model 2; 3 - experimental structure

has been obtained to be equal to $(0.2 \div 0.3) \cdot C$. According to the results obtained by Plaksin, one may also achieve $u = (0.7 \div 0.9) \cdot C$. The maximum possible amplitude of solitons, considering the nonlinear compressibility of the liquid, has been found to be lower. These results are in qualitative agreement with experiment. The velocity of the observed solitary wave is equal to 1100 m/s, which amounts to $0.7 \cdot C$. As the amplitude of the wave in question grows, the velocity of the wave increases and asymptotically approaches the sound velocity in the bubble-free liquid.

5. Conclusion

Therefore, solitary pressure waves stable to shape variation in the region of strong nonlinearity and dispersion, "resonance" solitons, are experimentally found. Description of the solitons thus found is only possible within the framework of a complete nondissipative system of equations, where hydrodynamic nonlinearity, nonlinearity of the bubble equations, nonlinearity of the liquid and gas component are taken into consideration.

References

Crighton DG, Ffowcs Williams JE (1969) Sound generation by turbulent two-phase flow. J. Fluid Mech. 36:585

Iordansky SV (1960) On equation of motion of a liquid containing gas bubbles. J. Prik. Mekh. i Tekn. Fiz. 3:102

Kedrinskii VK (1968) Disturbance propagation over a liquid with gas bubbles. J. Prik. Mekh. i Tekn. Fiz. 4:29

Malykh NV, Ogorodnikov IA (1979) Structure of pressure pulses in liquid with gas bubbles. J. de Physique 40:C8-300

Nakoryakov VE (1983) Wave propagation over gas- and- vapour liquid media. Inst. of Thermophys., Novosibirsk

Plaksin SJ (1983) On stationary solutions of the equations of motion of a bubbly liquid. J. Prik. Mekh. i Tekn. Fiz. 1:75

Analytical Techniques for the Problem of the Interaction of Nonlinear Sonic Waves with Nonuniform Media

A.A. Lugovtsov
Institute of Thermophysics, Siberian Branch of Russian Academy of Sciences, Lavrentiev Avenue 1, 630090 Novosibirsk, Russia

Abstract. Equations describing the propagation of waves with small but finite amplitude in the presence of bubbles with different radii are deduced, keeping account of the dependence of their distribution on space coordinates. The laws of conservation are shown to be significant in choosing the solutions coincident in basic physical properties with the solutions of the initial system. In some particular cases, the equations thus deduced are such known equation as KDV, Kadomtsev-Petviashvily and Khokhlov-Zabolotskaya.

Key words: Nonlinear waves, Nonuniform media

1. Introduction

Experimental and numerical investigations of wave processes in polyphasic media face difficulties associated with the large number of characteristic parameters. Exact and approximate solutions allow us to obtain the dependence on all parameters in an explicit form. The generalized KDV equations deduced below are of interest, because they represent a considerable progress in the analytical investigation of such type of equations.

The method of obtaining approximate equations proposed below may be used for investigations of wave processes in nonuniform media of diverse natures.

2. Approximate equations

The initial equations of motion for a liquid containing gas bubbles may be written as (Gavrilyuk 1991):

$$\frac{\partial}{\partial t}[(1-\varphi)\rho] + \frac{\partial}{\partial x_k}[(1-\varphi)\rho u_k] = 0 \tag{1}$$

$$\frac{\partial}{\partial t}[(1-\varphi)\rho u_i] + \frac{\partial}{\partial x_k}[p\delta_{ik} + (1-\varphi)\rho u_i u_k] = 0 \tag{2}$$

$$\frac{\partial N}{\partial t} + \frac{\partial}{\partial x_k}(N u_k) = 0 \tag{3}$$

$$R\frac{d^2R}{dt^2} + \frac{3}{2}\left(\frac{dR}{dt}\right)^2 + \frac{4\nu}{R}\frac{dR}{dt} = \frac{\left(p_* + \frac{2\sigma}{R_*}\right)\left(\frac{R_*}{R}\right)^{3\gamma} - \frac{2\sigma}{R} - p}{\rho_*} \tag{4}$$

$$\varphi = \frac{4}{3}\pi \int\limits_0^\infty N\left(R_*, t, \overrightarrow{x}\right) R^3\left(R_*, t, \overrightarrow{x}\right) dR_* \tag{5}$$

$$\rho = \rho_*\left[1 + n\frac{p - p_*}{\rho_* C_{1*}^2}\right]^{1/n} \tag{6}$$

where t is the time, $\boldsymbol{x}$ the space coordinate; $\rho(t,\boldsymbol{x}), \boldsymbol{u}(t,\boldsymbol{x})$ the density and velocity of the liquid, $p(t,\boldsymbol{x})$ the pressure, C_{1*} the velocity of sound in the liquid, R_* the undisturbed radius of the bubbles of the "R_*—fraction" ; $R(R_*,t,\boldsymbol{x})$ the curent radius of a bubble; $N(R_*,t,\boldsymbol{x})$ the density

Shock Waves @ Marseille III
Editors: R. Brun, L. Z. Dumitrescu

of distribution of bubbles on radius, $N(R_*,t,x)dR_*$ the number of bubbles of the "R_*—fraction" per unit of volume of mixture, φ the volume concentration of gas in the mixture.

Let ϵ be a small parameter. The unknown functions may be written as:

$$p = p_*(1+\epsilon p'), \quad u = \epsilon C_* u, \quad N = N_*(R_*)(1+\epsilon N'), \quad R = R_*(1+\epsilon R') \tag{7}$$

and the independent variables as:

$$x = \epsilon^{-1/2} L x', \quad t = \epsilon^{-1/2}(L/C_*)t', \quad \xi = x' - t', \quad \tau = \epsilon x' \tag{8}$$

Such a substitution means the transition into a system of coordinates, wich moves in the positive direction of the x axis with velocity C_*.

The chosen dependence of functions and variables on ϵ is typical for the KDV approximation.

Substituting Eqs.7 and 8 into Eqs.1-6 leads to the appearance of the small parameter ϵ in the initial equations. The presence of such a parameter permits us to search for unknown functions as:

$$F = \sum_{k=1}^{\infty} F_k(\xi,\tau)\epsilon^k, \qquad F = u, p, R, N \tag{9}$$

For the zero approximation it leads to:

$$N_0 = u_0, \quad p_0 = \frac{(1-\varphi_*)\rho_* C_*^2}{p_*} u_0; \quad R_0 = -\frac{(1-\varphi_*)\rho_* C_*^2}{3\gamma p_* + (3\gamma-1)\frac{2\sigma}{R_*}} u_0 \tag{10}$$

$$\left[1 - (1-\varphi_*)\rho_* C_*^2 \frac{4\pi}{3}\int_0^{\infty} \frac{N_*(R_*)R_*^3 dR_*}{\gamma p_* + \frac{3\gamma-1}{3}\frac{2\sigma}{R_*}} - \frac{(1-\varphi_*)^2 C_*^2}{C_{1*}^2}\right] u_0 = 0 \tag{11}$$

From Eq.11 and under the condition of the existence of an untrivial solution for u_0 it follows:

$$\frac{1}{(1-\varphi_*)\rho_* C_*^2} = \frac{1-\varphi_*}{\rho_* C_*^2} + \frac{4\pi}{3}\int_0^{\infty} \frac{N_*(R_*)R_*^3 dR_*}{\gamma p_* + \frac{3\gamma-1}{3}\frac{2\sigma}{R_*}} \tag{12}$$

where $\varphi_* = \frac{4\pi}{3}\int_0^{\infty} N_*(R_*)R_*^3 dR_*$. The formula (12) defines C_*. The function u_0 in the zero approximation is indefinite. The first approximation gives the equation for u_1, taking into account Eq.10:

$$A u_1 + \hat{L}(u_0) = 0$$

where A is the term in square brackets from Eq.11 (i.e. A is equal to zero) and the operator $\hat{L}(U_0)$ is the KDVB equation for u (index is omitted):

$$u_\tau + \alpha u u_\xi + \beta u_{\xi\xi\xi} - \mu u_{\xi\xi} = 0$$

The formulas for α, β, μ are the same, as were obtained in Lugovtsov (1991) if the sums present there, are changed for integrals.

Let N_* be a function of x. In the one-dimensional case the KDV approximation is possible if $N_*(\epsilon^{3/2}x, R_*)$, i.e., when the characteristic length of change of N_* is much larger than the length of the propagating wave.

What needs to be changed in Eq.8 is:

$$\xi = \int_{x_0'}^{x'} \frac{d\tau}{c(\tau)} - t' \tag{13}$$

which means transition to a system of coordinates, which moves along the x axis with variable velocity $C_*c(\tau)$, that is defined by the same formula (12) where one replaces: C_* by $C_*c(\tau)$ and φ_* by $\varphi_* = \int_0^\infty N_*(\tau, R_*)R_*^3 dR_*^3$.

The equation for u is:

$$u_\tau + \alpha(\tau)uu_\xi + \beta(\tau)u_{\xi\xi\xi} - \mu(\tau)u_{\xi\xi} + \left[\frac{k}{2\tau} + \delta(\tau)\right] u = 0 \tag{14}$$

where $\alpha(\tau), \beta(\tau)$ and $\mu(\tau)$ are defined by the same formulas with the replacements mentioned above, while $\delta(\tau)$ is:

$$\delta(\tau) = \frac{1}{2[1-\varphi(\tau)]c(\tau)} \frac{\partial}{\partial\tau} ([1 - \varphi_*(\tau)]c(\tau)) .$$

$k = 0, 1, 2$ for plane cylindrical and spherical cases respectively.

The laws of conservation for the initial system of equations for $\nu = 0$, written in the KDV approximation and valid at $u \to 0$ for $|\xi| \to \infty$, is

$$[1-\varphi_*(\tau)]\, \tau^k c(\tau) \int\limits_{-\infty}^{\infty} u(\tau,\xi)d\xi = C_1 + O(\epsilon) \tag{15}$$

$$[1-\varphi_*(\tau)]\, \tau^k c(\tau) \int\limits_{-\infty}^{\infty} u^2(\tau,\xi)d\xi = C_2 + O(\epsilon) \tag{16}$$

It is easy to show, that for $\nu = 0$ the conservation law of energy (16) is valid for any exact solution of Eq.14. The conservation law of momentum (15) for the case when $\delta(\tau)$ is not equal to zero, is valid only for $C_1 = 0$ even for $k = 0$ (plane case). In this case it is necessary to take into account the next order approximation. Solution of the equation for u_1 can include waves whose length is comparable to the characteristic length of the nonuniformities. Such waves make a contribution to momentum on account of their length, but do not make contribution to energy on account of their small amplitudes.

To investigate Eq.14, it is possible to write u in the form (for $\nu = 0$):

$$u_0 = f(\tau)U(\eta, T(\tau)), \eta = \Phi(\tau)\xi \tag{17}$$

The functions $f(\tau)$, $\Phi(\tau)$, $T(\tau)$ are determined from the conditions of separating out the main part of the equations for U in the form of the KDV classical equation and implementing the conservation laws in the following form:

$$\int_{-\infty}^{\infty} U^2(\eta,T)d\eta = C_1, \quad \int_{-\infty}^{\infty} U(\eta,T)d\eta = C_2$$

It follows from this:

$$\Phi = \frac{\alpha^{2/3}}{(1-\varphi_*)^{1/3}\tau^{k/3}\beta^{2/3}c^{1/3}}; \quad f = \frac{\alpha}{\beta}\Phi^2;$$

$$T = \int_{\tau_0}^{\tau} \frac{\alpha_2 d\tau}{(1-\varphi_*)\tau^k c\beta} \tag{18}$$

The equation for $U(\eta, T)$ is written as:

$$U_T + UU_\eta + U_{\eta\eta\eta} - F(\tau)\left(\frac{k}{\tau} + Q(\tau)\right)(2\eta U_\eta + U) = 0 \tag{19}$$

where $Q(\tau)$ is a known function. If the gas content distribution is such, that $\frac{k}{\tau} + Q(\tau)$ is equal to zero, Eq.18 becames the classical KDV equation, the solutions of which are well-known. The amplitude and width of the signal, and its propogation velocity are defined by the formulas (17). In some cases approximate solutions may be obtained, similarly as it is done in Lugovtsov (1980).

The most general equation within the KDV approximation may be deduced in the case, when $N_* = N_*(R_*, \tau) + \epsilon n_*(R_*, \tau, \epsilon y, \epsilon z)$. For this case the sound velocity is: $C_* c(\tau) + \epsilon c_1(\tau, \epsilon y, \epsilon z)$. Let us add to Eqs.7 and 8 the following: $v = \epsilon^{3/2} C_* c(\tau) v', w = \epsilon^{3/2} C_* c(\tau) w', y = \epsilon^{-1} L y', z = \epsilon^{-1} L z'$. It is possible to repeat the procedure of expansion in ϵ. The result is:

$$u_\tau + \alpha(\tau) u u_\xi + \beta(\tau) u_{\xi\xi\xi} - \mu(\tau) u_{\xi\xi} + \delta(\tau) u - c_1(\tau, y, z) u + \frac{1}{2}(v_y + w_z) = 0$$

$u_y = v_\xi, u_z = w_\xi$ The exclusion of v and w leads to:

$$\frac{\partial}{\partial \xi}(u_\tau + \alpha u u_\xi + \beta u_{\xi\xi\xi} - \mu u_{\xi\xi} + \delta u - c_1 u_\xi) + \frac{1}{2}(u_{yy} + u_{zz}) = 0 \qquad (20)$$

For the first time an analogous equation was obtained for surface waves by Lugovtsov and Lugovtsov (1969). Later on it became known as the Kadomtsev-Petviashvily equation.

If one changes the sign of ξ and poses the problems with boundary conditions, Eq.19 is a generalization of the known Chochlov-Zabolotskaya equation, which describes the propagation of a sonic wave.

It should be noted, that the method proposed above permits us to take into account such processes as the relative motion of bubbles and liquid, mass and heat exchange between phases etc.

Acknowledgements

The author is grateful to the Soros Fund for sponsoring this work.

References

Gavrilyuk SL (1991) Preprint N1, Institute of Hydrodynamics, Siberian Branch of Russian Academy of Sciences, Novosibirsk

Lugovtsov AA (1991) Dinamika Sploshnoy Sredy 100:148 (Institute of Hydrodynamics, Novosibirsk).

Lugovtsov AA, Lugovtsov BA(1969) Dinamika Sploshnoy Sredy, 1: 195 (Institute of Hydrodynamics, Novosibirsk)

Lugovtsov AA (1980) Jurnal Prikladnoy Mechaniki i Technicheskoy Fiziki (PMTF) 6: 24

An Adaptive Gridding Approach to the Computation of Reactive Two-Phase Flows in Two Dimensions

W. Speares and E.F. Toro
Department of Aerospace Science, College of Aeronautics, Cranfield Institute of Technology, Cranfield, Bedfordshire MK43 0AL, England

Abstract. In this paper we outline the application of adaptive gridding techniques to the calculation of reactive two-phase flows in two dimensions, in the context of Interior Ballistics. Computations of these flows on a single mesh, using a higher order shock-capturing Riemann problem based scheme, are presented and compared with experimental data. While encouraging agreement between the two is achieved the need for an adaptive approach to the problem is recognised. It is proposed that a scheme based on Adaptive Mesh Refinement is desirable. The algorithm for the adaptive scheme is described and validation tests are performed for the gas-phase (Euler) equations. The implementation of the adaptive scheme in the case of the full two-phase system is discussed.

Key words: Reactive two-phase, Shock-capturing, Adaptive mesh

1. Introduction

The reactive two-phase flows encountered in the study of Interior Ballistics make a challenging subject for investigation, from the viewpoints of both the modeling problem and the techniques employed for their computation. The modeling of two-phase processes in general is not straightforward and is reviewed in Stewart and Wendroff (1984).

The system of equations we study were derived in Fitt (1988) building on earlier work from Gough (1979). The equations seek to model the two-phase reactive flows which occur within the interior of a gun upon firing, and are of mixed-hyperbolic elliptic type. The solution strategy we adopt here consists of placing the system in a "pseudo-hyperbolic" form (Toro 1989) which leads to a strictly hyperbolic homogeneous problem. The elliptic type terms responsible for the mixed character of the system now appear as source terms in the non-homogeneous problem. The homogeneous problem is solved using the high resolution Riemann Problem (RP) -based Weighted Average Flux method (Toro 1992), with the source terms being integrated using the Euler step method. While results obtained on single mesh calculations in realistic geometries (Speares and Toro 1992) compare well with experiment, it is recognised that the relative coarseness of the meshes employed might compromise reactive processes present in the simulation. In order to perform fine mesh calculations but not incur prohibitive computational penalties, it is desirable to implement an adaptive gridding strategy.

The adaptive scheme we study is based on the so-called Adaptive Mesh Refinement (AMR) approach (Burger and Colella 1989, Quirk 1990). This involves adaption in both space and time based around a hierarchical arrangement of structured meshes, which dynamically track important flow features (for example flow discontinuities) and thus maintain high resolution in those regions. The performance of this type of scheme can be impressive and has not yet been widely exploited, chiefly due to the more involved nature of the coding problem. The use of powerful programming languages such as "C", significantly reduces this difficulty.

In this paper we present comparisons between single mesh two-phase calculations and experimental data, and demonstrate the adaptive algorithm on an idealised gas-phase problem. The full two-phase calculation is discussed. The paper is organised as follows: In Section 2 we describe the system of equations and in Section 3 we give a brief overview of the numerical integration scheme.

In Section 4 we show results of single mesh two-phase calculations compared with experimental data. In Section 5 we describe the adaptive algorithm, and in Section 6 we present a validation test of the algorithm for the Euler (gas-phase) system. Section 7 is the conclusion.

2. The governing equations

The mathematical model proposed by Fitt (1988) seeks to model the flow of a burning solid granular propellant bed interspersed with compressible gaseous products of combustion inside a gun combustion chamber. The propellant bed is initially at rest within the combustion chamber, and is then ignited by the injection of hot gases from a primer. As the propellant burns, it releases gaseous combustion products. The resulting high pressures within the combustion chamber cause the projectile to be accelerated down the gun barrel and eventually be released.

It is assumed throughout that cylindrical symmetry exists. Variables are functions of (z, r, t), r denotes the radial distance from the axis of symmetry and z the axial distance from the breech (back) end of the gun, t is time. The first order system we study is written in the form:

$$\boldsymbol{U}_t + [\boldsymbol{F}(\boldsymbol{U})]_z + [\boldsymbol{G}(\boldsymbol{U})]_r = \boldsymbol{S} \tag{1}$$

where the conserved variables and fluxes are:

$$\boldsymbol{U} = \begin{bmatrix} \alpha_1\rho_1 \\ \alpha_1\rho_1 u_1 \\ \alpha_1\rho_1 v_1 \\ \alpha_1 E_1 \\ \alpha_2\rho_2 \\ \alpha_2\rho_2 u_2 \\ \alpha_2\rho_2 v_2 \\ N \end{bmatrix}, \quad \boldsymbol{F} = \begin{bmatrix} \alpha_1\rho_1 u_1 \\ \alpha_1(\rho_1 u_1^2 + p) \\ \alpha_1\rho_1 u_1 v_1 \\ \alpha_1 u_1(E_1 + p) \\ \alpha_2\rho_2 u_2 \\ \alpha_2(\rho_2 u_2^2 + R + p) \\ \alpha_2\rho_2 u_2 v_2 \\ u_2 N \end{bmatrix}, \quad \boldsymbol{G} = \begin{bmatrix} \alpha_1\rho_1 v_1 \\ \alpha_1\rho_1 u_1 v_1 \\ \alpha_1(\rho_1 v_1^2 + p) \\ \alpha_1 v_1(E_1 + p) \\ \alpha_2\rho_2 v_2 \\ \alpha_2\rho_2 u_2 v_2 \\ \alpha_2(\rho_2 v_2^2 + R + p) \\ v_2 N \end{bmatrix} \tag{2}$$

Here subscripts 1 and 2 refer to the gas and particle phase variables respectively. ρ_i is for density; u_i and v_i axial and radial velocities; p is the pressure which is assumed the same in both phases; E_1 is the total gas phase energy and N is the particle number density. R is the intergranular stress, which is taken to be a function of α_2. The variables α_i satisfy $\alpha_1 + \alpha_2 = 1$, with $\alpha_i = \alpha_i(z, r, t)$ being a normalised measure of the cross sectional areas occupied by the gas and particle phases respectively. α_1 is called the porosity. The source vector $\boldsymbol{S}$ includes contributions from chemical sources, heat loss and heat transfer and also interphase drag. There are also terms which are due to the "hyperbolisation" of the system. These give:

$$\boldsymbol{S} = \begin{bmatrix} -\frac{\alpha_1\rho_1 v_1}{r} + M_{pr} + M_{ig} \\ p\frac{\partial\alpha_1}{\partial z} - \frac{\alpha_1\rho_1 u_1 v_1}{r} + u_2 M_{pr} - D_z \\ p\frac{\partial\alpha_1}{\partial r} - \frac{\alpha_1\rho_1 v_1^2}{r} + v_2 M_{pr} - D_z \\ p\left(\frac{\partial\alpha_2 u_2}{\partial z} + \frac{\partial\alpha_2 v_2}{\partial r}\right) - \frac{v_1(\alpha_1 E_1 + \alpha_1 p_1)}{r} - \frac{p_1\alpha_2 v_2}{r} + M_{pr}E_2 + M_{ig}e_2 - u_2 D_z - v_2 D_r - Q - W \\ -\frac{\alpha_2\rho_2 v_2}{r} - M_{pr} \\ p\frac{\partial\alpha_2}{\partial z} - \frac{\alpha_2\rho_2 u_2 v_2}{r} - u_2 M_{pr} + D_z \\ p\frac{\partial\alpha_2}{\partial r} - \frac{\alpha_2\rho_2 v_2^2}{r} + v_2 M_{pr} + D_r \\ -\frac{v_2 N}{r} \end{bmatrix} \tag{3}$$

The terms involving axial and radial derivatives of α_i are due to the process of 'hyperbolisation'. M_{pr} is the rate of mass addition due to the combustion of the solid propellant and M_{ig} is the rate of mass addition due to the igniter (Speares and Toro 1992). The terms D_z and D_r are the axial and radial drag respectively; Q is the interphase heat transfer and W is the heat loss to the wall. In the analysis below these terms are set to zero. The quantity $E_2 = \frac{1}{2}\left(u_2^2 + v_2^2\right) + e_2$ is the specific total energy of the solid phase, where e_2 is the chemical energy of the propellant. The system of equations is closed by giving the specific internal energy e_1 of the gaseous phase for a covolume equation of state to be:

$$e_1 = p\frac{(1 - b\rho_1)}{(\gamma - 1)\rho_1} \tag{4}$$

b is the covolume parameter. This is a constant with dimensions $m^3.kg^{-1}$ which is determined experimentally.

3. Numerical methods

For the solution of the homogeneous part of Eq.1:

$$\boldsymbol{U}_t + [\boldsymbol{F}(\boldsymbol{U})]_z + [\boldsymbol{G}(\boldsymbol{U})]_r = \boldsymbol{0} \tag{5}$$

we employ an operator-split finite-volume formulation of the WAF method (Toro 1992), to calculate a second order accurate intercell flux. The method achieves higher order accuracy by performing a spatial integral through the solution of the cell interface Riemann problem. The numerical flux is defined:

$$\boldsymbol{F}_N^{waf} = \frac{1}{\Delta n_{CML}} \int_{-1/2\Delta n_{CML}}^{0} \boldsymbol{F}_n^* dn + \frac{1}{\Delta n_{CMR}} \int_0^{1/2\Delta n_{CMR}} \boldsymbol{F}_n^* dn \tag{6}$$

where now $\boldsymbol{F}_n^* = \boldsymbol{F}_n[\boldsymbol{U}^*(x/\Delta t/2)]$ is the solution to the RP along the line $t = \Delta t/2$ normal to each cell face, with n_{CML} and n_{CMR} representing the normal distance from the face to the left and right cell centroids on the grid. An approximate solution to the RP is obtained for the system by a phase splitting approach. Having obtained an intercell flux the cell update is performed separately for each phase using the standard discretisation of the integral form of the conservation laws (Speares and Toro 1992). The source term is included by solving the homogeneous system, using data $\boldsymbol{U}^n$ at time level n for cell ij. This results in data $\hat{u}_{ij}^{n+1}$ after the homogeneous update has been performed which is then used as the initial data for the system of ODEs:

$$\frac{d\boldsymbol{U}}{dt} = S(\hat{\boldsymbol{U}}^{n+1}) \tag{7}$$

The new time state is now found by a simple Euler step update. We employ central differencing to calculate the elliptic derivative terms within the source vector.

4. Single mesh two-phase calculations

In Fig.1 we show comparisons between single mesh calculations and available experimental data for two propellant configurations in a 76 mm gun. In the first configuration the propellant is constrained to lie in the first two thirds of the combustion chamber from the breech (left hand picture), whereas in the second configuration the propellant is evenly distributed throughout. In both cases the initial mesh was of 51 × 4 cells which, due to the travel of the projectile, increases to 422 × 4 cells over the calculation. In both cases there is encouraging agreement between the calculated pressure variation and the experimental data provided by our sponsors.

5. The adaptive algorithm

The adaptive algorithm we employ here is based on the AMR approach (Burger and Colella 1989, Quirk 1990) and which involves adaption in both space and time. A hierarchy of structured refined meshes is constructed above flow features of interest such as shock waves and vortices. This structure is adapted dynamically as the flow evolves, with each successive level in the hierarchy being integrated a number of times and with a time step consistent with the factor of refinement.

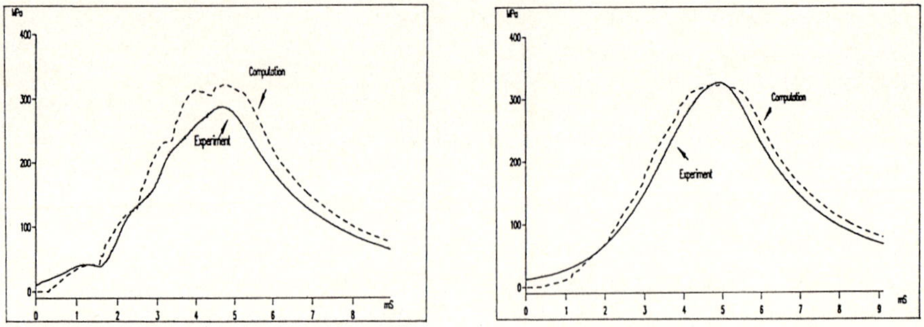

Fig. 1. Results for two propellant configurations

This approach can result in savings in computational effort in excess of 99%, depending on the nature of the problem being calculated. The algorithm may be viewed in a completely modular way, as being a data structure acted upon by an adaption function and an integration function. By employing a powerful language such as "C" it is possible to code the algorithm in an elegant manner. Below we briefly describe details of the algorithm.

5.1. The data structure

The data structure is constructed from various groupings of variables called structures (here we employ terminology from "C", although other languages such as PASCAL and FORTRAN 90 have similar facilities). Take, for example, the *Mesh* structure:

```
struct Mesh {
struct Mesh * parent;
struct Tags M_Tag;
struct Cell * MC_ptr;
struct Boundary * MB_ptr[4];
struct Mesh * prior;
struct Mesh * next; } ;
```

These Mesh structures may be used to construct the hierarchical data structure in a simple way. The Mesh structure contains the position of data connected to the individual meshes, stored in a doubly linked list joined via the Mesh pointers *prior* and *next*. Via the pointer MC_ptr the individual cells in the mesh may be accessed; these will be "Cell" structures containing the conserved variable data, refinement flags, cell geometry data and so on. The four "Boundary" structure pointers MB_ptr allow access to the four boundaries of each mesh. These structures will contain "ghost cell" storage for the mesh as well as mesh connectivity information. The "Tags" structure contains information about the size and coordinate position of the mesh, and the Mesh pointer *parent* contains the address of the parent mesh above which the (refined) Mesh was constructed. Using this type of data grouping it is possible to construct the data structure for the adaptive algorithm in a straightforward manner.

5.2. The adaption process

The success of the adaption process relies on the successful detection of flow features. This is currently done by the computation of a simple *Grad* operator of the density variable, Other differential operators such as *Curl* are clearly appropriate for the detection of skew-symmetric quantities like vorticity. In the future it is expected that information from the local cell Riemann problems will be used.

Once the flow detection has been achieved, cells affected are 'flagged' for refinement. A mesh of fine cells sufficient to cover the flagged area of the domain is then constructed, its position and connectivity are recorded, and it is added to the linked list of meshes. Upon adaption of the grid hierarchy to account for the changing structure of the flow, a new data structure is constructed in the same way. Where there is an overlap between the two grid hierarchies, data is transferred directly from the old list to the new. In parts of meshes where there is no overlap, the cells are primed with interpolated data from the coarse meshes on the level below (Quirk 1990). In this way strong features like shocks are always captured, and remain captured, as the flow evolves.

5.3. The integration process

The integration process is the third component to the adaptive algorithm. The nature of the numerical algorithm being applied to the system of equation under study can (should) be made completely separate from the adaptive machinery. By maintaining modularity of design it is a simple matter to change numerical methods, systems of equations and so on. One source of difficulty with this type of adaptive scheme is the procedure followed on boundaries between fine and coarse areas of the grid. Not only is it necessary to ensure conservation across such boundaries, but the "ghost cells" required for the integration of the fine meshes in the grid hierarchy must be primed with data for times in between the coarse mesh time discretisation. This implies that both time and space interpolations must be carried out. In our treatment here we follow (Quirk 1990), and apply a slope-limited spatial interpolation and a linear time interpolation between successive values in space and time, respectively, of the underlying coarse mesh discretisation. The processes of flow adaption and integration are then interleaved in such a way to ensure that strong flow features, such as shocks, may never escape across a fine-coarse boundary. This is an important requirement of the scheme.

6. Adaptive mesh gas-phase calculations

Below we demonstrate the operation of the adaptive code on a subset of the system (1), that of the gas-phase (Euler) equations, with the ideal gas equation of state (covolume parameter $b = 0$). The validation test shown in Fig.2 is that of the single Mach reflection of a Mach 2.12 shock wave impinging on a 30 degree wedge. The upper diagrams, left to right, show experimental data (courtesy of Dr W. Heilig, Fraunhofer Institut) and the computation respectively. The computation is presented as a contour plot. The lower diagrams show the physical features present in the experiment and the computation mesh structure at output (cell centroids). The base mesh is 50×50 with two levels of refinement, refining by factors of four and two respectively. This gives the calculation the resolution equivalent to a single mesh of 400×400 cells. The speed-up resulting from the use of adaptive meshing, for this problem, over the equivalent single mesh calculation is of the order of a factor of 25. As can be seen the agreement between the experiment and the computation is good for the inviscid flow regimes.

7. Discussion and conclusions

Above we have outlined a numerical approach and an adaptive gridding algorithm for the calculation of reactive two-phase flows within Interior Ballistics. The computation of high resolution numerical solutions to these systems is valuable in order to test both the current generation of numerical methods in use against these problems as well as the mathematical models themselves. Unsatisfactory features, such as the ill-posedness of the initial value problem, indicate that current models are not complete. Exhaustive testing of existing models using high-resolution adaptive methods is a necessary step in finding more satisfactory formulations. It should be recognised that very high grid resolutions of the granular scale and below for the type of model described above,

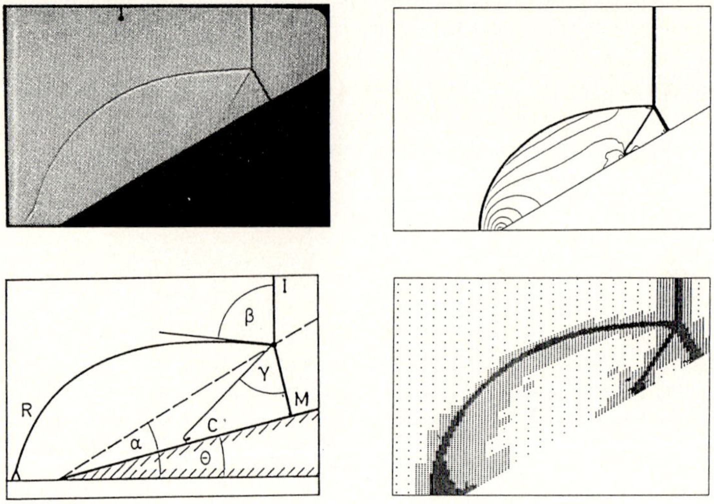

Fig. 2. Numerical calculation vs. experiment for single Mach reflection

while being of interest from the numerical viewpoint, would almost certainly lack physical meaning. However, the experience from single phase combustion model calculations suggests that to include chemical reactive terms such resolutions will be necessary. An adaptive approach is currently probably the only way of achieving this in 2-D and 3-D. There are a number of possible problems associated with applying the adaptive algorithm to the full two-phase system. An important first step is to determine satisfactory criteria for the detection of flow features of interest. In addition the problem we discuss here involves a moving boundary. An adaptive algorithm for the moving boundary is currently under development. It is hoped that preliminary results for the validation of the moving boundary for the gas-phase will be available soon. The inclusion of the full two-phase system will follow.

Acknowledgement

This work has been carried out under the sponsorship of the DRA Fort Halstead, U.K. The authors would also like to thank Dr W. Heilig of the Fraunhofer Institut für Kurzzeit-Dynamik, Freiburg, Germany, for providing experimental data.

References

Burger MJ, Colella P (1989) J. Computational Physics 82: 64-84

Fitt AD (1988) Some aspects of Internal Ballistics theory. 5th Anglo-German Ballistics Meeting, Unterluss, Germany

Gough P (1979) in: Krier and Sommerfeld (eds.) Progress in Astronautics and Aeronautics Series Vol. 66 pp 177-196

Toro EF (1992) Phil. Trans. Roy. Soc. 241: 499-530

Toro EF (1989) Lecture notes in Physics No. 351, Springer, pp 472-481

Speares W, Toro EF (1992) In: Hirsch et al. (eds.) ECCOMAS CFD Conference Proceedings, Elsevier, pp 565-573

Stewart H, Wendroff B (1984) J. Computational Physics 56: 363-409

Quirk JJ (1990) An adaptive grid algorithm for computational shock hydrodynamics. PhD thesis, Cranfield

Part 2: Shocks in Porous Media

On the Modelling of Wave Phenomena in Permeable Foam

M.E.H. van Dongen*†, D.M.J. Smeulders*, T. Kitamura† and K. Takayama†
* Faculty of Applied Physics, Eindhoven University of Technology, Eindhoven, The Netherlands
† Shock Wave Research Center, Institute of Fluid Science, Tohoku University, Sendai, Japan

Abstract. The linear Biot theory for wave propagation in porous media is applied to waves in open-cell permeable foam. The reflection of a weak shock wave from a semi-infinite elastic foam is considered. It is shown that the deformation of the foam and the related effective stress are caused by friction. The non-linear permeability and stress-strain relation of a flexible polyurethane foam is investigated. An example is shown of the reflection of a weak shock wave with Mach number 1.04 from a slab of foam adjacent to a solid wall. Both end-wall pressure and effective stress are measured. A partial mechanical collapse of the foam is observed.

Key words: Permeable foam, Wave phenomena

1. Introduction

Wave phenomena in permeable foams are quite interesting from a gasdynamic point of view. Such waves are strongly dispersive: a weak shock wave reflecting from a foam surface generates a gradual compression wave. When a shock wave reflects from a slab of foam adjacent to a solid wall, the end-wall pressures observed are much higher than expected. A recent survey on experimental and theoretical investigations of wave propagation in foams is given by Skews (1991). Until recently, the modelling of these wave phenomena was based on the assumption that foam and gas could be considered as effectively one phase. An example is the pseudo-gas model of Gel'fand et al. (1983), that was later modified by Bazhenova and Gvozdeva (1988). Such models are rather successful, although not all phenomena can be explained, the differences between observation and theory being often attributed to the two-phase character of the wave phenomena. Baer (1992) recently studied wave phenomena in foam by applying a non-linear two-phase model, with viscous and visco-elastic interaction terms. By varying the parameters he was able to obtain a rather satisfactory numerical simulation of the experiments of Skews (1991). Although permeable foams may show a strong non-linear mechanical behaviour, we shall apply the linear two-phase theory of Biot (1956) to waves in foams in order to investigate the basic principles. The deformation of the foam is attributed to the effective stress σ, which is the compressive stress acting on the porous structure in addition to the pressure p, positive in case of compression.

2. Dispersion relations, wave propagation and reflection

We start with the conservation laws of mass and momentum for two interacting media, porous material or foam, indicated with subscript p, and pore fluid, with subscript f (Bear and Bachmat 1990):

$$\partial/\partial t(\phi_a \rho_a) + \partial/\partial x(\phi_a \rho_a v_a) = 0 \ , \tag{1}$$

$$\partial/\partial t(\phi_a \rho_a v_a) + \partial/\partial x(\phi_a p_a + \phi_a \rho_a v_a^2) = m_a \ . \tag{2}$$

Here the subscript a is either p or f, ϕ_a is the volume fraction of component a, p_a the pressure or compressive stress, v_a the velocity, ρ_a the density, and m_a the interaction force. We shall assume that the solid temperature does not change. $\phi_f = 1 - \phi_p$ is also known as the porosity.

Shock Waves @ Marseille III
Editors: R. Brun, L. Z. Dumitrescu

The compressive stress acting in the cross section of a porous material consists of two contributions: $p_p = p_f + \sigma/\phi_p$. The so-called effective stress σ is related to the change in porosity: $\phi_f = \phi_f(\sigma)$. The density of the solid material is considered as constant. We specify the interaction term m_a in its linear form:

$$m_f = -m_p = p_f \partial/\partial x(\phi_f) + b_0(v_p - v_f) + \phi_f(\alpha - 1)\rho_f \partial/\partial t(v_p - v_f) \ . \quad (3)$$

The Darcy friction factor b_0 is related to the steady state permeability k_0 by $b_0 = \eta\phi_f^2/k_0$, with η the viscosity. The last term represents the added mass effect, with α the added mass parameter or tortuosity (Biot 1956).

We proceed as follows: Eqs.1-2 are linearized with respect to the steady state. Then, harmonic waves are considered: $p_f = \hat{p}_f \exp[i(\omega t - \kappa x)]$, etc. We eliminate the complex amplitudes of porosity and fluid density by introducing the effective bulk moduli K_f and K_p: $\hat{\rho}_f = (1/K_f)\rho_f\hat{p}_f$, $\hat{\phi}_p = -\hat{\phi}_f = (1/K_p)\phi_p\hat{\sigma}$. We then have obtained four linear homogeneous algebraic equations for the unknowns $\hat{p}_f, \hat{\sigma}, \hat{v}_f$ and $\hat{v}_p$, which results into the dispersion relation $\kappa(\omega)$. For an extensive description of the theory we refer to Van Dongen et al. (1993). It follows that in general two different wave modes exist, each with its own frequency-dependent wave speed and damping. A characteristic frequency ω_c is found, namely that frequency at which friction terms and inertia terms are of equal importance: $\omega_c = b_0/(\phi_f\alpha\rho_f) = \eta\phi_f/(k_0\alpha\rho_f)$. As a reference value for the wave speed we will take the low-frequency limit: $c_{ref} = ((K_p + K_f/\phi_f)/\rho)^{1/2}$. The reference length is $L_{ref} = c_{ref}/\omega_c$.

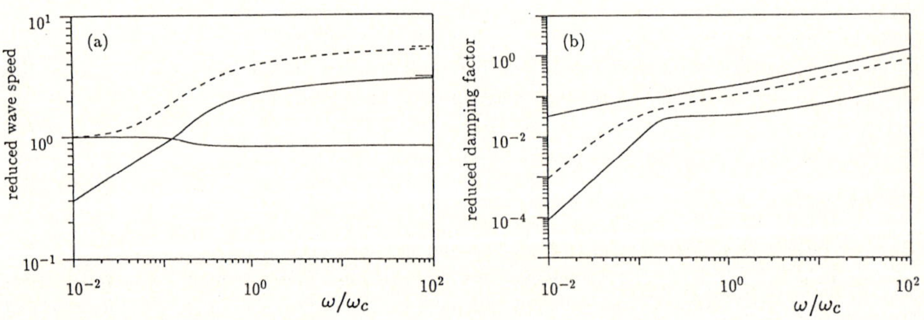

Fig. 1. Reduced wave speed and damping factors versus reduced frequency for permeable foam. $\phi_f = 0.98$, $\alpha = 1$, $\rho_p/\rho_f = 1000$. Solid lines: $K_p/K_{f0} = 2$; dashed curves: $K_p/K_{f0} = 0$

The description given above is somewhat modified by introducing a more realistic frequency-dependent friction factor $b(\omega)$ (Smeulders et al. 1992). The heat conduction between pore gas and pore solid is taken into account by introducing a frequency-dependent gas bulk modulus: $K_f(\omega) = n(\omega)K_{f0}$. To evaluate the polytropic coefficient $n(\omega)$ we applied the theory of Devin (1959) for oscillating gas bubbles to the present problem.

In Fig.1 some results are given for wave speeds and damping factors for foam. Two different cases are considered: $K_p/K_{f0} = 2$ and 0, respectively. The latter case corresponds to foams without any mechanical strength. Then only one wave mode remains. Other parameters are: $\phi_f = 0.98, \alpha = 1$ and $\rho_p/\rho_f = 1000$, which is a realistic value according to Gibson and Ashby (1988). Wave speeds and damping factors are given in dimensionless form. The isentropic wave speeds of the pore gas are indicated along the high-frequency axis. High-frequency components are propagating with the isentropic sound speed, but they are strongly damped. Low-frequency components can propagate without damping with the much lower speed c_{ref}. The reflection of a weak shock wave from a semi-infinite foam and from a slab of foam adjacent to a solid wall has been evaluated on the basis of the linear Biot theory. To that end proper boundary conditions have to be applied. At the free surface the effective stress has to be zero, according to its definition, the foam remaining there undeformed. Further, continuity of pressure and mass flux has to be satisfied. At a solid boundary the velocities of foam and gas have to be zero. The wave problems were solved as follows (Van Dongen et al. 1993): the incident shock wave is decomposed into its Fourier components. For each harmonic the reflection problem is solved including multiple reflections in between foam surface and solid wall. The solution in the time domain is obtained by an inverse numerical Fourier transform.

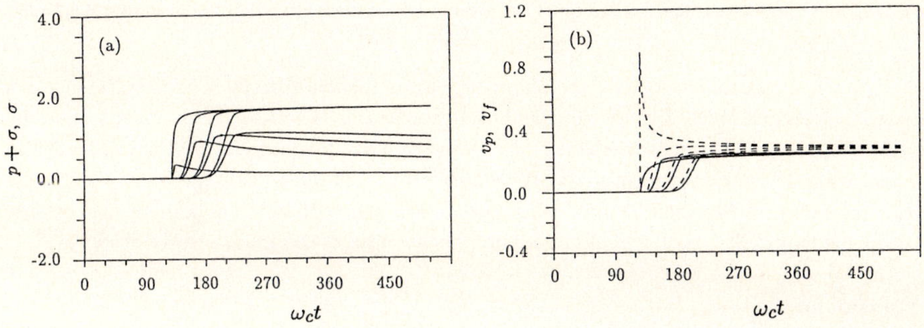

Fig. 2. Waves in a semi-infinite permeable foam after reflection of a weak shock wave from the interface. $K_p/K_{f0} = 2$. Other parameters: as in Fig.1. **a:** total stress (upper curves) and effective stress (lower curves) as functions of time at distances from the interface: x/L_{ref} = 5, 30, 55 and 80. **b:** velocities of gas (dashed lines) and solid at x/L_{ref} = 0, 25, 50 and 75

First, we consider the case of shock wave reflection from a semi- infinite permeable foam. The foam is assumed to behave elastically with $K_p/K_{f0} = 2$. In Fig.2 the total stress, $p + \sigma$, the effective stress σ and the velocities of gas and solid are shown as functions of reduced time for four different equidistant positions in the foam. The total stress behaves as a damped propagating wave. The steepness of the wave front clearly decays due to internal friction. The transmitted total stress attains a constant value after some time: T_r for an incident step wave of unit strength. The value of T_r can be evaluated: $T_r = [2(1+\phi_f K_p/K_{f0})^{1/2}]/[(1+\phi_f K_p/K_{f0})^{1/2} + (\gamma\phi_f\rho_f/\rho)^{1/2}]$. The effective stress shows a rather different behaviour. At the free boundary σ remains zero, according to its definition. The foam material will not be compressed there, in agreement with the observation by Skews et al. (1992). At some distance from the interface the stress first increases and then slowly decreases again. The maximum stress value increases with increasing distance from the interface and attains an asymptotic value. The ratio between maximum stress and total stress T_r is, according to linear theory: $\sigma_{max}/T_r = \phi_f(K_p/K_{f0})/(1+\phi_f K_p/K_{f0})$. When studying waves in collapsible foams it is important to compare the maximum value of σ with the collapse stress σ^*. For sufficiently weak shocks, $\sigma_{max} < \sigma^*$, and the foam will respond in an elastic way. For stronger shocks, if $\sigma_{max} >> \sigma^*$, the foam will collapse and will behave as if it has no strength at all. In that case there is a strong similarity between waves in foams and waves in dusty gases. In Fig.2b the velocities of gas and foam are shown, again at four equidistant positions, starting at the interface. The velocities have been reduced here with respect to the velocity induced by the incident wave.

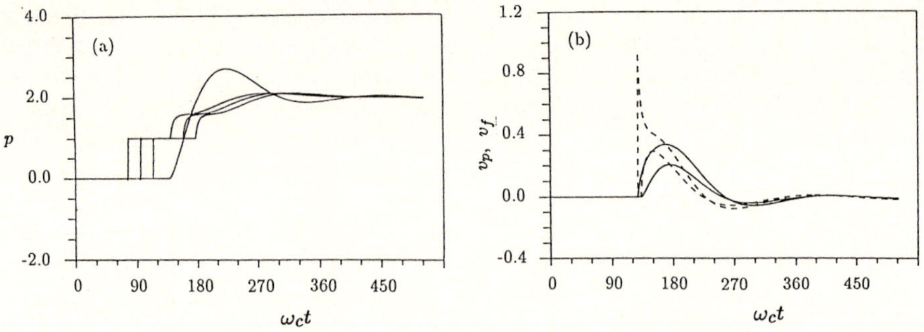

Fig. 3. Reflection of a weak shock wave from a permeable foam adjacent to a solid wall. The thickness of the foam in reduced cordinates is 50. **a:** gas pressures at distances of 270, 170 and 70 from the interface, and the total stress at the solid wall. **b:** velocities of gas (dashed lines) and solid at the interface and inside the foam at x/L_{ref}=25

In Fig.3 pressures and velocities are shown for wave reflection from a foam adjacent to a rigid wall. The length of the foam is 50 in reduced coordinates. The foam is assumed to be fully collapsible: $K_p/K_{f0} = 0$. The time histories are shown at three positions in front of the foam and at the end-wall. The incident wave front propagates with a relatively high speed, somewhat less than c_g. The wall pressure shows a maximum value, much higher than the maximum gas pressure in front of the foam. The exchange of momentum between solid and gas is clearly observed.

3. Experiments

The properties of the flexible polyurethane foam Moltopren Pu-Sc, with density $\rho = 28.8$ kg/m^3 and porosity $\phi_f = 0.977$, were investigated. The permeation properties of the foam were determined in a one-dimensional flow set-up for foam specimens with a length of 50 cm. The relation between the applied pressure gradient and the resulting air mass flow was measured in the velocity range between 0 and 0.1 m/s. This relation satisfies Forchheimer's law:$-dp/dx = \eta\phi_f v_f/k_0 + b_f\rho_f\phi_f^2 v_f^2$, with $k_0 = 8.2\ 10^{-9}$m^2 and $b_f = 3.4\ 10^4$m^{-1}.

The stress-strain relations were investigated in a test facility (Shimadzu AG5000c Autograph) at a constant atmospheric pore pressure for foam specimens with lengths of 7 and 14 cm, either unconstrained, or constrained in a tube, similar to the one that was used later for the wave experiments. A typical result is shown in Fig.4. As discussed by Gibson and Ashby (1988), the foam behaves elastically for a small compressive strain, with an effective bulk modulus K_p. At some critical stress the foam collapses with a plateau value σ^*, more or less independent of the compression ratio. When the compression ratio increases, the foam enters the densification regime, and σ increases again. The average values for the effective bulk modulus and collapse stress are: $K_p = 220 \pm 40$ kPa, and $\sigma^* = 7.8 \pm 0.4$ kPa. The Young modulus is 99 ± 4 kPa and the corresponding Poisson ratio is 0.40 ± 0.03.

In order to investigate the effect of the non-linear stress-strain relation on wave reflection, a shock tube experiment was performed. A piece of foam with a length of 7 cm was cut to fit precisely in an aluminium shock tube with a cross-section of 6 × 15 cm^2. The foam was placed in the shock tube such that the rear face was flush with the end-wall. In this end-wall two piezo-electric transducers were mounted, one in direct mechanical contact with the foam material, the other receded over a distance of 3 mm, having a diameter of 5 mm. In this way a direct mechanical contact between foam and transducer was avoided, so that the local gas pressure could be determined. In the side wall of the tube pressure transducers were mounted at distances of 15.6 and 25.6 cm from the end plate. A typical and illustrative result of a shock wave reflection experiment is shown in Fig.5. The shock Mach number is 1.04, the test gas is atmospheric air.

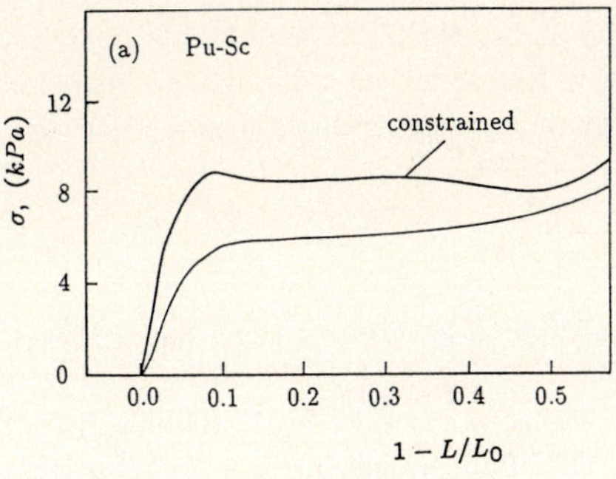

Fig. 4. Example of effective stress versus compression ratio for the flexible polyurethane foam Pu-Sc. Upper curve: constrained; lower curve: unconstrained

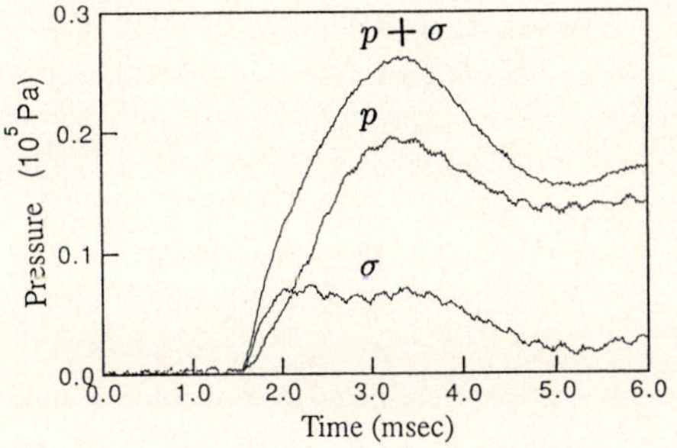

Fig. 5. Reflection of a shock wave with Mach number 1.04 from polyurethane foam (Pu-Sc) adjacent to the end-wall of a shock tube. Length of the foam: 7 cm. Upper trace: Total stress at the end-wall of the shock tube. Middle trace: End-wall pressure. Lower trace: Effective stress

The speed of propagation of the foot of the wave is 160 m/s, which is, as expected, somewhat larger than its low elastic low-frequency limit of 106 m/s. The upper signal is the total stress, the second signal is the pressure, and the lower signal is the difference between the upper two and should correspond to the effective stress of the foam. It is clear that the effective stress shows a constant value during 2 ms at a level of 0.07 bar, which corresponds very well with the collapse value shown in Fig.4.

4. Conclusions

Linear Biot theory, applied to the problem of the interaction of a weak shock wave with a semi-infinite or a finite permeable foam, shows that only the low-frequency limit of one of two possible wave modes can propagate without damping with a wave speed: $c_{ref} = ((K_p + K_{f0}/\phi_f)/\rho)^{1/2}$. High frequency components may propagate, strongly damped, at the isentropic sound speed of the pore gas. When a shock wave hits a high-porosity permeable foam there is almost no instantaneous reflection. The reflected wave is generated because the gas that enters the foam decelerates due to friction.

For the flexible polyurethane foam Pu-Sc it was found that the relation between a pressure gradient and the gas velocity satisfies a quadratic dependence. The quadratic term equals the linear term at a velocity of approximately 7 cm/s, so that the non-linear term will be the dominant one for most shock wave reflection experiments. The reflection of a weak shock wave, Mach number 1.04, from a slab of polyurethane foam adjacent to a solid wall was studied experimentally. The effective stress of the foam at the end-wall of the shock tube could be determined as the difference between total stress and gas pressure. Even for this very weak shock wave a partial mechanical collapse of the foam was observed. The wave foot propagates with a velocity in between the limits predicted by linear theory.

Acknowledgement

The first author is much indebted to Tohoku University and to the Institute of Fluid Science for the invitation to be a visiting professor at the Shock Wave Research Center during the second half of 1992. The valuable advice of Dr. M. Olim on the numerical implementation of the reflection problems is gratefully acknowledged.

References

Baer MR (1992) A numerical study of shock wave reflection on low density foam. Shock Waves, 2:121

Bazhenova TV, Gvozdeva LG (1988) Unsteady interactions of shock waves. In: Grönig H (ed) Proc. 17th Intl. Symp. on Shock Tubes and Waves, VCH Publ, Weinheim/New York, pp 41-49

Biot MA (1956) Theory of propagation of elastic waves in a fluid saturated solid, I. Low-frequency range. J. Acoust. Soc. Am. 28:168; II. Higher frequency range, ibid. 28:179

Bear J, Bachmat Y (1990) Introduction to Modeling of Porous Media. Kluwer

Devin C (1959) Survey of thermal radiation and viscous damping of pulsating air bubbles in water. J. Acoust. Soc. Am. 31:1654

Gel'fand, BE, Gubanov AV, Timofeev EI (1983) Interaction of shock waves in air with a porous screen. Izv. Akad. Nauk SSSR, Mekh. Zhidkosti i Gaza 4 :79

Gibson LJ, Ashby M (1988) Cellular Solids-Structure and Properties. Pergamom Press

Skews BW (1991) The reflected pressure field in the interaction of weak shock waves with a compressible foam. Shock Waves 1:205

Skews BW, Atkins MD, Seitz MW (1992) Gas dynamic and physical behaviour of compressible porous foams struck by a weak shock wave. In: Takayama K (ed) Proc. 18th Intl. Symp. on Shock Waves, Springer, New York, pp 511-520

Smeulders DMJ, Eggels RLGM, Van Dongen MEH (1992) Dynamic permeability: reformulation of theory and new experimental and theoretical data. J. Fluid Mech. 245:211

Van Dongen MEH, Smeulders DMW, Kitamura T, Takayama K (1993) On wave phenomena in permeable foam. In: Rep. Inst. Fluid Sci., Tohoku University, Sendai, 5, pp 55-68

A Multiphase Model for Shock-Induced Flow in Low Density Foam

M.R. Baer
Energetic Materials and Fluid Mechanics Dept 1512, Sandia National Laboratories Albuquerque, NM

Abstract. A multiphase mixture model is applied to describe shocked-induced flow in deformable low-density open-cell foam. This model includes interphase drag and heat transfer and all phases are treated as compressible. Volume fraction is represented as an independent kinematic variable and the entropy inequality suggests a thermodynamically-admissible evolutionary equation to describe rate-dependent compaction. This multiphase model has been applied to shock tube experiments conducted by B. W. Skews and colleagues in the study of normal shock impingement on a wall-supported low-density porous layer. Numerical solution of the multiphase flow equations employs a high-resolution adaptive finite-element method which accurately resolves contact surfaces and shock interactions.

Key words: Foams, Multiphase shock flow.

1. Introduction

In contrast to single-phase materials, porous materials contain internal boundaries which greatly influence the thermal and mechanical response to shock loading. In spite of the wealth of study in multiphase flow, much remains to be understood regarding the complex nature of shock-induced flow in distended materials. Continuum mechanics has provided a foundation for theoretical study and mixture theory has successfully described various multiphase waves (Baer et al. 1986a, 1989); however, much of this analysis has focussed on the strong shock wave response of high density porous materials. This study extends the application of mixture theory to shock-induced flow in low-density open-cell foam. In recent years, B.W. Skews and colleagues (1989,1992) have reported shock tube experiments with wall-supported low-density foams. Pressure measurements verified that dispersive waves are produced in the foam material following shock impingement and stagnation of material causes pressure amplification at the wall. As the foam material reflects off the wall, additional waves are observed. This study provides a theoretical description which replicates the observations of these experiments and complements a study of shock interactions on multiple foam layers.

2. Theoretical description

The continuum mixture model used in this study is taken, in its entirety, from the work of Baer and Nunziato (1986 a); only the final forms of this model are given here. For simplicity, the balance equations are given for one-dimensional two-phase flow. Thus, the effects of material strength and wall boundary friction are not included. Each phase is denoted with a subscript "a" to identify solid ($a = s$) and gas phase ($a = g$). Associated with each phase are the material densities, ρ_a, volume fractions, ϕ_a, particle velocities, v_a, pressures, p_a, temperatures, T_a, and internal energies e_a. The balance laws in Eulerian form, excluding the effects of mass exchange, are given as:

$$(Mass) \qquad \frac{\partial}{\partial t}(\phi_a\rho_a) + \frac{\partial}{\partial x}(\phi_a\rho_a v_a) = 0 \tag{1}$$

$$(Momentum) \qquad \frac{\partial}{\partial t}(\phi_a\rho_a v_a) + \frac{\partial}{\partial x}(\phi_a p_a + \phi_a\rho_a v_a^2) = m_a^{\ddagger} \tag{2}$$

Shock Waves @ Marseille III
Editors: R. Brun, L. Z. Dumitrescu

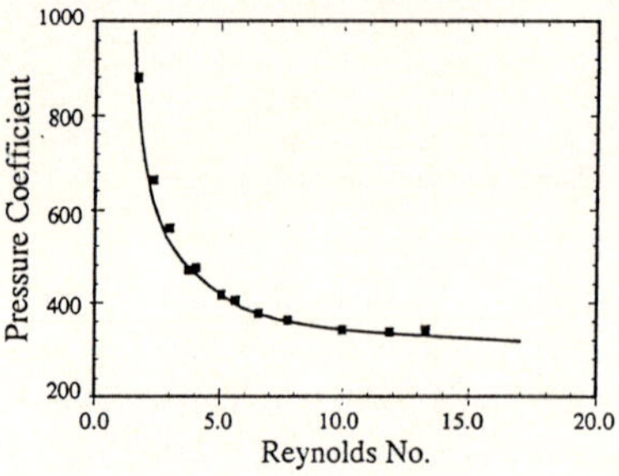

Fig. 1. C_p vs Re for polyester foam

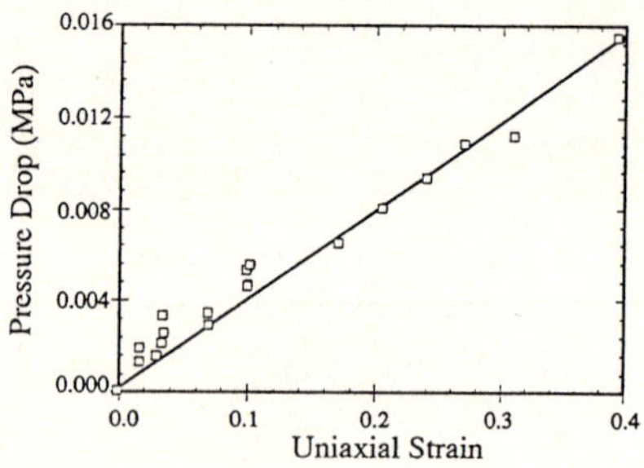

Fig. 2. Compaction data for polyester foam

$$(Energy) \qquad \frac{\partial}{\partial t}(\phi_a \rho_a E_a) + \frac{\partial}{\partial x}((\phi_a \rho_a E_a + \phi_a p_a) v_a) = e_a^{\ddagger} \tag{3}$$

where $E_a = e_a + v_a^2/2$ and the "$\ddagger$"superscript denotes a phase exchange quantity.

To describe the rate of change of solid volume fraction, a compaction model is derived from the entropy inequality given as:

$$\frac{\partial \phi_s}{\partial t} + v_s \frac{\partial \phi_s}{\partial x} = \frac{\phi_s \phi_g}{\mu_c}(p_s - p_g - \beta_s) \tag{4}$$

and $\phi_s + \phi_g = 1$ is the saturation constraint. The configurational effective stress, β_s, is the stress associated with the contact forces of the solid phase pore structure and is a different stress state than the bulk stress associated with the material compressibility. The compaction viscosity, μ_c , dictates the rate at which the volume fraction adjusts to an equilibrium stress state: $p_s = p_g + \beta_s$, and its value is estimated to be $O(10^2 - 10^4)$ [poise] as determined in prior dynamic compaction studies.

Consistent with the constraints of the Second Law of Thermodynamics, the interaction of momentum and energy are represented by:

$$m_s^{\ddagger} = -m_g^{\ddagger} = p_g \frac{\partial \phi_s}{\partial x} - C_d(v_s - v_g) \tag{5}$$

$$e_s^{\ddagger} = -e_g^{\ddagger} = m_s^{\ddagger} v_s - h(T_s - T_g) - (p_s - \beta_s)\frac{\phi_s \phi_g}{\mu_c}(p_s - p_g - \beta_s) \tag{6}$$

The drag coefficient, C_d, and the interphase heat transfer coefficient, h, are exchange coefficients for momentum and energy phase interactions and depend on local flow conditions based on Reynolds number scaling. These constitutive laws are discussed in the next section. Model closure is obtained using a nonporous solid thermoelastic equation of state fit to shock Hugoniot data. Additionally, the ideal gas law is used for the gas phase equation of state.

3. Constitutive equations

Critical constitutive relationships to the mixture model are the drag and heat transfer phase interactions and the compaction rate law for the low-density foam. Polyester and polyurethane foam cylindrical plug sections, provided by B. Skews (University of Witwaterstrand), were tested in a transparent flow channel. Pressure drop measurements at low to high flow rates were taken in the determination of permeability of the foams and the Reynolds number scaling of the interphase drag. At high air flow rate, drag forces induce compaction of the porous material and visual measurement of uniaxial strain is sufficient for estimating the effective mechanical properties of the foam materials.

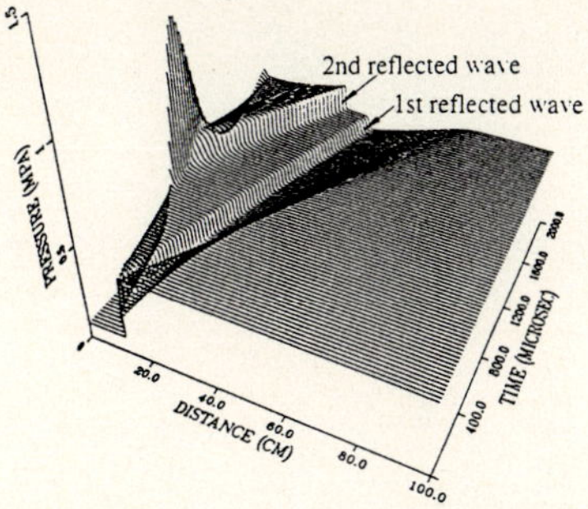

Fig. 3. Pressure wave fields during shock interaction

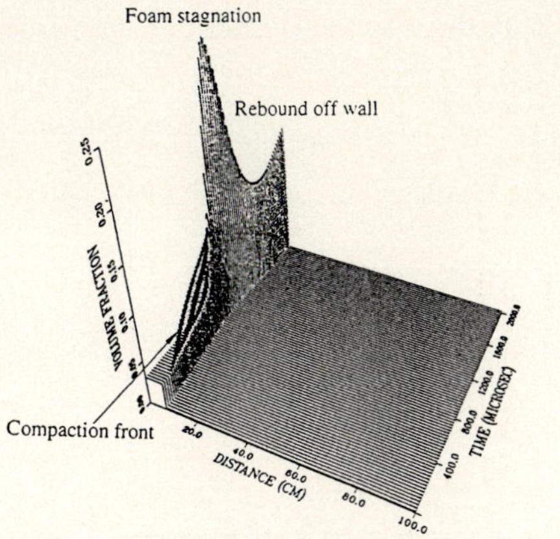

Fig. 4. Solid volume fraction during shock interaction

For nonshocked flow, pressure drop in rigid porous materials is related to flow velocity by the Forcheimer equation (Collins 1976):

$$\frac{\Delta p}{\delta} = \frac{\mu}{\kappa}V + \frac{c\rho_g}{\sqrt{\kappa}}V^2, \tag{7}$$

where δ is the length over which pressure drop occurs, μ is the gas viscosity, ρ_g is the gas density, κ is the foam permeability and c is an inertial constant (typically a value $\sim$ 1.). The fundamental length scale associated with the gas flow is $\sim \sqrt{\kappa}$ and a local Reynolds number is defined as $Re = \rho_g V\sqrt{\kappa}/\mu$; thus the pressure drop is recast in terms of a pressure coefficient $C_p = \Delta p/(\rho_g V^2/2)$:

$$C_p = \frac{2c\delta}{\sqrt{\kappa}} + \frac{2\delta}{Re\sqrt{\kappa}} \tag{8}$$

Fig.1 displays a representative data fit of equation 8 for gas flow into polyester foam (0.024 g/cc). The permeability of this material was measured as $\kappa = 0.29 \times 10^{-4}$ cm^2 and the inertial constant was determined to be c=0.31. Similar data was obtained for the polyurethane foam (0.027 g/cc) and the permeability was estimated to be $\kappa = 0.29 \times 10^{-5}$ cm^2 and $c \sim 10.4$. The interphase momentum was then determined as outlined in Baer et al. (1986 a). Having established appropriate Reynolds number scaling, the interphase heat transfer is represented by a fluidized bed correlation (see Baer et al. 1986 a).

The mechanical properties of the low-density foams were estimated by measurement of the uniaxial strain of the plug subjected to high gas flow rate. As described by Gibson and Ashby(1988), the mechanical behavior of these foam materials consists of a linear elastic region (within a few percent of strain), an elastic buckling range to 80-90% of strain and the upper limit of material densification. In the flow experiments, strains were measured in the elastic buckling region; hence the effective stress of the porous material is approximated by a constant yield stress, Y_{eff}. Using a force balance and mass conservation, this stress state is inferred from uniaxial compression data (induced by the drag forces of gas flow) by the following relationship:

$$Y_{eff} = \Delta p / 2\phi_s^\circ \frac{\Delta\delta}{\delta} \tag{9}$$

Fig.2 displays compaction data for a polyester foam material. The effective elastic buckling stress is estimated to be $\sim$ 1.07 MPa and for polyurethane foam Y_{eff} is estimated to be $\sim$ 0.8 MPa. It is noted that the effective bulk modulus is $K_{eff} \sim Y_{eff}/3 \sim 0.25$ MPa, which is consistent with shock Hugoniot measurement by Henderson et al. (1990).

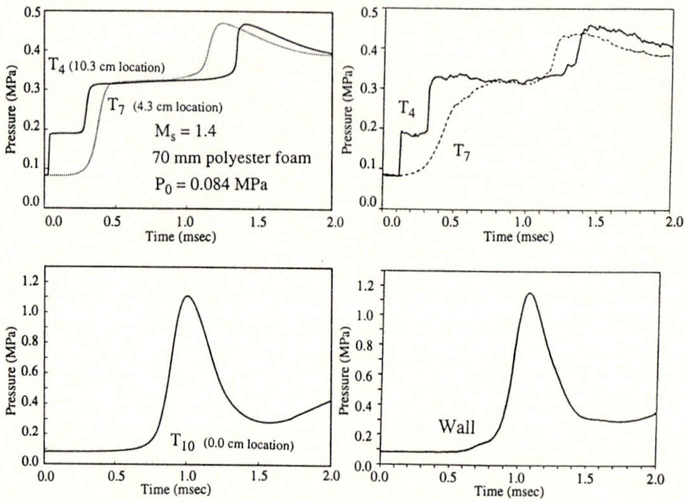

Fig. 5. Comparison of experimental and calculated pressure histories

4. Numerical solution

Appropriate numerical methods to resolve these highly stiff multiphase equations have been previously reviewed by Baer et al.(1986 b). This study uses an adaptive finite element numerical method. In this Eulerian approach, spatial derivatives are discretized using central differences and the resulting set of ordinary differential equations is solved using a stiff ODE solver. To stabilize the hydrodynamic calculation, a minimal amount of numerical viscosity is required to prevent numerical dispersion. An insert/delete adaptive noding method has been incorporated to enhance the numerical accuracy for resolving contact surfaces and shock fronts. This greatly reduces the effects of numerical diffusion. Up to eight levels of refinement (each level is a mesh reduction of a factor of two) are sufficient to capture the desired accuracy with approximately 200 adaptive elements.

5. Numerical calculations and comparison to experiment

Using the aforementioned model, one-dimensional numerical simulations of shock tube experiments by B.W. Skews and colleagues (1989,1992) were conducted. The computational domain was taken to be 100 cm long and the initial high pressure air driver pressure was specified consistent with experimental incident shock Mach number, M_s. Numerical calculations were done for $M_s = 1.25$ interactions on wall-supported foams of varied length. Preliminary calculations and a comparison to experiment are reported in Baer and al. (1986 b). Revised numerical calculations, with greater element refinement, have demonstrated that all of the wave features are well-resolved and modeled as described by multiphase mixture theory.

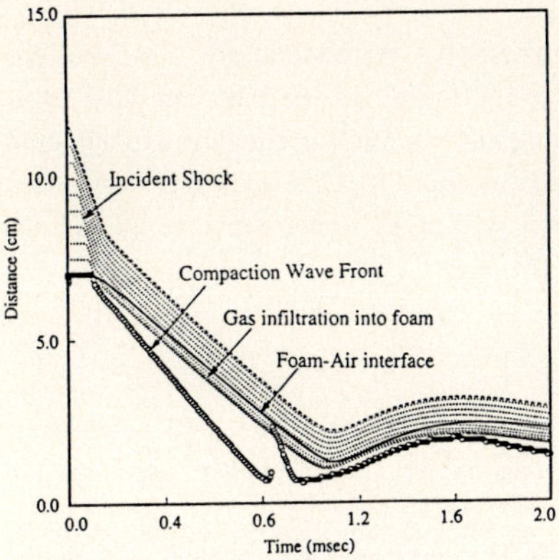

Fig. 6. Wave characteristics of shock interaction

Numerical calculations for a $M_s = 1.4$ shock interacting with a low-density (0.038 g/cc) polyester foam of length 7 cm are shown in Figs.3 and 4, which display the temporal and spatial variations of pressure and solid volume fraction. After the shock hits the foam, a reflected wave passes back into the driver gas and a compaction wave occurs in the foam as gas permeates into the porous material imparting momentum to the solid phase. Compaction wave behavior is clearly seen in Fig.4. As the compaction wave interacts with the rigid boundary, stagnation of material produces a large pressure rise. Reflection of this pressure pulse produces a second reflected wave. This wave initially has a dispersive front but steepens to a reflected shock as supported by the rebounding foam. Its amplitude is greater than the first reflection off of the foam/air interface. Reflections within the foam layer produce a third pressure wave of low amplitude.

In the shock tube experiments by B. Skews et al. (1993), several pressure transducers were used to probe the wave fields. Fig.5 is a direct comparison of the multiphase theory to the experimental measurements at locations $T_4 = 10.3$ cm and $T_7 = 4.3$ cm from the wall, representing conditions exterior and interior to the foam layer,respectively. Clearly, the pressure histories are well described by the theoretical model. The calculated vs. measured pressures at the wall are also compared in the bottom plots of Fig.5. At this shock condition, the reflected wall pressure is amplified by a factor of $\sim$10.

To better characterize the wave fields, particle paths are shown in Fig.6. The foam/air interface and the compaction wave head are also displayed. Note that the gas pathlines penetrate into the foam layer at a depth of $\sim$2mm from the foam/air interface. The compaction wave proceeds into the foam layer faster than the motion of the foam/air interface. At a later time, relaxation of the foam material is clearly evident.

6. Summary

In this study, a multiphase mixture theory has been successfully applied to describe shock tube experiments. A variety of shock states and experimental configurations have been replicated and continuum mixture theory has been shown to describe well the shock-induced flow in low-density open-cell foam. Numerical modeling can provide details of the complex wave behavior in highly distended materials.

Acknowledgments

I would like to thank Prof B. Skews and coworkers (University of Witwatersrand, S.Africa) for sharing their shock tube experimental data and providing the foam materials for the flow measurements. I also thank F. Chavez (SNL/NM) for assembling and conducting the flow experiments and A. Kraynik (SNL/NM) for his assistance with the foam mechanics. This work was performed at Sandia National Laboratories, supported by the US Dept. of Energy under contract DE-AC04-76DP00709.

References

Baer MR, Nunziato JW (1986 a) A two-phase mixture theory for the deflagration-to-detonation transition (DDT) in reactive granular materials. International Journal of Multiphase Flow 12, 6: 861-889

Baer MR, Benner RE, Gross RJ, Nunziato JW (1986 b) Modeling and computation of deflagration to - detonation transition (DDT) in reactive granular materials. Lectures in Applied Mathematics, American Mathematical Society 24: 479-498

Baer MR (1988) Numerical studies of dynamic compaction of inert and energetic granular materials. Journal of Applied Mechanics 55: 36-43

Baer MR, Nunziato JW (1989) Compressive combustion of granular materials induced by low-velocity impact. Ninth Symposium (International) on Detonation, OCNR 113291-7, August 1989, pp. 293-305

Baer MR (1992) A numerical study of shock wave reflection on low-density foam. Shock Waves 2: 121-124.

Collins RE (1976) Flow of Fluids Through Porous Materials. Penn Well Books and Pub. Company, Tulsa OK

Gibson LJ, Ashby M (1988) Cellular Solids-Structure and Properties. Pergamon Press

Henderson LF, Virgona RJ, Di J, Gvozdeva LG (1990) Refraction of a normal shock wave from nitrogen into polyurethane foam. In: Kim YW (ed) Proc. 17th Intl. Symp. on Shock Waves and Shock Tubes, American Institute of Physics, New York, NY, pp. 814-818

Skews BW (1989) The reflected pressure field in the interaction of weak shock waves with a compressible foam. Shock Waves 1, 3: 205-211

Skews BW, Atkins MD, Seitz M (1992) Gas dynamic and physical behavior of compressible porous foams struck by weak shock waves.In: Takayama K (ed) Proc. 18th Intl. Symp. on Shock Waves, Springer-Verlag, Heidelberg, pp 511-516

Structure of Deformation Waves in Porous Compressible Materials Under the Influence of Shock Waves

L.G. Gvozdeva, Yu.M. Faresov and Yu.L. Sharov
Institute for High Temperatures, Russian Academy of Sciences - Moscow, Russia

Abstract. Paper describes new experimental results concerning compression waves in porous compressible foams due to loading with shock waves generated in gas. A shock tube of 100 × 100 mm cross-section has been used. In addition to piezoelectric measurements and flow visualization, which were described elsewhere, new techniques for measuring the deformation waves in porous compressible media were developed. The density distribution in the deformation wave is determined, based on photodetection of a reference grating on the side face of the foam with short exposure times. The structure of the deformation wave was found to depend substantially on the permeability of the front face of the foam. In particular, the deformation wave in open-pore polyurethane foam was shown to be transformed into a wave-like isolated compression wave with a rarefaction region between the compression wave and the foam surface. The deformation wave in a foam with the front face covered with a thin mylar film is of a different structure, more like that expected in polyurethane if taken to behave like a pseudogas. The rarefaction region disappeared. The effect of pressure magnitude of the incident shock wave on the deformation wave structure was studied. The dependence of the maximum compression of porous medium on impinging wave intensity was obtained. Pressure distributions on the side wall of the shock tube under the foam were taken. The dependence of pressure distribution on the Mach number of the incident shock wave has been studied.

Key words: Porous compressible foam, Deformation wave, Shock wave, Optical methods

1. Introduction

Studies of the loading of a compressible foam with shock waves in air were started by Monti (1970). The effect of an anomalous increase of pressure at the shock-tube end wall covered with layers of compressible plastic foams was described first by Gelfand and al. (1975, 1983) and Gvozdeva and Faresov (1984, 1985). In recent years, many experimental and numerical studies on this phenomenon have been done, describing the entire process; the refraction of the shock wave at the surface of the sample of the foam and the formation of a refracted deformation wave inside the sample. Many striking features of the process were found and described by different authors recently. A list of references is given at the end of the paper.

In this paper some new results are given concerning the propagation and the structure of the deformation wave into polyurethane foam under shock wave loading. A special optical method is described for studying the deformation wave inside the foam.

2. Experimental facility

A sample of polyurethane foam was installed into the shock tube near its closed end as was described elsewhere (Gvozdeva and Faresov 1984, 1985). The shock tube has a square cross-section along its whole length - 100 × 100 mm, including the high-pressure chamber and the sample of polyurethane foam had also a strictly equal cross-section (100 × 100 mm). Attention was especially given that there should be no gap between the foam and the wall surface. The length of the sample was 40 mm, the value of apparent density ρ_H was 38 kg/m^3 and the porosity coefficient was equal to 0.98. Air was used in the low pressure chamber at initial pressure 10^5

Shock Waves @ Marseille III
Editors: R. Brun, L. Z. Dumitrescu

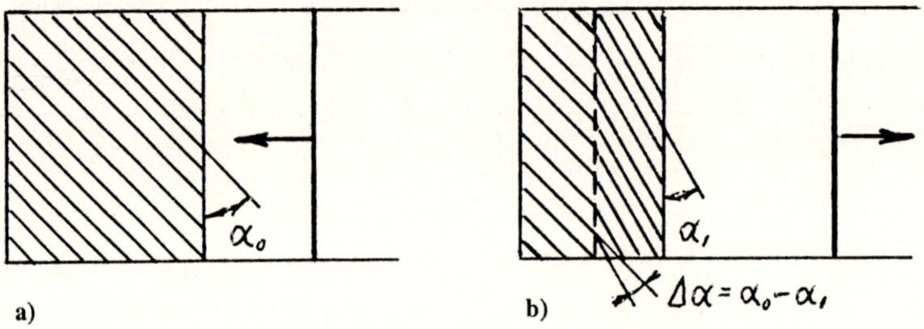

Fig. 1. Reformation of line pattern: a) before interaction; b) after interaction

Pa. Mach numbers of the incident waves were varied in a range $1.3 < M < 1.8$ by changing the pressure of the gas (helium) in the high-pressure chamber. On the side of the sample a set of parallel lines were drawn with a flowmaster pen, which formed an angle $\alpha_0 \approx 45^\circ$ with the front face of the sample (Fig.1a). For accurate experiments, it is better to reduce the width of the lines, δ and the distance between them, ΔX_0. Both these values should be greater than the diameter of the foam pores ($d \sim 0.5$ mm). In our case δ and ΔX_0 were 1 mm and 10 mm.

When the deformation wave propagates in the sample, the system of lines is deformed. A plane shock wave will change the angle of lines by some angle $\Delta\alpha$ (Fig.1b) and lead to a change of distance between any pair of lines in the disturbed region. It is easy to show that the density ratio across the shock wave ρ_1/ρ_0, is given by the formula: $\rho_1/\rho_0 = \tan\alpha_0/\tan\alpha_1 = \Delta X_0/\Delta X_1$, where $\Delta X_0, \Delta X_1$ are distances between pairs of lines. Measurement of the angles α_0, α_1 or the differences $\Delta X_0, \Delta X_1$ allow the calculation of the density ρ_1 behind the shock wave for a known value of ρ_0. When no shock wave but a compression wave exists in the compressible media, if one wants to determine the density distribution $\rho(X)$, one should first draw the distribution of angles $\alpha(X)$ and then use the equation $\rho(X)/\rho_0 = \tan\alpha(X)/\tan\alpha_0$. Strictly, the method gives the density distribution at the side of the foam and not the compression wave inside the foam. But, it was found by Bazhenova and al. (1988) that the deformation of a sample of polyurethane foam in a shock tube is one-dimensional and the distortion of the side of the foam by the glass windows was not noticed. Those findings formed a basis allowing the determination of the instantaneous density distribution in a layer of dynamically compressed polyurethane foam.

Fig.2 presents the optical device for density measurements. A source of light (1) supplies short (~ 10 μs) and very intensive light impulses. Light from the light source falls on the side of the foam with a system of optical fibres (3). They end in a circle on the glass window of the shock tube. In this manner the uniformity of illumination was achieved and there were no light reflection in the field of view of the camera (4).

Pressure measurements were taken with pressure transducers whose diameter were less than 3 mm. Measurements were made on the side-walls of the shock tube, the gauges were under the polyurethane foam.

3. Experimental results

3.1. Density measurements

The aim of this study was not only to find the structure and the dynamics of the propagation of compression waves in layer of polyurethane foams with open porosity under shock wave loading, but also to try to find the role of the gas flow inside pores on the compression wave structure. Bearing this in mind a series of experiments was done in which the side of the foam was covered with a thin (~ 20 μm) layer of mylar film which prevented gas leakage through the side from the compressed gas behind the reflected shock wave.

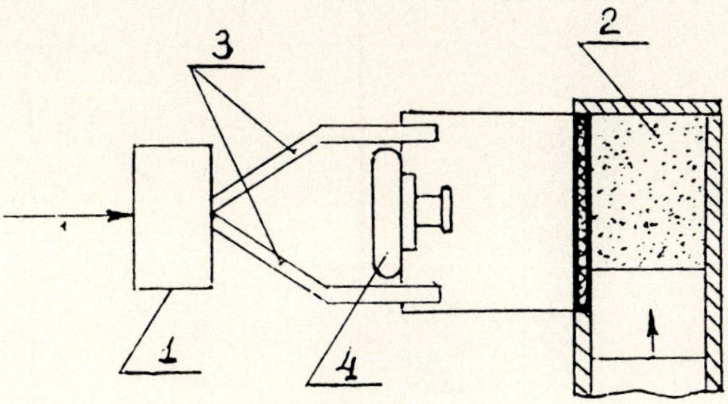

Fig. 2. Optical setup for density measurement: 1 light source; 2 polyurethane foam; 3 optical fibres; 4 camera

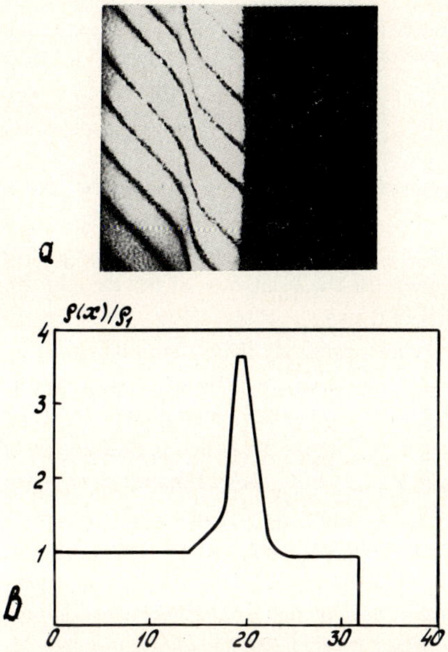

Fig. 3. Interaction of a shock wave ($M_s = 1.5$) with a foam (non-protected front surface): a) Deformation of line pattern; b) Density distribution

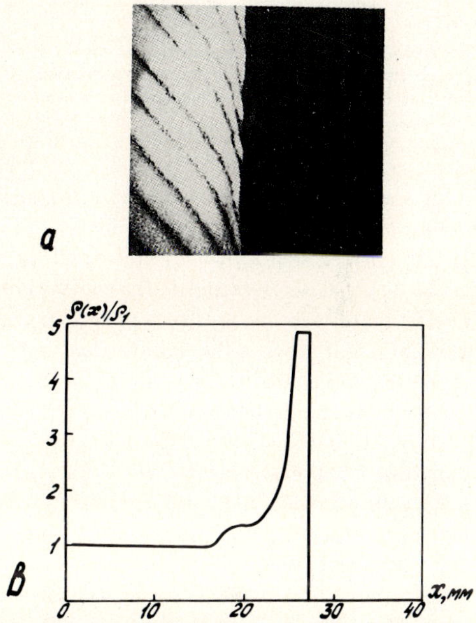

Fig. 4. Deformation of the system of lines (a) and density distribution (b) in the case of protected front face of polyurethane foam

A photo of the system of lines at the side of the foam deformed by the formation of a compression wave inside the unprotected sample of foam is given in Fig.3a. The Mach number of the shock wave in air was $M_1 = 1.5$, the delay between the beginning of the reflection of the shock wave from the side and the moment when the photo was taken was $\Delta t = 205$ μs.

In Fig.3b the instantaneous profile of density $\rho(X)/\rho_1$ is given. We see that in compressible polyurethane foam the compression wave is practically a shock wave with a relatively steep rise in density which is followed by a region of constant density and a rarefaction wave. The width of the leading side of the shock wave is approximately 2.0 mm, the width of the region of the shock wave with constant density is equal of 1.5 mm and the width of the whole wave is 8 mm.

Fig.4a presents a photo of the deformed set of lines in the experiment with the same sample of polyurethane foam as in Fig.3a, but with the side of the foam covered with a mylar sheet.

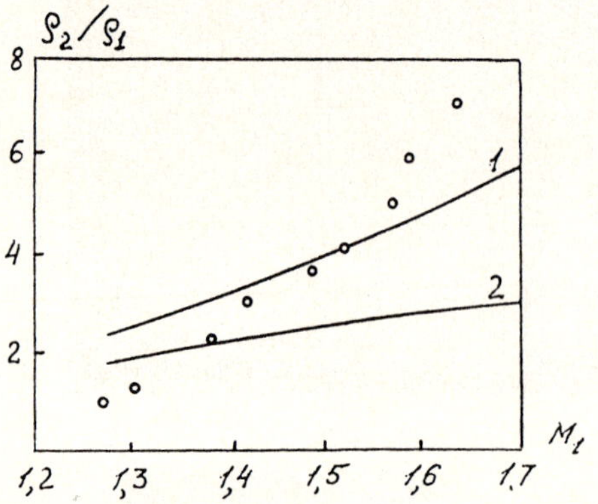

Fig. 5. Density ratio over compression wave in polyurethane foam: o measurements; 1 calculation (without heat exchange); 2 calculation (with heat exchange)

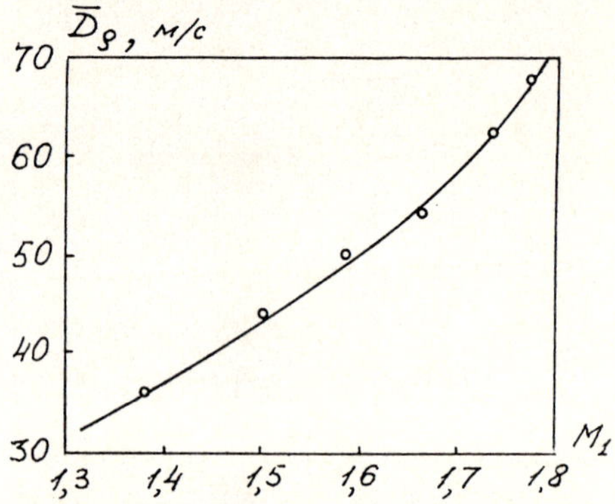

Fig. 6. Mean velocity in polyurethane foam: o measurements with curve fit

It was shown that in this case, the structure of the deformation wave in the polyurethane foam is different. Fig.4b gives the flow profile for this case. From a comparison of profiles in Fig.3b and Fig.4b it is seen that artificially cutting off the gas flow through the front face of the foam leads to the fact that the rear end of the shock wave coincides with the front face of the sample. No rarefaction wave exists, as is seen in Fig.4b. Comparison of density profiles in Fig.3b and Fig.4b shows also that gas flow through the front face influences comparatively little the maximum rate of compression behind the shock wave in the porous material. It is interesting to note that the density in the rarefaction wave, which we can see in Fig.3b, is practically constant and is about 0.9 times the initial density of the foam. Fig.5 presents results of measuring the maximum rate of compression in porous polyurethane foam as a function of the Mach number of the incident shock wave in air. Curves 1 and 2 in Fig.5 were calculated with the theory developed by Gvozdeva and Faresov (1986), where the model of shock wave refraction at the gas-polyurethane boundary was used. The polyurethane foam was tested as a two-phase media in equilibrium (1) and two phase media without heat exchange between phases (2). As can be seen from Fig.5, experimental points are in better agreement with model (1). Nevertheless, when we increase the Mach number M, the experimental values of density ratio show a steeper rise than the calculated curve. Probably, this is connected with the influence of gas flow through the skeleton of polyurethane foam which should be proportional to the pressure difference across the incident shock wave and so increase with the Mach number M_1. Gas filtration was not taken into account in the theory proposed by Gvozdeva and Faresov (1986).

Photos of the deformation of the system of inclined lines on the side wall of the foam taken at successive moments of time allowed information about the velocity of the compression wave, $\bar{D}_\rho$ in permeable polyurethane foam to be obtained. In Fig.6 results of the measurement of wave speed are given in the same sample of polyurethane foam ($d = 40$ mm, $\alpha = 0.98, \rho_r = 385$ kg/m^3 without mylar a covering layer, as function of the Mach number of the shock wave in air at initial atmospheric pressure. As seen from Fig.6 the values of the mean velocity of shock waves in polyurethane foam are relatively low. The solid curve in Fig.6 represents experimental results.

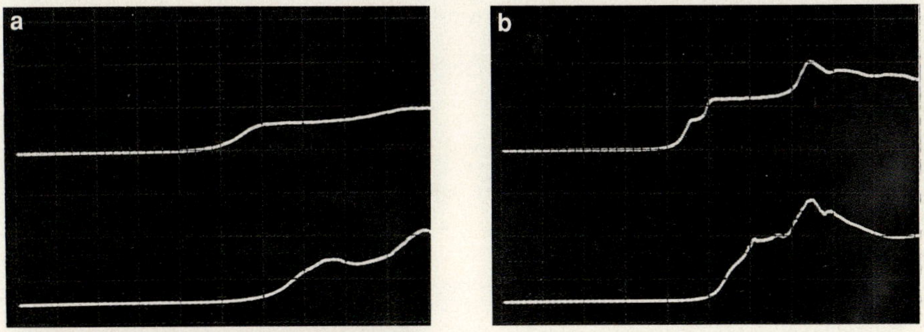

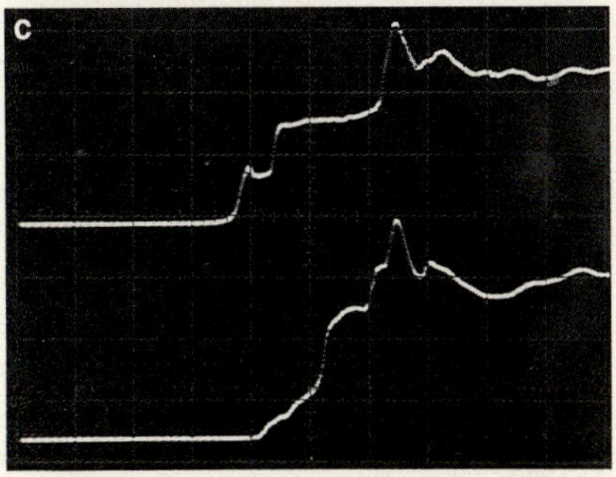

Fig. 7. Typical wall pressure records for polyurethane foam (unprotected front surface): a) $M_s = 1.38$; b) $M_s = 1.57$; c) $M_s = 1.73$

3.2. Pressure measurements

These measurements were taken with polyurethane foam of the same density and porosity but the length of sample was 90 mm.

This was done intentionally, because the aim of these experiments was to study the pressure change on the side wall of the shock tube under the sample of polyurethane foam. So the length of the sample should be long enough that after its compression the gauges remained under the polyurethane foam. The Mach number of the shock wave in air varied from $1.3 < M_1 < 1.9$ at initial atmospheric pressure. Pressure tranducers were set at 10 and 20 mm from the closed end of the shock tube. Fig.7 presents pressure records of these two gauges (lower trace and upper trace) at different Mach numbers of the loading shock waves. Fig.1a is for the Mach number $M_1 = 1.38$, Fig.7b for $M_1 = 1.57$ and Fig.7c for $M_1 = 1.73$. The apparent density was 38.5 kg/m^3. It is seen from Fig.7 that the main front of the shock wave (or compression wave) in the polyurethane foam is coincident with the second rise in the pressure records. The first pressure recorded seems to belong to the pressure wave in air in the narrow channels inside the porous material and the gap between the sample and side walls of shock tube. At low Mach numbers, the wave is a compression wave like a sound wave and the two rises look like one.

The third rise in Fig.7b and 7c is due to the reflected wave from the closed end of the shock tube. It is seen that increasing the Mach number of the incident shock wave in air leads to a steeper and faster shock wave in the polyurethane foam.

At the Mach number $M_1 = 1.73$ the length of the pressure wave front is about 10 ms and its velocity ~63 m/s which coincides perfectly with data taken from density measurements.

With the purpose of looking at the influence of the gas flow through the pores on the shock wave structure, experiments were taken with a sample made from 20 different pieces of polyurethane foam separated by thin mylar films. Experiments showed that the pressure in the reflected wave under the sample becomes higher. In the case when only one thin mylar film was attached to the front face of the sample, it turned out that there was practically no difference with pressure records taken without the film.

4. Conclusion

Density profiles in shock waves in a compressible porous material under loading by a shock wave generated in air have been obtained. The velocity of the shock wave in the porous media and rate of compression were estimated as function of the Mach number of the incident shock wave in air. Pressure records on the side of the material were taken. It is shown that the permeability of the skeleton of porous material has a great influence on shock wave structure.

References

Bazhenova TV, Gvozdeva LG, Faresov YuM et al. (1988). Unsteady interaction of shock and detonation waves in gases. Rus. ed. Korobeinikov VP, Eng. ed. Urtiew PA, Hemisphere Publishing Corporation, New-York, Washington, Philadelphia, London

Baer MR (1992). A numerical study of shock wave reflection on low density foam. Shock Waves 2: 121-124

Gelfand BE, Gubin SA, Kogarko SU et al. (1975). Study of special features of propagation and reflection of pressure waves in porous medium. Sov. Phys. Appl. Math. Tech. Phys. 6:74-77

Gelfand BE, Gubanov AV, Timofeev AI (1983) Interaction of shock waves in air with a porous screen. Izv. Akad Nauk SSSR, Mekh Zhidk. Gaza 4: 79-84

Gvozdeva LG, Faresov YuM (1984) Interaction of air shock waves with porous compressible material. Pisma v Journal Tekh. Fiz. 19,10: 1153-1156

Gvozdeva LG, Faresov YuM (1985) Calculation of parameters of steady shock waves in porous compressible media. Zh. Tekh. Fiz. 55, 4: 773-775

Gvozdeva LG, Faresov YuM (1985) Approximate calculation of the parameters of stationary shock waves in porous materials. Soviet Journal of Appl. Mech. and Techn. Phys. 1, 1: 120-121

Gvozdeva LG, Faresov YuM, Fokeev VP (1985) Interaction between air shock waves and porous compressible materials. Soviet Physics Applied Mathematics and Technical Physics 3: 111-115

Gvozdeva LG, Lyakhov VN, Rayevsky DK, Faresov YuM (1987) Numerical study of propagation of shock wave in gases and porous media. Fiz. Gorenia i Vzryva 4: 125-129

Henderson LF, Virgora RD, Di J, Gvozdeva LG (1989) Refraction of a normal shock wave from nitrogen into polyurethane foam. In: Kim YW (ed) Proc.17th Intl. Symp. Shock Waves and Shock Tubes. American Institute of Physics Conferences Proceedings 208, pp 814-818

Monti R (1970) Normal shock wave reflection on deformable wall. Mecanica 4, 5: 285-296

Rayevsky DK, Gvozdeva LG, Faresov YuM, Bressard J, Bailly P (1989) Reflection of shock and explosion waves from surfaces covered with layers of polyurethane foam. 11th Intl. Colloq. on Dynamics of Exp. and Reactive Systems

Seitz MW, Skews BW (1992) Three dimensional effects in the study of shock wave loading of porous compressible foams. In: Takayama K (ed) Proc. 18th Intl. Symp. on Shock Waves, Sendai, Springer Verlag, Heidelberg, pp 517-520

Skews BW, Altkink MD, Seitz MW (1992) Gas dynamics and physical behaviour of compressible porous foams struck by a weak shock wave. Shock waves. In: Takayama K (ed) Proc. 18th Intl. Symp. on Shock Waves, Sendai, Springer Verlag, Heidelberg, pp 511-516

Skews BW (1991). The reflected pressure field in the interaction of weak shock wave with a compressible foam. Shock Waves 1: 205-211

Transmission of a Blast Wave Through a Deformable Layer

J.J. Lee, D.L. Frost and J.H.S. Lee
Department of Mechanical Engineering, McGill University, Montreal, Quebec, Canada

Abstract. Experiments have been carried out to investigate the transmission of a blast wave through a deformable layer consisting of a variety of different types of foam. The shape of the stress-strain curve for a foam has a strong influence on the pressure wave transmitted through the foam. At low blast wave amplitudes, the transmitted pressure is attenuated because the strains induced in the foam fall within a relatively linear elastic region of the stress-strain curve. For high blast wave amplitudes, the transmitted pressure is strongly amplified because the foam is compressed into the densification region where the foam cells collapse almost completely and the foam material itself is compressed. A nonlinear spring-mass-damper model which incorporates the stress-strain curve as the spring force relation qualitatively reproduces the blast loading behavior observed experimentally for different blast wave amplitudes.

Key words: Shock-foam interaction, Shock attenuation, Deformable layer

1. Introduction

There are important practical applications where it is desirable to attenuate the peak pressure and impulse of a reflected blast wave on a surface. For example, the design of bomb suits requires a clothing material that will mitigate the blast effects as well as injury due to fragments and thermal radiation. Complicated composite materials comprised of layers of different materials are used for protection against the different threats. Perhaps the most difficult one to protect against is the blast effect. The aim is to limit the transmitted peak overpressure and impulse to the body to below the lung damage threshold. It is well established that foam is extremely effective in absorbing pressure wave energy, and providing significant impulse and peak pressure attenuation. Hence foam is widely used in crash padding and cushioning. The design guide for the use of foam for packaging and crash padding is well established. However the use of foam for the protection against air blasts has not been studied. A number of investigations have been carried out in shock tubes where shock reflection and transmission into a foam layer is studied (Gel'fand et al. 1975; Gvozdeva et al. 1986; Monti 1970; Henderson et al. 1989; Skews 1991). A number of attempts to model this phenomenon have also been made (e.g., Baer 1992). Of particular interest is the finding that the transmitted peak pressure to the rigid wall behind the foam layer can be significantly higher than the normal reflection pressure. This effect was credited to the momentum imparted to the solid foam material by the shock, requiring a larger impulse from the wall to bring the foam material to rest. This amplification of the transmitted pressure by the foam layer is contrary to the damping properties of foam for impact loading and suggests that the foam sample has been loaded beyond its plateau stress value for energy absorption. Furthermore, the previous studies were conducted in shock tubes where the loading function is a step input. In a blast wave, the loading function is typically a triangular pressure pulse of finite duration. It is well known in modelling a viscoplastic material with a spring-mass-damper system, that the response is characteristic of the input function. Thus the finite triangular pressure pulse of a blast wave may induce a different response than the step input in a shock tube. The objective of this study is to gain some insight into the effect of blast and foam parameters on the transmitted peak pressure and impulse through a foam layer. The approach is to consider the foam as a material characterized by its standard stress - strain curve and its response to a specific loading

Shock Waves @ Marseille III
Editors: R. Brun, L. Z. Dumitrescu

function is modeled by a mechanical spring-mass system. The detailed wave processes have not been considered in this formulation. Experiments on thin foam samples are carried out by actual air blasts using small explosive charges (up to 170 g of C-4) in a blast chamber.

2. Experimental

Blast wave tests on the foam samples were conducted using a large aluminum plate (61 cm in diameter and 3.2 cm thick). The explosive charges consisted of from 10 g to 170 g of C-4 explosive suspended at different distances above the plate to obtain the desired blast wave impulses and amplitudes. The duration of the positive phase of the blast wave pressure was maintained at approximately 600 μs for the different blast wave strengths. The foam samples were circular discs (48 mm in diameter) placed at the center of the plate, and a pressure transducer was flush-mounted in the plate below the sample to record the transmitted pressure pulse. Another transducer was mounted in a vertical assembly to measure the incident side-on pressure.

Three types of foams were tested in this work: a low density open cell foam (density = 30.2 kg/m^3), a closed cell neoprene foam (density = 157.2 kg/m^3), and a closed cell cross-linked polyethylene foam (density = 70.6 kg/m^3). The thickness of the samples varied from 6.2 mm for the neoprene foam to 25 mm for the low-density foam.

3 Results and discussion

3.1. Stress-strain curves of the foams

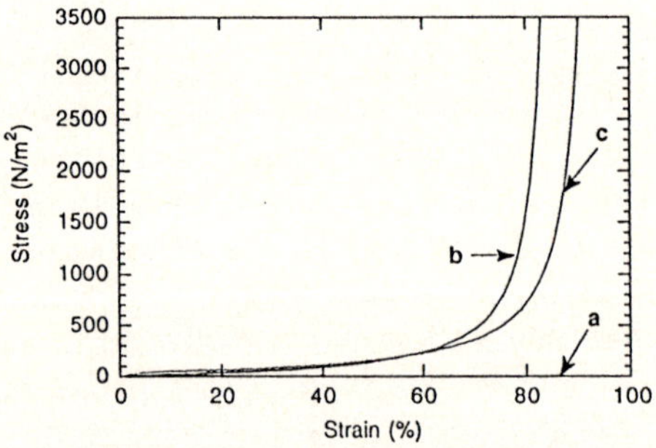

Fig. 1. Stress-strain curves for (a) low density open cell (thickness = 25 mm), (b) closed cell neoprene (thickness = 6.3 mm), (c) closed cell neoprene (thickness = 12.32 mm), and (d) closed cell polyethylene (thickness = 13.78 mm)

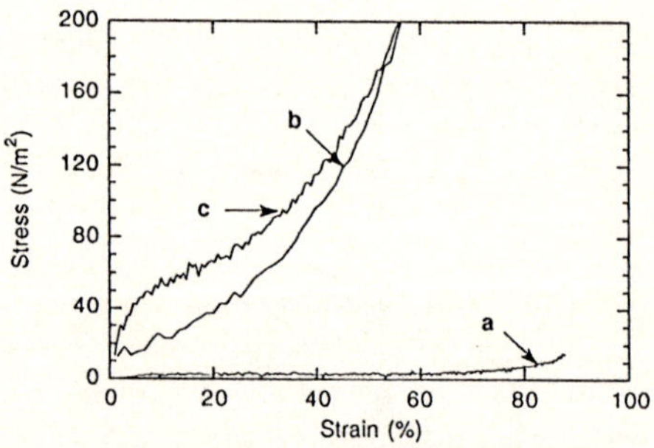

Fig. 2. Expanded scale view showing the stress-strain curve (a) of the low density open-cell foam (thickness = 25 mm)

Stress-strain curves are useful in characterizing the behavior of materials under pressure loading. By measuring the stress-strain curves of the various foams, the different regions of elastic behavior can be identified and related to the blast loading effects. Two distinct regions of behavior can be distinguished in the stress-strain curves of the foams used in this study as shown in Figs.1,2. For small strains, there is a long region of gradual stress increase, and for large strains (above 80%), there is a region where the stress increases sharply with strain. In the slowly rising region, compression results in elastic bending and buckling of the foam cells and elastic compression of the air in the cells. The rapidly rising region corresponds to densification of the foam as the foam cells collapse almost completely and the foam material itself is compressed. Blast waves of different amplitudes impacting on the foams produce different degrees of loading, resulting in different responses in the foam depending on the region of behavior on the stress-strain curve. The pressure-time profile of the pressure pulse transmitted through the foam is therefore highly dependent on the blast wave amplitude. It should be noted that due to the small thickness of the foam samples tested in this work, the densification region is reached with a relatively low blast wave amplitude. For a larger foam thickness, the finite duration impulse of the blast wave only compresses the foam by a certain distance equal to a small fraction of its original thickness. However, for thin samples, the duration of the blast wave impulse is long enough to compress the foam over a substantial distance relative to the original thickness.

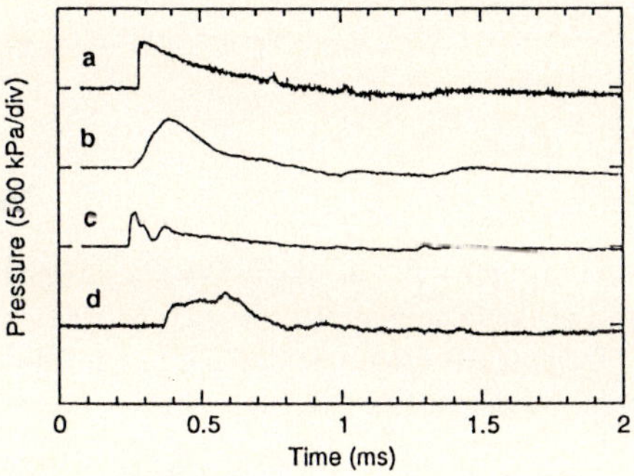

Fig. 3. Pressure-time profiles of (a) the reflected blast wave pressure with no sample, the transmitted wave with (b) low density open cell foam, (c) closed cell neoprene foam, and (d) closed cell polyethylene foam

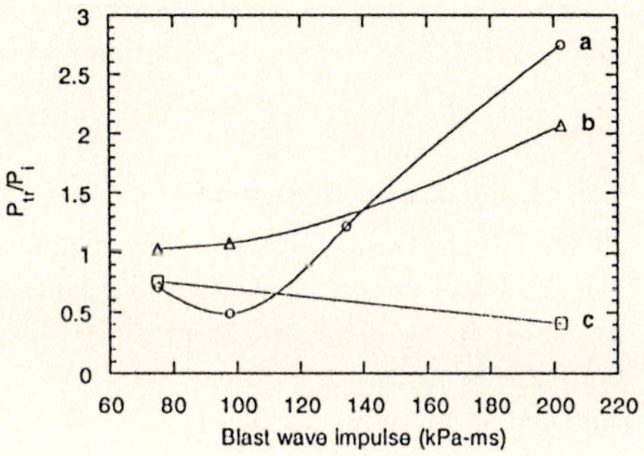

Fig. 4. Ratio of the peak transmitted pressure to the incident blast wave pressure (P_{tr}/P_i) at different incident blast wave impulses for (a) closed cell neoprene foam, (b) low density open cell foam, and (c) closed cell polyethylene foam

3.2. Transmission of blast wave pressure through the foams

The pressure-time profiles of the transmitted waves through the three tested foams each underwent a change in shape as shown in Fig.3. For the low density open cell foam (trace b), the rate of change of the blast wave pressure was slowed, resulting in a smoother profile. This behavior has been described by Henderson et al. (1989) as a spreading of the incident shock into compression waves. For the closed cell neoprene foam (trace c), the transmitted pulse resembles the incident blast wave although an oscillatory behavior is present, possibly due to the higher elasticity of the foam. As the "stiffness" of the elasticity increases (i.e., the slope of the stress strain curve increases), the oscillation frequency of the material increases. As the strain of the foam increases, the region of stiffer elasticity is approached (Fig.1) and the period of the oscillations becomes comparable to the duration of the blast wave impulse and consequently begins to appear in the transmitted wave. For the closed cell polyethylene foam (trace d), the transmitted pulse appears approximately square, suggesting the influence of a pressure limiting effect. This pressure cut-off can be attributed to the relatively flat portion of the stress-strain curve prior to the densification region. If the blast wave impulse produces a strain below $\sim$75%, then the densification region is not reached and the pressure of the transmitted wave is limited by the stress of the foam which is low in the pre-densification region. The peak pressures of the transmitted waves are shown in Fig.4. Since the pressure of the transmitted wave depends on the strain induced in the foam by the incident blast wave impulse, large transmitted pressures can be expected if the foam is compressed in the densification region on the stress-strain curve. For the closed cell neoprene foam (curve a), a peak pressure *attenuation* is observed at low blast wave impulses. As the impulse increases, causing the strain to approach the densification region, the transmitted pressure begins to be amplified. For the low density open cell foam (curve b), the densification region is reached with a low blast wave impulse due to the high compressibility of the foam. As a result, the incident blast wave pressure is amplified over the range of blast wave impulses used and the amplification increases with increasing impulse. For the closed cell polyethylene foam, attenuation is always obtained, indicating that the foam is not compressed into the densification region. This is consistent with the stress-strain curves (Fig.1) since the densification region begins at a strain of ≈85% compared to ≈75% for the neoprene foam, indicating that larger blast wave impulses are needed to reach the region where peak pressure amplification would occur. The transmitted pulse shape as well as the pressure depends strongly on the characteristics of the stress-strain curve as the blast loading is increased. For example, for the neoprene foam, a change in the pressure profile of the transmitted wave is observed for high blast wave amplitudes, as shown in Fig.5 for a blast wave impulse of about 200 kPa-ms.

3.3. Spring mass damper model

When a blast wave impacts a foam sample placed on a rigid wall, an impulse is transmitted onto the wall due to the acceleration and compression of the foam material towards the wall by the blast wave pressure. The pressure-time profile of this transmitted impulse can be modeled by a purely mechanical system. The system used to model the various foam samples is a simple damped spring mass oscillator.

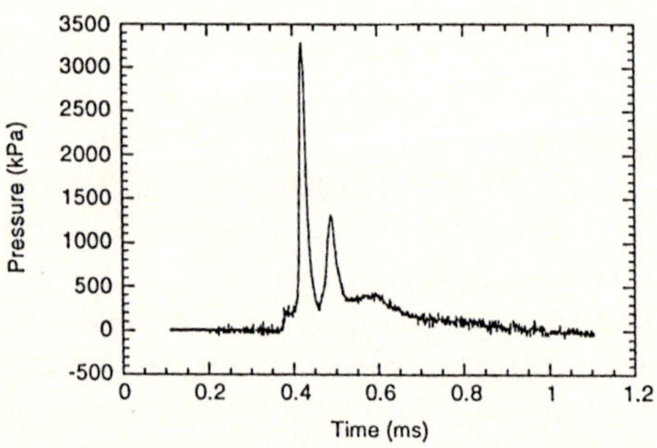

Fig. 5. Pressure-time profile of transmitted wave for a large blast wave amplitude (≈1200 kPa peak pressure with a 170 g charge at a 1 m distance) on closed cell neoprene foam

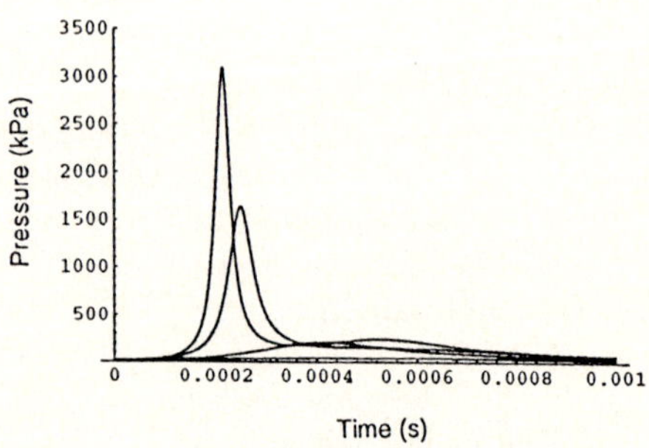

Fig. 6. Transmitted pressure profiles calculated with the spring-mass-damper model for four different input blast waves with peak pressures of 300, 559, 1007, and 1200 kPa

The three principal parameters of this system are mass m, damping $F_d(x)$, and spring force $F_s(x)$, which can be measured experimentally for the foam material. The damping relation $F_d(x)$ is governed mainly by plastic deformation of the foam material and viscous dissipation in the case of open-cell foam (Gibson and Ashby 1988). In the present study, closed cell foams with comparatively low damping were used. The damping relation was assumed to be constant and its value was estimated by measuring the decay of the oscillatory behavior of the transmitted pressure through a foam sample when a relatively weak shock ($M \approx 1$) is applied to it.

The spring force relation $F_s(x)$ was determined by measuring the stress-strain curves of the foams. A very good curve fit for the stress-strain curve was obtained using the relation:

$$F_s(x) = a_0(a_1 - x)^{-a_2} \tag{1}$$

The equation of motion for the damped-spring-mass system is:

$$mx''(t) + cx'(t) + F_s(t) = E(t) \tag{2}$$

where m is the mass, c is the damping constant and $E(t)$ is an excitation function which corresponds to the pressure wave incident upon the foam sample. In the case of a blast wave, it has the form of a decaying exponential function. Due to the highly non-linear spring force relation, the equation of motion was integrated numerically. The pressure against the rigid wall was found by substituting the time-dependent displacement $x(t)$ back into the spring force relation.

Fig.6 shows the transmitted pressure profiles calculated with the model for four different input blast waves with peak pressures of 300, 559, 1007, and 1200 kPa (with the same pressure profile as that obtained experimentally). Qualitatively, the transmitted blast pressure for the theoretical model agrees with the experimental results, in that both show a peak pressure attenuation at lower blast wave amplitudes, and then a sudden rise, leading to a strong peak pressure amplification, for stronger blast amplitudes. The general shape of the calculated pressure profiles also reproduces the general features of the actual experimental profiles, particularly for strong blast waves.

4. Conclusions

The present experiments demonstrate that the utility of a deformable material such as foam for attenuating a blast wave is strongly dependent on the amplitude and duration of the blast wave and the details of the compressibility of the foam (as described by the stress-strain curve for the foam). For weak blast waves on foams with a relatively long elastic region in the stress-strain curve, the peak pressure is attenuated. If the foam cells are completely collapsed so that the foam material itself is compressed, the stress-strain curve rises sharply and the incident blast wave is actually amplified. The shape of the transmitted pressure pulse is also governed by the elastic and dissipative properties of the foam. A nonlinear spring-mass-damper model, while it doesn't consider wave effects, is still useful for reproducing the general features observed in the experiments.

Acknowledgements

The authors gratefully acknowledge CANMET for the use of their blast testing facilities and technical assistance as well as A. El Shareif and W. Chatila for their help with the experimental work.

References

Baer MR (1992) A numerical study of shock wave reflections on low density foam. Shock Waves 2: 121-124

Gel'fand B, Gubin, S, Kogarko SM, Popov OE (1975) Investigation of propagation and reflection of pressure waves in porous media. Zh. Prikl. Mekh. Tekh. Fiz. 16, 6: 74-77

Gibson LJ, Ashby FA (1988) Cellular Solids, Structure and Properties. Pergamon Press, Oxford, England

Gvozdeva LG, Faresov YM, Brossard J, Charpentier N (1986) Normal shock reflections on porous compressible material. Progress in Astronautics and Aeronautics 106:155-165

Henderson LF, Virgona RJ, Di J, Gvozdeva LG (1989) Refraction of a normal shock wave from nitrogen into poly-urethane foam. In: Kim YW (ed) Proc. 17th Intl. Symp. on Shock Waves and Shock Tubes. American Institute of Physics Conference Proceedings 208, pp 814-818

Monti R (1970) Normal shock wave reflections on deformable solid walls. Meccanica: 285-296

Skews BW (1991) The reflected pressure field in the interaction of weak shock waves with a compressible foam. Shock Waves 1: 205-211.

Peculiarities of Weak Shock Wave Propagation in a Gas-Liquid Foam

I.I. Goldfarb*, **V.P. Melnikov***, **Z.M. Orenbakh**†, **I.R. Shreiber**† **and F.I. Vafina***
*Institute of Earth Cryosphere, 625000 Tyumen, p.o.b.1230, Russia
†Ben-Gurion University of the Negev, 84105 Beer-Sheva, p.o.b.653, Israel

Abstract. A theoretical model of weak shock wave propagation in gas-liquid foams of different structure, and experimental investigation results are presented in the paper. It is concluded that the general influence on the evolution of the initial perturbation has been exerted by the liquid flow through the system of Plateau-Gibbs channels and the heat transfer process between carrying and dispersion phases.

Key words: Gas-liquid foam, Shock wave

1. Introduction

The object of investigations in many papers devoted to studying the peculiarities of the propagation of weak shock waves in gas-liquid systems is that of bubbly media, i.e., two-phase structures, the volumetric liquid content of which is in the range of 95% and above. The other limiting case, the gas-liquid in which this range covers the volumetric void fraction, is studied insufficiently in the experiments. A great number of publications devoted to foams contains different experimental data on the propagation of strong shock waves in foams ($\Delta P/P \gg 1$). In this case, when such a wave passes through the the foam, the latter's internal structure is destroyed and the parameters of the two-phase system change rather significantly. The number of experimental studies devoted to investigations in such a range of wave amplitudes is rather limited.

According to the definition given by Manegold (1953), foam is a gas-liquid medium, consisting of a large number of gas bubbles separated by thin liquid films. To characterize a foam the parameter K - expansion factor - is usually applied because of the great change in foam properties during small changes in volumetric gas content. This expansion factor is the ratio of foam volume to the liquid volume contained in it and is equal to the inverse moisture content ($K = 1/\alpha_L$).

2. Theory

The problem statement is reduced to the formulation of interphase heat transfer conditions and the derivation of the equations of liquid motion through the channels. A number of general physical assumptions are made to specify each problem separately. The consideration is restricted only to such weak shock waves, and consequently to such foam motions, that any destruction of the internal foam structure does not occur. The effect of gravity forces is neglected. The isobaric condition is valid for the gas in a bubble. The liquid is considered compressible and for taking into account liquid compressibility various hypotheses are used and analysed (suggested by Gilmore-Kirkwood-Bethe, Herring-Trilling and Shagapov-Nigmatulin).

The model suggested in the paper considers diverse physical mechanisms bearing an influence on the signal evolution in the foam. Some of them, which were investigated earlier (interphase heat transfer processes, surface tension, effects of added liquid mass and viscous deformable boundary), are accounted for in a standard way.

The statement of the heat transfer problem is considered within the framework of the general cellular approximation of a two-phase medium (Nigmatulin 1990), which was specified as applied to foam structure by Shagapov (1985). The foam is formulated as a set of ideal spherical gas

Shock Waves @ Marseille III
Editors: R. Brun, L. Z. Dumitrescu

bubbles, which are surounded by a thin liquid layer. Under the standard boundary conditions for the cell model the system of equations may be written in the form

$$\left.\begin{aligned}
&(0<r<r_b): && \rho_G\, c_{pG}\frac{\partial T_G}{\partial t}=\frac{1}{r^2}\frac{\partial}{\partial r}\left(\lambda_G r^2\frac{\partial T_G}{\partial r}\right)+\frac{\partial P_G}{\partial t},\\
&(r_b<r<r_b+r_1): && c_L\rho_L\frac{\partial T_L}{\partial t}=\lambda_L\frac{\partial^2 T_L}{\partial r^2},\\
&r=r_b: && T_L=T_G,\qquad \lambda_L\frac{\partial T_L}{\partial r}=\lambda_G\frac{\partial T_G}{\partial r};\qquad P_G=\rho_G B T_G,\\
&r=0: && \frac{\partial T_G}{\partial r}=0,\\
&r=r_b+r_1: && \frac{\partial T_L}{\partial r}=0,\\
& && \frac{\partial P_G}{\partial t}=-\frac{3\gamma P_{G0}}{r_{b0}}\frac{\partial r_b}{\partial t}+3(\gamma-1)\frac{\lambda_G}{r_{b0}}\left[\frac{\partial T_G}{\partial r}\right]_{r=r_b}
\end{aligned}\right\}\qquad (1)$$

Here T is the temperature; P the pressure; ρ the density; λ the thermal conductivity coefficient; c the thermal capacity; γ the ratio of specific heat capacities; B the universal gas constant. Subscripts are related to: L liquid, G gas; 0 unperturbed state; b bubble, l liquid surrounding a gas bubble, f foam; subscript p means under constant pressure; r is a microcoordinate for the distance from the bubble center.

The additional (in comparison with small gas content suspensions) hydrodynamical mechanism influencing initial profile evolution is under consideration also (Goldfarb, Shreiber and Vafina 1992). This effect is caused by the resistance, which takes place during the liquid flow along the Plateau channels (elements of the internal foam structure). One supposes, that the intensity of this friction is defined by two factors: foam cell surface dynamics and liquid motion degree of freedom. The physical equivalent of the first is bubble radius dependence upon time, of the second - foam hydroconductivity. The mathematical expression of the last property depends strongly on the particular foam structure type (cellular, spherical or polyhedral) and is taken in the form suggested by Kann (1989). On the basis of these suggestions one may obtain an analog of the Rayleigh-Lamb-Plesset equation in the framework of the suggested model, which may be written in the form (K_f is a foam hydroconductivity function)

$$\rho_L r_{b0}\frac{\partial^2 r_b}{\partial t^2}+\frac{4\mu_L}{r_{b0}}\frac{\partial r_b}{\partial t}+\frac{2\sigma}{r_{b0}}+\frac{\mu_L r_{b0}}{K_f}\frac{\partial r_b}{\partial t}+\frac{\rho_L\alpha_L r_{b0}}{K_f}\sqrt{\frac{S\nu_L}{\pi}}\int_{-\infty}^{t}\frac{\partial^2 r_b}{\partial\tau^2}\frac{d\tau}{\sqrt{t-\tau}}=P_G-P_L \qquad (2)$$

Here S is the cross section of the Plateau-Gibbs channels; σ the surface tension coefficient; α the volume phase content; ν and μ the kinematic and dynamic viscosities respectively.

Eq.2 differs from the standard Rayleigh-Lamb-Plesset equation (Nigmatulin 1990) by two new terms containing the foam hydroconductivity function. The first of these terms (the first derivative on time) corresponds to the Stokes component in the friction force, the second one (containing the Duhamel integral) to the Basset component. The relation between these components is defined by the formula presented, in particular, by Landau (1988).

Analysis of the system of equations describing the suggested model allows to make the following conclusions (Goldfarb, Shreiber and Vafina 1992; Vafina 1993): Dissipation availability leads to the disappearance of the "opacity window" to the right of the Minnaert frequency, as in usual gas-liquid mixtures with small gas content. This "window seal" process in foam occurs rather intensively, due to the presence of friction, caused by liquid flow. The value of Minnaert's analog frequency is much higher than in an usual gas-liquid mixture, because of the low moisture content level in foam. The point of resonance (frequency, which is obtained through equating to nought the real part of the dispersion correlation denominator) in the present model is caused not by added liquid effect, but by the high-frequency friction and heat transfer. In the limit $\omega \to \infty$ the phase velocity is equal the to sound velocity in liquid divided by the square root of the moisture content. The frequency dependence of the attenuation decrement is defined by the friction and heat transfer, but in the limit $\omega \to \infty$ its behavior depends on the basic hypothesis: in the framework of Gilmore-Kirkwood-Bethe and Shagapov-Nigmatulin models the limit values of the imaginary part of the wave are identical constants, depending on foam parameters only; in the framework of the Herring-Trilling models $\mathcal{I}m[k(\omega)]$ decreases monotonously with frequency increase.

3. Experimental data

The experimental facility used by Orenbakh and Shushkov (1991) is a vertical shock tube (Orenbakh and Shushkov 1991; Orenbakh, Shreiber and Shushkov 1991). The pressure disturbance in the experiments amounted to 3-15 kPa, i.e. $\Delta P/P \ll 1$. The upper level of the main harmonics was about 600-800 Hz. The average radius of the foam cells was about 0.1 mm. The expansion ratio of the foams was in the range 35-180. 1% and 2% sulphanol solutions were used for foam generation. The test section is made of organic glass and equipped with 6 piezoelectric sensors placed through 0.05 m. The distance from the diaphragm to the first sensor and from the last sensor to the bottom of the test section is larger than 1 m, which helps to exclude reflected waves from consideration. Experiments with foams containing different gas components (air, helium, and carbonic acid, which are distinguished by their thermophysical properties) were conducted to ascertain the degree of heat transfer effect on weak shock wave evolution in foam.

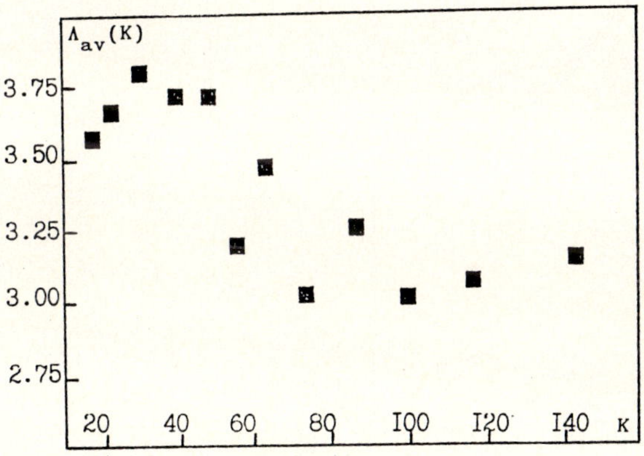

Fig. 1. Average attenuation coefficient vs. expansion factor

Experiments showed that measured velocity values lie between the curves corresponding to adiabatic and isothermal sound velocities in two-phase media. It testifies to the polytrope character of the heat transfer in foam bubbles. No nonlinear effects of the type of steepening the leading front of the wave were registered in the experiments and no dispersion of sound velocity in the form of any oscillations was evident. The average attenuation factor (averaged harmonics of all frequencies contained in the signal) has a weak tendency to decrease with the expansion factor growth (Fig.1). Comparison of the experimental data obtained on the foams with gases of different thermophysical properties allows one to state that dissipation of pressure perturbation depends to a small extent only on the heat transfer process between phases. The developed

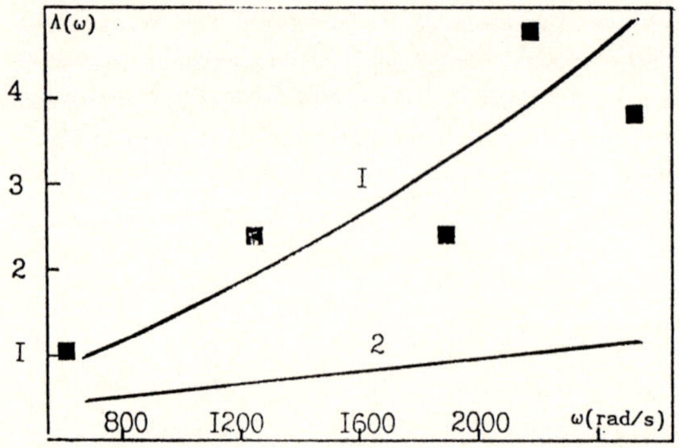

Fig. 2. Decrement of attenuation vs. frequency. 1 - calculation in accordance with the model suggested by the authors, 2 - calculation without taking into account liquid flow through the channnels (Shagapov, 1985), ■ - experimental points. Air-water foam parameters: 1% sulphanol solution, $K = 143$, $r_b = 0.075$ mm

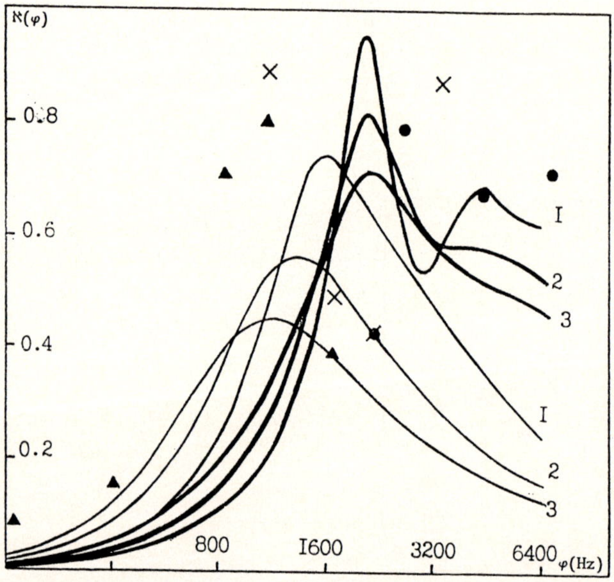

Fig. 3. Experimental and theoretical absorption coefficient vs. frequency for the conditions of Run#119 conducted by Manson, Lieberman and Burge (1971). Thin lines - calculations taking into account both Bass and Stokes components of friction; thick lines - calculations taking into account Stokes component of friction only; ▲, •, x - experimental points. Numerals designate the value of the minimal expansion factor: 1 - Kmin = 4; 2 - Kmin = 8; 3 - Kmin = 12

software permits to determine the phase velocity and attenuation decrement dependency on frequency. Experimental results allow to reach the conclusion that, in the region of foam, thermal dissipation mechanism is not a determining physical effect influencing signal evolution in foam, and one should take into account the friction caused by the liquid flow through the elements of the internal foam structure (Fig.2).

A weak shock wave propagation may be described in the acoustical approximation and the experimental data obtained in the acoustical studies be used for the comparison with the theory. Both the standing wave and reverberation room methods were used in one of the most detailed fundamental investigations of the acoustical properties of gas-liquid foams (Manson, Lieberman and Burge 1971). The value of the absorbtion coefficient of the foam layer was the main characteristic which was measured in these experiments. In this work the foam expansion factor was changed in the range 50-150, the sound signal frequency did not exceed 6400 Hz, the length of foam samples was varied in the range 0.5-5 inches, the foams were rather polydispersive and contained up to 15 fractions with different bubble sizes (depending on foaming agent, gas in the bubbles, the way of foam production, etc.) in the interval 0.04-2 mm. Comparison between the experimental data obtained in Run #119 and the model suggested by the authors is presented in Fig.3. In this Run the foam contained 8 fractions (from 0.08 mm to 0.51 mm in radius), the expansion factor was 85 (•) and 72 (▲; × - aged foam), the 10% protein foaming agent was for air-water foam production. For a detailed description the hydroconductivity of polydispersive foam the additional empirical parameter - minimal expansion factor (Kann 1989) - was used. Its value may be determined by experiment only, so in numerical calculations the authors were compelled to use several different values of this parameter.

Analysis of the experimental results allows to conclude that the relation between two components of the friction changes from the ideal model used in this work and the boundary frequency is moved to the high-frequency region. A similar conclusion can be drawn on the basis of experimental data obtained by Tolstov (1992).

4. Conclusion

Both theoretical and experimental investigations have shown that the gas-liquid foam is a medium with strong dissipation. The basic reason for shock wave attenuation in foam is the process of interphase heat transfer and the presence of liquid motion through the system of interrelated Plateau-Gibbs channels. When the external pressure perturbation is superimposed in the Plateau-Gibbs channels, secondary flows appear, and the energy of the wave is spent on the formation of these flows.

Acknowledgements

The authors express their sincere gratitude to the International Science (J. Soros) Foundation for the possibility given to present the results obtained at the ISSW-19. This work was supported also, in part, by a Soros Humanitarian Foundations Grant awarded by the American Physical Society.

References

Goldfarb II, Shreiber IR, Vafina FI (1992) Heat transfer effect on sound propagation in foam. J. of Acoust. Soc. of Amer. 92:2756-2769

Kann KB (1989) Capillar Hydrodynamics of Foams. Nauka, Siberian Branch, Novosibirsk

Landau LD, Lifshits EM (1988) Hydrodynamics. Nauka, Moscow

Manegold E (1953) Schaum. Center, Heidelberg

Manson L, Lieberman S, Burge HL (1971) A study of the use of liquid based foams for jet noise reduction. Report NASA CR-1695

Nigmatulin RI (1990) Dynamics of Multiphase Media. Hemisphere, New-York

Orenbakh ZM, Shreiber IR, Shushkov GA (1991) Pressure waves propagation in gas-liquid foam. In: Leroy O, Breazeale MA (eds) Physical Acoustics: Fundamentals and Applications. Plenum Press, New-York, pp 563-567

Orenbakh ZM, Shushkov GA (1991) Experimental determination of velocity and attenuation factor of acoustic perturbation in a gas-liquid foam. Acoustical Journal (USSR) 37: 403-405

Shagapov VS (1985) The influence of heat transfer processes between phases on small disturbances propagation in foam. Thermophysics of High Temperature (USSR). 23:126-132

Tolstov GS (1993) Acoustical waves propagation in foams. Acoustical Journal (USSR) 38: 1091-1099

Vafina FI (1993) Heat transfer role in wave dynamics of gas-liquid foams. Tyumen State University (Ph.D. thesis), Tyumen

Pressure Waves in Gas-Liquid Foams

T. Elperin, O. Igra and Z. Orenbakh

Pearlstone Center for Aeronautical Engineering Studies, Department of Mechanical Engineering, Ben-Gurion University of the Negev, Beer-Sheva, Israel

Abstract. The homogeneous mixture model of a two-phase media (Miksis and Ting (1991)) is modified to describe pressure waves propagation in foams. It is shown that the principal physical phenomena which determine shock wave propagation and their structure in foams are the fluid creeping flow through the Plateau-Gibbs channels and heat transfer between gas bubbles and liquid in foams.

Key words: Pressure waves, Liquid foams

1. Introduction

Foams differ from bubbly liquids mainly by their close to unity gas content and are composed of large number of gas bubbles separated by thin liquid films. Foams with liquid content which does not exceed 0.1, are of particular interest. In such foams the thickness of the liquid film is significantly smaller than the radius of the gas bubbles. Stability of such a delicate structure is achieved due to the presence of surfactants in a foam-forming liquid. Absence of this small amount of surfactants (about 1%) will lead to rapid bubble coagulation. Because they are in very small amount, stabilizing surfactants have practically no effect on the properties of the foam-forming liquid.

The purpose of the present investigation is to present results of an analysis of a complete system of equations describing propagation of pressure waves in gas-liquid foams.

2. Theoretical background

A foam is modeled by a cellular structure; each cell consists of a spherical gas bubble of radius a, which is surrounded by a liquid layer with thickness b. In such a case the volumetric liquid content α_L is: $\alpha_L = 1 - (a/(a+b))^3$. Foams with liquid content which does not exceed 0.1 are of particular interest. In such foams $b \ll a$ and the equation of thermal conductivity for the liquid can be written using a plane boundary approximation.

Assuming a homogeneous pressure distribution in a single gas bubble the equation of energy conservation for the gaseous phase, which takes into account compressibility effects, reads (Nigmatulin 1990):

$$\rho_G c_{pG} \left(\frac{\partial T_G}{\partial t} + u_G \frac{\partial T_G}{\partial r} \right) = \frac{1}{r^2} \frac{\partial}{\partial r} \left(\lambda_G r^2 \frac{\partial T_G}{\partial r} \right) + \frac{\partial P_G}{\partial t} \qquad \text{for } 0 < r < a \qquad (1)$$

The liquid phase is considered to be incompressible and the equation of energy conservation for the liquid phase can be formulated as follows (Nigmatulin 1990):

$$\rho_L c_{pL} \left(\frac{\partial T_L}{\partial t} + u_L \frac{\partial T_L}{\partial r} \right) = \frac{1}{r^2} \frac{\partial}{\partial r} \left(\lambda_L r^2 \frac{\partial T_L}{\partial r} \right) \qquad \text{for } a < r < a+b \qquad (2)$$

In the above equations T is the temperature, c_p is the specific heat capacity at constant pressure, ρ is the density, u is the velocity, λ is the thermal conductivity coefficient; subscript L denotes liquid and subscript G denotes gaseous phase, r is a distance measured from the bubble center.

Shock Waves @ Marseille III
Editors: R. Brun, L. Z. Dumitrescu

The boundary conditions at the gas-liquid interface are:

$$\lambda_L \frac{\partial T_L}{\partial r} = \lambda_G \frac{\partial T_G}{\partial r}, \qquad T_G = T_L \qquad \text{at } r = a \tag{3}$$

Symmetry conditions at the center of the gas bubble and at the cell boundary yield,

$$\frac{\partial T_G}{\partial r} = 0 \qquad \text{at } r = 0 \tag{4}$$

and

$$\frac{\partial T_L}{\partial r} = 0 \qquad \text{at } r = a + b \tag{5}$$

In the following we assume that the equation of state of an ideal gas holds for the gas contained inside a bubble:

$$P_G = \rho_G R_G T_G \tag{6}$$

where R_G is the gas constant.

Using the continuity equation for the gaseous phase yields,

$$\frac{\partial \rho_G}{\partial t} + u_G \frac{\partial \rho_G}{\partial r} + \frac{\rho_G}{r^2} \frac{\partial}{\partial r} \left(r^2 u_G\right) = 0 \tag{7}$$

and following the procedure described by Nigmatulin (1990) we arrive at the following relation:

$$\frac{\partial P_G}{\partial t} = -3\gamma P_G \frac{1}{a} \frac{\partial a}{\partial t} + 3(\gamma - 1)\frac{\lambda_G}{a} \left(\frac{\partial T_G}{\partial r}\right)_{r=a} \tag{8}$$

where $\gamma = c_{p_G}/c_{v_G}$. In a case where no heat transfer takes place equation (8) reduces to the following adiabatic relation, P_G/ρ_G^{γ}=Const.

Combining the continuity equation for the liquid phase with Darcy's equation, which describes fluid motion in the Plateau-Gibbs channels, yields the modified Rayleigh equation (for details see Elperin et al. 1993) which describes oscillating gas bubble in a foam:

$$\rho_L a \frac{d^2 a}{dt^2} + \mu_L \left(4 + \frac{\alpha_L a^2}{K_F}\right) \frac{1}{a} \frac{da}{dt} + \rho_L \frac{3}{2} \left(\frac{da}{dt}\right)^2 + \frac{2\sigma}{a} = P_G - P_L \tag{9}$$

where $K_F(\alpha_L, a)$ is the foam permeability, μ_L is the liquid viscosity and σ is the coefficient of surface tension.

In the following we employ the conservation equations of mass and momentum in a homogeneous mixture approximation to describe the propagation of one-dimensional pressure waves in a foam (Miksis and Ting 1991).

Define density, pressure and velocity of a foam as:

$$\begin{aligned} \rho_F &= \rho_L \alpha_L + \rho_G \alpha_G \\ P_F &= P_L \alpha_L + P_G \alpha_G \\ u_F &= \frac{(\rho_L u_L \alpha_L + \rho_G u_G \alpha_G)}{(\rho_L u_L + \rho_G u_G)} \end{aligned} \tag{10}$$

Based upon these definitions, the continuity equation for the foam is:

$$\frac{\partial \rho_F}{\partial t} + \frac{\partial (u_F \rho_F)}{\partial x} = 0 \tag{11}$$

In the one-velocity approximation, i. e. when $u_F = u_G = u_L$, and neglecting inertial terms, the equation of momentum conservation is:

$$\rho_F \frac{\partial u_F}{\partial t} = -\frac{\partial P_F}{\partial x} \tag{12}$$

Eqs.1, 2, 6-12 and the condition of constant mass gas content $\delta = 4/3 \cdot \pi \cdot a^3 \cdot \rho_G \cdot N = \text{Const.}$ (δ is the mass gas content and N is the constant number of gas bubbles per unit mass in the foam), govern the propagation of pressure waves in foams.

Note that, although the above conservation equations of mass and momentum for foams do not contain dissipative and dispersive terms, dissipation and dispersion present themselves through the equation of state of the foam. The first-order time derivatives of density give rise to dissipation while the second order time derivatives of density are responsible for dispersion. It can be shown that in cases where the pressure depends on both density and its time derivative, the system of Euler equations is equivalent to the Navier-Stokes equation with the first and second viscosity terms (Elperin et al. 1993). The dissipation and dispersion of pressure waves in foams arise due to bubble oscillations and fluid squeezing through the network of Plateau-Gibbs channels.

Differentiation with respect to time of the linearized continuity equation for foam (Eq.11) allows to transform Eq.12 to the following form:

$$\frac{\partial^2 \rho_F}{\partial t^2} = \frac{\partial^2 P_F}{\partial x^2} \tag{13}$$

The wave equation (13) describes propagation of weak pressure waves in foams.

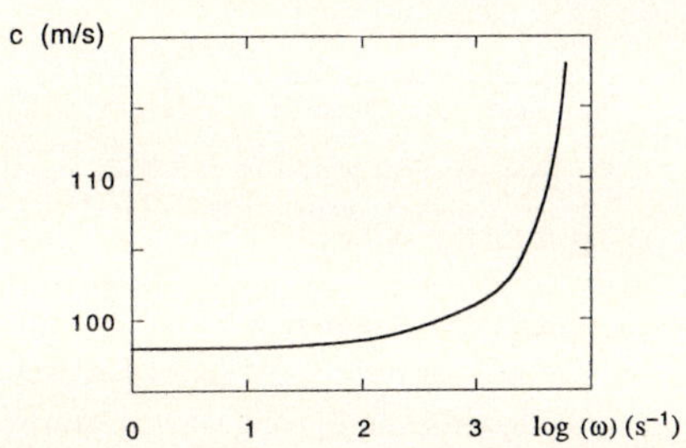

Fig. 1. Phase velocity $c = \frac{\omega}{k}$, ($\alpha_L = 0.01$, $a_0 = 10^{-4}$ m)

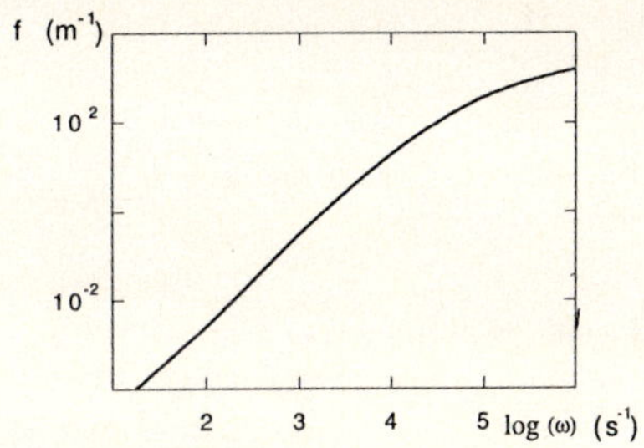

Fig. 2. Attenuation coefficient $f = \mathcal{I}m\,[k(\omega)]$ ($\alpha_L = 0.01$, $a_0 = 10^{-4}$ m)

3. Results and discussion

Sound propagation in foam was studied on the basis of the linearized system of the conservation equations presented above. Let us look for a solution to this system of equations in the form of a decaying traveling wave, i. e.,

$$\begin{gathered} P_L,\ P_G,\ P_F,\ a,\ b,\ \alpha_L,\ \alpha_G \sim \exp(i\,kx - i\,\omega t) \\ T(x,r,t) \sim T(r)\exp(i\,kx - i\,\omega t) \end{gathered} \tag{14}$$

The dispersion relation characterizing this model is found from the condition of the existence of nontrivial solution in the form (14) of the linearized system of conservation equations:

$$\frac{k^2}{\omega^2} = \frac{\alpha_{L_0}(\rho_L - \rho_{G_0}) + \alpha_{G_0}\,\rho_{G_0}\left(1 + \xi_1\xi_2^{-1}(y_G)\right)(1+\xi_1)^{-1}}{\gamma\,P_{G_0}(1+\xi_1)^{-1} - \frac{4\sigma}{3a_0} - i\,\omega\frac{\alpha_{L_0}\,\rho_L\,\nu_L}{3} - \omega^2\rho_L\frac{a_0\,b_0}{3}} \tag{15}$$

where

$$\xi_1 = (\gamma - 1)\left[\xi_2^{-1}(y_G) + (\eta\, \alpha_{L_0})^{-1}\ \coth(y_L)\right]^{-1},$$

$$\xi_1(x) = 3\ x^{-2}\ [x \cdot \coth(x) - 1]\,,$$

$$y_L = \sqrt{i\,\omega/\omega_{TL}}, \qquad y_G = \sqrt{i\,\omega/\omega_{TG}}, \qquad \omega_{TL} = k_L/b_0^2, \qquad \omega_{TG} = k_G/a_0^2,$$

$$\eta = \frac{\rho_L\ c_{p_L}}{\rho_G\ c_{p_G}}, \qquad k_L = \frac{\lambda_L}{\rho_L\ c_{p_L}}, \qquad k_G = \frac{\lambda_G}{\rho_G\ c_{p_G}}$$

The dispersion relation, Eq.15, describes the propagation of pressure waves in a gas-liquid foam in the linear approximation. Eq.15 takes into account all physical mechanisms incorporated into the model and determining the evolution of the initial disturbance: effect of added mass; motion of the liquid-gas interface; surface tension; fluid flow inside the network of Plateau-Gibbs channels and heat transfer between phases. The phase velocity $c = \omega/k$ and attenuation coefficient, $f = \mathcal{I}m[k(\omega)]$ calculated with the aid of Eq.15, are presented in Fig.1 and Fig.2.

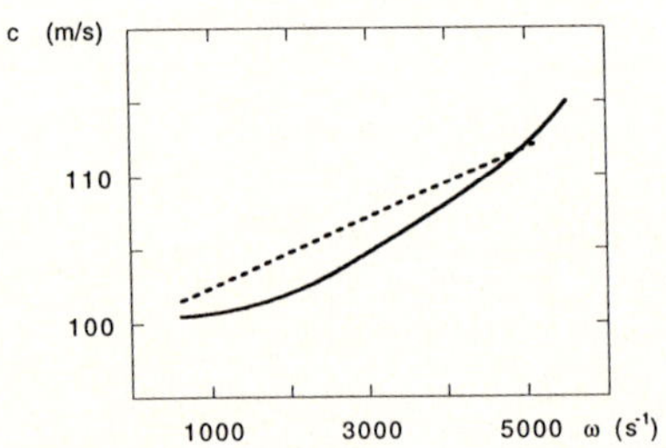

Fig. 3. Phase velocity ($\alpha_L = 0.0085$, $a_0 = 7 \cdot 10^{-5}$ m). — theory, - - experiment

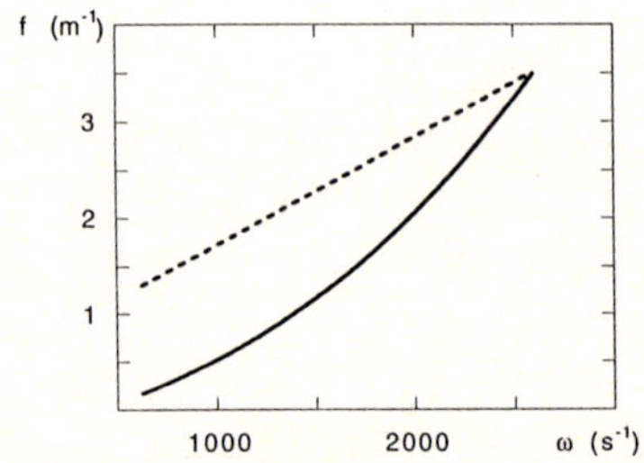

Fig. 4. Attenuation coefficient, ($\alpha_L = 0.0085$, $a_0 = 7 \cdot 10^{-5}$ m). — theory, - - experiment

The curves presented in Fig. 1 and Fig. 2 show that the dispersion of pressure waves is negligibly small up to frequencies close to the resonance frequency of the gas bubbles. The attenuation coefficient is large and increases monotonically with frequency. Therefore oscillating wave patterns are not formed in foams with large gas content, contrary to the bubbly liquid case.

Variation of phase velocity and attenuation coefficient as functions of frequency are compared with experimental results of Orenbakh and Shushkov (1991) in Fig.3 and Fig.4. The experimental results were obtained by processing a large number of experimental points using the least square method. A fairly good agreement between the theoretical predictions and the experimental findings is evident.

References

Elperin T, Igra O, Shreiber I (1993) Formation of shock waves in gas-liquid foams. These Proceedings

Miksis MJ, Ting L (1991) Effective equations for multiphase flows- waves in a bubbly liquid. Advances in Applied Mechanics 28:143

Nigmatulin RI (1990) Dynamics of Multiphase Media. Hemisphere Publ. Co., New York

Orenbakh Z, Shushkov G (1991) Experimental determination of velocity and attenuation of acoustics waves in gas-liquid foam. J. Soviet Acoustics 37:403

Formation of Shock Waves in Gas-Liquid Foams

T. Elperin*, **O. Igra*** and **I. Shreiber**†
*Pearlstone Center for Aeronautical Engineering Studies Department of Mechanical Engineering
†Institute for Industrial Mathematics, Ben-Gurion University of the Negev P.O.Box 653, Beer Sheva 84105, Israel

Abstract. The purpose of the present investigation is to analyze the phenomenon of shock wave formation in foams using the Burgers equation which is derived from the conservation laws for bubbly liquids. It is shown that the ratio of parameters of nonlinearity in the Burgers equation describing wave propagation in bubbly liquids to that in foams is of order 40, while the coefficient of bulk viscosity in foams is by a factor 10^3 higher than that in bubbly liquids. This explains why oscillatory shock waves are not detected in shock tube experiments with foams while they are observed under similar conditions in bubbly liquids.

Key words: Gas-liquid foams, Shock waves, Bulk viscosity

1. Introduction

The main difference between gas-liquid foams and the well known and investigated bubbly liquids is in the high (close to unity) gas content which characterizes gas-liquid foams. Foams are composed of a large number of gas bubbles separated by thin liquid films and have a cellular quasi-ordered structure. The structure of foams can be well described by a polyhedral model whereby an elementary foam cell represents an irregular polyhedron (Bikerman 1973, (§41, §42) and Kraynik 1988). The liquid film junction regions, called Plateau borders, determine the boundaries of polyhedral gas bubbles. Foams with a liquid content which does not exceed 0.1 are of particular interest. In such foams the thickness of the liquid film is much less than the radius of the gas bubbles. Stability of such a delicate structure is achieved due to the presence of surfactants in the foam-forming liquid. Absence of this small amount of surfactants (about 1%) will lead to rupture of the thin liquid films and rapid bubble coagulation. Due to the very small amount of stabilizing surfactants they have practically no effect on the properties of the foam-forming liquid. Propagation of pressure disturbances in foams can be described by the method developed for analyzing bubbly liquid flows (Miksis and Ting 1991). However, it is essential to take into account the creeping liquid flow in the capillary network formed by the Plateau borders. This flow is caused by changes in the volume of the gas bubbles resulting from the disturbances induced by the incident pressure wave. The latter process is complicated by other phenomena, e.g., viscous effects at the gas-liquid interface, inertial effects, surface tension, etc.

It is well known that oscillatory pressure waves in bubbly liquids may reach an amplitude twice as large as that of the original pressure impulse. However, experiments showed that pressure disturbances in foams always attenuate without significant change in the wave pressure profile. Shock waves are formed due to interaction of two competing effects: nonlinearity and dissipation (and dispersion). However, it is showed in this investigation that the parameter of nonlinearity is much higher in bubbly liquids than in foams while the dissipation in foams is significantly higher than that in the bubbly liquid. Therefore shock waves in bubbly liquids can be observed in shock tubes of ~1 m length. The latter is impossible for foams since the length of the shock wave formation in foams and the initial amplitude of the incident pressure disturbance which results in shock wave formation, are considerably higher than in the bubbly liquid case.

Shock Waves @ Marseille III
Editors: R. Brun, L. Z. Dumitrescu

2. Derivation of conservation equations for gas-liquid foams

In the present investigation the cellular model of the multiphase media (see, e. g., Miksis and Ting 1991) is employed, whereby the foam is modeled by a regular cellular structure. Each cell consists of a spherical gas bubble surrounded by a thin liquid film. The foam density ρ may be expressed as:

$$\rho = \rho_f(1-\beta) + \beta\rho_g \tag{1}$$

where ρ_f and ρ_g are the densities of liquid and gas, respectively and β is the gas volume fraction in the foam.

The incident pressure wave, which travels through the foam, causes deformation of each of the foam cells and thereby initiates creeping fluid flow in the channels between Plateau borders. The network of these channels can be viewed as a kind of a porous medium. The fluid filtration flow in this porous medium causes dissipation and damping of the incident pressure wave. Since the fluid slip velocity at the gas-liquid interface is zero, due to the presence of surfactants (Bikerman 1973), in the following we assume that gas and liquid flows have the same velocity. We further assume that the entire foam structure is not damaged by the incident wave, i.e. there is no annihilation or formation of cells. Then, since $\rho_g \ll \rho_f$ the velocity of sound in the foam c, is given by the following relation (see Miksis and Ting 1991):

$$c^2 = \frac{\gamma P}{\rho\beta} \tag{2}$$

where γ_g is the ratio of a specific heats in the gas, P the pressure. In the derivation of Eq.2, the liquid phase is considered incompressible. Consider a variation of expression (2) with respect to parameters P, ρ, β. Then, after some algebra, using basic relations from the theory of homogeneous mixtures (Miksis and Ting 1991), we arrive at the following relation for the velocity of sound in foams:

$$c = c_0\left(1 + \frac{\delta P}{P_0}\frac{1}{2}\frac{\gamma+1-2\beta}{\gamma}\right) \tag{3}$$

where subscript 0 indicates an unperturbed state. The propagation of a pressure wave in foam is governed by the following system of mass and momentum conservation equations:

$$\frac{\partial\rho}{\partial t} + \frac{\partial}{\partial x}(u\rho) = 0 \tag{4}$$

$$\frac{\partial u}{\partial t} + u\frac{\partial u}{\partial x} = -\frac{1}{\rho}\frac{\partial P}{\partial x} \tag{5}$$

where u is the velocity.

In order to close the system of conservation equations it is necessary to determine the equation of state of foam, i.e., $\rho = \rho(P)$. Due to the fluid inertia and the gas compressibility, the gas bubble in a foam cell experiences radially symmetrical oscillations. These oscillations cause a creeping fluid flow in the network of channels between the Plateau borders. The liquid flow in the channels between Plateau borders is governed by the equation of continuity:

$$\frac{\partial}{\partial r}\left(r^2 v\right) = 0 \tag{6}$$

and the Navier-Stokes equation

$$\frac{\partial v}{\partial t} + v\frac{\partial v}{\partial r} = -\frac{1}{\rho_f}\frac{\partial P}{\partial r} + \nu_f\left[\frac{1}{r^2}\frac{\partial}{\partial r}\left(r^2\frac{\partial v}{\partial r}\right) - \frac{2v}{r^2}\right] \tag{7}$$

where v is the velocity of the liquid, ν_f is the kinematic viscosity of the liquid. Eqs.6-7 must be supplemented with regular boundary conditions at the gas-liquid interface:

$$P_f = P_g - \frac{2\sigma}{R} - 4\mu R \frac{dR}{dt} \tag{8}$$

$$v(R,t) = \frac{dR}{dt} \tag{9}$$

where $R(t)$ is the radius of the gas bubble, σ the coefficient of surface tension, μ the dynamic viscosity of the liquid. Integrating Eqs.6-7 with boundary conditions (8-9) yields the regular Rayleigh equation (see, e.g., Miksis and Ting 1991):

$$\rho_f\, R\frac{d^2R}{dt^2} + \frac{4\mu}{R}\frac{dR}{dt} + \rho_f \frac{3}{2}\left(\frac{dR}{dt}\right)^2 + \frac{2\sigma}{R} = P_g(R) - P \tag{10}$$

The pressure in the the foam, P, can be found from Darcy's equation (Dullien 1992):

$$-\frac{\partial P}{\partial r} = \frac{\nu_f \rho_f V}{k} \tag{11}$$

where k is the foam permeability and V is the liquid flow rate velocity. The integral mass balance equation yields:

$$4\pi r^2 V = 4\pi R^2 \frac{dR}{dt}\alpha \tag{12}$$

where α is the fraction of a unit cross-section of foam which is occupied by liquid.

Note that the presence of surface porosity α in the right-hand side of Eq.12 is due to the deformation of the walls of gas bubbles when their oscillations squeeze liquid through the channels between Plateau borders. It is essential that Darcy's filtration law which is described by Eq.11 does not involve net flow through the foam. This expression describes a local flow in the vicinity of an oscillating bubble with zero volume average flow. Propagation of shock waves through foams is associated with compressibility of gas bubbles. The attenuation of shock waves occurs due to local friction during filtration flow through the Plateau borders and stretching and oscillations of liquid films. All these effects are accounted for in the modified Rayleigh equation (Eq.10).

Integrating Eqs.11 and 12 allows to determine the pressure distribution in the liquid film surrounding the gas bubble:

$$P(r) = P_f + \alpha\,\frac{\mu_f}{k}\,\frac{dR}{dt}\,\frac{R^2}{r} \tag{13}$$

Substituting Eq. 13 into Eq.10 yields the modified Rayleigh equation:

$$\rho_f \frac{3}{2}\,\frac{d^2R}{dt^2} + \frac{4\mu_f}{R}\,\frac{dR}{dt}\left[1 + \frac{1}{4}\frac{\alpha R^2}{k}\right] + \rho_f \frac{3}{2}\left(\frac{dR}{dt}\right)^2 + \frac{2\sigma}{R} = P_g(R) - P \tag{14}$$

Eqs.4-5 together with the modified Rayleigh equation (14) describe the propagation of pressure waves in the foam. In order to further simplify the analysis we neglect in Eq.14 the second-order inertial and surface tension terms:

$$4\mu_f\left[1 + \frac{1}{4}\frac{\alpha R_0^2}{k}\right]\,\frac{1}{R}\,\frac{dR}{dt} = P_g - P \tag{15}$$

where R_0 is the average equilibrium size of the gas bubble. The inertial term in the modified Rayleigh equation can be safely neglected since the mass of liquid surrounding gas bubbles in foams is very small. It is of interest to note that this inertial term in bubbly liquids is much higher (by a factor $\propto R/\delta$, where δ is the width of the liquid film) and causes wave dispersion and amplification of pressure in the oscillating wave front (Karpman 1974).

Using equation

$$\frac{1}{R}\frac{dR}{dt} = -\frac{1}{3\rho\beta}\frac{\partial\rho}{\partial t} \qquad (16)$$

together with Eqs. 15 and 16, we arrive at the nonlocal equation of state for foams (see, e.g., Karpman 1974, §9) $P = P(\rho, \dot{\rho})$ which can be written as follows:

$$\delta P = \left(\frac{\partial P}{\partial \rho}\right)_{S,\dot{\rho}} \delta\rho + \left(\frac{\partial P}{\partial \dot{\rho}}\right)_{S,\rho} \delta\dot{\rho} = c^2\delta\rho + \nu_b\delta\dot{\rho} \qquad (17)$$

where ν_b is the bulk viscosity of the foam:

$$\nu_b = \frac{4\mu_f}{3\rho\beta}\left[1 + \frac{1}{4}\frac{\alpha R_0^2}{k}\right] \qquad (18)$$

It is essential for further analysis to estimate the bulk viscosities, given by Eq.18, for gas-liquid foams ($\beta \sim 0.9$) and for bubbly liquids ($\beta \sim 0.1$). For estimating the permeability of foam we use the Karman-Kozeny equation (Dullien 1992, p. 243):

$$k = \frac{R_0^2(1-\beta^3)}{720\ \beta^2} \qquad (19)$$

The surface porosity α can be estimated as $\alpha \cong (1-\beta)^{2/3}$. Then, from Eq.18 one obtains $\nu_b(\beta = 0.9)/\nu_b(\beta = 0.1) \sim 10^3$, i.e., the bulk viscosity of foam is 10^3 times greater than that of bubbly liquids.

3. Derivation of Burgers equation

The simplified Rayleigh equation (15) yields the equation of state for foams in the form of $P = P(\rho, \dot{\rho})$. Using the regular definition for the speed of sound, i. e.,

$$c^2 = \left(\frac{\partial P}{\partial \rho}\right)_{S,\dot{\rho}} \qquad (20)$$

the conservation equations (4-5) can be written as follows:

$$\frac{\partial \rho}{\partial t} + \frac{\partial}{\partial x}(\rho u) = 0 \qquad (21)$$

$$\frac{\partial u}{\partial t} + u\frac{\partial u}{\partial x} + \frac{c}{\rho^2}\frac{\partial \rho}{\partial x} = \left(\frac{\partial P}{\partial \dot{\rho}}\right)_{\rho}\frac{\partial^2 u}{\partial x^2} \qquad (22)$$

Eqs. 21-22 have a solution in the form of a quasi-simple wave (Karpman 1974, §13):

$$\rho(x,t) = \rho(u) + \psi(x,t) \qquad (23)$$

where

$$\frac{d\rho}{du} = \frac{\rho(u)}{c(u)} \qquad (24)$$

and

$$\frac{dP}{du} = c(u)\rho(u) \qquad (25)$$

and $\psi(x,t)$ is the solution of the linear wave equation:

$$\frac{\partial \psi}{\partial t} + c_0\frac{\partial \psi}{\partial x} = 0$$

Substituting Eqs.24 and 25 into Eqs.21-22 results, after some algebra, in the following equation for a quasi-simple wave:

$$\frac{\partial u}{\partial t} + u\frac{\partial u}{\partial x} + c\frac{\partial u}{\partial x} = \frac{1}{2}\left(\frac{\partial P}{\partial \dot{\rho}}\right)_{\rho} \frac{\partial^2 u}{\partial x^2} \tag{26}$$

Since in a simple wave $\Delta P = c\rho\Delta u$ expressions (2) and (3) yield:

$$c = c_0 + \frac{\gamma + 1 - 2\beta}{2\beta}\, u \tag{27}$$

where c_0 is the sound velocity in an unperturbed foam.

By combining Eqs.15-18 and 26-27 we arrive at the Burgers equation describing the wave propagation in a gas-liquid foam:

$$\frac{\partial u}{\partial t} + u\frac{\partial u}{\partial x} + (c_0 + \Gamma u)\frac{\partial u}{\partial x} = \frac{\nu_b}{2}\frac{\partial^2 u}{\partial x^2} \tag{28}$$

where ν_b is a bulk viscosity of the foam; it can be determined from Eq.18, while Γ is a nonlinearity parameter defined as:

$$\Gamma = \frac{\gamma + 1 - 2\beta}{2\beta} \tag{29}$$

The above Burgers equation describes the propagation of a pressure wave with steepening profile in foam, i.e., the formation of a shock wave. Shock waves are formed due to interaction of two competitive mechanisms: nonlinearity and dissipation. The velocity of sound c, in foams and in bubbly liquids is determined by the following expression:

$$\frac{1}{c^2} = \frac{(1-\phi)\rho}{\rho_f^2\, c_f^2} + \frac{\phi\rho}{\rho_g^2\, c_g^2} \tag{30}$$

where $\phi = \rho_g\beta/\rho$ is the gas mass fraction. In the cases when $\beta \sim$0.1 or $\beta \sim$0.9, i.e., in bubbly liquids and foams, respectively, the sound velocities are practically equal. However, the nonlinearity parameter, Γ, is about 40 times higher in bubbly liquids than in foams. It is of interest to note that when heat transfer is taken into account the parameter, γ is effectively reduced and the nonlinearity parameter Γ , for foams, becomes even smaller than that of a pure gas. Therefore, under such circumstances the length required for shock waves formation in foams is 40 times longer than in a bubbly liquid. At the same time the bulk viscosity of a foam is about 10^3 times greater than that of a bubbly liquid. The above analysis explains why in shock tube experiments with foams shock waves are not detected, while they are easily observed when bubbly liquids are used under similar conditions. In contrast to a pure gas, or bubbly liquid, shock waves cannot propagate in gas-liquid foams. Therefore, propagation of strong shock waves in foams is accompanied by their destruction. The latter is the reason why foams are widely used for attenuation of strong pressure waves (Krasinski 1992).

4. Conclusions

It is shown that propagation of pressure waves is governed by the Burgers equation, as in the case of a bubbly liquid. However, the parameter of nonlinearity which determines the length of formation of a shock wave, is significantly higher in foams than in a bubbly liquid. The dissipation coefficient (bulk viscosity) of foams is about 10^3 times higher than that of a bubbly liquid. The latter is caused by high viscous losses during liquid filtration through the porous structure formed by the gas bubbles. A model of bubble oscillations in foams is suggested and the modified Rayleigh equation, which describes these oscillations, is derived. The presented analysis explains why shock waves are not formed in foams while they are observed in bubbly liquids under similar experimental conditions.

References

Bikerman JJ (1973) Foams. Springer-Verlag, New York

Dullien FAL (1992) Porous Media. Fluid Transport and Pore Structure. Academic Press, New York

Karpman VI (1974) Nonlinear Waves in Dispersive Media. Pergamon Press, Oxford

Krasinski SJ de (1992) Some aspects of the fluid dynamics of liquid-air foams of high dryness fraction. Prog. Aerospace Sci. 29:125

Kraynik AM (1988) Foam flows. Ann. Rev. Fluid Mech. 20:325

Miksis MJ, Ting L (1991) Effective equations for multiphase flows waves in bubbly liquid. Adv. Appl. Mech. 28:143

Jump Conditions Across Strong Compaction Waves in Gas-Saturated Rigid Porous Media

A. Levy*, **G. Ben-Dor***, **S. Sorek*** **and J. Bear**†

*Pearlstone Center for Aeronautical Engineering Studies, Department of Mechanical Engineering, Ben-Gurion University of the Negev, Beer Sheva, Israel
†Albert and Anne Mansfield Chair in Water Resources, Technion-Israel Institute of Technology, Haifa, Israel

Abstract. Based on experimental results and some additional simplifying assumptions, the general macroscopic two-phase equations governing the flow field which is developed in a gas-saturated rigid porous medium domain were simplified to a form which enabled us to develop two analytical models for calculating the jump conditions across a strong compaction wave. Predictions obtained by these two simplified analytical models are compared to the experimental results of Sandusky and Liddiard (1985) and to predictions of another more complicated model which was proposed by Powers et al. (1989). Fairly good to excellent agreements are evident.

Key words: Compaction waves, Porous materials

1. Introduction

In general, two approaches have been employed by researchers for treating gas-saturated flows in a porous medium domain: the single-phase approach, which is based on an oversimplified mixture theory and the two-phase approach, which is in the authors' opinion, a better approach for analyzing transport phenomena in porous media. For this reason only the two-phase approach is presented in this paper. Readers interested in more details regarding the single-phase approach are referred to Gibson and Ashby (1988). In the two-phase approach, the fluid filling the pores and the porous medium are considered as two phases which interact with each other. A detailed description of this approach was given by Baer and Nunziato (1986). A one-dimensional two-phase analysis of air as the fluid phase was presented by Baer (1988) and Powers et al. (1989).

In addition to numerical solutions Baer (1988) and Powers et al. (1989) also presented simplified analytical models. In their studies, they showed that compaction waves having velocities smaller than the speed of sound of the solid material are possible. Powers et al. (1989), who referred to these waves as subsonic compaction waves, showed that they arise from non-ideal effects in the material. Furthermore, they showed that the minimal compaction wave velocity when Tait's equation is ideal is the speed of sound in the solid. In addition, by assuming no change in the porosity across the compaction wave, they showed that the supersonic compaction waves are indeed shock waves. From their calculations, Powers et al. (1989) realized that the change in the density of the solid is only fractions of one percent, therefore, they concluded that the solid phase can practically be assumed to be incompressible. In view of their finding regarding the compressibility of the solid phase, Powers et al. (1989) introduced a constitutive relation to address the unknown porosity, which they regard as the compaction equation. This equation relates the effective stress to a function of the solid volume fraction i.e., $P_s - P_g = f(\phi_s)$.

Finally, both the analyses of Baer (1988) and Powers et al. (1989) indicated that the attenuation of the compaction waves is negligibly small, and, hence, the compaction waves can be assumed to propagate with a constant velocity. Based on these findings, a Galilean transformation could be preformed on the governing equations and they could be simplified from a set of time-dependent equations to a set of equations which is independent of time. The simplified unsteady set of governing equations was further simplified by us resulting in two analytical models capable of predicting the jump conditions across strong compaction waves in rigid porous media.

Shock Waves @ Marseille III
Editors: R. Brun, L. Z. Dumitrescu

2. Present study

The macroscopic mass, linear momentum and energy balance equations, for a two-phase saturated porous medium, as obtained by the dimensional analysis which Levy et al. (1992) conducted on the macroscopic balance equations of Bear et al. (1992) and Sorek et al. (1992), were adopted. The macroscopic conservation equations were obtained by an averaging process over a representative elementary volume (REV) of the microscopic balance equations (Bear and Bachmat 1990). In addition, the solid phase was assumed to be rigid and thermoelastic and the fluid phase was considered as a Newtonian air.

Combining the same balance equations for the two phases and non-dimensionalizing them, results in a situation in which the gaseous-phase terms are multiplied by coefficients whose values are orders of magnitude smaller than the coefficients multiplying the solid-phase terms. Consequently, it turns out that the gaseous-phase terms in the overall conservations equations can be neglected in comparison with the solid phase terms (see Levy et al. (1993). Based on the findings of Baer (1988) and Powers et al. (1989) that the velocity of the compaction waves is almost constant the resulted equations can be further simplified, to obtain the following set of time-independent equations:

$$\frac{d}{d\xi}(\phi\rho v) = 0 \tag{1}$$

$$\frac{d}{d\xi}(\phi\rho v^2 + \phi P) = 0 \tag{2}$$

$$\frac{d}{d\xi}\left(e + \frac{v^2}{2} + \frac{P}{\rho}\right) = 0 \tag{3}$$

The above set of equations consists of three equations with five non-dimensional unknowns, namely: the porosity, ϕ, density, ρ, velocity, v, pressure, P, and internal energy, e. In order to have a solvable set of equations, two complementary equations are needed. One could be derived from the caloric equation for the solid phase (Powers et al. (1989):

$$e + \frac{P}{\rho} = \frac{\gamma}{(\gamma - 1)} \frac{P + \sigma}{\rho} \qquad (4).$$

Here γ and σ are parameters that define Tait's equation of state. The value of γ is chosen to match the Hugoniot data. It is analogous to the specific heat ratio for an ideal equation of state; σ is the nonideal solid parameter. It is an adjustable parameter which enables one to vary the equation of state in a simple way in order to show how the results are sensitive to nonideal effects. When $\sigma = 0$, the equation of state is referred to as ideal.

At this stage we have a set of four equations, (Eqs.1-4) and five unknowns, ϕ, ρ, v, P and e. In order to have a solvable set, Powers et al. (1989) introduced an empirical equation which was based on their experiments. In order to avoid dependence on experimental results we propose two simplified models in which one of the five general unknowns is chosen to be a known parameter. Hence, the problem is reduced to four unknowns and a set of four equations (Eqs.1-4). In the first proposed model we assume that the solid phase density, ρ, remains constant while in the second proposed model we assumed that the post-compaction wave porosity, ϕ, is equal to unity. The relations as obtained from the two models are presented in the form $X_{21} = X_2/X_1$ where X is a property of the solid phase (e.g., ϕ, P, v, and c.) and the indices 1 and 2 denote locations ahead and behind the compaction wave, respectively.

3. Model I: Post-compaction wave volume fraction (porosity) is sufficiently less than unity

It can be assumed that the solid phase starts to be compressed only when its volume fraction ϕ_2 approaches unity. Hence, when ϕ_2 is sufficiently smaller than unity, the solid phase can be assumed to be incompressible and hence $\rho_2 = \rho_1 = \rho$. Under these assumptions Eqs.1-4 result in the following relations:

$$P_{21} = (\alpha + \phi_{12})/\left(\alpha + \phi_{21}\right), \qquad \text{where} \qquad \alpha = (1+\gamma)/(1-\gamma) \tag{5}$$

Eq.5 can be rewritten to read

$$\phi_{21} = \left(-\alpha(P_{21}-1) + \sqrt{\alpha^2(P_{21}-1)^2 + 4\,P_{21}}\right)\Big/ 2\,P_{21} \tag{6}$$

In order to relate the pressure ratio across the compaction wave to the compaction wave Mach number let us first define $M_s = D/c_1$, where D is the compaction wave velocity and c_1 is the speed of sound in the solid ahead of the compaction wave. The speed of sound, c, could be calculated from: $c^2 = \partial P/\partial\rho|_S$. In order to calculate the speed of sound we introduce the following thermal equation of the state of the solid phase (Powers et al. (1989): $P = (\gamma-1)\rho C_v T - \sigma$. Combining this relation with Eq.4 results in:

$$c^2 = \gamma(\gamma-1)C_v T = \frac{\gamma(P+\sigma)}{\rho} \tag{7a}$$

The ideal speed of sound, c^*, can be obtained from Eq.7a by setting $\sigma = 0$, i.e.,

$$\left(c^*\right)^2 = \frac{\gamma P}{\rho}. \tag{7b}$$

By applying some algebra we get,

$$M_s = \left\{ \frac{-1 - \frac{\alpha}{2}(P_{21}-1) + \sqrt{\left(\frac{\alpha}{2}\right)^2 (P_{21}-1)^2 + P_{21}}}{\gamma(1+\bar{\sigma})\left(1 - \frac{P_{21}}{-\frac{\alpha}{2}(P_{21}-1)+\sqrt{\left(\frac{\alpha}{2}\right)^2(P_{21}-1)^2+P_{21}}}\right)} \right\}^{1/2} \tag{8}$$

where $\bar{\sigma} = \sigma/P_1$.

The non-dimensional velocity of the solid phase, $U_{21} = u_2/c_1^*$ is:

$$U_{21} = \left\{ \frac{1}{\gamma}\left(-1 - \frac{\alpha}{2}(P_{21}-1) + \sqrt{\left(\frac{\alpha}{2}\right)^2 (P_{21}-1)^2 + P_{21}}\right) \times \right.$$
$$\left. \left(1 - \frac{P_{21}}{-\frac{\alpha}{2}(P_{21}-1) + \sqrt{\left(\frac{\alpha}{2}\right)^2 (P_{21}-1)^2 + P_{21}}}\right) \right\}^{1/2} \tag{9}$$

Note that U_{21} is independent of the non-ideal behavior of the material.

4. Model II: Post-compaction wave volume fraction (porosity) is very close to unity ($\phi_2 = 1 - \varepsilon$ and $\varepsilon \to 0$)

When the post-compaction wave volume fraction, ϕ_2 is very close to unity, the assumption that the solid phase is incompressible can no longer be justified, i.e., $\rho_1 \neq \rho_2$. Under this additional fact Eqs.1-4 can be manipulated to result in

$$\rho_{21} = \left(\frac{2\gamma}{(\gamma-1}(P_{21}+\bar{\sigma}) - (P_{21}-\phi_{12})\right) \Big/ \left(\frac{2\gamma}{(\gamma-1)}\,(1+\bar{\sigma}) + \phi_{21}P_{21} - 1\right) \tag{10}$$

Following a procedure similar to the one outlined earlier, the compaction wave Mach number is:

$$M_s = \frac{D}{c_1} = \left\{ \frac{(\phi_{21}P_{21} - 1)\left(\frac{2\gamma}{\gamma-1}(P_{21}+\bar{\sigma}) - (P_{21} - \phi_{12})\right)}{\frac{2\gamma(1+\bar{\sigma})}{\gamma-1}(P_{21} + \gamma\bar{\sigma} - (1+\gamma\bar{\sigma})\phi_{12})} \right\}^{1/2} \tag{11}$$

The non-dimensional velocity of the solid, U_{21}, becomes:

$$U_{21} = \frac{u}{c_1^*} = \left\{ \frac{\frac{2}{\gamma-1}(\phi_{21}P_{21} - 1)\ (P_{21} + \gamma\bar{\sigma} - (1-\gamma\bar{\sigma})\phi_{12})}{\gamma\left(\frac{2\gamma}{\gamma-1}(P_{21}+\bar{\sigma}) - (P_{21} - \phi_{12})\right)} \right\}^{1/2} \tag{12}$$

Table 1. Comparison of analytical models with experimental results

	Sandusky and Liddiard Experimental data (1985)	Powers et al.'s Model (1989)	Present Model I
ϕ_2	0.91	0.83	0.83
D	300	340	365
$(\phi P)_2$	3.4×10^7	4.5×10^7	4.74×10^7
ρ_{21}	not available	not available	1*

T_1=300 K, ρ_1=1900 kgm/m^3, C_ν=1500 J/(kgK), S =8.976$\times10^6$ m^2/s^2, γ=5, ϕ_1 = 0.6, u_1 = 100 m/s
*the model assumes an incompressible solid.

Table 2. Comparison of analytical models with experimental results

	Sandusky and Liddiard Experimental data (1985)	Powers and al's Model (1989)	Present Model I	Present Model II
ϕ_2	0.95	0.97	1	1**
D	370	410	365	370
$(\phi P)_2$	5.9×10^7	6.2×10^7	5.71×10^7	5.8×10^7
ρ_{21}	not available	1901/1900	1*	1901/1900

T_1=300 K, ρ_1=1900 kgm/m^3, C_ν=1500 J/(kgK), S =8.976$\times10^6$ m^2/s^2, γ=5, ϕ_1 = 0.73, u_1 = 100 m/s
*the model assumes an incompressible solid
** this was imposed as an input

5. Results and discussion

The experimental results of Sandusky and Liddiard (1985), the predictions of the empirical model proposed by Powers et al. (1989) and those of our further simplified analytical models are shown in Tables 1 and 2 for two cases of different initial conditions.

In the first experiment, the post-compaction wave volume fraction of the solid phase was sufficiently less than unity, i.e., ϕ_2=0.91, and hence only Model I could be used. The results shown in Table 1 indicate that the predictions of our proposed Model I are slightly worse than those of the empirical model proposed by Powers et al. (1989).

In the second experiment, the post-compaction wave volume fraction of the solid phase was 0.95. First we applied Model I. The post-compaction wave volume fraction, ϕ_2, as calculated by Model I was $\phi_2 \approx 1$. For this reason we decided to apply our Model II. The results, shown in Table 2, indicate that for this case the predictions of Model I are very good for D and relatively good for $(\phi P)_2$. However, the predictions of Model II are excellent for D and very good for $(\phi P)_2$. Note that both Models I and II fail to predict ϕ_2 well enough. The empirical model of Powers et

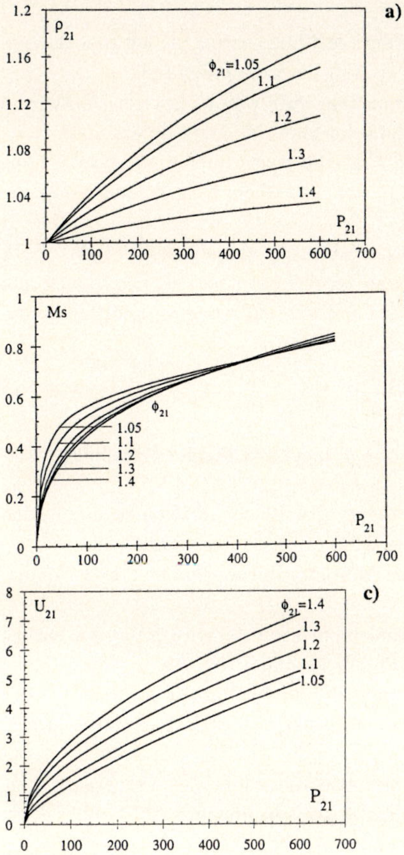

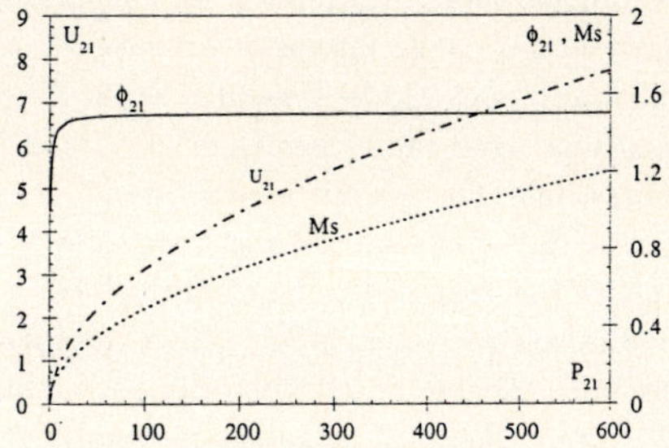

Fig. 1. Possible solutions of Model I, which assumes the the post-compaction wave volume fraction is significantly less than 1.

Fig.2. (left)Possible solutions of Model II, which assumes that the post-compaction wave volume fraction is very close to unity
a)ρ_{21} vs. P_{21}; b)M_s vs. P_{21}; c)U_{21} vs. P_{21}.

al. (1989) on the other hand is much better in predicting ϕ_2. However, its predictions of both D and $(\phi P)_2$ are much worse than those of Model II and even those of Model I.

Based on the comparisons shown in Tables 1 and 2, it could be concluded that:

1. Model II could be practically applied for $\phi_2 \geq 0.95$;
2. Using the compaction equation is much more important when ϕ_2 is sufficiently less than unity. Dropping this equation when $\phi_2 \rightarrow 1$ seems to be justified;
3. The present simplified models are sufficient for estimating the post-compaction wave properties behind strong compaction waves propagating in air-saturated rigid porous materials.

Once these conclusions were reached it is worthwhile to graphically present possible solutions as predicted by both Models I and II. The values of ϕ_{21}, M_s and U_{21} as a function of P_{21} are shown in Fig.1, for $\bar{\sigma} = 374$ and $\gamma = 5$. While M_s and U_{21} are seen to continuously increase with P_{21} , ϕ_{21} is seen to asymptotically approach the value of $\alpha = 1.5$.

Possible solutions as obtained from our Model II are shown in Figs.2(a) to 2(c), in which ϕ_{21} serves a parameter. These figures are in fact graphical presentations of Eqs.10, 11 and 12, respectively. It is evident that ρ_{21}, M_s and U_{21} increase as P_{21} increases. Furthermore, while ϕ_{21} is seen to strongly affect both ρ_{21} and U_{21}, its effect on M_s is minor.

6. Conclusions

Two simplified analytical models for calculating the jump conditions across strong planar compaction waves propagating inside air-saturated rigid porous media were presented.

The first model, Model I, assumed that the post-compaction wave volume fraction of the solid material was significantly less than unity and hence the solid material could be assumed to be incompressible. The second model, Model II, assumed that the post-compaction wave volume fraction of the solid material was very close to unity ($\phi_2 = 1 - \varepsilon$, where $\varepsilon \to 0$), and hence the solid phase was assumed to be compressible.

Predictions were compared to the experimental data of Sandusky and Liddiard (1985). Both models were capable of satisfactorily predicting the jump conditions across strong compaction waves without the need to numerically solve the flowfield and without using an additional constitutive compaction equation, as proposed by Powers et al. (1989).

References

Baer MR (1988) Numerical studies of dynamic compaction of inert and energetic granular materials. ASME J. Appl. Mech. 55:36-43

Baer MR, Nunziato JW (1986) A two-phase mixture theory for the deflagration to detonation transition (DDT) in reactive granular materials. Intl. J. Multiphase Flow 12:861-889

Bear J, Bachmat Y (1990) Introduction to Modeling of Transport Phenomena in Porous Media. Kluwer Academic Publishers, Dordrecht

Bear J, Sorek S (1990) Evolution of governing mass and momentum balance following abrupt pressure impact in a porous medium. Transport in Porous Media. 5:169-185

Gibson LJ, Ashby MF (1988) Cellular Solids: Structure and Properties. Pergamon Press, Oxford

Levy A, Ben-Dor G, Sorek S, Bear J (1993a) Jump conditions across compaction waves in gas-saturated porous media. Shock Waves 3, 2: 105-111

Levy A, Sorek S, Ben-Dor G, Bear J (1993b) Evolution of the balance equations in saturated thermoelastic porous media following abrupt simultaneous changes in pressure and temperature. Submitted to Advances in Water Resources

Olim M, van Dongen MEH, Kitamura K, Takayama K (1993) On numerical simulation of propagation of shock waves in porous foams. In: Takayama K (ed) Proc. 18th Symposium on Shock Waves, Sendai, Japan, pp 261-264

Powers JM, Stewart DS, Krier H (1989) Analysis of steady compaction waves in porous materials. ASME J. Appl. Mech. 56:15-24

Sandusky HW, Liddiard TP (1985) Dynamic compaction of porous beds. NSWC TR 83-256,Naval Surface Weapons Center, White Oak, Md

Sorek S, Bear J, Ben-Dor G, Mazor G (1992) Shock wave in saturated thermoelastic porous media. Transport in Porous Media 9:3-13

Head-on Collision of a Normal Shock Wave with a Polyurethane Foam

E. Zaretsky and O. Igra
Pearlstone Center for Aeronautical Engineering Studies, Department of Mechanical Engineering, Ben-Gurion University of the Negev, Beer-Sheva, Israel

Abstract. The head-on collision of weak ($M = 1.28$) shock waves with polyurethane foam with water content up to densities of about 700 kg/m^3 was studied experimentally in a shock tube. Analysis of the experimental pressure-time curves allows to suggest a simple one-phase equation of state for water-containing foams.

Key words: Shock wave interactions, Porous materials

1. Introduction

In the recent past the head-on collision of shock waves, propagating in a gaseous phase, with a compressible porous media (such as foams) were studied. In these studies the interaction of the shock wave with the porous media was treated as shock waves propagating in a two-phase media. Among the published works the following should be cited: the review paper of Gvozdeva et al. (1986) provides a summary of both experimental and theoretical works done untill 1986. Gelfand et al. (1983) and Gvozdeva and Faresov (1985) examine the gas - foam system as a gas mixture in which the foam is treated as a pseudo-gas. Later, Gvozdeva et al. (1987) studied the gas-foam combination as a two-temperature gas system. Recently, Baer (1992) reanalyzed numerically the findings of Gelfand et al. (1983). He used the theoretical model proposed by Baer and Nunziato (1986) in which a gas - foam combination was treated as a mixture of gas with solid particles. Each phase had its own thermodynamic parameters. Based upon the past works of Skews et al. (1992) and Seitz and Skews (1992), in which the velocity of the shock wave transmitted into a foam was found to be constant, Ben-Dor and Zaretsky (1993) showed that the transmitted shock wave in the foam could be described in a manner similar to shock propagation in a one-phase solid. The present work adopts this approach and shows that for low Mach numbers of the incident shock wave, the one-phase equation of state provides satisfactory description of the obtained experimental results. This equation, which relates the shock velocity in the foam to the foam's particle velocity holds even for wetted foams, up to water concentration of about 70% by volume.

2. Experimental investigation

Experiments were conducted in the 90-mm diameter shock tube of the Shock Wave Laboratory of Ben-Gurion University of the Negev. The driver length is 2 m and the driven part is 8 m long. Kistler pressure transducers were used for monitoring the incident shock wave velocity and the pressure at the shock tube end-wall. The transducers output signals were fed into a Tektronix 212 digital oscilloscope. The investigated samples, disks of 5 or 10 mm thickness polyurethane foam, were positioned just before the shock tube end-wall. They were fixed to the wall by 3 steel pins. Both dry and water-containing (up to volume concentration of about 70%) foams were studied. Several (1 to 9) foam layers were used in order to obtain the required homogeneous distribution of water in the samples cross-sections (equal to the initial concentration of absorbed water) and for studying the shock wave propagation through the foam layers. Special attention was given to obtaining a homogeneous water distribution in the built-up samples both parallel and perpendicular to the shock wave propagation direction. The difference in water concentrations for different

Shock Waves @ Marseille III
Editors: R. Brun, L. Z. Dumitrescu

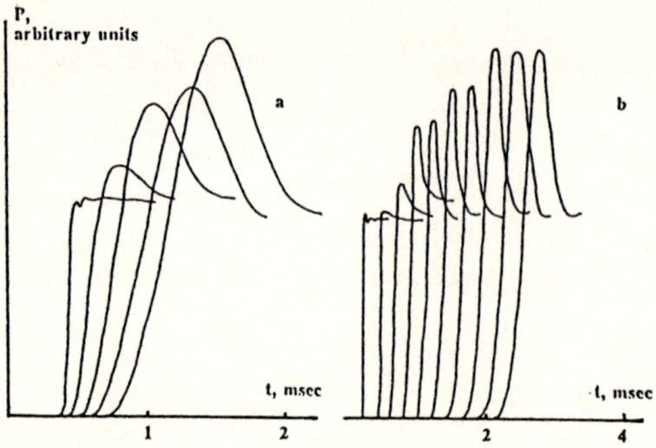

Fig. 1. Pressure readings of: a) water-saturated foam (1 to 5 5-mm foam sheets, ~30 vol.% water);b)(0 to 9 5-mm sheets)

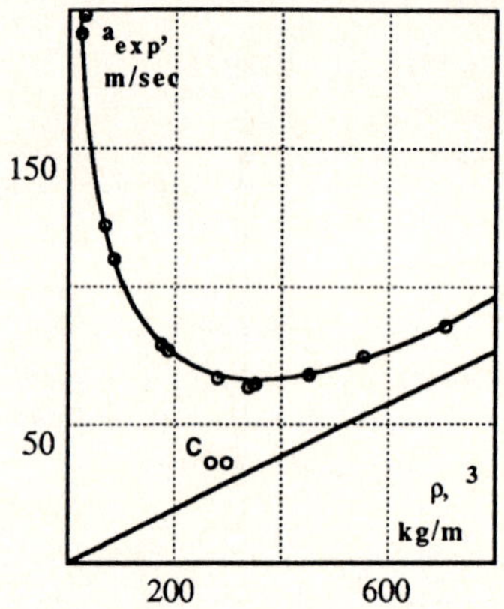

Fig. 2. Experimental and calculated (solid line) values of $a_{exp} = 2\,L_0/t$. C_{00} as a function of sample density

layers, in the same built-up sample, was less than 0.5%. The water desorbtion rate was studied by measuring the amount of desorbed water from a sample sheet which was placed edgewise on a digital balance pan. Maximal water losses during 4 to 5 minutes, the time required for sample preparation and installation in the shock tube, were less than 1% for water concentration of about 70% by volume. The end-wall pressure profiles obtained for two sets of experiments are shown on Fig.1a,b.

3. Results and discussion

In Fig.2 the relation betweens a_{exp} and the density of the investigated water-containing foams is shown. The sample density was defined as the mass of water contained in the sample divided by its volume, and a_{exp} was defined as: $a_{exp} = 2\ L_0/t$, where L_0 is the thickness of the foam sample and t is the time passed from the beginning of the pressure pulse recording until its maximum value is reached. Let us consider the shock wave behavior in the $u-P$ (particle velocity - pressure) and $x-t$ (distance - time) diagrams. These are shown in Figs.3a, b, and c respectively: In Fig.3a the head-on collision of the incident shock wave with a rigid wall is shown, while in Fig.3b,c the case appropriate to the presence of foam layers, near the wall, is described. The incident (Fig.3a) shock wave propagates into a quiescent gas and changes the gas pressure from P_0 (upstream of the shock wave) to P_1 (behind the shock front). For relatively weak shock waves, (in the present work M was equal to 1.28), the pressure, the gas particle velocity behind the shock front, and the shock wave Mach number can be related as follows (Landau and Lifshitz 1987):

$$P_1 - P_0 = \rho_0\ C_0\ u_1, \tag{1}$$

$$M = 1 + \frac{\gamma + 1}{4\gamma}\ \frac{P_1 - P_0}{P_0}, \tag{2}$$

where $\rho_0\ C_0$ is the acoustic impedance of the gas, ρ_0 and C_0 are the density and speed of sound of the pre-shocked gas (1.22 kg/m^3 and 330 m/sec), respectively, u_1 the air particle velocity and γ the adiabatic exponent of the gas; $\gamma = 1.4$. In the same approximation, the gas state after reflection of the incident shock wave, of pressure P_1, from the shock tube end-wall is:

$$P - P_1 = \rho_1\ C_1\ (u_1 - u), \tag{3}$$

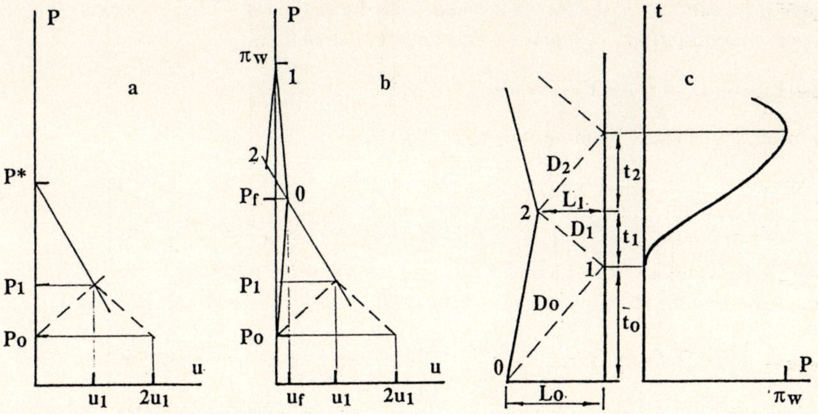

Fig. 3. $P-u$ diagrams of gas shock wave reflection from rigid wall (a), reflections in presence of a foam sample (b) and $t-x(-P)$ diagram of wave propagation in the foam (c)

where u is the gas particle velocity behind the reflected shock wave front. From Eqs.1 and 2 u_1 could be expressed as:

$$u_1 = \frac{P_1 - P_0}{\rho_0 \, C_0} = \frac{0.653 \, P_0}{\rho_0 \, C_0}.$$

The acoustic impedance behind the reflected shock wave, $\rho_1 \, C_1$, could be expressed in term of $\rho_0 \, C_0$ as (Landau and Lifshitz 1987),

$$\rho_1 \, C_1 = \rho_0 \, C_0 \, (2 \, M - 1) = 1.560 \, \rho_0 \, C_0,$$

which gives, in the $P-u$ plane, for the gas state behind the reflected shock wave

$$P = 2.672 \, P_0 - 1.560 \, \rho_0 \, C_0 \, u. \tag{4}$$

For the case of a rigid wall a simple boundary condition prevails, i.e., $u \equiv 0$, and for such a case the gas pressure is $P = 2.672 \, P_0 \equiv P^*$. For the case when a sample, composed of foam layers, is attached to the rigid wall, $u \not\equiv 0$. When the sample thickness L_0 is relatively small ($L_0 \ll 2\Lambda$, Λ being the driver section length) the pressure acting on the shock tube end-wall, after a number of shock reflections within the foam, approaches the value P^*. This can be verified from pressure records, obtained by a Kistler gauge installed at the shock tube end-wall; see Fig.1. The head-on collision of the incident shock wave with the foam sample is analyzed below. The shock wave reflections from the foam leading and trailing interfaces are shown in Fig.3b. For the shock wave propagating into the foam (with shock front velocity D) we use an equation of state which is normally used for shock waves in solids, (Zel'dovich and Raizer 1966),

$$D = C_{00} + s \, u, \tag{5}$$

where C_{00} is close to the speed of sound in the undisturbed material, s is a constant which depends on the material and u is the particle velocity in the material. Combining Eq.5 with the equations of mass and momentum conservation across the shock front, leads to the following expression for the pressure behind the shock front (in the foam sample),

$$P - P_0 = \rho_{00} \, D \, u = \rho_{00}(C_{00} + s \, u) \, u, \tag{6}$$

where P_0 is the initial pressure in the shock tube; (in the present case it was equal to the atmospheric pressure) and ρ_{00} is the foam's initial density. Since the same pressure exists on both sides of the gas - foam interface, by equating Eqs.4 and 6, one obtains

$$1.672\ P_0 - 1.560\,\rho_0\,C_0\,u = \rho_{00}(C_{00} + s\,u)u. \tag{7}$$

The solution of Eq.7 yields the foam particle velocity; it is

$$u_f = \frac{C_{00}}{2\ s}\left\{\left[\left(1 + 1.560\ \frac{\rho_0\,C_0}{\rho_{00}\,C_{00}}\right)^2 + \frac{4\ \pi^*\ s}{\rho_{00}\ C_{00}^2}\right]^{1/2} - 1\right\}, \tag{8}$$

where $\pi^* = P^* - P_0$. Combining Eqs.5 and 8 provides the following relation for the shock wave velocity in the foam and for the pressure π_w, acting on the shock tube end-wall (see Fig.3b):

$$D = C_{00}+s\,u = C_{00}+\frac{C_{00}}{2}\left(1+1.560\frac{\rho_0 C_0}{\rho_{00}C_{00}}\right)\times\left\{\left[\left(1 + 1.560\frac{\rho_0 C_0}{\rho_{00}C_{00}}\right)^2 + \frac{4\ \pi^*\ s}{\rho_{00}\ C_{00}^2}\right]^{1/2} - 1\right\} \tag{9}$$

$$\pi_w = 2\ \rho_{00}(C_{00} + s\,u)\ u = 2\ \frac{\rho_{00}\ C_{00}^2}{s} \times \left[\left(1 + 1.560\frac{\rho_0\ C_0}{\rho_{00}\ C_{00}}\right)^2 + \frac{4\ \pi^*\ s}{\rho_{00}\ C_{00}^2}\right]^{1/2} \times$$

$$\times\left\{\left[\left(1 + 1.560\frac{\rho_0\ C_0}{\rho_{00}\ C_{00}}\right)^2 + \frac{4\ \pi^*\ s}{\rho_{00}\ C_{00}^2}\right]^{1/2} - 1\right\}. \tag{10}$$

It is apparent from the $x - t$ diagram, showing the reflections of the shock wave travelling in the foam (Fig.3c), that the measured velocity $a_{exp} = 2L_0/t$ may be evaluated as

$$a_{exp} = \frac{2\ L_0}{t_1 + t_2} \tag{11}$$

where $t_0 = L_0/D_0$, $t_1 = L_1/D_1$ and $t_2 = L_1/D_2$ are durations of the first, the second and the third shock wave passages through the sample, respectively. For one-dimensional compression of the gas-foam interface, its velocity u (sample material particle velocity in laboratory coordinate system) may be deduced from

$$L_1 = L_0 + u\ (t_0 + t_1) = L_0 - u\ L_0/D_0 - u\ L_1/D_1 = L_0\ \frac{D_0 - u}{D_1 - u}\ D_1/D_0. \tag{12}$$

Assuming $D_0 \approx D_2$, see Fig.3b, for the experimentally measured values of a_{exp} one has,

$$a_{exp} = \frac{2\ L_0}{t_1 + t_2} = \frac{2\ L_0}{L_1/D_1 + L_1/D_2} = \frac{2\ L_0}{L_1/D_1 + L_1/D_2}, \tag{13}$$

and substituting Eq. 12 into Eq. 13 yields

$$a_{exp} = \frac{2\ (C_{00} + s\,u)^2\ (C_{00} + 2\,s\,u + u)}{(C_{00} + s\,u - u)\ (2\,C_{00} + 3\,u)}. \tag{14}$$

It is of interest to compare the experimental values of a_{exp} with the predictions of Eq.14. It can be done only when C_{00} and s are known. It can be shown, (Zel'dovich and Raizer 1966), that the parameter s, which is used in equations of state of type (5) is the inverse value of the maximum material compressibility (for infinitely large pressure values). For weak shock waves it is safe to assume that the solid matrix of the porous material is incompressible, i.e. ρ_s =Const., and, therefore, the parameter s could be expressed as:

$$s = \frac{V_{00}}{V_{00} - V} = \frac{1}{1 - \rho_{00}/\rho} = \frac{1}{1 - \rho_{00}/\rho_s}, \tag{15}$$

where ρ_{00} and ρ are the initial and final densities of the shock-compressed material. Proposing that the speed of sound C_{00} is a linear function of initial material density ρ_{00}, $C_{00} = k\ \rho_{00}/\rho_s$,, we found that the best approximation of experimentally measured values of a_{exp} may be achieved when

$$C_{00} = (96 \mp 4)\ \frac{\rho_{00}}{\rho_s} \tag{16}$$

The obtained values for a_{exp} calculated from Eq.14 with s from Eq.15 and C_{00} from Eq.16 are shown on Fig.2. It appears from Fig.2 that for a wide range of water concentrations the proposed equation of state of water-saturated porous polyurethane describes the material behavior with an accuracy of better than 5%. The velocity of the gas-sample interface, calculated from Eq.8 and shown Fig.4, is in good agreement with the experimental results of Skews (1991) and the numerical data of Olim et al. (1993).

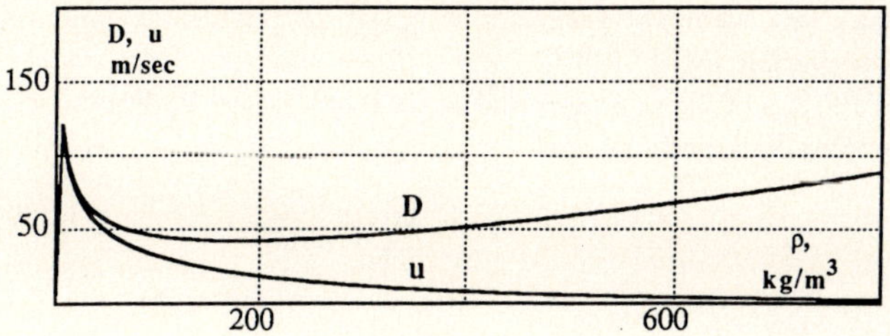

Fig. 4. Calculated concentration dependences of front (D) and particle (u) velocities in the water-containing foam

4. Conclusions

The shock equation of state for water-saturated polyurethane foam was studied for a wide range of water concentrations. It was shown that a simple type of equation of state (e.g., Eq.5) with concentration-dependent parameters: $C_{00} = 96\ \rho_{00}/\rho_s$ m/sec and $s = \rho_s/(\rho_s - \rho_{00})$ provides a satisfactory description of the obtained experimental results. The C_{00} values for dry foam ($\rho_{00} = 30$ kg/m^3) and for water-saturated foam ($\rho_{00} = 1000$ kg/m$^3 = \rho_s$) are 2.9 and 96 m/sec. This is smaller by a factor of 1/2 than the longitudinal sound velocity in these materials.

References

Baer MR, Nunziato JW (1986) A two-phase mixture theory for the deflagration-to-detonation transition (DDT) in reactive granular materials. Intl. J. Multiphase Flow 12: 861-889

Baer MR (1992) A numerical study of shock wave reflection on low density foam. Shock Waves 2: 121-124

Ben-Dor G, Zaretsky E (1993) Head-on interaction of planar shock wave with polyurethane foam - a semi-empirical model. Archive of Applied Mechanics (to be published)

Gelfand BE, Gubanov AV, Timofeev EJ (1983) Interactions of shock wave in air with a porous screen. Izv. Akademii Nauk SSSR, Mechanika Zhidkosti i Gaza 4: 78-84

Gvozdeva LG, Faresov YM (1985) Calculating the parameters of steady shock wave in porous compressible media. Zhurn. Tech. Fiz. 55: 773-775

Gvozdeva LG, Faresov YM, Brossard J, Charpentier N (1986) Normal shock wave reflection on porous compressible material. In: Bowen JR, Leyer JC, Soloukhin RI (eds) Dynamics of Explosions. Progress in Astronautics and Aeronautics 106: 155-165

Gvozdeva LG, Liakhov VN, Raevsky DK, Faresov YM (1987) Numerical study of shock wave propagation in gas and porous media. Fiz. Gorenya i Vzryva 4: 125-129

Landau LD, Lifshitz EM (1987) Fluid Mechanics, second edition. Pergamon Press, Oxford, New York, Beijing, Frankfurt, Sao Paulo, Sydney, Tokio, Toronto

Olim M, van Dongen MEH, Kitamura T, Takayama K (1993) Numerical simulation of shock waves in porous foams. In: Takayama K (ed) Japanese Natl. Symposium on Shock Waves, Sendai, pp. 261-264

Seitz MW, Skews BW (1992) Three-dimensional effects in the study of shock wave loading of porous compressible foams. In: Takayama K (ed) Proc. 18th Intl. Symp. on Shock Waves, Sendai. Springer -Verlag, pp 517-520

Skews BW (1991) The reflected pressure field in the interaction of weak shock waves with a compressible foam. Shock Waves 1: 205-211

Skews BW, Atkins MD, Seitz MW (1992) Gas dynamic and physical behavior of compressible porous foam struck by a weak shock wave. In: Takayama K (ed) Proc. 18th Intl. Symp. on Shock Waves, Sendai. Springer -Verlag, pp 511-516

Zeldovich YB, Raizer YP (1966) Physics of Shock Waves and High-Temperature Hydrodynamic Phenomena. Academic Press, New-York and London

Head-on Collision of Shock Waves with Porous Materials: Experimental and Numerical Investigation

G. Ben-Dor*, O. Igra*, G. Mazor*, H. Onodera† and K. Takayama‡
* Pearlstone Center for Aeronautical Engineering Studies, Department of Mechanical Engineering, Ben-Gurion University of the Negev, Beer Sheva, Israel
† Department of Mechanical Engineering, Faculty of Engineering, Iwate University, Morioka, Japan
‡ Shock Wave Research Center, Institute of Fluid Science, Tohoku University, Sendai, Japan

Abstract. The equations governing the phenomenon in which a planar shock wave collides head-on with a cellular material and interacts with it were developed by Mazor et al. (1993) using a Lagrangian approach. In addition, the numerical approach for solving the equations was briefly outlined there. In this study we report on experimental and numerical results of the head-on reflection of a planar shock wave with an open-cell polyurethane foam.

Key words: Porous materials, Numerical simulation

1. Introduction

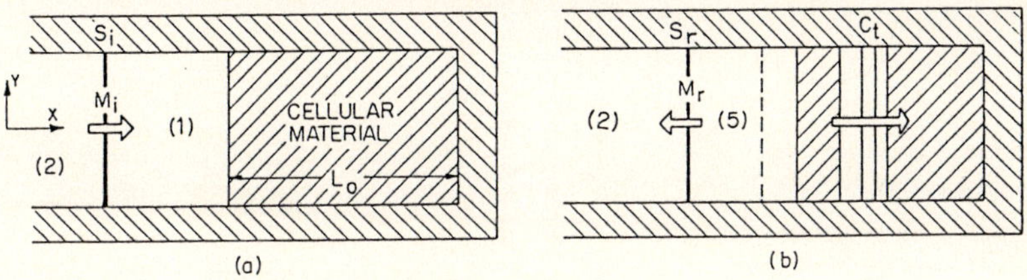

Fig. 1. Schematic illustration of the investigated interaction

The problem under consideration is shown schematically in Fig.1. Since the foam cannot expand in the $y-$ and $z-$ directions, it experiences a uni-axial strain compression, i.e.: $\epsilon_x \neq 0$, $\epsilon_y = 0$ and $\epsilon_z = 0$. As a result, $\sigma_x \neq 0$, $\sigma_y \neq 0$ and $\sigma_z \neq 0$.

The analytical investigation resulted in a set of ten governing equations. For details, see Mazor et al. (1993). The interest in better understanding the head-on reflection of a planar shock wave from a foam has grown quite intensively in the past decade. However, most of the reported researches related to this problem were experimental, e.g., Gvozdeva et al. (1985) and Skews (1991). It was found in these studies that: (i) The peak pressure acting on the shock-tube end-wall is higher than the pressure which would have been obtained, had the incident shock wave reflected head-on from the shock-tube end-wall without the presence of a foam; (ii) the peak pressure increases as the initial length of the foam increases. In addition, as the initial length of the foam increases the increase in the peak pressure seems to approach an asymptotic value; (iii) the front edge of the foam moves at a constant velocity. These results, which are not intuitive and partially yet unexplained, motivated us to undertake the present study. The main purposes of this study were: 1) to experimentally investigate the problem under consideration in order to better understand the phenomenon and the parameters affecting it, and 2) to develop a simulation code capable of reproducing the above-presented results.

Shock Waves @ Marseille III
Editors: R. Brun, L. Z. Dumitrescu

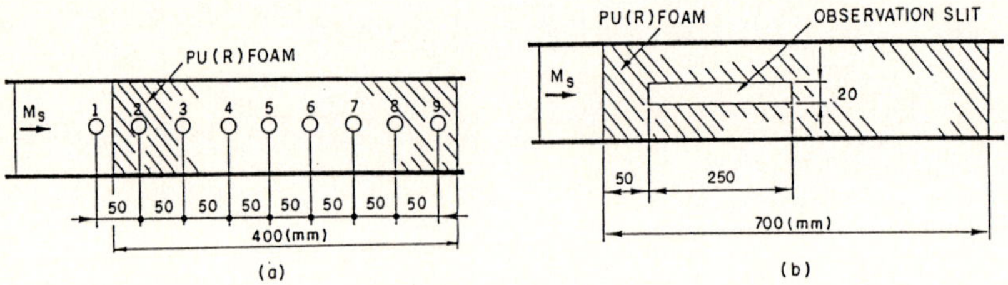

Fig. 2. Experimental set-ups used in the present study: a) for pressure and front edge displacement measurements, b) for holographic interferometry measurements of the waves propagating inside the foam

2. Experimental study

Open-cell flexible polyurethane foam was used. The foam density was $\rho_c = 31.5$ Kg/m^3 while the pure flexible polyurethane density is 1100 Kg/m^3. These values result in a relative density of about $\rho_c/\rho_s = 0.02625$, i.e., a porosity of $\phi = 0.97375$.

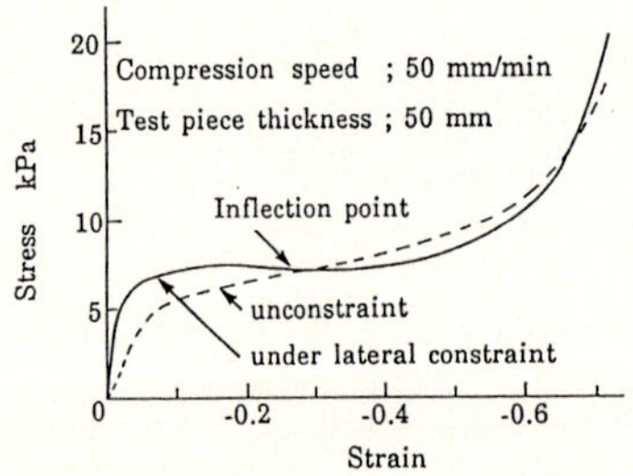

Fig. 3. Stress-strain curve of the foam

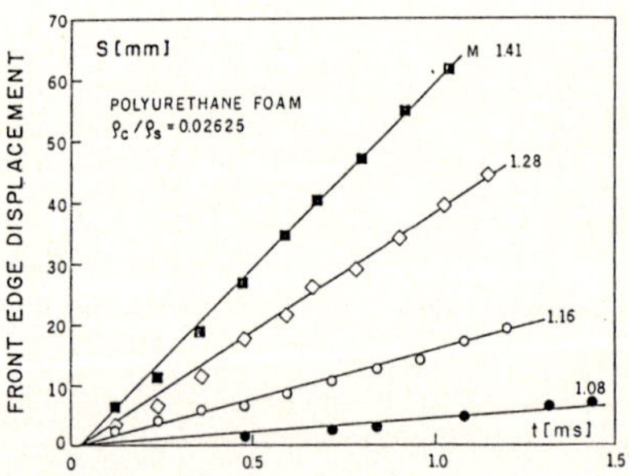

Fig. 4. Displacement of the foam front-edge for different incident shock wave Mach numbers

The experiments were conducted in the 60 mm×150 mm shock tube of the Shock Wave Research Center of the Institute of Fluid Science of Tohoku University. The major experimental set-up is shown in Fig.2a. The foam was 400 mm long. Nine Kistler-type pressure transducers were flush mounted, 50 mm apart, on the shock-tube side-walls. The first gauge, located 25 mm ahead of the foam front-edge recorded the pressure changes across the incident shock wave and the shock wave which reflected head-on from the foam front-edge. The experimental set-up shown in Fig.2b was used for flow visualization of the waves propagating inside the foam. For this case a 20 mm×250 mm slit was cut along the foam-centerline. Its leading edge was located 50 mm behind the foam front-edge. The foam length in these experiments was 700 mm.

The foam stress-strain relation was measured using a SHIMAZU material testing machine (Auto-Graph AG 5000-C). A typical stress-strain curve is shown in Fig.3. It is evident that the foam stress-strain relation depends on the compression mode. By measuring the slope of the stress-strain curve in its elastic regime the Young modulus could be deduced. It was found that the Young modulus depends on both the compression speed and on the foam initial length.

The time dependence of the displacement of the foam front-edge as obtained by streak photography, for four different incident shock wave Mach numbers, is shown in Fig.4. The streak photographs were obtained using an image converter camera (Hadland Photonics IMACON 790) with an argon-ion laser (Spectra Physics 2030) as a light source. The almost perfectly straight trajectories of the foam front-edge indicate that its velocity is almost constant, in agreement with the experimental results of Gvozdeva et al. (1985).

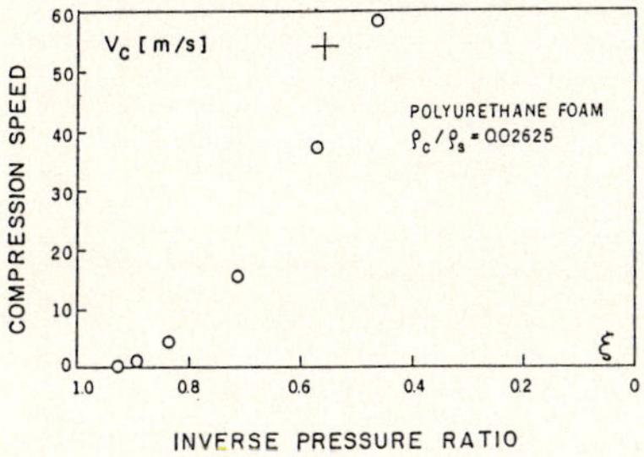

Fig. 5. Dependence of the foam front-edge velocity on the inverse shock wave pressure ratio

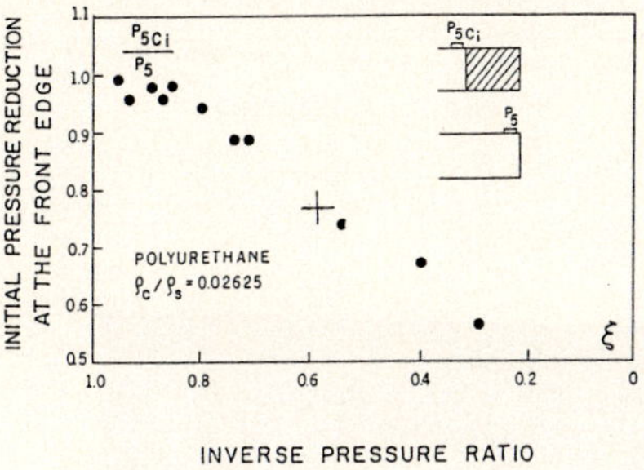

Fig. 6. Dependence of the initial pressure relative reduction at the foam front-edge on the inverse shock wave pressure ratio

The dependence of compression speed, as deduced from Fig.4, on the inverse pressure ratio across the incident shock wave, ξ, is shown in Fig.5. In spite of the fact that the pressure induced by the reflected shock wave on the foam front-edge increases as the incident shock wave Mach number increases, its value is smaller than the pressure that would have been obtained had the incident shock wave reflected head-on from a solid surface. This is shown in Fig.6.

The recorded wave velocity along a 2 m long foam, as obtained by time counters, is shown in Fig.7. It is evident that the transmitted wave attenuates. However, in the case of relatively short foams and moderate incident shock wave Mach numbers it could be assumed that the transmitted wave velocity remains nearly constant.

Holographic interferometry was applied in order to study the waves propagating inside the foam. This was achieved by cutting a slit along the foam centerline as shown in Fig.2b. A typical interferogram is shown in Fig.8a. The relative density difference, $\Delta\rho/\rho_0$, between neighboring fringes is 0.038. If one assumes that the entropy is constant, then the relative pressure difference, $\Delta P/P_0$ between the neighboring fringes is 0.0536. The pressure distributions as obtained from the interferogram and by Kistler pressure transducers are shown in Fig.8b. Good agreement between the two pressure measurements is evident.

3. Numerical study

The basic details of the numerical method and approach are outlined in Mazor et al. (1992). The governing equations of the considered problem consist of ten equations of which only the constitutive relation, i.e., $\sigma = \sigma(\epsilon)$, depends on the porous material specifics. Consequently, the appropriate stress-strain relation of an open-cell polyurethane foam is required. A well-tailored set of empirical stress-strain relations was developed by Ben-Dor et al. (1994)

When the incident shock wave collides head-on with the foam front-edge it transmits into it compression waves which can converge to a shock wave provided the stress-strain relation fulfills

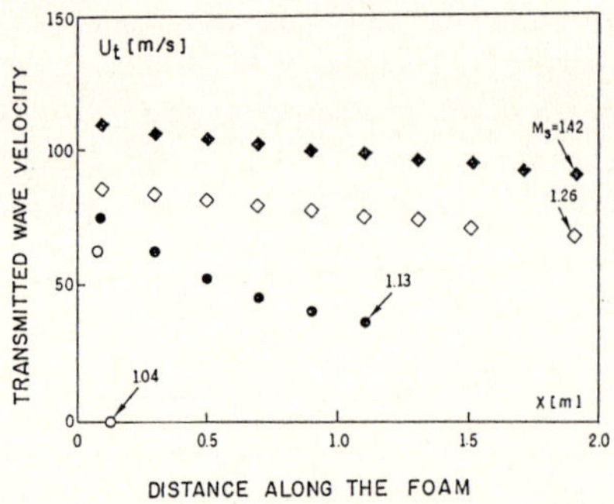

Fig. 7. Dependence of the wave velocity on its propagation distance inside the foam for different incident shock wave Mach numbers

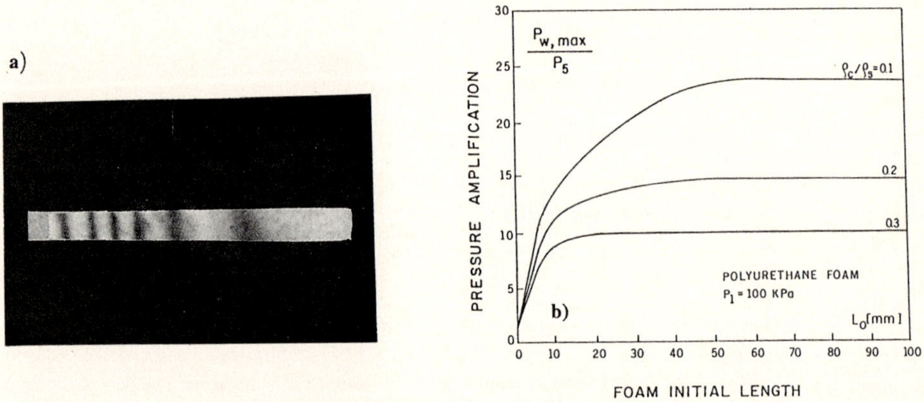

Fig. 8. a) Holographic interferogram of the wave propagating inside the foam, b) Comparison of the pressure distribution inside the foam as obtained from pressure and optical measurements. $M_s = 1.28$

the following conditions:

$$\frac{\partial \sigma}{\partial \lambda} > 0 \quad \text{and} \quad \frac{\partial^2 \sigma}{\partial \lambda^2} > 0$$

Applying these convergence conditions on the stress-strain relations indicates that foams are the only known materials in which a shock wave can be generated by an uni-axial strain compression.

During the course of the present numerical study the governing equations were solved for a flexible open-cell polyurethane foam having the following relative densities; ρ_c/ρ_s =0.1, 0.2 and 0.3. In order to verify and better understand the experimental results of Gvozdeva et al. (1985) four different initial lengths were assigned to the foam, i.e., L =10, 20, 50 and 100 mm. The interaction was simulated for the above parameters and two incident shock Mach numbers, i.e., M_s =2 and 5. The initial conditions ahead of the incident shock wave were: $u_{g_1} = 0, P_1 = 100$ KPa and $T_1 = 300$ K. The foam cross section was 0.01 m^2 and the properties of the flexible polyurethane foam as taken from Gibson and Ashby (1988) were $\rho_s = 1100$ Kg/m^3, $E_s = 45$ MPa and $\nu_s = 0.45$. In the following only some typical results are presented.

The peak pressure developed at the shock-tube end-wall, $P_{w,max}$, relative to the pressure which would have been developed there, had the incident shock wave reflected directly from the shock-tube end-wall as a function of the initial length of the foam, L_0, for different relative

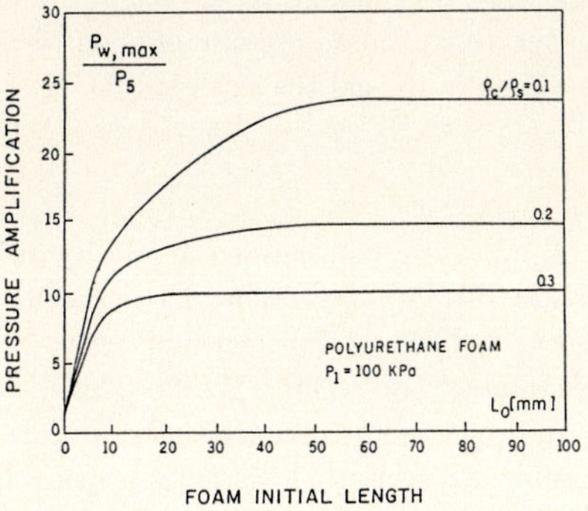

Fig. 9. Dependence of the peak pressure at the shock-tube end-wall on the foam initial lengh and its relative density

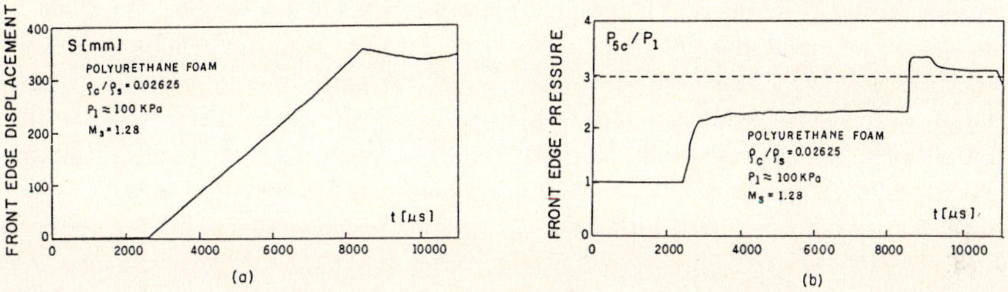

Fig. 10. Numerical simulation of an actual experiment for $M_s = 1.28$. a) Displacement of the foam front-edge. b) Pressure history at the foam front-edge

densities is shown in Fig.9 for the case $M_s = 2$. The results ressemble an increase in $P_{w,max}$ as L_0 increases until a certain critical value of L_0 after which $P_{w,max}$ reaches an asymptotic value. This critical length increases as the foam relative density decreases. It is also evident that the smaller the relative density, the larger is the peak pressure at the end-wall. This could be attributed to the fact that foams having smaller relative densities are easier to accelerate and hence they reach higher material velocities, which eventually translate into higher end-wall "stagnation" pressures.

4. Comparison with experiments

Using the manufacturer's data together with $M_s = 1.28$, $P_1 = 100$ KPa, $T_1 = 300$ K, $L_0 = 400$ mm, and a foam cross section of 90 cm^2, resulted in the front edge displacement shown in Fig.10a, and the pressure history at the foam front-edge shown in Fig.10b. Based on this figure, the foam compression speed is about 55 m/s and the pressure acting on the front edge is about 0.77 P_5. These values, which have been added (as cross signs) to Figs.5 and 6, respectively, seem to show a poor agreement with the measured compression speeds and a very good agreement with the measured pressures. The poor agreement with the compression speed could be attributed to the unknown correct value of the Young modulus. As mentioned earlier the foam Young modulus depends on the compression speed. The value used in the computation was deduced from Fig.3 in which the compression speed is 0.833×10^{-3} m/s while the actual compression speed is of the order of 50-150 m/s. This difference of about five orders of magnitude could explain the over-predicted value of the compression speed. Note that the value of the Young modulus as deduced by us was $E_s = 194.5$ MPa, however, as quoted by Gibson and Ashby (1988) the Young modulus of polyurethane foam is about five time larger, i.e., 1100 MPa. Replacing only this value in the computation results in a compression speed of about 48 m/s (as compared to about 55 m/s in the previous computation and to about 38 m/s in the experiment) and a presssure of about 0.79 P_5 (as compared to about 0.77 P_5 both in the previous computation and the experiment).

5. Summary and conclusions

The head-on interaction of a planar shock wave with an open-cell foam has been investigated both experimentally and numerically for the case of a uni-axial strain compression. The numerical predictions were found to qualitatively agree both with our and others' experimental results. A quantitative comparison could not be made since the computer code requires the knowledge of some properties of the foam which unfortunately could not be measured. Consequently, in order to overcome this shortcoming there is a need to develop some experimental techniques which will enable one to obtain the required mechanical properties of the foam.

The major conclusions regarding the investigated interaction are: (i) the foam stress-strain relation depends on its: a) mechanical behavior, b) internal structure, c) compression mode, d) compression speed, and e) initial length; (ii) due to the presence of the foam the induced pressure on the shock-tube end-wall is enhanced as compared to the pressure which is obtained without a foam. The pressure enhancement increases as the relative density of the foam decreases. The pressure enhancement increases as the initial length of the foam increases until a critical length is reached, beyond which the pressure enhancement remains unchanged. The critical length increases as the relative density decreases; (iii) during the compression phase, the foam front-edge moves with an almost constant velocity. The velocity is larger when the relative density is smaller. The velocity is larger when the incident shock wave is stronger. The time duration of the almost constant velocity is shorter when the incident shock wave Mach number is larger; (iv) during the relaxation (decompression) phase, the foam front-edge moves with an almost constant velocity in a direction opposite to that during the compression phase; (v) at the early stages of the interaction, the pressure induced on the foam front-edge is smaller than the pressure induced by a shock wave reflecting head-on from a solid wall. At a later stage of the interaction, the pressure induced on the foam front-edge increases beyond the pressure induced by a shock wave reflecting head-on from a solid wall; (vi) the wave propagating inside the foam attenuates.

References

Ben-Dor G, Cederbaum G, Mazor G, Igra O (1993) Well-tailored compressive stress-strain relations for electromeric foams in a uni-axial stress compression. Submitted for publication to J. Mat. Sci.

Gibson BE, Ashby MF (1988) Cellular Solids: Structure and Properties, Pergamon Press, 1st Ed.

Gvozdeva LG, Faresov YuM, Fokeev VP (1985), Interaction between air shock waves and porous compressible materials. Sov. Phys. Appl. Math. and Tech. Phys. 3: 111-115

Mazor G, Igra O, Ben-Dor G, Mond M, Reichenbach H (1992) Head-on collision of normal shock waves with a rubber supported rod. Trans. Roy. Soc. Lond. Ser. A338: 237-269

Mazor G, Ben-Dor G, Igra O, Sorek S, Takayama K (1993) Shock wave interaction with cellular materials. Part: I Analytical investigation and governing equations. Shock Waves, to be published

Skews BW (1991) The reflected pressure field in the interaction of weak shock waves with a compressible foam. Shock Waves 1: 205-211.

Head-on Collision of Normal Shock Waves with Rigid Porous Materials

A. Levy*, **B.W. Skews**†, **G. Ben-Dor*** **and S. Sorek***
*Pearlstone Center for Aeronautical Engineering Studies, Department of Mechanical Engineering, Ben-Gurion University of the Negev, Beer Sheva, Israel
†School of Mechanical Engineering, University of Witwatersrand, Johannesburg, South Africa

Abstract. The head-on collision of a planar shock wave with a rigid porous material has been investigated experimentally. The study indicated that unlike the reflection from a flexible porous material, where the transmitted compression waves do not converge to a sharp shock wave, in the case of a rigid porous material the transmitted compression waves do converge to a sharp shock wave, which decays as it propagates along the porous material.

Key words: Porous materials, Compression wave, Reflection/interaction process

1. Introduction

The investigation of the head-on collision of normal shock waves with porous materials has increased quite intensively in the past decade due to the application of this phenomenon to many engineering fields.

Table 1. The experimental models

Material	Model	Pore per inch (ppi)	Air porosity	Model length (mm)
Silicon Carbide*	I	10	0.728±0.016	40, 60, 81
(SiC)	II	20	0.745±0.001	41, 62, 83
Alumina*	III	30	0.814±0.010	48, 93
(Al_2O_3)	IV	40	0.821±0.007	50, 99

*The structure of the material was open-cell

In the present study air-saturated rigid porous materials rather then flexible ones were used. The motivation for this study was to experimentally investigate the extent of the differences in the entire interaction/reflection process when the flexible porous material is replaced by a rigid one. In addition, Bear et al. (1992), Sorek et al. (1992), Levy et al. (1993a), have recently initiated analytical models for predicting the flowfield which is developed inside a rigid porous material under a variety of initial conditions and constrains. The study reported in this paper is also aimed at providing an experimental data base for validating their and hopefully others' analytical models.

2. Experimental set-up

The experiments were conducted in the 75 mm×75 mm shock tube of the School of Mechanical Engineering of the University of Witwatersrand in Johannesburg, South Africa. The incident shock wave Mach number range was $1.2 \leq M_i \leq 1.7$, the initial pressures and temperatures throughout this study were about 830 mbar and about 288 K, respectively.

The rigid porous materials were made of silicon carbide (SiC) and alumina (Al_2O_3). The SiC-manufactured porous material had either 10 or 20 pores per inch and the Al_2O_3-manufactured porous material has either 30 or 40 pores per inch. The air porosity of these porous materials as

Shock Waves @ Marseille III
Editors: R. Brun, L. Z. Dumitrescu

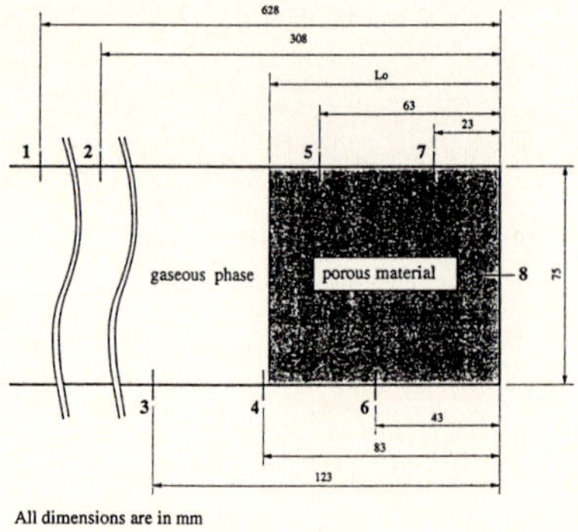

Fig. 1. Schematic illustration of the experimental set-up and the location of the pressure transducers (numbered 1 to 8)

well as the initial length, L_0, of the models which were cut from these materials are given in Table 1. The cross-section of the models was 75 mm×75 mm and hence they experienced a uni-axial strain compression. The models were mounted in such a way that their back edge was supported by the shock-tube end-wall. In order not to scratch the high-quality shock-tube windows and walls the models were enveloped by a very thin and mechanically strong cardboard.

A schematic drawing of the experimental set-up used during the present experimental study is shown in Fig.1. Although there were eight pressure transducer-ports, only seven of them were actually used in each experiment. The pressure transducers in ports 1 and 2 were used to measure the incident shock wave velocity. Pressure transducers were mounted in ports 4, 5, 6, 7 and 8 for the models having an initial length up to (inclusive) 83 mm (see Table 1). For the two alumina models with initial lengths of 93 mm and 99 mm the pressure transducer of port 4 was moved to port 3. This was done in order to have a measurement of the pressure changes that were caused by the shock wave which reflected backwards from the front edge of the porous material.

3. Experimental results

3.1. Gas pressure inside the porous material

Fig.2 illustrates the pressure histories of the gas in the pores as recorded by the various pressure transducers for an experiment with an 81 mm long model made of SiC having 10 pores per inch. The incident shock wave Mach number was M_i=1.542. The four pressure histories shown in Fig.2 were recorded by the pressure transducers which were mounted in the ports numbered 5, 6, 7 and 8 in Fig.1. It is evident from these pressure histories that:

a) The pressure jump across the compaction wave is sharp and resembled a typical pressure jump across a shock wave (note, the sharp pressure jumps are indicated by arrowheads).
b) The sharp pressure jumps are followed by continuous pressure increases which finally reach asymptotic values.
c) As the compaction wave propagates the pressure jump across it decreases, but the final pressure remains almost unchanged.

The decrease in the pressure jump across the compaction wave as measured from Fig.2 is shown as a function of distance propagated by the compaction wave in Fig.3. The distance propagated by the compaction wave as function of time is shown in Fig.4, which indicates that the compaction wave velocity has a nearly constant value of 330 m/s. In spite of this fact, Fig.3 indicates that the initial pressure jump across the compaction wave decreases. These supposedly contradicting experimental observations suggest that the compaction wave broadens as it propagates along the porous material and develops a dispersed structure. Note, that although the initial jump across the compaction wave decreases, the overall pressure increase remains almost unchanged. It should also be noted here that the compaction wave velocity, as deduced from Fig.4, is slightly smaller than the speed of sound of the air filling the pores. (This will be further discussed in the next section). As a result of the observation that the sharp pressure jump across the transmitted compaction wave decays, it is clear that if the porous model is short enough, then the compaction wave could reach the shock-tube end-wall prior to its degeneration to either a fully dispersed or a sound wave.

The pressure histories shown in Fig.2 indicate that although a compaction wave having a sharp jump of about 105 KPa (as estimated from Fig.3) collided head-on with the shock-tube end-wall, there is no noticeable evidence of a sharp reflected compaction wave. Furthermore, the pressure history, as recorded by the pressure transducer which was mounted in port 8, at the shock-tube end-wall, clearly indicates that the pressure there does not remain uniform, as it should have been had a compaction wave reflected there. Instead, it is seen to continuously increase until it levels off at a value slightly higher than the pressure, P_5, that would have been obtained had the incident shock wave (M_i=1.542) collided head-on with the shock-tube end-wall without the presence of the porous material.

3.2. Gas pressure outside the porous material

The pressure histories, as recorded in the same experiment by the pressure transducers which were mounted in the ports numbered 1, 2 and 4 in Fig.1, are shown in Fig.5. The pressure histories, as recorded by the pressure transducers mounted in ports 1 and 2, indicate the passage of two shock waves across them. The first step jump is due to the incident shock wave, and the second one is due to the shock wave which reflects head-on from the front edge of the porous material. Since the center of the pressure transducer mounted in port 4 was only 2 mm ahead of the front edge of the porous material, and since the diameter of the transducer was 3 mm, it did not have enough resolution to clearly separate between the pressure jumps across the incident and the reflected shock waves.

The pressure jump across the reflected shock wave, S_r, is clearly seen to increase as the reflected shock wave moves away from the front edge of the porous material. In addition, following the above-mentioned pressure jump across the reflected shock wave, the pressure is seen to continuously rise to values slightly larger than the values, P_5, which would have been obtained had the incident shock wave (M_i=1.542) reflected head-on from a rigid wall.

Although Figs.2 to 5 describe the results of only one experiment, similar results were obtained in all the experiments. However, it should be mentioned that a comparison between Models I and II (see Table 1) revealed that the initial strengths of the transmitted compaction wave were almost the same when identical incident shock waves were used, in spite of the fact that the number of pores per inch in Model II was twice the number in Model I.

This was not the case when the results of Models III and IV were compared. Here the compaction wave transmitted into Model III was clearly stronger than that transmitted into Model IV. The reason for this peculiar behavior could be attributed to the fact that while the porosity of Model II is clearly larger than that of Model I, the porosities of Models III and IV are similar. Hence the larger porosity of Model II, together with the fact that the pores were smaller, could yield a response equivalent to that of Model I, where the porosity was smaller but the pores were larger.

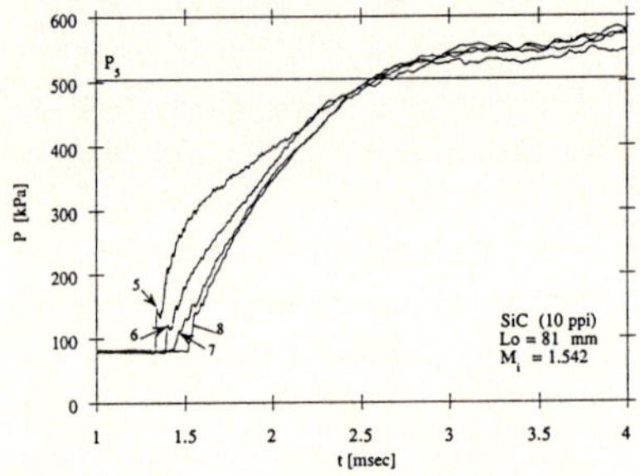

Fig. 2. The pressure history of the gas in the pores of the porous material, as recorded by the pressure transducers mounted in ports 5, 6, 7 and 8 of Fig.1

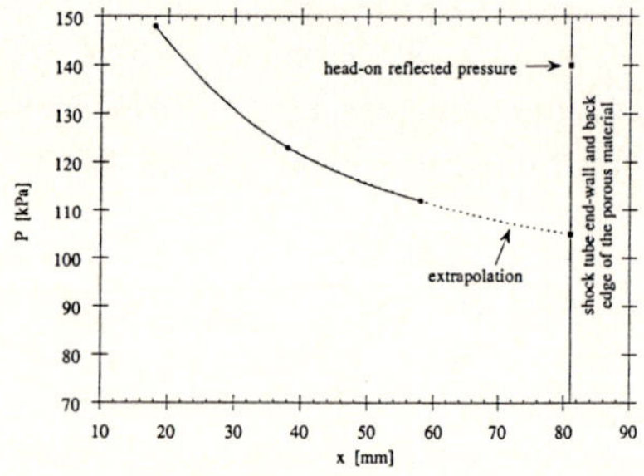

Fig. 3. The sharp pressure jump across the transmitted compaction wave as it propagates inside the porous material, as deduced from Fig.2

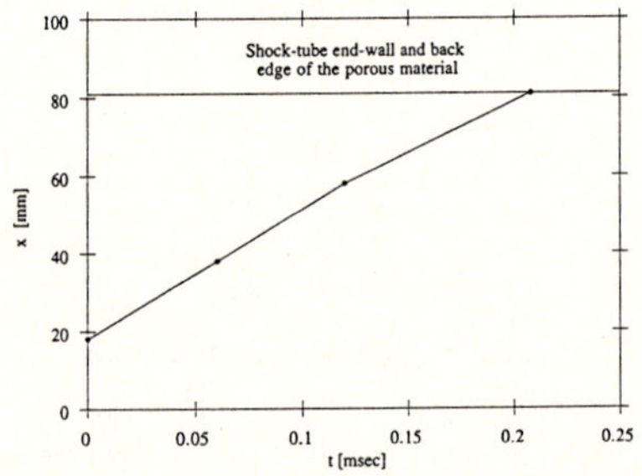

Fig. 4. The distance travelled by the transmitted compaction wave as a function of time, as deduced from Fig.2

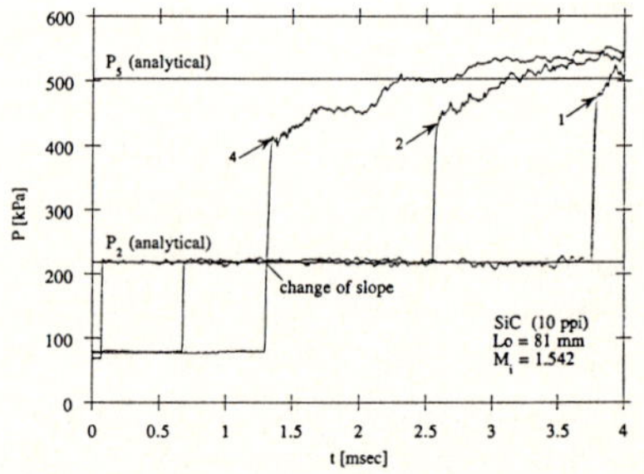

Fig. 5. The pressure history of the gaseous phase as recorded by the pressure transducers mounted in ports 1, 2, and 4 of Fig.1

4. Suggested modified phenomenological model

Based on the above-described observations, it is clear that the existing commonly accepted model for describing the investigated interaction phenomenon is insufficient, because it cannot account for the fact that the transmitted compaction wave, C_t, develops a dispersed structure, and the reflected shock wave, S_r, accelerates. Consequently, a different phenomenological model is suggested. The presently accepted model (see e.g., Skews et al. 1992) is shown in the (x,t) diagram of Fig.6. According to this model, which assumes constant velocities for all the discontinuities, the pressure, P_5, induced by the reflected shock wave, S_r, at the front edge of the porous material should remain constant until an additional shock wave, S_t, is transmitted into the gaseous phase, following the head-on reflection of the reflected compaction wave, C_r, from the front edge of the porous material. Similarly, the pressure, P_{II}, induced by the transmitted compaction wave, C_t, should also remain constant until the passage of the reflected compaction wave, C_r, which changes it from P_{II} to P_V. Furthermore, P_V should also remain constant until the reflected rarefaction wave, R_r, changes its value. The above-described model obviously cannot explain the present experimental observations in which, as mentioned earlier, the initial pressure jump across the transmitted compaction wave, C_t, decays, the reflected shock wave, S_r, accelerates, the reflected compaction wave, C_r, is not noticeable, and the pressures in states (5) and (II) are not uniform, but increase continuously until they reach a constant value long times after the initiation of the process.

Based on the observed experimental results, the continuously increasing pressure jump across the reflected shock wave, S_r, and the continuously decreasing initial pressure jump across the transmitted compaction wave, C_t, could be attributed to two processes:

a) For short times, following the instance when the incident shock wave collided head-on with the front edge of the porous material, multiple reflections, diffractions and interactions which occur inside the pores that constitute the front part of the porous material are dominant.

b) At longer times, the friction between the gaseous phase and the skeleton of the porous material becomes dominant.

The proposed phenomenological model is based on the fact that the porous materials which were investigated during the course of the present experimental study have relatively very large pores. For example the average diameter of the pores of Model I is about 2.5 mm (10 pores per inch). As a consequence, unlike the case of a head-on collision from the front-edge of the porous materials in which the size of the pores is smaller than 0.1 mm, where filtration processes are dominant, here the transmitted compaction wave propagates in a very complex geometry which forces it to undergo a variety of interaction processes at the early stages of its penetration into the porous material.

The idea of the multiple reflection process which the incident shock wave undergoes when it penetrates the front edge of the porous material is described by Levy et al. (1993b). The penetration of the shock wave through the front layer of the porous material can be viewed as a sharp front, which sweeps through the porous material, while parts of it are left behind. These parts bounce back and forth inside the cavities of the pores, parts of them interact with each other, and parts of them even find a way to emerge from the front edge of the porous material as weak compression waves, which eventually catch-up with the main reflected shock wave and increase the pressure field behind it. While penetrating the porous material, the compaction wave becomes more and more dispersed and the sharp jump across its front vanishes. Simultaneously, the friction between the gaseous phase and the rigid skeleton of the porous material becomes more and more pronounced, until it takes over and becomes the dominant mechanism. As a result of the friction interaction, the gas flowing through the open cell pores is slowed down and in turn its pressure builds up. Note that the slow-down process of the gas which flows through the pores might generate compression waves, which can also emerge from the front edge of the foam and catch up with the reflected shock wave, S_r, and further increase the pressure behind it.

Based on the above-described multi-interaction process we propose the detailed (x, t) diagram shown in Fig.7. Unlike the general (x, t) diagram shown in Fig.6, compression waves are seen to emerge from the front edge of the porous material. They eventually catch-up with the reflected shock wave, strengthen it and also increase the pressure behind it.

5. Conclusion

An experimental investigation of the head-on interaction of a planar shock wave with a rigid porous material having a large permeability has revealed that:

1. Unlike flexible porous materials, where the available experimental data indicated that the transmitted compression waves did not converge to a sharp discontinuity, here there was evidence that sharp-front transmitted waves were formed in the rigid porous models.
2. The transmitted wave velocity was found to be nearly constant and about 3-5% smaller than the speed of sound of the air filling the pores.
3. The sharp jump across the transmitted waves was found to decay as the compaction wave propagates.
4. The pressure at the shock tube end-wall, behind the reflected wave, did not remain uniform, as expected, but slightly increased. The pressure rise increased as the air porosity increased. In addition, this phenomenon became more pronounced as the length of the model increased.

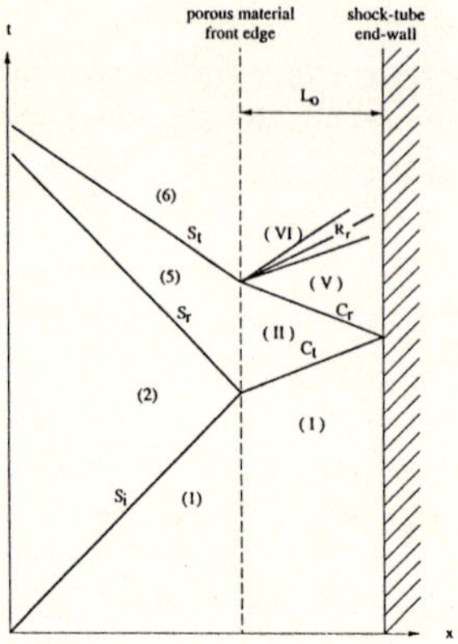

Fig. 6. The commonly accepted (x, t) diagram of the various waves involved in the interaction under consideration (see Skews et al. 1992)

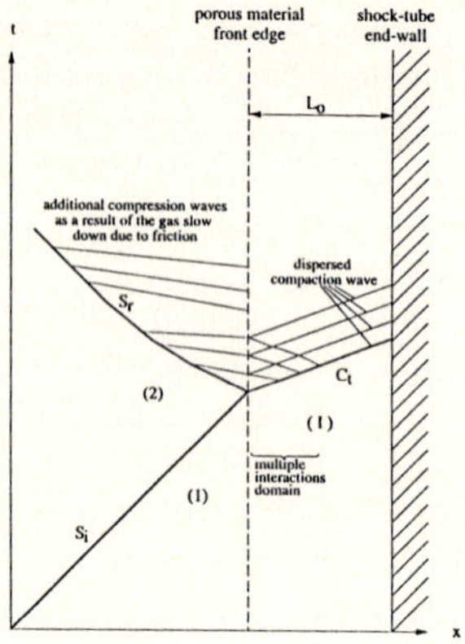

Fig. 7. The proposed detailed (x, t) diagram of the main discontinuities involved in the interaction under consideration, as suggested by our modified proposed model

5. The above-mentioned last three observations suggest that the compaction wave developed a dispersed wave structure.
6. A comparison between Models I and II (see Table 1) indicated that the strength of the transmitted compaction waves were almost identical when identical incident shock waves were used. This was not the case when the results of Models III and IV were compared.
7. The pressure jump across the reflected shock wave was found to increase as it propagated backwards.

All the above-listed observations led us to modify the generally accepted phenomenological model of this reflection/interaction process. The proposed modified model implies that the entire interaction phenomenon is probably governed at it early stages by multiple interactions of the transmitted wave with the internal walls of the pores of the porous material. Only at a later time does friction become the dominant process.

References

Bear J, Sorek S, Ben-Dor G, Mazor G (1992) Displacement waves in saturated thermoelastic porous media. Fluid Dyn. Res. 9:155-164

Levy A, Ben-Dor G, Sorek S, Bear J (1993a) Jump conditions across compaction waves in porous media. Shock Waves 3, 2: 105-111

Levy A, Ben-Dor G, Skews B, Sorek S (1993b) Head-on collision of normal shock waves with rigid porous materials. Experiments in Fluids, to be published

Skews BW, Atkins MD, Seitz MW (1992) Gas dynamic and physical behavior of compressible porous foams struck by a weak shock wave. In: Takayama (ed) Proc. 18th Intl. Symp. on Shock Waves, Sendai. Springer Verlag, pp. 511-516

Sorek S, Bear J, Ben-Dor G, Mazor G (1992) Shock wave in saturated thermoelastic porous media. Transport in Porous Media 9:3-13

Slow Detonation in Filtration Combustion

Vladimir Gol'dshtein*, **Isaak Shreiber**[†] **and Gregory Sivashinsky**[‡]
* Department of Mathematics and Computer Sciences

[†]Institute for Industrial Mathematics, Ben Gurion University of the Negev, Beer Sheva, Israel

[‡]School of Mathematical Sciences, Tel Aviv University, Tel Aviv, Israel

Abstract. In porous media the local elevation of pressure slowly penerates to the adjacent layers of the gas causing the temperature to rise there. In the presence of explosive gases this mechanism may lead to the formation of self- sustaining combustion waves propagating at a constant speed.

Key words: Detonation, Filtration, Combustion

1. Introduction

The objective of this paper is to demonstrate the possibility of self-sustaining combustion waves in porous media, whose propagation is controlled not by heat and mass diffusivities but rather by the adiabatic compression. The latter is in turn maintained by the energy release provided by the chemical reaction. The phenomenon is quite analogous to the classical detonation occuring in gaseous combustion in open systems. In contrast to the conventional detonation, however, in the case of a porous medium the reaction wave spreads at an essentially subsonic speed controlled by the viscous friction.

2. Mathematical model

In the formulation of the pertinent model the following basic asumptions are employed:

a. Deformation of the porous skeleton is ignored.
b. Gas flow through the skeleton obeys Darcy's law.
c. The local temperatures of the solid and gaseous phases are identical. Indeed, in many realistic systems the characteristic time for temperature equalizing between the phases is much shorter then the time associated with the overall heating-up.
d. Porosity of the skeleton is sufficiently high. As a result, the effective features of the gas-solid system are controlled mainly by its gaseous phase.
e. In order to elucidate the impact of barodiffusion, the thermal and molecular diffusivities are regarded as negligibly small compared to the former.

In light of the above assumptions the system of governing equations reads as follows:
Energy equation:

$$c_p\rho\left(\frac{\partial T}{\partial t}+u\,\frac{\partial T}{\partial x}\right)-\left(\frac{\partial P}{\partial t}+u\,\frac{\partial P}{\partial x}\right)=QW \qquad (1)$$

Concentration equation:

$$\rho\left(\frac{\partial C}{\partial t}+u\,\frac{\partial C}{\partial x}\right)=-W \qquad (2)$$

Momentum equation:

$$\rho u=-\frac{K}{\nu}\frac{\partial P}{\partial x} \qquad (3)$$

Continuity equation:

$$\frac{\partial \rho}{\partial t} + \frac{\partial \rho u}{\partial x} = 0 \tag{4}$$

State equation:

$$P = (c_p - c_v)\rho T \tag{5}$$

Chemical kinetics equation:

$$W = Z\rho C \exp(-E/RT) \tag{6}$$

Eqs.1-4 are writen in a frame of reference attached to the skeleton. u is the gas velocity relative to the skeleton. C the concentration of the deficient reactant controlling the termination of the reaction, W the chemical reaction rate, ν the kinematic viscosity of the gas, K the permeability of the porous medium, Z the frequency factor, E the activation energy, R the universal gas constant, Q the heat release. Other notations are conventional.

Model (1-6) is similar to that discussed in Aldushin and Merjanov (1988); Laevsky and Babkin (1988) where, however, effects due to pressure, central to the present study, were omitted.

3. Travelling wave solution

The solution of the system (Eqs.1-6) is sought in the form of a travelling wave moving at a speed V from left to right, i.e.

$$T = T(x - Vt), C = C(x - Vt), P = P(x - Vt),$$

$$\rho = \rho(x - Vt), u = u(x - Vt) \tag{7}$$

Eqs.1-4 are thus reduced to the following system of ordinary differential equations:

$$c_p\rho(u - V)\frac{dT}{dy} - (u - V)\frac{dP}{dy} = QW, \quad \rho(u - V)\frac{dC}{dy} = -W$$

$$\rho u = -\frac{K}{\nu}\frac{dP}{dy}, \quad \frac{d}{dy}[P(u - V)] = 0. \tag{8}$$

Here $y = x - Vt$.

Eqs.8 are solved subject to the boundary conditions:

$$T(+\infty) = T_0, \ P(+\infty) = P_0, \ \rho(+\infty) = \rho_0 = P_0/(c_p - C_v)T_0,$$

$$C(+\infty) = C_0, \ C(-\infty) = 0, \ T_y(-\infty) = P_y(-\infty) = \rho_y(-\infty) = 0 \tag{9}$$

Eqs.8 and condition (9), as it may be easily shown, yield the following relation:

$$\frac{K}{\nu}\frac{dP}{dy} = \frac{V}{c_p - c_v}\left(\frac{P_0}{T_0} - \frac{P}{T}\right) \tag{10}$$

Introduce the following reference parameters: $Q = \sqrt{\gamma(c_p - c_v)T_0}$, (Here $\gamma = c_p/c_v$, and a is the speed of sound in the fresh mixture); $D_b = Ka^2/\gamma\nu$ the barodiffusivity; $L_b = D_b/V$ the reaction wave width; $\tau = Z^{-1}\exp(E/RT_0)$ the reaction time.

Introduce the following dimensionless parameters and variables:

$\Lambda = V\sqrt{\tau/D_b}$, the dimensionless speed of combustion wave; $\Delta = QC_0/c_pT_0 = (T_b^* - T_0)/T_0$ where T_b^* is the adiabatic temperature of combustion products for the freely propagating flame (in the absence of a porous medium);

$$\eta = y/L_b, \ \Pi = P/P_0, \ \Theta = T/T_0, \ \Phi = C/C_0, \ \Omega = \tau W/\rho_0 C_0 \tag{11}$$

In new units Eqs.8,10 and the boundary conditions (9) become

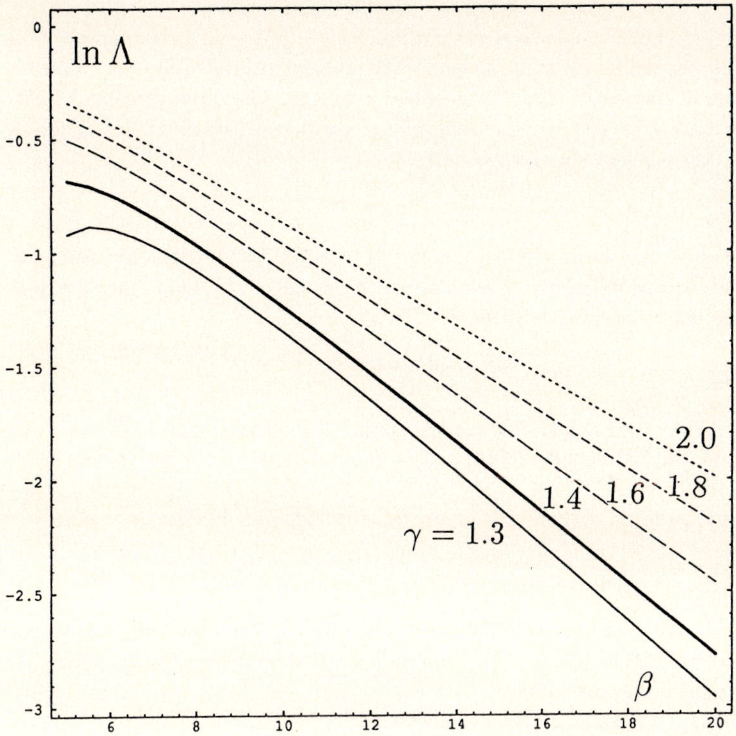

Fig. 1.

$$\frac{\Theta}{\Pi}\frac{d\Pi}{d\eta} - \frac{\gamma}{\gamma-1}\frac{d\Theta}{d\eta} = \frac{\gamma}{\gamma-1}\frac{\Delta}{\Lambda^2}\Omega,$$

$$\frac{d\Pi}{d\eta} = 1 - \frac{\Pi}{\Theta}, \quad \frac{d\Phi}{d\eta} = \frac{1}{\Lambda^2}\Omega, \tag{12}$$

$$\Pi(+\infty) = \Theta(+\infty) = \Phi(+\infty) = 1$$

$$\Pi_\eta(-\infty) = \Theta_\eta(-\infty) = \Phi_n(-\infty) = 0. \tag{13}$$

In the limit of high activation energy, E/RT_0, and small heat release, Δ, the problem may be considerably simplified.

Introduce the reduced temperature $\mathcal{U}$ and pressure $\mathcal{V}$ according to the following relations:

$$\Theta = 1 + \gamma\Delta\mathcal{U}, \quad \Pi = 1 + \gamma\Delta\mathcal{V}. \tag{14}$$

At $\Delta \ll 1$ and $E/RT_0 \gg 1$ the problem (12), (13) then becomes

$$\gamma\mathcal{V}_{\eta\eta} + \mathcal{V}_\eta + \Lambda^{-2}(1 - \mathcal{V} - \gamma\mathcal{V}_\eta)\exp[\beta(\mathcal{V} + \mathcal{V}_\eta)] = 0 \tag{15}$$

$$\mathcal{V}(+\infty) = 0, \quad \mathcal{V}(-\infty) = 1, \tag{16}$$

$$\mathcal{U} = \mathcal{V}_\eta + \mathcal{V}, \quad \Phi = 1 - \mathcal{V} - \gamma\mathcal{V}_\eta, \quad \beta = \gamma E/RT_0. \tag{17}$$

To circumvent the cold boundary difficulty, typical for these kind of systems (Frank-Kamenetsky 1969), the factor $\exp[\beta(\mathcal{V} + \mathcal{V}_\eta)]$ will be replaced by zero at $\mathcal{U} = \mathcal{V}_\eta + \mathcal{V} < 1/\beta$, i.e. at low temperatures.

Analysis of the corresponding nonlinear eigenvalue problem, with Λ being the parameter to be determined, shows that for sufficiently high values of β the problem is solvable and the solution is unique. The formal proof, however, is rather involved and will be presented elsewhere. Fig.1 shows the dependency of $\ln \Lambda$ on the scaled activation energy β (for various values of the parameter γ), which appears to be nearly linear. For example,

$$\Lambda \simeq 0.17 \exp(-0.166\beta) \quad \text{at} \quad \gamma = 1.4.$$

Thus barodiffusion indeed may provide a constant speed, $V \sim (D_b\tau)^{1/2}$, of the reaction wave even in the absence of thermal diffusivity. This mode of self-sustained convolution therefore may well be regarded as a slow or creeping detonation.

Acknowledgements

This work was supported by the Israel Ministry of Science and Technology, by the US Department of Energy under Grant No. DE-FG2-88ER1382 and by the National Science Foundation under Grant No CTS-9213414.

The authors would like to thank Prof. E. Gelman for an enlightning discussion.

References

Aldushin AP, Merjanov AG (1988) Theory of filtration combustion: general concepts and the current state of research. In: Matros YuSh (ed) Propagation of Thermal Waves in Heterogeneous Media, Nauka, Novosibirsk, pp 9-15.

Laevsky Yu M , Babkin V S (1988) Filtration combustion of gases. In: Matros YuSh (ed), Propagation of Thermal Waves in Heterogeneous Media, Nauka, Novosibirsk, pp 108-144.

Frank-Kamenetsky DA (1969) Diffusion and Heat Transfer in Chemical Kinetics. Plenum, New York

Part 3: Shocks in Condensed Matter

Liquid Fracture at Explosive Loading

V. Kedrinskii
Lavrentyev Institute of Hydrodynamics, Novosibirsk, RUSSIA

Abstract. Paper presents a short review of some results on the cavitation liquid fracture at explosive loading. The problems of the real liquid state (from the viewpoint of its inhomogeneity) and limit tensile stress, as well as the mechanics of the cavitation development excited by intense rarefaction waves and the dynamic feature of breaking of a spherical liquid drop layer under the action of detonation products are considered.

Key words: Explosive loading, Cavitation fracture, Two-phase liquid

1. Introduction

The problem of liquid fracture in intense rarefaction waves at the explosive loading of a liquid with a free surface is often connected with the concept of liquid strength i.e. a critical tensile stress p_* admitted by the cavitating liquid. One of the recent studies of liquid strength was performed by measuring the dome velocity v at shallow underwater explosions when it tends to zero: $v = (2p - p_*)/\rho_o U$ (Wilson et al. 1975). The value of p_* for settled tap water turned out to be -0.85 MPa, for deionized and vacuumized water - 1.5 MPa. At high velocity of loading they obtained p_* = -60 MPa (Carlson and Henry 1973).

Practically in all the experiments these data are associated with visible breaks in liquid (cavities), which develop on cavitation nuclei when affected by intense rarefaction waves. However the above-mentioned strength concept restricts essentially the understanding of the complicated process of fracture. It may be defined as an effect of inversion of a two-phase medium stage consisting in the transformation of the cavitating liquid into a gas-droplet system. The inversion scheme involves a number of stages such as formation of bubble clusters, its transformation in a structure of a foam type followed by fracture into cavitating fragments, transition to the droplet state and its evolution.

2. Bubble cavitation

Being the most investigated, this section deals with such problems as the state of a real liquid, the concept of liquid strength, the formation mechanism of bubble clusters as well as its mathematical simulation. Real liquid structure in macroscale is such that even after special purifications, distillation and deionization there is a lot of microinhomogeneities, which play the role of cavitation nuclei: fluctuating holes, $R_o = \sqrt{kT/\sigma}$ (Ya. Frenkel'), hydrophobic particles with nuclei in crevices $R_o = 2\sigma/p_o$ (E.Harvey), nuclei as solid particles (M. Plesset), combination structures (experimental result, Fig. 1, A.Besov, V.Kedrinskii, E.Pal'chikov) (Kedrinskii 1993).

The experimental data show that in fresh distilled water the maximum of the bubble size spectrum is approximately at the level of 4 μm, in settled water 0.85 μm (Kedrinskii 1993). Experimental results on bubble spectrum in settled water (Hammit et al. 1974) and their generalization (Kedrinskii 1985) allowed to estimate the total density of microinhomogeneities: $N_o = 1.5 \cdot 10^5$ cm^{-3} . This magnitude correlates with the estimation $10^5 - 10^6$ cm^{-3}, obtained from track recording of diffraction spots arising from light scattering on microinhomogeneities of any nature.

The results on acoustic diagnostics of free gas microbubbles, denoting their extremely low density served as the basis for the mechanism of avalanche-type "settlement" of the developing bubble cavitating zone by the nuclei (Sirotyuk 1969). It is caused by the instability of the nuclei

Shock Waves @ Marseille III
Editors: R. Brun, L. Z. Dumitrescu

shape, their intense growth and collapse. However, the development of a dense cavitation zone in the field of a single rarefaction pulse (for example at an underwater explosion near the free surface) cannot be explained by means of the above-mentioned scheme.

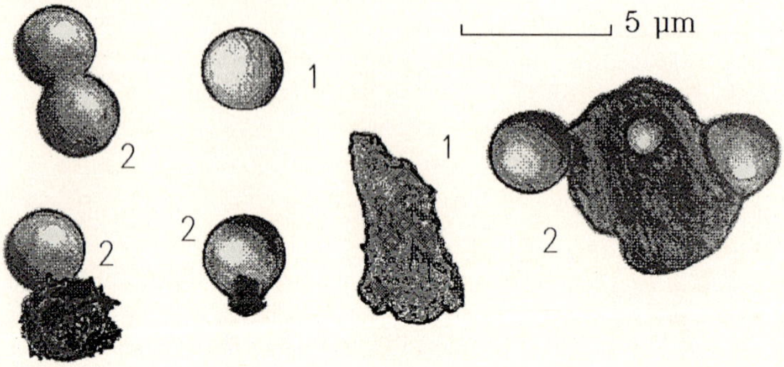

Fig. 1. Individual particles (1) and combination structures (2) of microinhomogeneities in real liquids

According to a new mechanism of cavitation zone development (Kedrinskii 1986) a real liquid is considered to contain a wide spectrum of cavitation nuclei in the range of sizes $10^{-7} - 10^{-3}$ cm at density of $10^5 - 10^6$ cm^{-3}. The concept of visible, i.e. detectable size of cavitation bubble is introduced; the mentioned effect of cavitation zone "settlement" in relatively weak ultrasonic fields is explained by gradual saturation of the zone containing bubbles, which attained the visible size over different time intervals τ_v according to their initial position in the spectrum.

To formulate last result on the dependence of nucleus growth time up to a visible size the first integral of Rayleigh's equation was analysed in the simplest case of constant tensile stresses (Kedrinskii 1985)

$$V^{-1/3}\left(\frac{dV}{d\tau}\right)^2 = 6\eta F(V, V_o, p),$$

where $$F = \frac{1 + W_e V_o^{-1/3}}{\gamma - 1} V_o^\gamma \left(V_o^{1-\gamma} - V^{1-\gamma}\right) - p(V - V_o) - 1.5\mathrm{We}\left(V^{2/3} - V_o^{2/3}\right),$$

$V = (R/R_v)^3$, R_v - the visible size of the bubble, R_o - the nucleus size, $p = p_\infty/p_o = \mathrm{const.}$; p_∞ - the pressure given at infinity; We$= 2\sigma/p_o R_v$; $\eta = p_o/\rho_o c_o^2$; $\tau = tc_o/R_v$. Qualitative analysis has shown that three types of solutions are possible at different values of p: unbounded bubble growth, asymptotic attainment of a finite size during an infinite time, or periodical pulsations. For the smaller sizes of nuclei the value τ_v was found to be practically constant and at $R_o \ll R_v$ to be determined by a simple dependence

$$\tau_v \approx 3/\sqrt{6|p|\eta}$$

The data presented interprete the real liquid as a two-phase medium despite the insignificant initial gas content with volume concentration in the range $10^{-8} - 10^{-12}$ cm^{-3}. In this case it is natural to suppose that the transformation mechanism of rarefaction waves in a cavitating liquid is similar to the known effects of shock wave propagation in bubbly media and to use the numerical model of these media for the description of cavitation effects (Kedrinskii 1976). This model is a system of conservation laws, written for the average density ρ, pressure p and mass velocity u. The state of the medium is described by Rayleigh's equation for the volume concentration of the vapor-gas phase k and by the dependence of $\rho(k)$. If the liquid component is incompressible ($\rho_{liq} = \rho_o$) the above-mentioned system can be reduced to the form

$$\Delta p = -\rho_o k_o \frac{\partial^2 k}{\partial t^2} \tag{1}$$

$$\frac{\partial^2 k}{\partial t^2} = 3k^{1/3}(p_o k^{-\gamma} - p)/\rho_o R_o^2 + (\frac{\partial k}{\partial t})^2/6k \tag{1'}$$

The dynamics of concentration growth in a cluster is rather adequately described by the equation

$$\frac{\partial^2 k}{\partial t^2} \simeq -3k^{1/3} p/\rho_o R_o^2.$$

This enables to obtain the equation for the pressure in a cavitating liquid of the Helmholtz type:

$$\Delta p \simeq p \tag{2}$$

for the new spatial variable $\eta = \alpha r k^{1/6}$ where $\alpha = \sqrt{3k_o/R_o^2}$ and some assumptions (Kedrinskii 1976; Hansson et al. 1982). Soluving simultaneously Eqs.2 and 1' determines the parameters of the rarefaction waves and the dynamics of the cavitation process (Kedrinskii 1978). This model allowed to solve a number of principal questions and, first of all, to introduce the concept of limit tensile stresses (Kedrinskii 1976) which may be measured in a liquid.

It is known that the stress in a rarefaction wave front has a certain steepness t_*. The computation of the axisymmetric problem for the cavitation zone development at an underwater explosion near the free surface has shown that this fact is of fundamental importance for the problem of limit stresses (Kedrinskii 1976).

Let us consider this effect by the example of a tube filled with a real liquid being accelerated vertically downward by an impact (Hansson et al. 1982). For simplicity the liquid is assumed to occupy a half-space $z \geq 0$. Under the boundary condition at the tube bottom (z=0): $\partial p/\partial z = -\rho_o a(t)$, where $a(t)$ - the tube acceleration, z - the vertical coordinate, the solution of Eq.2 gives the analytical dependence $p(k)$:

$$p = -\rho_o \; |a(t)| \; exp(-\alpha k^{1/6} z) \; / \; \alpha k^{1/6} \tag{3}$$

If the gas pressure inside the bubbles may be neglected then, substituting Eq.3 in Eq.1', the latter, for the case $z = 0$, takes the form

$$\frac{d}{dt}\,(k^{-1/6}\,\frac{dk}{dt}) = \; 3a(t) \; / \; \alpha R_o \tag{4}$$

The last equation is solved analytically, if $a(t)$ is known. For example, if $a(t) = a_{max} \exp(-t/t_*)$, the solution of Eq.4 presents the dynamics of concentration growth in the following form:

$$k^{5/6} = 1 + \frac{5t_* a_{max}}{2\alpha R_o^2} \; [t - t_*(1 - \exp(-t/t_*))] \tag{5}$$

The calculation by Eqs.3 and 5 for $a_{max} = 5 \cdot 10^5$ m/s^2 (corresponding to -30 MPa in the rarefaction wave), $k = 10^{-10}$, $R_o = 1$ μm and $t_* = 10$ μs shows that by the instant of time $t = t_*$ the amplitude of the rarefaction wave in a cavitating liquid is reduced 20-fold (in a one-phase liquid only by a factor of e). To estimate the influence of front steepness on the limit tensile stresses $-p_{max}$, which are admitted in a cavitating liquid it will suffice to consider $a(t)$ in the form of the linear dependence $a(t) = a_{max} t/t_*$, where $1/t_*$ determines the front steepness. At $t = t_*$ the wave amplitude attains the negative maximum and is determined by the following expression:

$$p = -\rho_o\left(\frac{a_{max}^2 R_o^3}{3k_o}\right)^{2/5} t_*^{-2/5} \tag{6}$$

Under the conditions $k_o = 10^{-11}$ and $R_o = 0.5$ μm for the time of front increase $t_* = 1\mu$s and $a_{max} = 3.28\ 10^5$ m/s^2 we will obtain $p_{max} \approx -3$ MPa instead of -30 MPa for a one-phase model. Calculations of underwater explosions (Kedrinskii 1976) give the same order of magnitude for the same external conditions.

3. Fracture, sandy model

As it was noted above, in reasonably intense rarefaction waves the development of bubble cavitation is characterized by the unrestricted growth of nuclei of the total theoretically possible range of sizes. Unfortunately, so far there is no complete understanding of the significantly nonlinear processes of unbounded growth of bubbles in cavity clusters and their hydrodynamic interaction with formation of dense packages with volume concentration of 0.5-0.75 and transition through foam structure to spallation and droplet phase separation. At the same time the experimental research of the structure of the flow of a cavitating liquid at an underwater explosion near the free surface has shown that at its destruction in intense rarefaction waves cavitating spallings are recorded (Fig.2) (Kedrinskii 1974). Thus, a cavitating liquid manifests both plastic and brittle properties, that is not characteristic of a one-phase liquid. It is possible that the prerequisite for their formation is initially "established" by the wave field.

a) b)

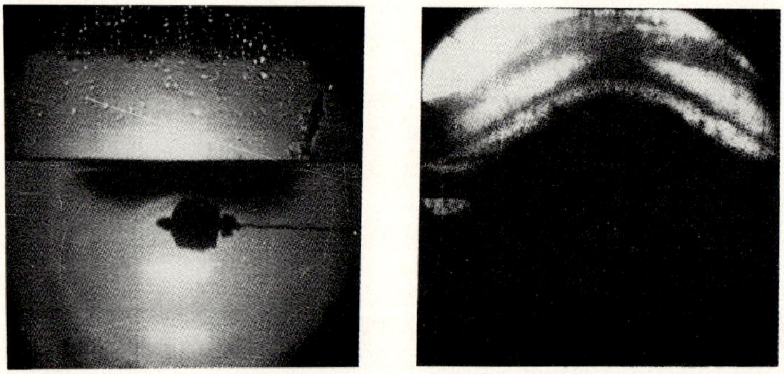

Fig. 2 a,b. Underwater explosion near the free surface. Two stages of fracture process: (a) cavitation zone; (b) spallings

Disregarding yet unknown details of the transition process from the foam structure into a droplet one, let us consider this process to happen instantaneously as soon as the structure of the cavitation zone attained the state of densely packed bubbles. In doing so we assume that the dense package of bubbles was instantaneously transformed into a dense package of elastic spherical liquid droplets, which do not flow together. This model may be called "sandy". The experiments confirmed that such an approach is positive.

High-speed photography of the fracture process of both liquid and natural sandy cylindrical shells under axial loading has shown the identity of the basic structural peculiarities of two-phase flows: analogous streamer structures characteristic of thin shells and essential stratification of flows to the moment of process completion (small particles take its central parts, while coarser particles are at the periphery).

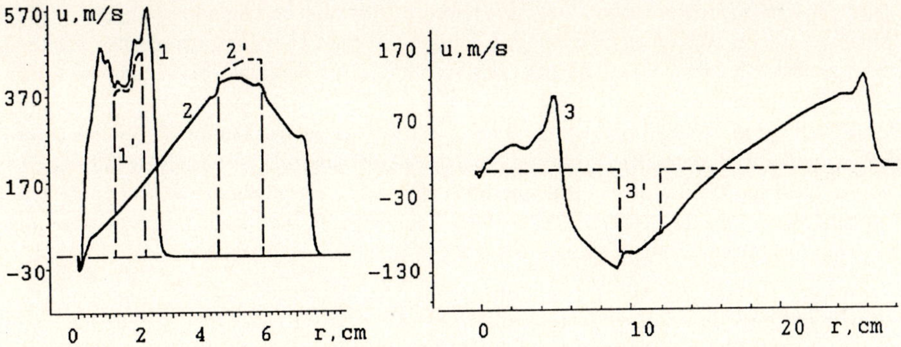

Fig. 3. Breaking of a spherical shell. Distributions of mass velocities of gas and disperse (dotted lines) phases: $d =$ $= 6\ \mu$m - $t_1 = 20\ \mu$s, $t_2 = 100\ \mu$s; $d = 1\mu$m - $t_3 = 500\ \mu$s

The above model with instantaneous transformation of the foam structure into a droplet one was analyzed numerically by Getts and Kedrinskii (1989). A spherical charge of HE with the density ρ^o and radius r_1 is surrounded by a shell with outer radius r_2 which is a two-phase mixture "liquid particles - air" with a volume fraction of the dispersed phase of 0.74, equal to the concentration of the dense package of drops. A charge detonation is simulated by an instantaneous explosion at a constant volume, with the detonation products having the same density ρ^o. Spherically symmetrical motion of such a two-phase mixture may be described by the well-known system of equations for heterogeneous medium mechanics written for each component separately (Nigmatulin 1987). The right-hand parts of the pulse conservation laws include a condition of phase interaction determined by the formula:

$$f \ = \ 0.75\ \alpha_2\ \rho_1^o\ C_d\ (u_1 - u_2)\ |u_1 - u_2|/d$$

where α_2 is the volume concentration of the disperse phase, $u_{1,2}$ - the mass velocities of phases, d - the diameter of drops, ρ_1 - the gas component density, C_d - the drag coefficient depending on α_2 and Re= $\rho_1^o\ |u_1 - u_2|\ d/\ \mu_1$. The system is closed by the condition of joint phase deformation (Getts and Kedrinskii 1989). Numerical calculations were carried out by the method of coarse particles.

The shock wave in the disperse phase is initiated as a result of discontinuity decay at the boundary "detonation products - drop shell" (Fig.3). When it reaches the outer boundary the shock wave appears in air and a rarefaction wave propagates into the liquid particles. This wave, along with divergent effects, leads to a fast decrease of tensions in the particles, and their density after 20 μs becomes lower than the bulk one (the particles are separated, the shell becomes penetrable).

Analysis of numerical calculations carried out for an explosive of RDX type and water drops at $r_2/r_1 \ = \ 5$ showed that it is possible to set apart three stages of the process development. The third stage is characterized by focusing of the rarefaction wave at the center and rather intense reverse gas flow. This flow decelerates, and later carries away small particles to the center thus determining the mechanism of particle stratification by sizes, observed in the experiments with sandy shells. Wave processes in gaseous phase noticeably influence the dynamics of the inner boundary which oscillates with the frequency characteristic of these processes.

4. Conclusion

The above analysis of some essentially nonlinear effects, responsible for the behavior of real liquids under explosive loading, has shown that, despite the considerable complexity, adequate physical and mathematical models describing wave processes in cavitation and liquid fracture, as well as the dynamics of their state, may be developed. Results discussed above and approaches proposed allowed to solve a number of crucial questions concerning the development mechanism of the process of liquid fracture.

From the scope of unsolved problems it is necessary to note the mechanism of "brittle" failure of foam structure and transition "foam - droplets", the development of techniques allowing to resolve the total spectrum of nuclei, the questions of stability of combinations of the type "gas nuclei - solid particles", the problem of metastable liquid state in the "deep" negative phase and the formation kinetics of vapor centers at the front of an intense rarefaction wave.

Acknowledgement

This work was supported in part by the Russian Foundation of Fundamental Research, Grant 93-013-16383.

References

Carlson GA, Henry KW (1973) Technique for studying tension failure in application to glycerol. J. Appl. Physics 42(5)

Getts IG, Kedrinskii VK (1989) Dynamics of explosive loading for a finite volume of a dense two-phase mixture. J. Appl. Mechanics and Tech. Physics 2: 120-125

Hammitt FG, Koller A, Ahmed O, Pyun J, Yilmaz E (1976) Cavitation threshold and superheat in various fluids. In: Cavitation, Proc. Intl. Conf., Mech. Eng. Publ. Ltd, London, NY, Edinburg, September 3-5

Hansson I, Kedrinskii V, Morch K (1982) On the dynamics of cavity clusters. J. Physics D: Appl. Physics 15: 1725-1734

Kedrinskii VK (1974) The experimental research and hydrodynamical models of a "sultan". Arch. Mech. 26 (3): 535-540

Kedrinskii VK (1976) Negative pressure profile in cavitation zone at underwater explosion near a free surface. Acta Astronautica 3: 623-632

Kedrinskii VK (1978) Surface effects at underwater explosions (review). J. Appl. Mechanics and Tech. Physics 4: 66-87

Kedrinskii VK (1985) Peculiarities of bubble spectrum behavior in cavitation zone and its effect on wave field parameters. In: Ultrasonics Intl. 85, Proc. Intl. Conf., Butterworth Sci. Ltd London: 225-230

Kedrinskii VK (1986) On multiplication mechanism of cavitation nuclei. In: ICA, Proc. 12th Intl. Congr. on Acoust., Toronto, July 24-31: 14-8

Kedrinskii VK (1993) Nonlinear problems of cavity liquid fracture at explosive loading (review). J. Appl. Mechanics and Tech. Physics 3: 74-91

Nigmatulin RI (1987) Dynamics of Many-Phase Media. Nauka, Moscow, Vol. 2 Chap. 6

Sirotyuk MG (1968) Experimental study of ultrasonic cavitation. In: Power ultrasonic fields, Nauka, Moscow: 167-220

Wilson DA, Hoyt JW, McKune JW (1975) Measurement of tensile strength of liquid by explosion technique. Nature 253: 723-5

Characterization of Pressure Waves in Liquids Using an Interferometric Method

A.P. Alloncle, D. Dufresne and M. Autric
Institute of Fluid Mechanics - Laser-Matter Interaction Group - Parc Scientifique et Technologique de Luminy, 163 Avenue de Luminy, 13009 Marseille, France

Abstract. Paper describes experimental studies on shock waves associated with optical cavitation phenomena in liquids. Direct radial pressure profiles of the transients are obtained using an optical method. Results on the initial step of the phenomenon and on the collapse of the cavity are given.

Key words: Optical cavitation, Laser, Bubble dynamics, Shock wave

1. Introduction

The interaction of a high power radiation with a transparent liquid induces, beyond a certain threshold of power density, the dielectric breakdown of the medium. This is essentially characterized by the formation of a luminous plasma, absorbing the incident radiation. This instantaneous deposit of energy leads to the emission of a shock wave and to the rapid heating and vaporization of a small volume of liquid. This cavity of high temperature is going first to expand and then its properties become very similar to those of cavitation bubbles formed in high-velocity flows, due to fluctuations of the pressure. This method of laser-induced cavitation or "optical cavitation" (Lauterborn 1980) has been used in many experimental researches on the understanding of the basic mechanisms of cavitation. It offers the possibility of studying cavitation phenomena under many boundary conditions. For example, studying the interaction between an isolated bubble and a solid wall under controlled conditions has shown a deformation of the bubble shape leading to the appearance of a liquid jet, going through the bubble and directed towards the wall. This jet, associated with the pressure waves emitted when the bubble is collapsing, can have an important role in the erosion cavitation phenomenon (Testud 1985, Vogel 1989, Tomita 1991). Collective effects have also been observed using this experimental method and the results show different sorts of interaction (change of pulse duration, coalescence, formation of jets, forced collapse) (Testud 1990). Moreover, these researches present an increasing interest for the medical fields like ophthalmology and lithotripsy, in which the same mechanisms are encountered.

This paper contributes to the study of vapour bubbles and laser-generated spherical shock waves, and it is most particularly concerned with direct pressure measurements using high-speed visualizations by an interferometric method. The final objective of our studies is to get an evaluation of the distribution of the total energy in its thermal, acoustic and mechanical components under different experimental (laser energy, hydrostatic pressure) and boundary conditions (near solid walls, free surfaces, or in the presence of other cavitating bubbles).

2. Experimental set-up

A schematic diagram of the experimental set-up is presented in Fig.1. Experiments are performed using a Nd:YAG laser, fitted with a frequency doubler, and delivering a pulse of 1.3 J energy during 10 ns. The second harmonic is isolated using two dichroic mirrors (MD). The absorption length of this radiation in water is in the range of 100 m; consequently, every other absorbing mechanism, except the dielectric breakdown, can be disregarded. First, the beam is enlarged by a telescope (T) system made of two confocal lenses. It is then focused into a cylindrical tank

Shock Waves @ Marseille III
Editors: R. Brun, L. Z. Dumitrescu

of 20 cm diameter filled with demineralized water. The size of the spot at the focus is 50 μm (referring to $1/e^2$). The breakdown occurs for an intensity of 5×10^9 W/cm^2. The quantity of energy actually deposited in water is determined by direct measurement using two calorimeters (C1, C2).

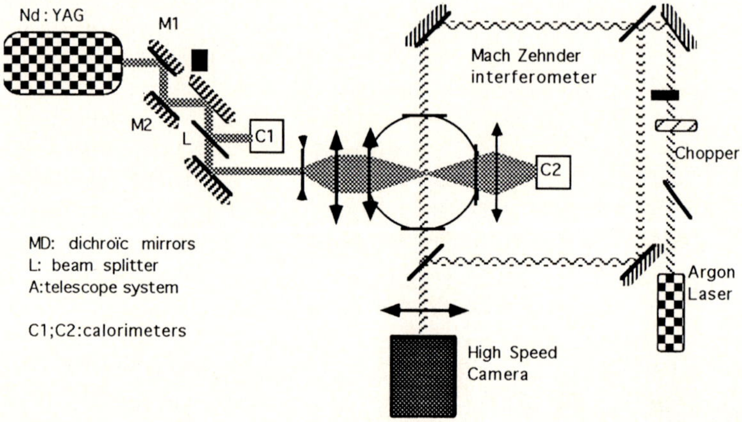

Fig. 1. Experimental set-up

Shadowgraph and interferometric visualizations have been taken with a high-speed camera (Imacon 790 - Hadland) allowing a framing rate of up to 2×10^7 frames/sec and a streak speed up to 1 ns/mm. A CW argon laser is used as an intense and monochromatic light source. The beam is divided in two equal parts. One of them is directed through the water, and records the disturbances of the medium induced by the breakdown, the pressure waves and the bubbles. The other one is used as a reference. Both beams are then collected by a beam splitter and directed onto the photo cathode of the camera. Accurate adjustments of the mirrors and the beam splitters cause interferences to occur. The fringe pattern is then recorded by the camera. Every perturbation occurring in the visualized area leads to a fringe shift. Operation of such an apparatus needs an accurate triggering device between the modulation of the continuous laser (with a shutter and a chopper), the pulsed laser and the camera.

3. Interpretation of the interference pattern

Direct pressure data are obtained from the interference pattern. The use of an optical method brings two significant advantages compared to pressure measurements performed with pressure gauges. First, no disturbance is induced in the medium by probes, and consequently, the pressure is measured close to the centre of emission. The second one is the accurate knowledge of the location where the measurement is done. A similar technique was used by Ward et al. (1990) in optical cavitation experiments, with interesting results.

The pressure wave induces some variations of the refractive index along its propagation, consequently a shift of the fringes from the initial parallel pattern is recorded. The whole phenomenon follows a spherical symmetry, therefore, it is possible to link the fringe shift $y(x)$ to the variation of the refractive index $\Delta n(r)$ using Abel inversion (Deutsch 1983). The pressure P is deduced from:

- the Gladstone-Dale law which gives, according to some authors (Partington 1953), a good value of the refraction index vs. the density ρ for liquids: $(n-1)/r = \text{Const.}$
- a state equation for water which has to be valid in a large range of pressures: $(P+B)/(P_0+B) = (\rho/\rho_0)^m$ (B and m are constants, and the suffix "0" refers to the state of the liquid not yet disturbed by the shock wave) (Cole 1948).

4. Shock waves generated during the optical breakdown

The phenomenon can be described in three steps:

First step: The *optical breakdown* threshold is first reached in the smaller part of the caustic beam, and as a function of the incident power its location moves to a larger surface. The created plasma is very absorbing to the incident radiation (shielding effect), consequently, the breakdown moves towards the incident beam (Docchio 1988). The energy deposition is not spherical. That explains the dissymmetry recorded during the first nanoseconds on the high speed photographs (Fig.2). Actually, Fig.2 shows a streak representation of the first fifty nanoseconds of the interaction. The streak speed is 2 ns/mm. The beam is coming from the top of the picture. The plasma is quickly replaced by a small volume of vapour which is growing rapidly . 10 to 20 ns after the beginning of the breakdown, the shock wave can be distinguished from the bubble.

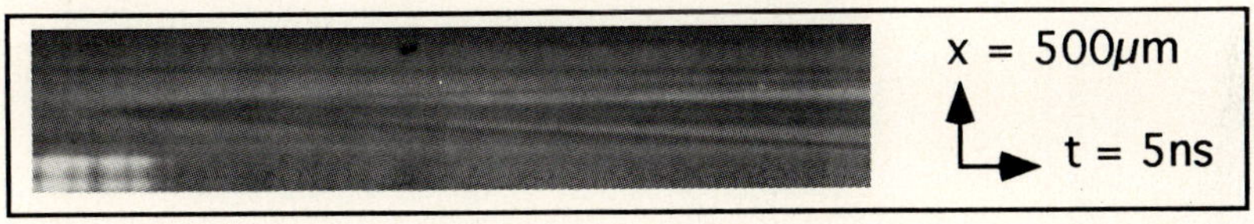

Fig. 2. Streak visualisation of the initial phase of the dynamics of the shock and bubble; incident energy 1 mJ

Second step: *supersonic propagation.* The shock wave seems to be separated from the bubble earlier on the side of the incident radiation (around 10-20 ns) than on the other side (30 ns). The initial propagation velocity of the shock wave is measured from the visualization and changes from 2100 m/s to 2600 m/s (Fig.3).

Using the Hugoniot relations for water, the pressure corresponding to these velocities is in the range $7 \cdots 15 \times 10^8$ Pa. This velocity quickly decreases and reaches the sound velocity around 100 ns after the beginning (Fig.3), that is also the characteristic time when the geometry becomes spherical.

Third step: *sonic propagation.* Fig.5 shows visualizations (framing rate 500 ns) and pressure profiles associated for 5mJ of incident energy. The small grey point at the left upper corner of each picture is the bubble and the shock wave is visualized by the fringe shift. The rising front is very short (60 ns). The measurements are performed at 2.1 mm and 1.3 mm from the centre of the cavity, and the maximum pressure amplitudes are respectively 50×10^5 Pa and 90×10^5 Pa.

The pressure is then decreasing slowly down to a value somewhat higher than the hydrostatic pressure (10^6 Pa for a radial travel propagation of the wave of 1.3 mm). The liquid is kept under pressure between the wave and the bubble interface, caused by the explosive expansion of the bubble.

On Fig.5, maximum pressure amplitudes are plotted. Measurements could be performed until 200 μm from the centre and pressures up to 500×10^5 Pa were recorded. As the wave propagates, the maximum pressure decreases as a function of $1/r$, characteristic of spherical waves. These results are in good agreement with measurements performed by PVDF probes located 1 cm far from the center of the cavity.

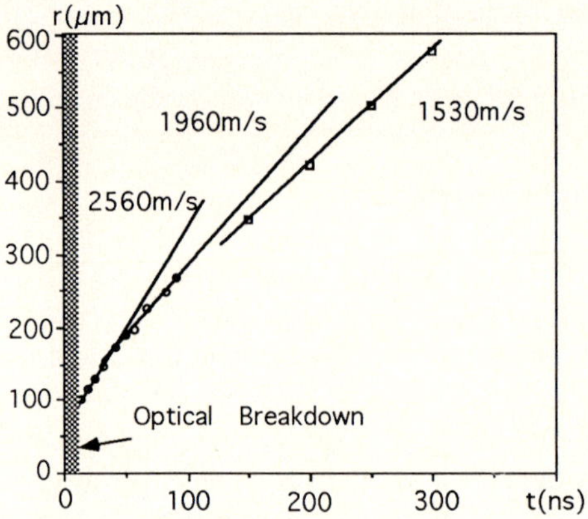

Fig. 3. Measured shock velocity during the first few nanoseconds

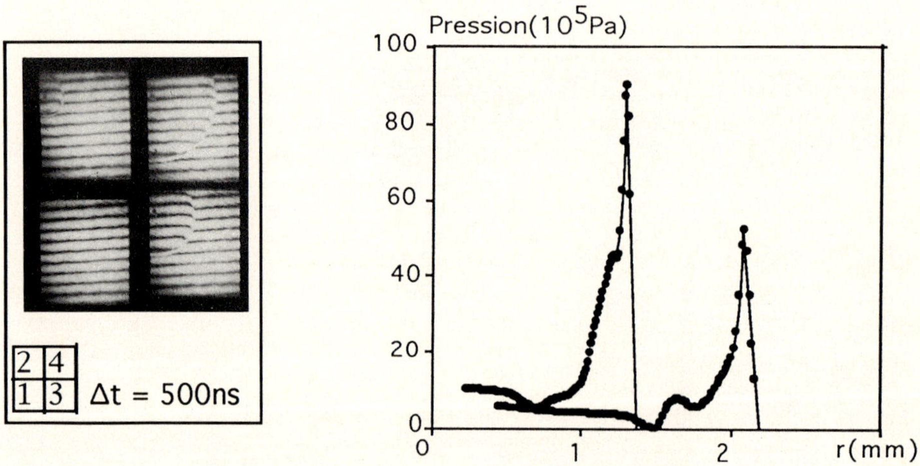

Fig. 4. Visualisation of the spherical transient and associated radial pressure profiles

5. Pressure waves generated during the collapse of the cavity

First this cavity expands until a maximum volume is reached, and then it begins to collapse. The range of incident laser energy used in this experiment was 1 to 5 mJ, and the characteristic radius of the bubbles produced can reach 800 μm. The velocity of the bubble interface near its maximum volume is first very low and increases strongly especially at the end of the collapse phase. It can reach values of up to 400 or 500 m/s. This motion is associated with a pressure increase around the bubble in the liquid, as it is shown on Fig.6 by the slight fringe shift. The bubble reaches a minimum volume and rebounds. A large part of its energy is dissipated in the medium as a spherical acoustic wave (Figs.6 and 7). The cavity can survive to several successive oscillations whose amplitudes are decreasing with time.

The acoustic transient produced during the collapse presents a symmetrical pressure profile. The rising time is around 400 ns, and pressure amplitudes up to 50 bars have been measured after

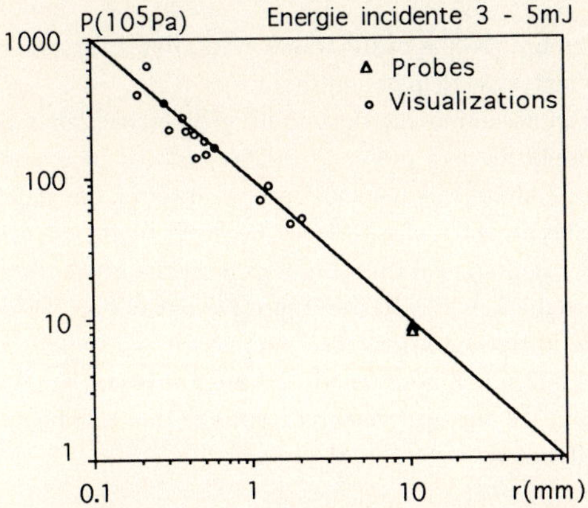

Fig. 5. Pressure wave amplitude vs. propagation distance

Fig. 6. Streak visualisation of the collapse of a cavitation bubble

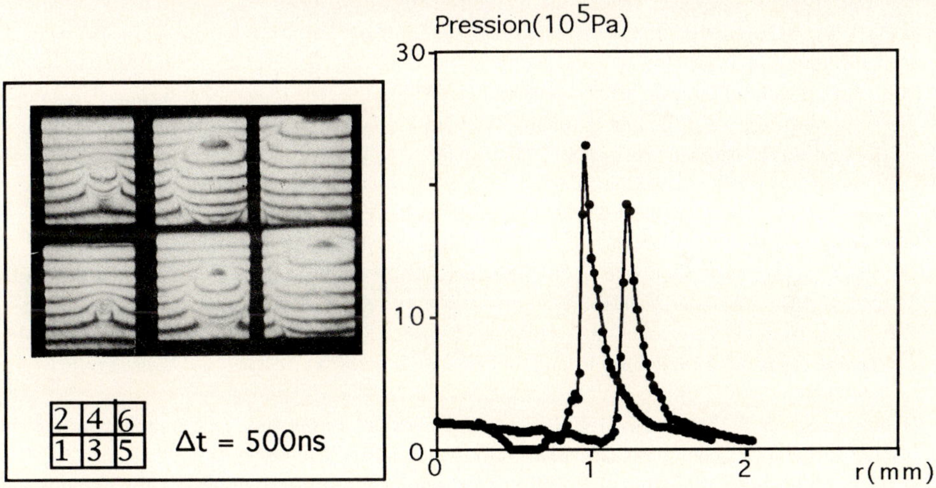

Fig. 7. Visualisation of the acoustic transient emitted during the collapse of the bubble and the associated radial pressure profiles

1 mm of propagation. Pressure amplitudes equal around half the values of the first shock wave. The propagation velocity of the spherical transient equals the sound velocity of the medium. The minimum value of the bubble which we can measure is around 100 μm diameter. Assuming a sonic and spherical propagation of the wave even at the earliest time (streak records up 10 ns/mm do not show different conditions of propagation) the initial pressure can be estimated to 10^8 Pa.

6. Conclusion

This method of vizualisation can give radial pressure profiles of the waves even after a very short time of propagation and without inducing any disturbance in the medium.

It will be very interesting to use this method while submitting the bubble to different boundary conditions in order to quantify the pressure amplitude and pressure profiles induced by a non-spherical collapse. High-speed vizualisations have shown a supersonic propagation of the shock wave during 100 ns. The results do not show the presence of a rarefaction wave, moreover, the liquid enclosed behind the shock wave is slightly compressed during 1 or 2 μs after the breakdown. This is probably caused by the presence of the bubble, which has an explosive behaviour. These results are very interesting because they provide initial conditions for the pressure and the initial volume of vaporized liquid, which are difficult to estimate, in order to perform numerical simulations. The next step of our studies will be to compare our experimental results with a theoretical analysis of the vapour bubble and shock wave dynamics.

Acknowledgements

The authors wish to acknowledge the financial assistance of the CEA for these studies. The authors wish to acknowledge the substantial assistance of Mr J.P.Fragassi in carrying out these experiments.

References

Cole RH (1948) Underwater Explosions. Princeton University Press

Deutsch M, Beniaminy I (1983) Inversion of Abel's integral equation for experimental data. J. Appl. Phys. 54-1: 137-142

Docchio F, Regondi P, Capon MRC, Mellerio J (1988) Study of the temporal and spacial dynamics of plasmas induced in liquids by nanosecond Nd:YAG laser pulses. 1: Analysis of the plasma starting times. Applied Optics 27,17: 3661-3667

Lauterborn W (1980) Cavitation and Inhomogenities in Underwater Acoustics. Springer Series in Electrophysics 4, Springer-Verlag Berlin, Heidelberg, New-York pp 3-12

Partington JR (1953) An Advance Treatise on Physical Chemistry. Longmans

Testud-Giovanneschi P, Dufresne D (1985) Experimental study of laser induced cavitation bubbles. J. Appl. Phys. 58, 2: 651-652

Testud-Giovanneschi P, Alloncle AP, Dufresne D (1990) Collective effects of cavitation, Experimental study of bubble-bubble and bubble-shock wave interaction. J. Appl. Phys. 67, 8: 3560-3564

Tomita Y, Shima A, Takahashi H (1991) The behaviour of a laser-produced bubble near a rigid wall with various configurations. FED vol.116 Cavitation' 91, ASME

Vogel A, Lauterborn W, Timm R (1989) Optical and acoustic investigations of the dynamics of laser-produced cavitation bubbles near solid boundary. J. Fluid Mech. 209: 299-338

Ward B, Emmony D (1990) The energies and pressures of acoustic transients associated with optical cavitation in water. J. Modern Optics 1 37, 4: 803-811

Laser-Generated Shock Waves

D.C. Emmony, M.K.B. Suaidi and B. Ward
Physics Department, Loughborough University of Technology, Loughborough, Leicestershire, LE11 3TU, UK

Abstract. A low-energy, Q-switched Nd YAG laser has been used to generate shock waves in air, liquids and solids and high time- and space-resolution optical interferometry then leads to the pressure field in the breakdown zone.

Key words: Laser ultrasonics, Cavitation, Shocks

1. Introduction

The optical breakdown of materials by intense laser radiation has been studied from the earliest stages of laser development. The interaction of a Q-switched, focused laser pulse with the free surface of a solid is particularly important from the point of view of laser damage in optical components for high power laser systems. If the peak optical power density at the focus exceeds the threshold for optical breakdown or if heating, vaporisation and ionisation of the material takes place, this leads to the generation of acoustic transients. If the pressure amplitude in these waves is large enough, shock waves are produced.

When dielectric breakdown occurs in a gas the expanding plasma or spark follows blast wave theory. In a liquid the event is much more complex. The high temperature plasma rapidly vaporises the surrounding liquid and the resulting vapour expands to give a spherical bubble. The event, often known as optical cavitation, is characterised by the large amplitude radial oscillations of the bubble walls on a time scale of 10^{-4} s to 10^{-3} s before oscillations cease with the breaking up of the bubble. The early stages of liquid breakdown include the generation of a shock wave which propagates into the surrounding liquid and may be involved in a number of complex interactions, for example with the walls of the vessel or with other solids. Further shock waves are generated by the collapse of the cavity at the end of each oscillation cycle.

In each case; gas, liquid or solid, the small scale, $\sim$ mm, and fast propagation velocities of the shock waves, $\sim 10^3$ m.s^{-1} (mm.μs^{-1}) imply difficult experimental diagnostics. A range of optical techniques have been developed using macro video recording. Focused shadowgraphy, where the object plane is imaged on the recording plane shows large density gradients as well as opacities in the field. Interferometry gives the full field optical path and where the resolution is sufficiently high and circular symmetry exists, can provide three-dimensional refractive index and hence pressure information. Shear stress is given by the addition of some form of photoelastic technique.

This paper is concerned with a study of laser-generated shock waves in air, water and solids and the associated effects when the waves interact with other media.

2. Experimental arrangement

The study of laser-generated shock waves was carried out using optical apparatus centred around a Nd-YAG Q-switched laser system. This laser was a J K Lasers 'Mini Q' system with a single transverse mode beam and a nominal output around 10 mJ in an 8 ns long pulse.

The light source for all forms of photography was a dye laser, PRA LN102 which, when used with the green dye, Coumarin 500, gave a pulse about 0.5 ns long with a coherence length of $\sim$ 20 μm and a beam divergence 4.4 mrad. The dye laser was pumped by a N_2 laser with a pulse length of 300 ps. The temporal jitter of the system was about 2 ns and the dye laser energy of around 15 μJ was sufficient to expose ASA 100 FP4 film in the earliest experiments. The experimental layout is shown in Fig.1. Recording of the visual events was by means of a

Shock Waves @ Marseille III
Editors: R. Brun, L. Z. Dumitrescu

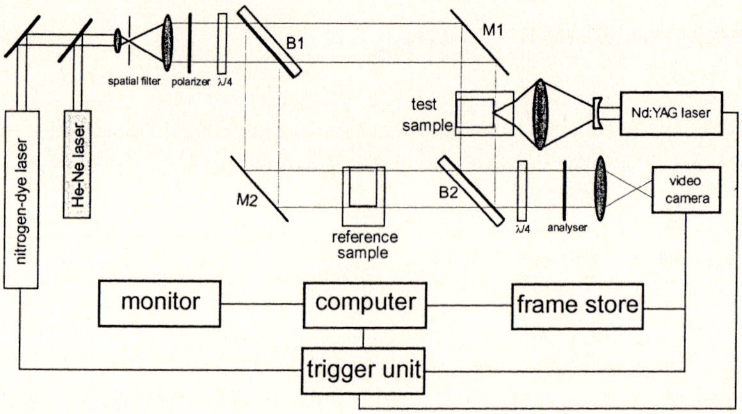

Fig. 1. Experimental arrangement

CCD video camera (NEC T1-22C) and a video frame grabber controlled by an IBM-compatible personal computer. A custom-built trigger unit controlled the video frame store, Nd-Yag laser and introduced a variable delay between the Nd laser and the illuminating dye laser.

Air breakdown studies involved the simplest optical set-up whereas laser-liquid and solid interactions required careful compensation of the two paths in the interferometer using identical glass-walled cells in each arm.

3. Results

3.1. Air breakdown

Fig.2a,b shows short sequences of Schlieren photographs and interferograms of air breakdown shocks. The central white elongated spot is the laser breakdown plasma which is not completely removed by the 1.06 μm blocking filter. The full horizontal frame dimension is 6.7 mm. The early stages (< 1 μs) are characterised by blast wave theory where the radius $r \propto t^n$ where $n = 0.66$ for the detonation regime and $n = 0.4$ later and t is the time. The energy in the shock wave, E_g can be estimated (Ready 1971) from:

$$r = (E_g/\rho_\infty)^{1/5}\, t^{2/5} \tag{1}$$

where ρ_∞ is the ambient gas density. In the shocks here, the energy is $\approx$ 40 μJ which is less than 1% of the laser pulse energy.

The shock wave pressure is obtained quantitatively from the interferograms using an Abel inversion matrix. The total optical path length is calculated from the fringe shift and the radial distribution of the refractive index, from the inversion, then gives, together with the Gladstone-Dale relationship, the pressure.

3.2. Air-solid breakdown

An important application of laser interactions is laser ultrasonics where laser light generates acoustic waves at a boundary. Different regimes have been identified including the thermoelastic source where no breakdown takes place. However when the laser flux density is high enough a plasma is formed accompanied by melting, ablation and vaporisation of the solid. Fig.3 shows the pressure profiles for waves generated in Perspex by air breakdown on the surface for a range of delays between the Nd laser and the illuminating dye laser. The peak pressure falls as r^{-1} as

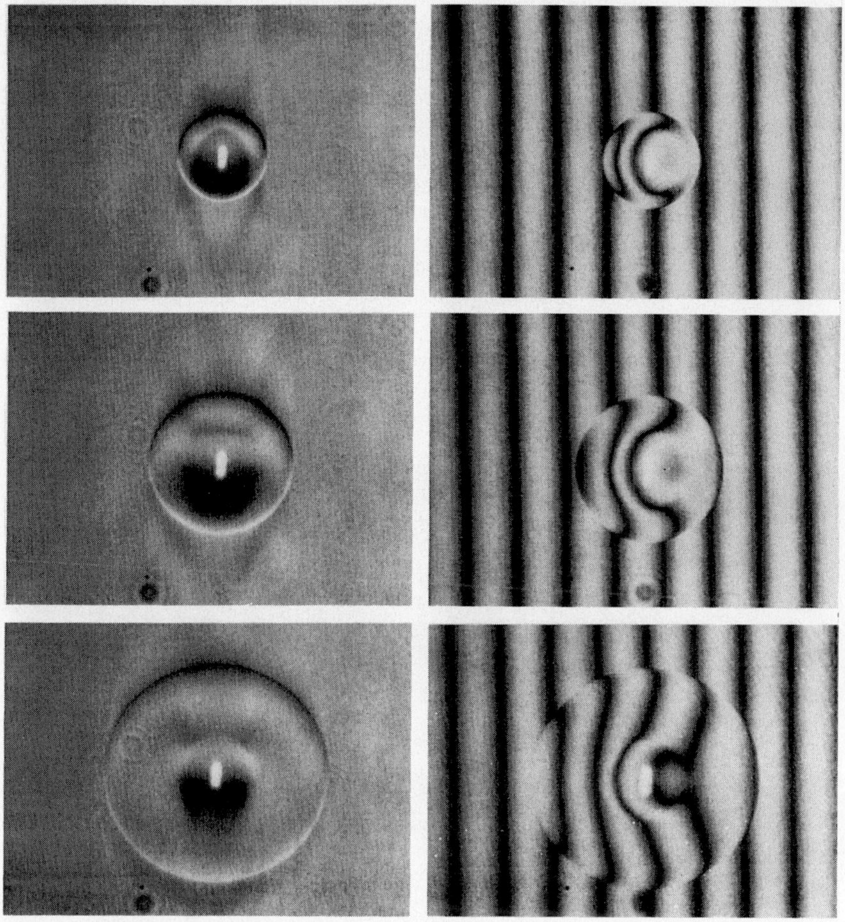

Fig. 2. Schlieren and Mach Zehnder interferograms of air shocks at 1.0 μs (top), 2.0 μs (middle), 4.0 μs (bottom). Horizontal frame length 8 mm

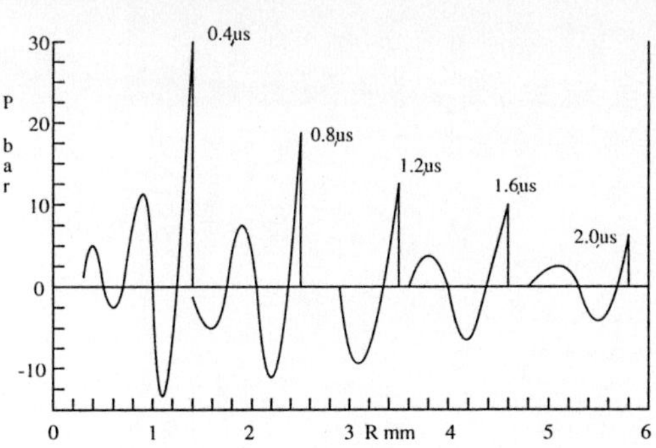

Fig. 3. Radial stress profiles for P-wave in Perspex

may be expected for a simple geometrical energy distribution where little attenuation takes place. Similar results have been obtained for breakdown in the body of the solid but in this case the shock wave duration is shorter (Ward and Emmony 1992).

3.3. Laser-liquid interactions

Laser breakdown in liquids leads to a complex sequence of phenomena with the general description of optical cavitation. When the pulse from a Q-switched laser is focused into the bulk of a transparent liquid, dielectric breakdown may occur. A plasma spark is formed, which heats the surrounding liquid to form a vapour bubble. The expanding wall of the bubble moving at a velocity greater than the speed of sound leads to the generation of a shock wave in the liquid. As the bubble expansion slows the bubble reaches some maximum radius before condensation of the vapour initiates a collapse of the bubble. During the collapse phase thermodynamics predicts that the pressure in the bubble again rises above the saturation vapour pressure of the liquid and eventually above that of the surrounding liquid. A stage is reached where the pressure and temperature both increase violently leading to a fast reversal in the direction of motion of the bubble wall and another shock wave. This 'rebound' is followed by a second expansion and then collapse which is repeated in a sequence of oscillations first discussed by Rayleigh. Laser breakdown is therefore responsible for the generation of a series of shock waves or acoustic transients at each bubble expansion stage.

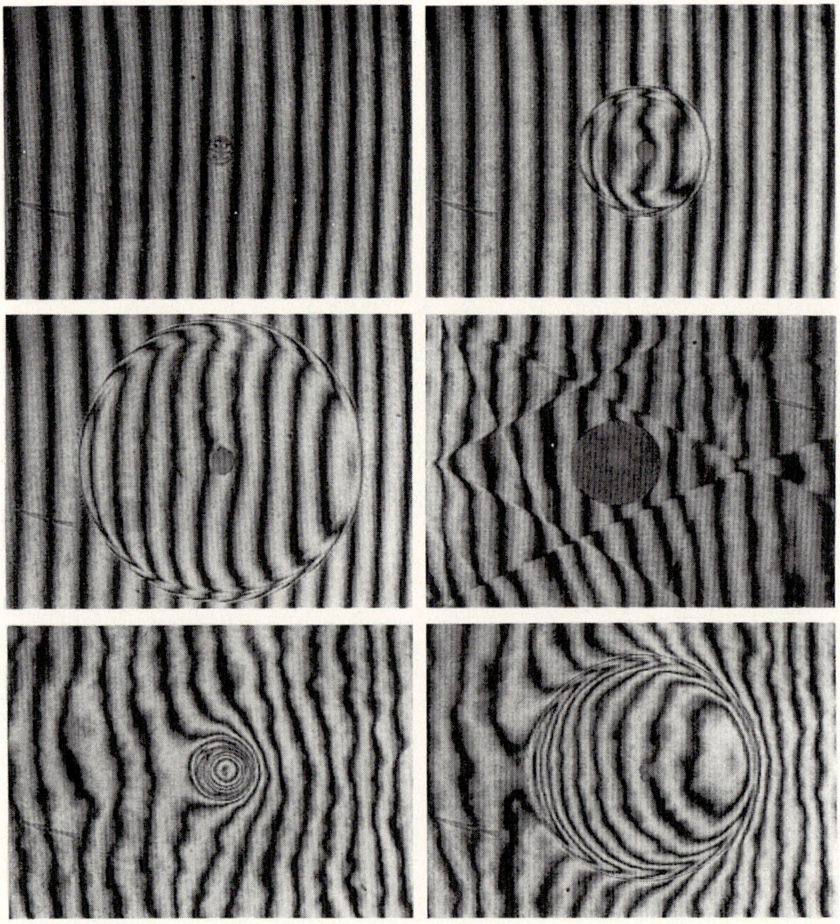

Fig.4. Water shock waves taken 0.2 μs (top left), 1.0 μs, 50 μs, 196 μs and 198 μs (bottom right)

Fig.4 shows a short sequence of water interferograms. In each case the laser radiation is focused from the top; (a), (b) and (c) were taken 0.2 μs, 1.0 μs and 2.0 μs after the Nd laser pulse. Fig.4d (50 μs) shows the growing bubble with a number of reflected waves from the free water surface and the cell walls. Fig.4e (196 μs) about 1 μs after rebound clearly shows the second shock wave which by (f) (198 μs) has grown to nearly fill the field of view. Figs.4(b) and (e) are similar except that in the rebound case the collapsed bubble leaves a low pressure region.

3.4. Laser-liquid-solid interactions

The laser generation of shock waves offers the opportunity to study a range of phenomena, particularly where optical recording provides information concerning the phase or optical path over the whole field. Optical cavitation can easily be induced near a solid using laser breakdown and the generation of acoustic waves at liquid-solid boundaries offers the possibility of characterising ultrasound effects such as diffraction around defects.

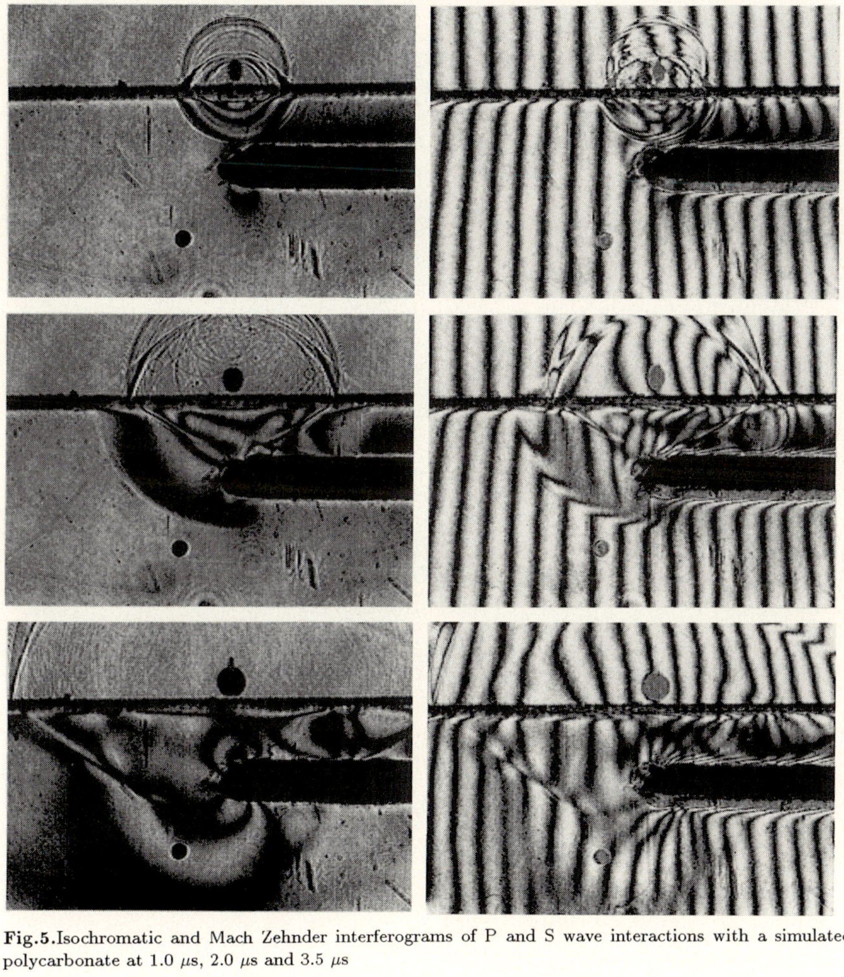

Fig.5. Isochromatic and Mach Zehnder interferograms of P and S wave interactions with a simulated crack in polycarbonate at 1.0 μs, 2.0 μs and 3.5 μs

4. Shock wave interactions in solids

The short rise-time of the laser-generated shock wave in a solid implies a broad bandwidth in the ultrasound. Consequently the shock wave is potentially a high-resolution source for ultrasonic investigations. The laser-generated shock wave at a liquid/solid interface has been used to study the interaction with mechanical faults in a model material. Fig.5 shows a short sequence of isochromatic fringe shadowgraphs and corresponding interferograms of the interaction of laser generated waves with a model crack in annealed polycarbonate.

Laser breakdown in the water immediately above the polymer surface leads to a shock wave which then is incident on the solid. Mode conversion produces P and S wave transients which travel at their respective sound speeds. Diffraction around the model crack tip is evident as well as the generation of a series of surface waves along the crack interfaces. This interaction lacks cylindrical symmetry and so a precise inversion procedure is not possible but the full field images allow the interaction to be studied semi-quantitatively.

5. Conclusions

Focused, pulsed laser radiation has been shown to be a reliable source of low total energy shock waves. The energy exchange rates and pulse duration of a small Nd-YAG laser are such that shock waves may be produced in a range of materials. Optical diagnostics have been developed using simple video recording techniques and interferometry allows quantitative measurements of the pressure field.

References

Ready JF (1971) Effects of High Power Laser Radiation. Academic Press, London

Ward B, Emmony DC (1992) Energy dissipation during laser breakdown in solids. 23rd Symposium on Laser Induced Damage in Optical Materials, SPIE, Washington

Laser Generation of Stress Waves in Metal

Jean-Eric Masse and Gérard Barreau
Laboratoire Mécasurf / ENSAM - 13617 Aix-en-Provence - France

Abstract. Laser-shock surface treatment can create plastic strains and induce compressive residual stresses in metals. This process consists in irradiating a metallic sample with a short but intense laser pulse. As a consequence, a high temperature and pressure plasma is created at the sample surface. An elastic-plastic wave is then propagating in the sample creating plastic strains. This paper presents the effects of laser-induced stress waves on the residual stresses of a standard steel: the surface residual stress induced is approximately 0.6 times the compressive yield strengh of the material for a plastically affected depth of one millimeter.

Key words: Plasma, Shock wave, Residual stresses

1. The physical principle of laser shocks

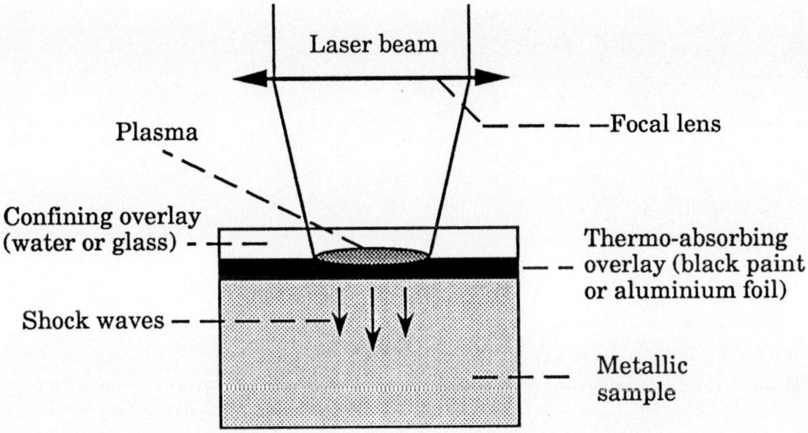

Fig. 1. Confined plasma configuration

This process, which was developped in the USA twenty years ago (Anderholm 1970), consists in exposing a metallic sample surface to an intense laser beam (80 J) during a very short time (< 50 ns). The heated zone vaporizes. The vapour rapidly achieves temperatures greater than 10,000° C, whereupon atoms are ionized and the vapour is transformed into a plasma. This continues to absorb the laser energy until the end of the deposition time. The plasma expansion can create a shock wave in metal. This type of interaction is called "direct ablation" and pressures of only some tenths of GPa are achieved.

To increase the shock pressures, it is necessary to delay the plasma expansion. To this purpose, we coat the surface sample with a material opaque to laser radiation (black paint or aluminium foil), then with a material transparent to the laser radiation (distilled water or BK7 glass). The "confined plasma configuration" is shown in Fig.1.

Shock Waves @ Marseille III
Editors: R. Brun, L. Z. Dumitrescu

The laser beam passes through the transparent overlay and is absorbed by the opaque overlay, partially vaporizing the latter. The expansion of the plasma in the confined region between the opaque and transparent overlays creates a high amplitude-short duration pressure pulse required for laser shock treatment. The shock pressures are magnified by a factor of 10 compared to a "direct ablation" configuration (Fabbro et al. 1990).

2. Generation of residual stresses by laser shock treatment

During the laser interaction with an uncovered suface sample, the treated zone is expanded by thermal effect, then compressed by the surrounding matter inducing a compressive stress field. After the deposition time, the treated zone is reset by the mechanical action of the undeformed matter inducing a tensile stress field.

If the surface sample is covered with a material opaque to laser light, the thermal effect affects only the protective covering. Only shock waves penetrate into the sample, inducing a tensile stress field. The surrounding matter reacts to this volume expansion by inducing a compressive stress field.

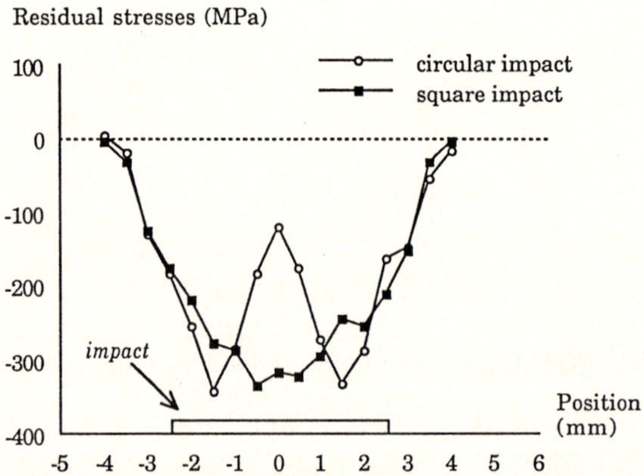

Fig. 2. Surface residual stresses for circular and square impacts

For circular spots, a lack of residual stresses exists at the center of the treated zone after the laser-shock treatment. It may be attributed to the simultaneous focusing of the waves emitted from the edges of the striked zone (Forget 1993). To avoid this defect, we must eliminate the circular symmetry, for example by using square-shaped impacts (Fig.2).

3. Material and shock conditions

The material studied is an hypoeutectoid steel. Its chemical composition is given underneath:

Table 1. Sample composition

Element	C	Mn	Si	S	P	Ni	Cr	Mo
% in weight	0.54	0.74	0.28	0.023	0.015	0.11	0.20	0.04

The yield strength is 600 MPa. In order to obtain an undamaged reference, the sample surface is mirror polished and the samples have been annealed to eliminate machining residual stresses.

The laser used for this work is the Q-switched neodymium:glass laser of the L.A.L.P. laboratory, operating at 1.06 μm wavelength. It basically consists of an oscillator and a Pockels cell followed by four amplifier stages. Maximum output energy is 80 J for nearly Gaussian pulses with a full width at half-maximum of about 25 ns. The laser spot is a 5 mm-side square.

We can obtain a maximum pressure of 3 GPa corresponding to a fluence of 10 GW/cm^2 in a water-confined mode (or 5 GPa in a glass-confined mode). Piezoelectric X-cut quartz gauges have been used for pressure measurements (Fabbro et al. 1990). The fluences used for the steel surface treatments are ranging between 1 and 10 GW/cm^2 in water- or glass-confined mode.

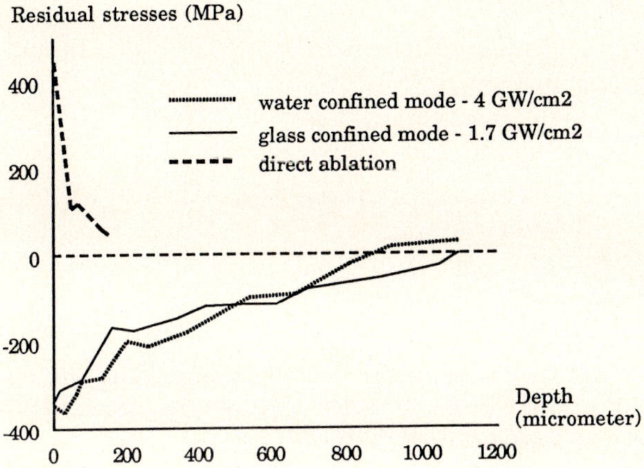

Fig. 3. In-depth residual stesses for different treatments

4. Residual stress measurements

X-ray diffraction technique was used to determine surface stress profiles along the treated zone. We also determined some in-depth profiles of residual stresses (after successively surface electropolishing).

Interaction without confining and absorptive overlays

In this case, the very surface of the metal sample forms the plasma and the thermal effect is significant. It results from this a tensile stress field. Fig.3 shows that the thermal effect induces a surface stress level of 450 MPa when treating a sample with a 5 GW/cm^2 fluence. The affected depth is 200 μm.

Confined plasma laser shocks

Residual stress analysis shows that the optimum impact pressure seems to be approximately 25 kbars i.e. 4 GW/cm^2 in a water-confined mode. This corresponds to a -350 MPa surface stress. In a glass-confined mode, we obtain the same results with a fluence of 1.7 GW/cm^2. Above those fluences, residual stresses saturate and the effect of the surface wave tends to decrease the value of residual stresses. The plastically affected depths are ranged between 800 μm and 1000 μm for these optimum shock conditions (Fig.3). With the optimized conditions (fluence of 4 GW/cm^2, in water-confined mode) we can produce an homogeneous cover strip (Fig.4).

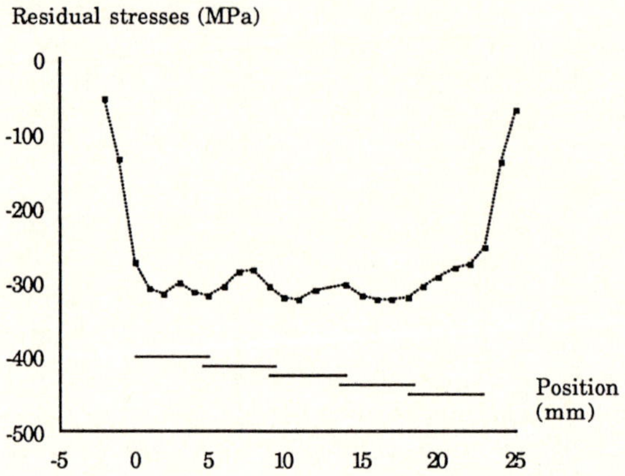

Fig. 4. Residual surface stresses of a cover strip. Five impacts with 10% overlap

5. Conclusion

A high-power pulsed laser can be used to induce compressive residual stresses in a metallic material. The plastically affected depth is higher than with other mechanical treatments. These reproducible results are obtained without deterioration of the treated zone (no surface pollution, homogeneous surface depression (less than 10 μm), low increase of roughness). Moreover, because of the precision of the laser beam, many regions inaccessible by other mechanical surface treatments can be treated.

In the following years, the industrialization of the laser shock process can be considered insofar as high frequency pulse lasers can be developed.

Acknowledgments

The authors would like to thank PEUGEOT S.A. for making available the material samples, and Remy Fabbro and the technical team of LALP (16 bis avenue Prieur de la Côte d'Or - 94114 Arcueil Cedex - France) for their part in the experimentation.

References

Anderholm NC (1970) Laser-generated stress waves. J. Appl. Phys. 16: 113

Fabbro R, Fournier J, Ballard P, Devaux D, Virmont J (1990) Physical study of laser-produced plasma in confined geometry. J. Appl. Phys. 68 (2): 775

Forget Ph.D (1993) Thesis, Ecole Nationale Supérieure des Mines de Paris (to be published)

High-Power Laser Radiation-Induced Shock Waves in Solids

Ch. Prat and M. Autric
Institute of Fluid Mechanics - Laser-Matter Interaction Group. Parc Scientifique et Technologique de Luminy, 13009 Marseille, France

Abstract. When a solid material is irradiated using high-power laser radiation, a fraction of the incident radiation is dissipated in the form of a shock wave in the sample. This article gives an account of shock measurements carried out using various pressure gauges on aluminium alloy and graphite samples irradiated using a CO_2 pulsed laser. These results are then used for comparison with a theoretical model.

Key words: Laser, Solid, Shock measurements

1. Introduction

During the first instants of the interaction, the optical properties of the material determine the energy fractions that it reflects and that penetrate it to a thickness generally defined by the optical penetration depth. The relative importance of specular reflection and diffusion is then linked to the surface condition of the material.

Rough areas on the sample surface very rapidly release electrons and are the site at which changes of state occur. Heating of the material by the absorbed energy is accelerated by the changes in its thermal properties which inevitably occur when its temperature rises. Depending on the energy density used, fusion and vaporization then affect the sample mass. The vapor produced undergoes more or less complete ionization, whose main mechanisms are generally the multiphotonic effect and inverse bremsstrahlung. Vaporization and plasma expansion transmit a fraction of the energy to the irradiated solid in the form of a shock wave and mechanical impulse.

A plasma therefore develops on the sample surface and spreads in the opposite direction to the radiation that created it, partially shielding the target. Over the duration of its existence the plasma absorbs, diffuses and if the critical electron density is reached, reflects the incident radiation; it also emits radiation in a large spectral range. In the ultra-violet region, this emission is particularly well absorbed by some materials and contributes to the surface heating and to vaporization and therefore to mechanical effects.

2. Theoretical considerations

The plasma flow created in front of the irradiated sample is now generally considered to contain (Phipps 1988, Puell 1970) :

- a plane one-dimensional part of length R (corresponding to the radius of the interaction spot)
- a part whose radial evolution is no longer negligible (two-dimensional flow).

The first region of the flow can therefore be considered to be entirely sonic (Nemchinov 1967, Pert 1986). According to Puell, overestimation of the plasma speed as a results of this approximation remains acceptable as long as $x < R$. According to Basov (1968), the instant t at which he expansion evolves under conditions dominated by the spherical expansion of the plasma now transparent is given by :

$$t = K_0^{-1/9} R^{8/9} \Phi^{-2/9} \qquad [\mathrm{s}] \qquad (1)$$

where $K_0 = 3.7 \times 10^{20} \cdot Z^3 \, (Z+1)^{3/2} \lambda^2 A^{-7/2}$ (in MKSA units) where Z represents the degree of ionization, A the atomic mass, λ and Φ are the wavelength and incident radiation power density.

Shock Waves @ Marseille III
Editors: R. Brun, L. Z. Dumitrescu

For graphite, and for $R = 0.6$ cm, this time varies from 400 ns at 10 MW/cm^2 for a degree of ionization $Z = 1$, to 90 ns for $\Phi = 10^3$ MW/cm^2 and $Z = 3$.

Basov also proposed an analytical expression of the pressure exerted on the target in the case of one-dimensional flow :

$$P_{1D} \approx \frac{1}{3}\left[3.7 \times 10^{20}\ Z^3(Z+1)^{3/2}A^{-7/2}\right]^{-1/8}\left(\Phi^{-3}\lambda\sqrt{\tau}\right)^{-1/4} \qquad [\mathrm{Pa}] \tag{2}$$

where the factor $(\Phi\lambda\tau^{1/2})$ is used to characterize the radiation conditions (τ is the pulse duration)

According to Caruso et al. (1968), the fraction of intensity dissipated in the solid in the form of a shock wave does not exceed $(\rho_{\mathrm{plasma}}/\rho_{\mathrm{solid}})^{1/2}$ where ρ is the density of the medium.

Work carried out by Astic (1990) under experimental conditions similar to those under which the results presented in the last part of this paper were obtained, shows that the maximum ionization degree for the species observed in the plasma plume was equal to 3. Furthermore, electron density values were between 1×10^{17} and 9×10^{17} electrons/cm^{-3} for graphite C5890 and the Al alloy (AU4G) studied. Under these conditions, the maximum power fraction transmitted to the solid in the form of a shock wave is approximately 10^{-2}.

3. Experimental conditions

Most of the pressure measurements were carried out using piezoresistive carbon gauges. Some tests were however carried out on assemblies fitted with quartz and PVDF piezoelectric gauges.

3.1. Carbon gauges

Pressure measurement series were carried out using Dynasen FC 300-50 Ekrte piezoresistive gauges in which the sensitive element, consisting of a 50 Ω nominal resistance and 25 μm-thick carbon sheet, is inserted between two kapton sheets, each 25 μm-thick. As the sensitive part is so small (0.8 mm^2) in comparison with the interaction surface (0.4 to 1 cm^2), the measurements taken are virtually pinpoint. The gauge is bonded between the target and a base of the same type in order to minimize shock reflections at the rear side of the sample. The gauge forms one of the branches of a Wheatstone-type assembly driven by a current pulse. The measurement consists in reading the bride unbalance voltage at the time of the laser shock.

3.2. Quartz gauges

The gauges used for our experiments consist of a 5 mm-thick quartz disk almost entirely metallized. A band free of metal coating delimits a surface area for measurements of 28 mm^2. The gauge is bonded to the rear side of the target, taking care to insulate it electrically from the target.

3.3. PVDF gauges

This gauge consists of a PVDF sheet 25 μm-thick whose surfaces have been coated by cathode sputtering to provide platinum or gold electrodes. The intersection of these electrodes forms the active surface of the sensor. (Bauer 1986, Wey et al. 1986)

4. Experimental comparison of the various gauges used

The rise-time and quality with which the gauge responds to stress are inherent to both its individual capability and the way it is set up (Bosca 1981). Given these restrictions, a piezoelectric sensor used in a current mode measurement configuration seems the most appropriate. However, with a quartz gauge the length of signal validity corresponds to the shock wave transit time in the dielectric ($\approx$875 ns). This sensor can therefore only be used with short laser pulses and is not suitable for our experimental conditions insofar as the pressure record is required during an entire μs laser pulse. In addition, due to its low amplitude (a few millivolts), the signal obtained during our experiments is disturbed by unavoidable electromagnetic interferences of the same order of magnitude. Using a PVDF gauge, the signal delivered by the sensor is around one volt and interference is relatively much lower. The main drawback of this signal lies in its analysis, which requires integration, and results in an information loss. In addition, at the time of our measurements, no calibration curves existed in the dynamic low pressure range.

Using a carbon gauge, it is possible to record pressure signals created by a laser pulse of several μs, while the amplitude of the signal delivered is of the same order of magnitude as the quartz gauge. However, the recorded signal reveals thermal drift in the gauge due to its heating by the Joule effect. This drift can be quantified and taken into account during computer analysis of the signals. This carbon gauge proved the most effective during our measurement series in spite of its delicate installation.

Fig.1 illustrates these comments using characteristic signals obtained under similar conditions, on AU4G, with each of the three gauges. On each recording, the upper trace represents the pressure signal and the lower trace represents the laser pulse. It can be seen that the measurement validity time is relatively short for the quartz gauge (in comparison with the carbon gauge) but their sensitivity levels are comparable (it should be noted that the quartz signal is positive whereas the carbon gauge signal is negative). The recordings also show the fundamental difference in appearance between the directly analysable quartz gauge signal on one hand and the signal derived from the PVDF gauge on the other hand.

5. Mechanical assembly

The signal given by the gauge reflects the part of pressure wave arriving at instant t. Strictly speaking, at each interface between two materials, the difference in acoustic impedance causes multiple reflections, likely to arrive in succession, which affect the sensitive element of the gauge. If the layers of glue or insulation are thin, these reflections are not generally noticeable but may increase the signal rise-time. These reflections in kapton sheets insulating the carbon gauge explain why the response time of this sensor is greater than the quartz and the non-insulated PVDF sensors.

At low dynamic pressures, the shock Hugoniot curve is approximated by a straight line whose slope corresponds to the acoustic impedance of the material Z_{ac}. This approximation is used to relate the stress transmitted from material 1 to material 2 using the relation : $\sigma_1 = (Z_{ac1} + Z_{ac2}/2Z_{ac2})\sigma_2$. Fig.1 illustrates the shock wave reflection problems for Al alloy; the pressure peak produced by the laser pulse appears several times on the signals obtained. The time separating two successive peaks corresponds to the time taken for the return trajectory of the shock wave in the target thickness. In the case of graphite targets, reflections cannot generally be observed on the recordings. This apparent absence of reflection can be explained by better acoustic impedance matching between the sample, gauge, glue and base with the graphite than with the AU4G. In addition, any reflections would be damped to a greater extent by the graphite than by the metal.

Assuming that the fraction of the shock energy that is not transfered to the gauge is reflected toward the target, the measurement of the amplitude of the first reflection makes it possible to estimate the value of Z_{ac2}. This way of estimation could be usefull if the reflection is due to a medium whose characteristics are not well known (glue, kapton sheets...). The values of Z_{ac2} we have calculated are close to the estimated one of the glue, used in such cases.

The Al alloy used in this study is usually known as AU4G or 2017A. Its chemical composition (Fe$\leq$ 0.7 %; Si$\leq$ 0.8 %; Cu= 4.2 %; Zn$\leq$ 0.25 %; Mg= 0.8 %; Mn= 0.7 %; Ti$\leq$ 0.2 %; Cr$\leq$ 0.1 %) is very close to that of alloy 2024, often quoted in the literature. Graphite C5890 produced by Carbone Lorraine, is a polycristalline form with very fine structure and porosity of 10 %. Its

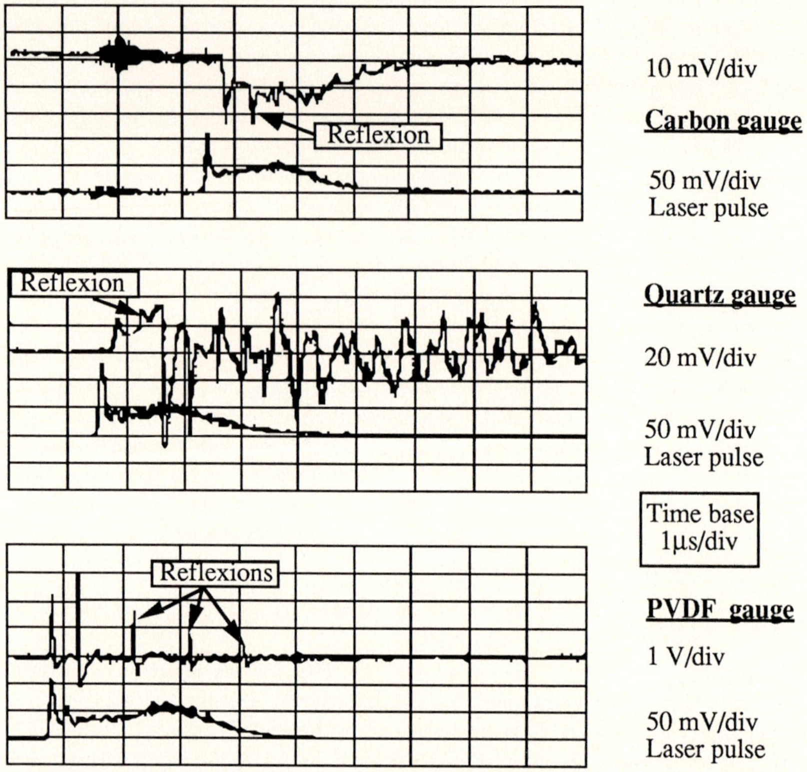

Fig. 1. Comparison of pressure signals obtained using different gauges

density and structure make it comparable to ATJ-S produced by Union Carbide (Laturelle 1986). The acoustic impedance levels of the graphite and Al alloy studied are 4.7×10^6 and 18.2×10^6 kg/s.m^2 respectively; and that of the glue can be estimated for several set-ups at approximately 5×10^6 kg/s.m^2.

6. Results and discussion

The results presented show the amplitude of the peak of the pressure signal to which it is possible to match a power density value read at the laser pulse, as long as the shape of the laser pulse and the pressure signals are quite similar.

6.1. Graphite

Pressure measurements were carried out using 1.5 mm-thick targets The dispersion of results can be attributed to the unevenness of the spread of energy over the surface of the target. The theoretical spread of intensity reveals that the width of the rings on which incident energy is distributed is of similar order of magnitude as the sensitive element of the carbon gauge. Depending on focusing conditions, a deviation of approximatively 0.5 mm in the incident radiation pointing to the surface can result in a considerable intensity variation from one shot to the next at a given point. However, as the distance between two adjacent intensity rings is of the same order of magnitude or less than the sample thickness, two basic shock waves, each created by one of these rings, mix together during their transit through the material to form an almost uniform shock front on the rear side of the sample. The slope, of around 0.7, of the mean curve through these results is very close to that given by Eq.2, 0.75 (fig 2.A).

The theoretical approximation of Basov's model for an ionization degree of 1 can be considered to give a good approximation of the stress level exerted by the plasma on the material. Studies by Astic provide no information concerning the degree of plasma ionization during a single laser pulse peak. These measurements were integrated over 1 μs at various instants from the start of the pulse. Strictly speaking, uncertainty therefore remains concerning the degree of ionization that can be used in the theoretical expression compared with experiment.

Moreover, examination of pressure signals provides a wealth of information concerning the evolution of plasma. Above an intensity of around 40 MW/cm^2, an abrupt pressure drop can be clearly observed at the center of the signal plateau, revealing total but momentary decoupling between the incident radiation and the material. A photodiode facing the surface simultaneously detects an intensified light emission in the visible spectrum. These two observations are sure signs of an increase in electron density in the plasma, thereby increasing its opacity, as is confirmed by Astic's results revealing a maximum absorption coefficient approximately 2 μs after the start of the laser pulse. According to Criterion 1 in Part 2 of this paper, the flow remains one-dimensional during the length of the laser pulse peak (80 ns).

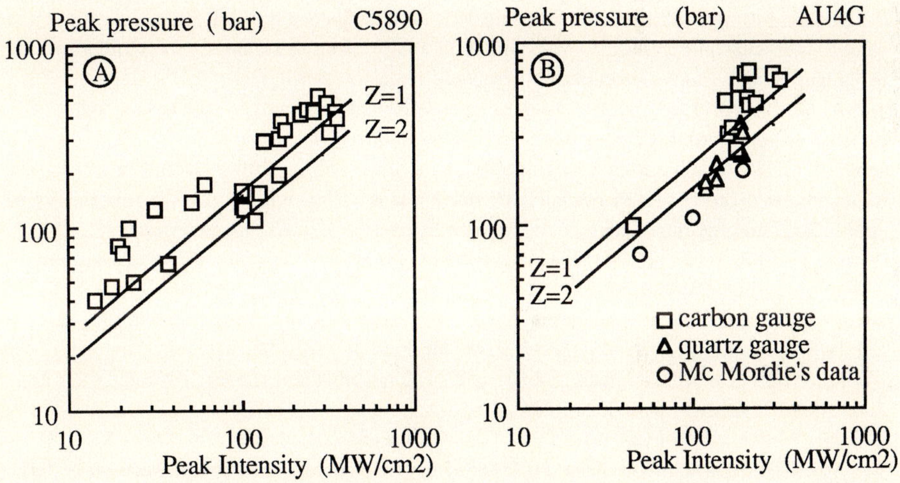

Fig. 2. Theoretical and experimental pressure results. Theoretical curves: 1D; t = 80 ns

6.2. Aluminium alloy

Pressure measurements were carried out using 1.5 mm-thick targets. The consistency of measurements obtained using carbon and quartz gauges (with an active area of 0.8 and 28 mm^2 respectively) shows that the flatness of the shock front on rear side of the target is satisfactory and that measurements carried out using the carbon gauges can be taken as representative of the mean pressure exerted on the sample. Although the acoustic impedance of quartz ($\approx 15 \times 10^6$ kg/s.m^2) is close to that of aluminium, shock reflection could occur. So the pressure induced in the sample is calculated taking into account the glue film at the rear side of the target.

These results are compared with those obtained by Mc Mordie et al. (1975) on a 0.66 mm-thick DTD-5104 Al alloy target using quartz gauges, for various rise times of the laser pulse peak. For the longest time (43 ns), his measurements prove to be in good agreements with ours (Fig.2B). As most of the experimental results obtained are between 100 and 300 MW/cm^2, a significant mean curve cannot be calculated for them. An acceptable theoretical approximate curve through these points can be drawn however. The same reservations as for graphite concerning the ionization degree, should be observed in this comparison.

Acknowledgements

The results presented here were obtained in the framework of studies supported by the Direction des Recherches Etudes et Techniques under contract No. 89-207. The authors would like to thank Dr. J.P. Fragassi for his technical assistance throughout the course of the experiments.

References

Astic D (1990) Caractérisation par spectroscopie d'émission d'un plasma créé par laser CO_2 impulsionnel sur matériaux solides. Thèse de Doctorat de l'Université d'Aix-Marseille II

Basov NG, Gribkov VA, Krokhin ON, Sklizkov GV (1968) High temperature effects of intense laser emission focused on a solid target. Soviet Physics, J.E.T.P. 27, 4: 575-582

Bauer F (1986) Propriétés ferroélectriques et réponse sous choc du polymère PVF2 et des copolymères VF_2/C_2F_3H polarisés. Rapport interne I.S.L.No. C 245/86

Bosca G, David J, Starynkevitch C, Wettling JC (1981) Influence des différents éléments d'une chaine de mesure piézorésistive sur la mesure d'une pression dynamique. 1er Symposium sur les jauges et matériaux piézorésistifs. Arcachon, France

Caruso A, Gratton R (1968) Some properties of the plasmas produced by irradiating light solids by laser pulses. Plasma Physics 10: 867-877, Pergamon Press

Laturelle F (1986) Cratérisation d'origine mécanique d'une cible irradiée par un faisceau laser impulsionnel : étude expérimentale, contribution à la modélisation des effets thermoélastiques et simulation numérique. Thèse de Doctorat d'Université, Ecole Centrale Paris

Mc Mordie JA, Roberts PD (1975) The interaction of pulsed CO_2 laser radiation with aluminium. J. Phys. D: Appl. Phys. 8:768-781

Nemchinov IV (1967) Steady-state motion of radiation-heated vapors of a material in the presence of lateral spreading flow. J. of Applied Math. Mechan. 31: 320-338

Pert G (1986) Models of laser-plasma ablation Part. 2, steady-state theory: self regulating flow. J. Plasma Physics, 36, part 3: 415-446

Phipps CR, Turner TP , Harrisson RF, York GW, Osborne WZ, Anderson GK, Corlis XF, Haynes LC, Steele HS, Spicochi KC, King TR (1988) Impulse coupling to targets in vacuum by KrF, HF and CO_2 single pulse lasers. J. Appl. Phys. 64, 3:1083-1096

Puell H (1970) Heating of laser produced plasmas generated at plane solid targets (I. Theory). Z. Naturforsch 25a: 1807-1815

Wey J, Baca W, Hugenschmidt M (1986) Mesure des pressions mécaniques induites par laser. Rapport interne I.S.L. No. R 123/86

Beam-Matter Experiments with High-Power Proton Beams on KALIF

K. Baumung*, H.J. Bluhm*, P. Hoppé*, H.U. Karow*, D. Rusch*, G.I. Kanel§, V. Licht†, S.V. Razorenov‡ and A.V. Utkin‡

*Kernforschungszentrum Karlsruhe, Karlsruhe, Germany
§High Energy Density Research Center, Moscow
†Karlsruhe University
‡Inst. for Chemical Physics, Chernogolovka, Russia

Abstract. The Karlsruhe Light Ion Facility ("KALIF") is a pulse power accelerator delivering up to ~40 kJ proton beam energy at power densities on target up to 1 TW/cm^2. For the 10- to 20 μm-thick energy deposition layer in a solid target, simulations predict temperatures of several tens of eV, pressures in the 10- to 100 GPa-range and densities which rapidly decay due to material ablation. Applying a laser Doppler velocimeter, we have utilized the dynamic response of 20- to 50 μm-thick planar metal targets to investigate the ion beam properties and to measure the ablation pressure. Thin metal foils were accelerated to velocities beyond 12 km/s. To demonstrate the potential of KALIF as a fast ablative shock-wave generator we have performed dynamic tensile strength measurements in the nanoseconds load duration regime.

Key words: Charged-particle beams, Hypervelocity projectiles, Spall strength, Shock waves

1. Introduction

The Karlsruhe Light Ion Facility (Bluhm et al. 1985) "KALIF" is a 1.7 MeV, 600 kA, 50 ns fwhm pulse power generator which is used to produce an intense proton beam and to focus it to a spot with high power density. The ion sources, the so-called "diodes", consist of an annular anode, on which a plasma layer is formed at the beginning of the accelerator pulse. Ions are extracted from this plasma and accelerated by the electrical field between the anode and an electron cloud oscillating just a few millimeters in front of the anode, the so-called "virtual cathode". To prevent the electrons from carrying a predominant portion of the diode current, a strong magnetic field must be applied in the anode-cathode-gap. This can be achieved either by the electron current itself like in the so-called "self-magnetically insulated" B_Θ-diode (Schimassek et al., 1991), or, in a better controllable way, by an externally applied magnetic field in the B_{appl}-diode (Bluhm et al. 1992). Behind the accelerating gap the ions essentially drift ballistically. Focusing is mainly effected by the shape of the accelerating electric field and requires both, current and space charge neutralization of the beam. This is realized by picking up electrons from a low-pressure background gas (B_{appl}-diode) or from metal vanes behind the virtual cathode (B_Θ-diode) which then drift together with the ions, thus providing the neutralization. The B_{appl}-diode provides a total beam energy of ~40 kJ, representing 20% of the primary energy store. The maximum power density is 1.05 TW/cm^2, the focal spot has a fwhm of ~6 mm. Most of our measurements were carried out using the B_Θ-diode. Although this diode delivers 0.25 TW/cm^2 only, it is convenient for experimenting since it allows up to 12 successive shots without major hardware restoring.

Beam-target interaction experiments performed on KALIF in the past investigated the enhancement of the ion stopping power in ionized matter (Goel and Bluhm 1988), filamentation of the ion beam in the focus (Karow et al. 1990), and the peak ablation pressure (Bachmann et al. 1988). Recently, a laser-Doppler-velocimeter was set up allowing to investigate the beam-target interaction by measurement of the hydrodynamic response of thin planar samples. These experiments will be presented in the following.

Shock Waves @ Marseille III
Editors: R. Brun, L. Z. Dumitrescu

2. Technique and results

2.1. Experimental set-up

For our experiments we have used an ORVIS-type velocimeter (Bloomquist and Sheffield, 1983) synchronized with KALIF by an electrical signal picked up from the pulse line. This trigger signal shows a jitter of ±7 ns relative to the ion beam impact on the target and allows to realize high streak rates which are essentially limited by the available laser power of 1 W. We have performed a series of measurements at a streak rate of 0.3 ns/mm and a total streak length of 21 ns achieving an estimated temporal resolution of <200 ps.

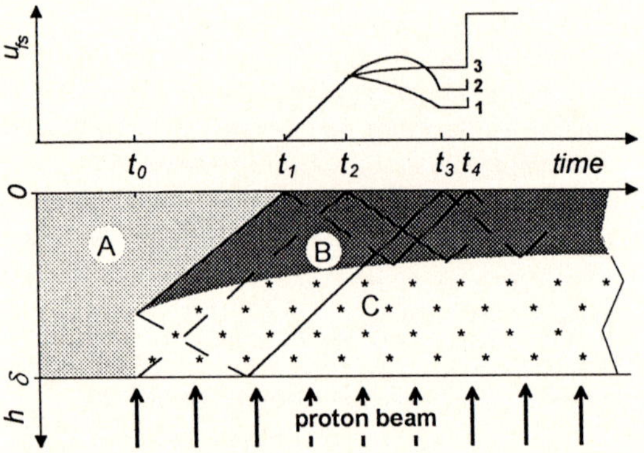

Fig. 1. Schematic h, t-diagram (h: Lagrange coordinate) of the hydrodynamic phenomena in the target in the early phase of energy deposition. At t_0, the proton beam of initial range R hits the target of thickness d. Pressure waves (solid lines) and release waves (dashed) circulate in the target and produce, at the rear target surface, the velocity u_{fs}; velocity histories (curves 1, 2, and 3) depend on the power histories. (A: undisturbed material; B: solid, compressed material; C: energy deposition zone)

2.2. Experiments and results

2.2.1. Pressure measurements

At present, only very scarce experimental information exists on the state of matter in the dense plasma which forms in solid targets exposed to the KALIF beam. For that reason, one of the main objectives of our target experiments consists in providing EOS-data in this regime. In a first step, we have determined the ion beam-induced ablation pressure in planar geometry, for the first time, quantitatively and time-resolved from measurements of particle velocity histories by applying the hydrodynamic relations used in shock-wave physics and the Hugoniot-EOS. In order to diagnose the ablation pressure waveform unsteepened by wave propagation, the velocity has to be measured in close vicinity ($\leq$ 10 μm) of the energy deposition zone. The depth of the zone is determined by the stopping range of the ion beam in the target and amounts to ~10 to 20 μm, depending on the material.

In the pressure range below 100 GPa encountered in our experiments, the particle velocity can be taken as 0.5 times the free surface velocity u_{fs}. However, as can be seen from Fig.1, the free surface velocity is an undisturbed function of the pressure only until the arrival, at t_4, of the reflected first wave front, i.e. during a few nanoseconds only. However, this time interval is sufficient to investigate the properties of the front of the ion beam. Because of the ~20 ns rise-time of the high voltage pulse delivered by the generator (cf. Fig.2) the first ions leaving the

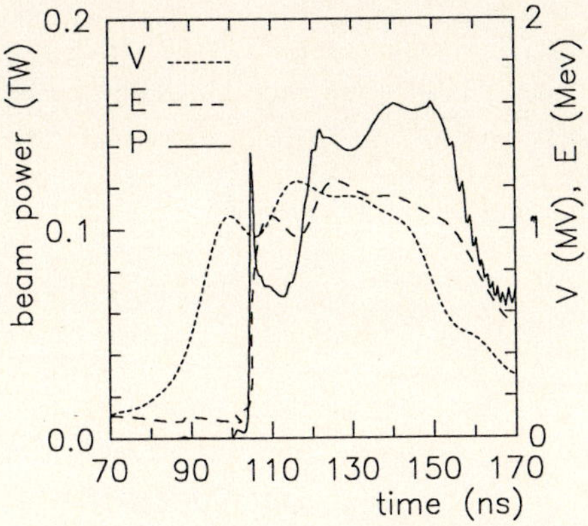

Fig. 2. Generator voltage (V), bunched ion energy (E), and bunched ion power on target (P)

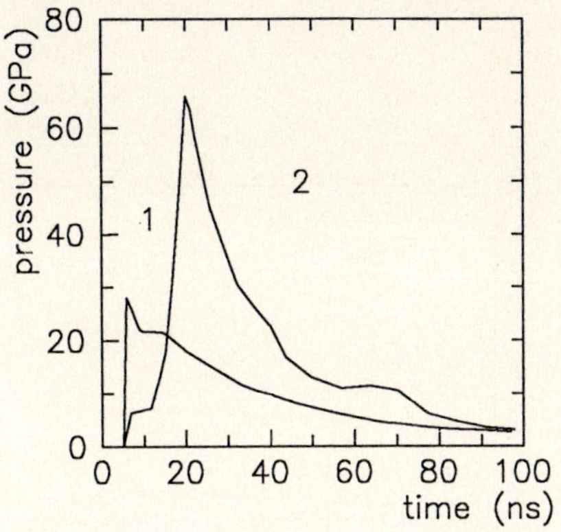

Fig. 3. Ablation pressure histories for the B_Θ (1) and the B_{appl} diode (2)

anode with low energies are overtaken on their 15 cm trajectory to the target by ions starting later but with higher energies. As a result of this time-of-flight compression or "bunching" of the beam front, the power P on target rises much faster than the generator voltage V and may even steepen up to a jump as shown in Fig.2 for an extreme case. Another effect of the bunching is that the energy E and thus the range R of the leading ions may reach half the maximum value and vary considerably with the respective voltage history.

From Fig.1 it follows that the initial range R depends on the rise time t_2-t_1 of the free surface velocity. For later times, the range can be deduced from the wave transit time t_4-t_1 which, in its turn, depends on the target thickness. Ranges derived from these data compare well with table data for corresponding ion energies. An analytical model based on an acoustic approximation allows to explain the details of the velocity profiles in the early phase of energy deposition, i.e. at times where the density in this zone is still close to the initial solid state density (Baumung et al. 1994). It allows to determine e.g. the power density and the increase rate of the power density and provides complementary information in a regime in which, due to the smallness of the signals, the accuracy of the electrical KALIF instrumentation is very limited.

For pressure measurements over longer time intervals, the disturbing wave reverberations have to be suppressed by use of a thick window in direct contact with the target. The window should have the same dynamic properties as the target, and remain optically homogeneous and transparent to the laser light of the interferometer up to high pressure states. For our experiments aluminium as target material and LiF single-crystal as a window was the appropriate choice. In this case, the measurement, after correction for the pressure effects on the refractive index of the windows material (Wise and Chhabildas, 1986), directly provides the particle velocity profile $u_p(t)$ of the Al-LiF- interface. Two ablation pressure histories evaluated from such measurements are displayed in Fig.3. Pressure amplitudes of $\sim$25 and 65 GPa are obtained with peak power densities of 0.25 and 1 TW/cm^2, respectively. With the B_Θ-diode, due to the bunching effect, there is an initial pressure jump to the maximum value. This level is then approximately maintained as long as the power density rises. As soon as the power increase rate reduces, the pressure falls off. In the case of the B_{appl}-diode there is a significant turn-on delay of some 15 ns during which the ion current stays very small. As a result, despite bunching, the power density and hence the pressure remain low. Only after reaching the nominal working point the diode delivers the full power density of 1 TW/cm^2 resulting in a corresponding pressure increase. The peak pressures approximately correspond to code predictions.

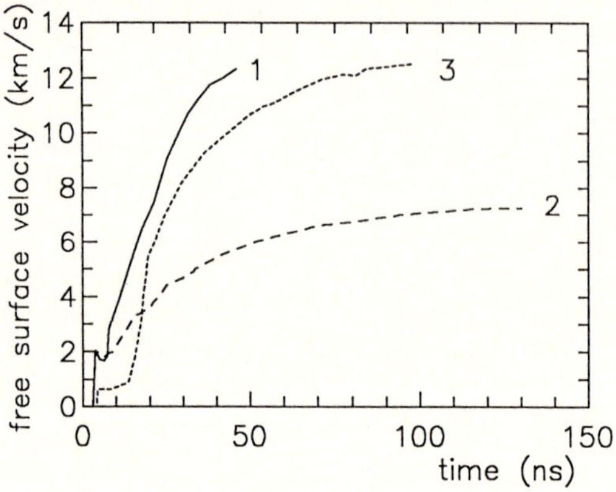

Fig. 4. Free surface velocities as a function of time. Curve 1: 33 μm aluminum, 0.25 TW/cm^2; curve 2: 50 μm Al, 0.25 TW/cm^2; curve 3: 50 μm Al, 1 TW/cm^2

2.2.2. Ablative acceleration of flyer foils:

Fig.4 shows three examples of ablative acceleration of thin plane sheets of matter. The acceleration takes place over a very short flight path of ~200 μm reaching values of more than 10^{11} g. With a beam power density of ~0.25 TW/cm^2 and an aluminum foil of ~10 μm residual thickness - i.e. the difference between the initial thickness of 33 μm and the thickness of the ablated layer of ~20 μm - a velocity of ~12 km/s was obtained (curve 1). The acceleration could be tracked up to a velocity of ~13 km/s. From the velocity profile one concludes that the acceleration still continues. Whether the signal loss is caused by hydrodynamic instability or a change of reflectivity due to material melting as predicted by simulations has to be clarified. For that purpose, we are going to improve the dynamic response of our system by using an intensified CCD-array and extend the velocity measurement from the actual 100-μm-diameter spot to a larger, spatially resolved area of the target.

With the same power density, an initially 50 μm-thick target remains stable and is accelerated to a final velocity of 7 km/s (curve 2). With a power density of 1 TW/cm^2 and 50 μm-thick foil, a final velocity of 12.5 km/s was reached (curve 3). With particle velocities in this range, interesting impact experiments in the 100- to 500 GPa-range are feasible. This was demonstrated in preliminary measurements with Al flyers impacting on optically transparent thick targets. In this case, it was possible to record with the ORVIS system two kinematic quantities in one and the same experiment: the acceleration and impact velocity of the flyer on the one hand, and, on the other hand, after impact, the particle velocity of the interface, or the shock front velocity in the target, respectively. This was verified by experiments with Al flyer foils impacting on LiF targets, and on PMMA targets, respectively.

2.2.3. Spall strength measurements at high strain rates:

We have used KALIF as a fast ablative driver for spall strength measurements at strain rates between 5×10^5 s^{-1} and 3×10^7 s^{-1} (Kanel et al, 1993). Using the ablation pressure pulse allows

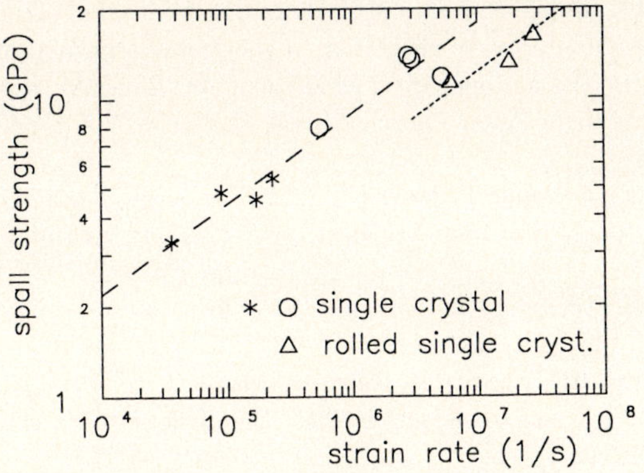

Fig. 5. Results of spall strength measurements on undeformed and rolled molybdenum single crystals (⋆ data from explosives drivers)

to realize strain rates between 10^5 s^{-1} and 3×10^6 s^{-1}. The acceleration of target foils of residual thicknesses between 10 μm and 50 μm to velocities between 4 km/s and 5 km/s allowed to obtain strain rates as high as 2.8×10^7 s^{-1} so far and load durations down to a few nanoseconds. The measurements were performed on samples of undeformed as well as rolled molybdenum single crystals. Fig.5 shows some of the results together with data obtained by explosives methods in the Institute for Chemical Physics at Chernogolovka, Russia. At the highest strain rates the dynamic strength reaches 30% of the ultimate theoretical strength.

3. Conclusions

The results presented above demonstrate the capabilities of the KALIF generator as a fast ablative driver in the domain of short (~30 ns) pulses in the 10 to 100 GPa range. With ablatively accelerated flyer plates of a few millimeter diameters, pressure pulses in the 100 to 1000 GPa range and pulse durations down to 2-3 ns can be realized for a variety of applications. The hydrodynamic response of thin targets can be utilized to diagnose the on-beam properties.

Acknowledgements

This work was performed with the support of the Bundesministerium für Forschung und Technologie in the framework of the Bilateral Scientific and Technical Co-operation, contract no. WTZ VI.20.

The authors appreciate the continuous interest and support of their joint work by G. Kessler, director of the Institut für Neutronenphysik und Reaktortechnik, KfK Karlsruhe, and V.E. Fortov, director of the High Energy Density Research Center, Moscow, of the Russian Academy of Sciences. The authors also would like to acknowledge the technical assistance by J. Singer and O. Stoltz (KfK/INR), and the KALIF operators.

References

Bachmann H, Bluhm H, Cardenes H, Goel B, Hoppé P, Karow HU, Munz CD, Müller F, Rusch D, Schub I, Singer J (1988) Experiments on ablative generation of shock waves in matter by intense proton beams: KALIF experiments and numerical simulations. In: Bauer W and Schmidt W (eds) Proc. 7th Intl. Conf. on High- Power Particle Beams, Karlsruhe, Vol. I, pp. 827

Baumung K, Karow HU, Rusch D, Bluhm HJ, Hoppé P, Kane GI, Utkin AV, Licht V (1994) High-power proton beam diagnostics by analysis of wave dynamics in solid targets. Accepted for publication in J. Appl. Phys.

Bloomquist DD, Sheffield SA (1983) Optically recording interferometer for velocity mesurements with subnanosecond resolution. J. Appl. Phys. 54: 1717

Bluhm H, Böhnel K, Buth L, Hoppé P, Karow HU, Klumpp A, Rusch D, Scherer T, Schülken U, Singer J (1985) Experiments on KfK's light ion accelerator KALIF. Dig. Tech. Papers, 5th IEEE Pulsed Power Conf., Arlington, VA, pp 114-117

Bluhm HJ, Hoppé PJW, Laqua HP, Rusch D (1992) Production and investigation of TW proton beams from an annular diode using strong radial magnetic insulation fields and a preformed anode plasma. Proc. IEEE 80: 995

Goel B, Bluhm H (1988) Determination of proton stopping power in hot plasma at KALIF. Journ. de Physique, Coll. C7, 49, C7-169

Kanel GI, Razorenov SV, Utkin AV, Fortov VE, Baumung K, Karow HU, Rusch D, Licht V (1993) Spall strength of molybdenum single crystals. J. Appl. Phys. 74, 12:7162

Karow HU, Bachmann H, Bauer W, Baumung K, Bluhm H, Guth H, Hoppé P, Ratajczak W, Rusch D, Schultheiß Ch, Singer J, Stoltz O (1990) In: Breizman BM, Knyazev BA (eds) Proc. 8th Intl. Conf. on High-Power Particle Beams Novosibirsk, p 450

Schimassek W, Bauer W, Stoltz O (1991) Generation of a focused proton beam with a self-magnetically B_Θ-insulated ion diode. Rev. Sci. Instrum. 62: 168

Wise JL, Chhabildas LC (1986) Laser interferometer measurements of refractive index in shock-compressed materials. In: Gupta YM (ed) Shock Waves in Condensed Matter - 1985, Plenum Publishing, PP 441-454

A Numerical Study of Explosive-Driven Cylindrical Imploding Shocks in Solids

T. Hiroe, H. Matsuo and K. Fujiwara
Department of Materials Science and Resource Engineering, Kumamoto University, Kumamoto 860, Japan

Abstract. A one-dimensional Random Choice code has been developed for a medium of layered structure including solids and gases, and applied to the numerical simulation of explosive-driven cylindrical imploding shocks in solids. In the numerical experiments, imploding shocks in solids (aluminium, PMMA) are supposed to be generated by simultaneous detonations of cylindrical explosive (PETN) shells. Detonations are assumed to be initiated simultaneously from the outer surface of the PETN shells. The method of characteristics is adopted only for the flow calculation. The numerical results show that for some cases a negative pressure caused by the interference of the strong expansion waves emerges in the period after the shock is reflected on the axis. The total energy contained in the center region of the solid cylinder is compared with the chemical energy of the explosive.

Key words: PETN, Imploding shock, Numerical simulation, Random Choice Method

1. Introduction

Explosive-driven shocks have been an interesting research topic in the field of applied physics and high energy engineering for the past two decades. In the numerical simulation of strong and steep shock waves it is essential to maintain the discontinuity at the shock front throughout the calculation. A typical method which possesses such a character is the Random Choice Method (RCM) (Glimm 1965, Sod 1977) which requires the solution of the Riemann problem. Cylindrical converging shock waves in atmospheric air have been numerically simulated using the RCM successfully (Matsuo et al. 1990).

On the other hand there have been only a few studies on imploding shocks in solids. There have been some attempts to apply the RCM to wave phenomena for solids with limited applications (Esser and Grönig 1985) and using the analytical-type solution of the Riemann problem (Nagayama 1989). A RCM for solids has been constructed solving isentropes numerically and applied to the cylindrical Riemann problem (Hiroe et al. 1992). The superiority of the RCM over the finite difference method (FDM) is demonstrated there. Recently, explosive-driven imploding shocks in solids have been produced (Hiroe et al. 1993) by the simultaneous detonation of cylindrical explosive shells using the wire explosion technique. In this study the developed RCM code for solids is extended for layered structures including solids and gases and applied to the numerical simulation of explosive-driven cylindrical imploding shocks in solids.

2. Construction of a 1D RCM code for solids and gases

In the explosive-driven imploding shock analysis, shock pressure rises soon enough to neglect rigidity effects. Then we consider inviscid and nonconducting fluid-like solids with a Grüneisen-type equation of state as follows:

$$P = P_H(\nu) + [\Gamma(\nu)/\nu]\,[E - E_H(\nu)], \tag{1}$$

where $P, \nu(= 1/\rho)$, ρ, E and Γ denote the pressure, specific volume, density, specific internal energy and Grüneisen parameter. The subscripts H and 0 represent the Hugoniot and the value at zero standard conditions. P_0 and E_0 are approximately set equal to zero. $P_H(\nu)$ and $E_H(\nu)$ are derived from a linear Hugoniot: $U_s = c_0 + su$, where U_s, u, c_0 and s are the shock velocity, particle velocity, zero-pressure intercept and the slope. The following familiar relation is adopted

Shock Waves @ Marseille III
Editors: R. Brun, L. Z. Dumitrescu

for $\Gamma(\nu) : \Gamma(\nu)/\nu = \Gamma(\nu_0)/\nu_0 = \Gamma_0/\nu_0 = K_0 = \text{const}$. On the other hand, assuming a polytropic gas, the equations of state for the combustion gas or the air are represented as

$$P = (\kappa - 1)E/\nu, \tag{2}$$

where κ is the ratio of specific heats. It is noticed that Eq.1 coincide with Eq.3 when the following equations are assumed as: $\Gamma(\nu) = \Gamma_0 = \kappa - 1,\ P_H(\nu) = 0$. The one-dimensional conservation laws for cylindrical flow of solids and gases in this study are written as

$$\boldsymbol{U}_t + \boldsymbol{F}_r(\boldsymbol{U}) + \boldsymbol{W}(\boldsymbol{U}) = 0, \tag{3}$$

with

$$\boldsymbol{U} = \begin{bmatrix} \rho \\ \rho u \\ e \end{bmatrix}, \quad \boldsymbol{F}(\boldsymbol{U}) = \begin{bmatrix} \rho u \\ \rho u^2 + P \\ u(e+P) \end{bmatrix}, \quad \boldsymbol{W}(\boldsymbol{U}) = \begin{bmatrix} \rho u/r \\ \rho u^2/r \\ u(e+P)/r \end{bmatrix}, \quad e = r(E + u^2/2)$$

where t and r are independent variables that represent the time and the space coordinate. Approximating $\frac{\partial U}{\partial t}$ in Eq.3 by the first-order difference quotient and applying operator-splitting (Sod 1997), we get a set of two-step difference equations. The equation is solved by the RCM (Glimm 1965) when the solver is constructed so as to applicable to the Riemann problems for solids, gases and a combination of solid and gas.

For a plane shock wave, we have the Rankine-Hugoniot (RH) equations. Then accurate relations of any two of the five variables P, v, E, u and U_s can be obtained from the iterative calculation of Eq.1 or Eq.2 and RH eqs. for solids and gases. For example, two Hugoniot curves (forward-facing and backward-facing) of $P = P_s(u)$ are drawn from a reference state in the $P - u$ plane.

In the isentropic process, we have three differential equations expressed using state variables P, v, E, u and c (sound velocity) derived from the conservation laws of Eq.3. We draw two expansion curves $P = P_e(u)$ by the accurate iterative calculation for gases and by numerically solving the following successive difference equations only for the solid:

$$P_i = \frac{P_H(\nu_i) - K_0(P_{i-1}\Delta\nu_i/2 + E_H(\nu_i) - E_{i-1})}{1 + K_0\Delta\nu_i/2}, \quad u_i = u_{i-1} \pm \left[(P_{i-1} - P_i)(\nu_i - \nu_{i-1})\right]^{1/2} \tag{4}$$

The accuracy of the above numerical approximation for solids was discussed in a previous study (Hiroe et al. 1992) and it was known that the above equations for solids produce sufficiently accurate solutions with fine difference divisions. The Riemann problem is now solved by the combination of $P_s(u)$ and $P_e(u)$ discriminating two types of flow.

Table 1. Conditions for numerical experiments

Cases	Solid materials	Cond. of Outer Region
A	Aluminium	Atmospheric air
B	Aluminium	Rigid wall
C	PMMA	Rigid wall

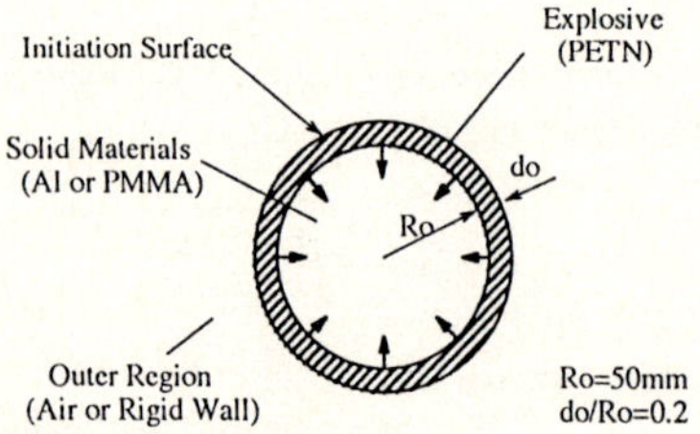

Fig.1. Schematic diagram of numerical experiments for explosive-driven cylindrical imploding shock generation

Table 2. Material constants tor Eq. of State

Materials	ρ_0 (kg/m^3)	c_0 (km/s)	s	Γ_0 or $\kappa - 1$
Aluminium	2729	5.25	1.372	2.0
PMMA	1181	2.26	1.816	0.75
PETN(C.G.)	—	—	—	1.98
Air	—	—	—	0.4

Table 3. C-J values of 1.0 g/cc PETN

ρ (kg/m^3)	u (km/s)	P (GPa)	U_s (km/s)
1340	1.40	7.68	5.55

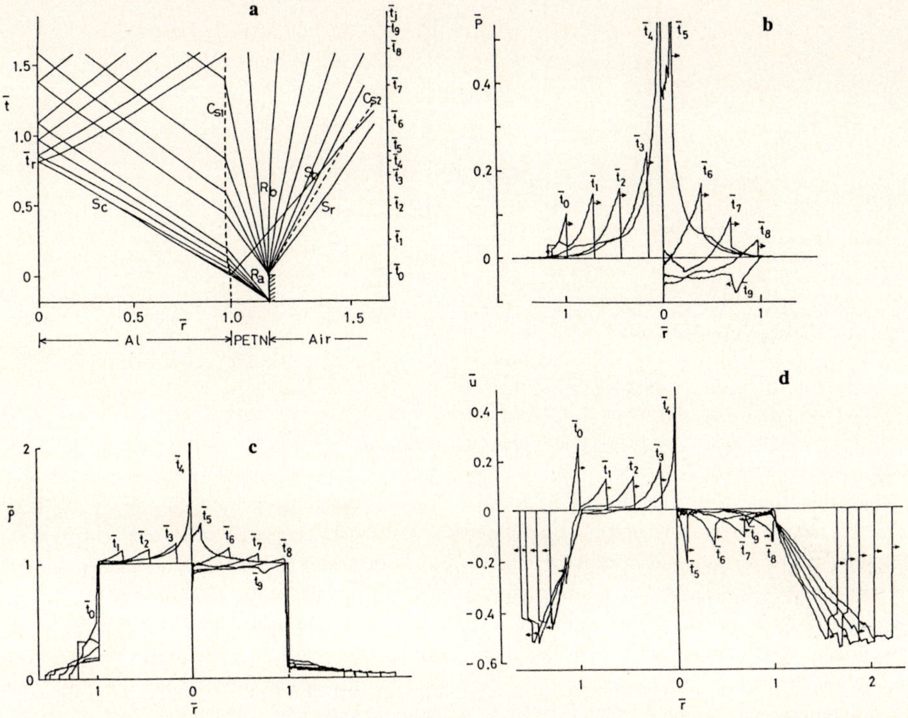

Fig. 2 a-d). Numerical results for case A (solid: aluminium; outer boundary: rigid). **a)** wave diagram in $\bar{t}-\bar{r}$ plane; **b, c, d)** Spatial distributions of pressure, density and particle velocity

3. Numerical experiments and discussion

Fig.1 shows the schematics of explosive-driven imploding shock generation. The cylindrical explosive shell of inner radius R_0 is made of PETN. A solid cylinder of aluminium (2024) or PMMA (or Lucite) is set inside the explosive shell. The explosive is initiated simultaneously at the outer surface, which is in contact with a rigid wall or a free outer boundary (air). It is assumed (Matsuo et al. 1990) that the flow associated with the detonation phenomenon in the explosive is planar and the detonation front with the Chapman - Jouguet (C-J) propagation velocity suddenly appears at the instant of initiation and propagates through the explosive with that constant speed. In this study, three cases: A, B and C are considered, as listed in Table 1. The outer region is also assumed to be a rigid wall until the detonation front reaches the solid surface (see Fig.2.a). C-J values for PETN are listed in Table 2. The method of characteristics is adopted for the flow calculation of the detonation assuming that the combustion gaseous product of PETN is a perfect gas. The numerical analysis by the RCM has started just after the detonation front has arrived at the outer surface of the solid cylinder, using the results by the characteristics as the initial condition. The material constants for the equation of state are listed in Table 3. The number of mesh divisions for RCM is 250 for the solid region. Nondimensional quantities are defined as follows: $\bar{r} = r/R_0$, $\bar{t} = c_0 t/R_0$, $\bar{P} = P/\rho_0 c_0^2$, $\bar{u} = u/c_0$.

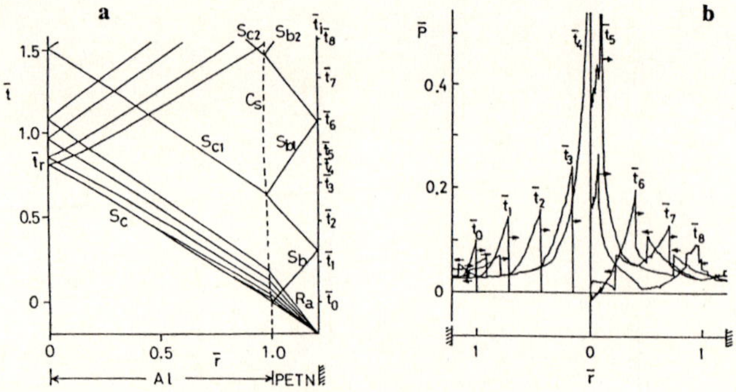

Fig. 3 a,b). Numerical results for case B (solid: aluminium; outer boundary: free). a) wave diagram in $\bar{t} - \bar{r}$ plane; b) Spatial distribution of pressure

Numerical results are shown in Fig.2a-d for case A. Fig.2a is the wave diagram in the $\bar{t}$ – $\bar{r}$ plane. The detonation wave propagates with the constant C-J velocity being accompanied by strong expansion waves R_a from behind. At the moment $t = t_0$, when the detonation front hits the inner boundary, a forward-facing S_c and a backward-facing shock wave S_b emerge. S_c first moves at a nearly constant velocity but accelerates as the shock front approaches the central axis. We assumed that a rigid wall in contact with the outer surface of the explosive suddenly change to a free outer boundary at $t = t_0$ and then a strong shock S_r and strong expansion waves R_b emerge at the outer surface of the explosive. Fig.2b shows the spatial distributions of pressure at 9 instances in the focusing (left side in the figure) and reflecting stages (right side). It is seen that a strong negative pressure emerges in the period after the shock has reflected on the axis. The wave diagram indicates that it is caused mainly by the interference of the strong expansion waves R_a and the expansion waves R_b coming from the outer surface of the explosive. Fig.1c shows that the density near the center is suddenly increased by the imploding shock and is soon decreased beneath the initial value by the expansion waves.

The wave diagram for case B is shown in Fig.3a, where the rigid wall outside the explosive is fixed all the time. The main difference between Fig.2a and Fig.3a is that the strong expansion waves R_b in Fig. 2a do not appear in Fig.3a. Then the interferences of the expansion waves are limited to those of R_a waves and their reflections. As a result, the magnitude of negative pressure which emerges in the central region is rather small, as shown in Fig.3b. The wave diagram for case C is shown in Fig.4a, where PMMA is compressed, instead of aluminium, as in case B. Since the C-J value of PETN is close to the Hugoniot value of PMMA, the reflected shock wave S_b is negligibly weak. PMMA is compressed much more than aluminium. Therefore most of the strong expansion waves R_a soon overtake the shock S_c during the shock implosion, reducing the self-amplifying effect of pressure. Then the interference effect of the expansion waves after the reflection is very small and the calculated pressure keeps positive in all the stages.

Finally, the energy concentration near the axis is discussed. Here, the total energy ratios E_1 and E_2 are defined as q_1/Q_0 and q_2/Q_0, where Q_0 ($= 1.95\text{GJ/m}^3$) is the chemical energy per unit volume of PETN and q_1 and q_2 are the total energies per unit volume in the region $r/R_0 < 1.0$ and $r/R_0 < 0.1$ respectively. Fig.5 shows the time-varying energy ratios for cases A - C. Energy ratios E_1 for overall solids continue increasing up to 0.24 for PMMA (case C) and 0.09, 0.12 for aluminium (cases A, B). The ratio for aluminum is smaller than for PMMA, since the aluminium has higher shock impedance and reflects larger amount of detonation energy. The difference of ratios between cases A and B comes from the influence of boundary conditions at the outer surface

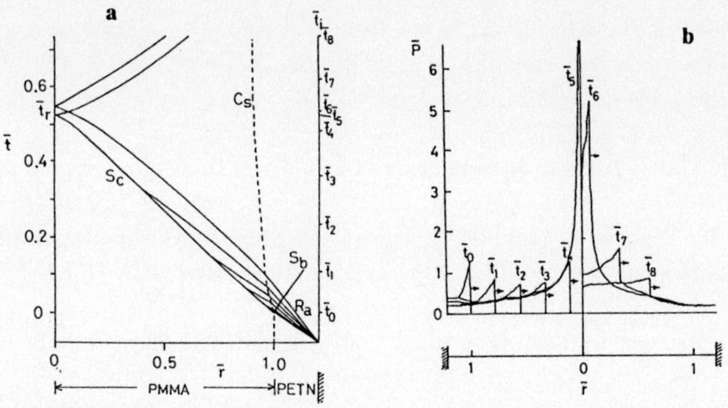

Fig. 4 a,b). Numerical results for case C (solid: PMMA; outer boundary: rigid). a) Wave diagram in $\bar{t} - \bar{r}$ plane; b) Spatial distribution of pressure)

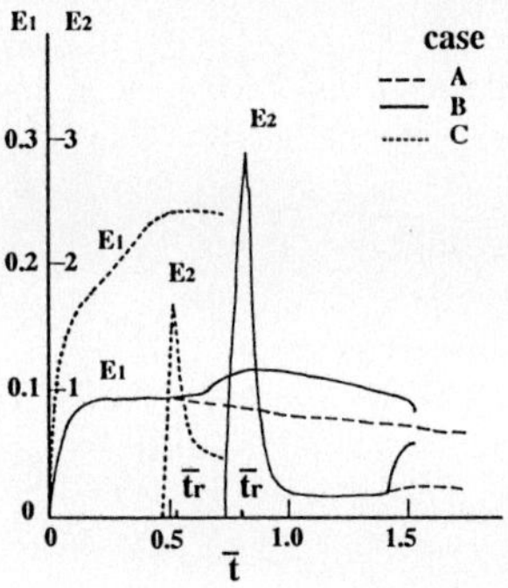

Fig. 5. Time-varying energy ratios E_1 and E_2 for cases A-C:E_1 and E_2 are q_1/Q_0 and q_2/Q_0. Q_0 is the chemical energy per unit volume of PETN. q_1 and q_2 are the total energies per unit volume in regions $r/R_0 < 1.0$ and $r/R_0 < 0.1$ respectively

of PETN. Energy ratios E_2 for the central region of solid start increasing after the imploding shocks are transmitted into the boundary $\bar{r}$ =0.1 and continue increasing up to 2.9 for aluminium and 1.7 for PMMA respectively, then they decrease rapidly. In this case boundary conditions of the outer surface of PETN have no influence on the energy concentration at the center.

4. Concluding remarks

The one-dimensional RCM for solids has been extended to layered systems including gases, solving isentropes for solids in the Riemann problem, using finite difference method with fine mesh divisions. A practical method to produce a one-dimensional imploding shock in solid is to simultaneously explode a long cylindrical explosive assembly charged outside the solid cylinder: in this study aluminium and PMMA are compressed by exploding PETN shells. The numerical results indicate that it is difficult to recover the solid medium compressed by this method since a negative pressure emerges. In the case of a low-impedance medium, pressure has kept positive but energy concentration at the central region is rather small. It is noticeable that the total energy per unit volume within the central region has increased up to around two or three times as large as the chemical energy of PETN.

References

Esser B, Grönig H (1985) Application of the Random Choice Method to wave propagation in elastic disks. Presented at the Sonderforschungsbereich 27, "Wellenfokussierung" RWTH Achen

Glimm J (1965) Solutions in the large for nonlinear systems of equations. Comm. Pure Appl. Math. 18:697-715

Hiroe T, Matsuo H, Fujiwara K (1992) Numerical simulation of cylindrical converging shocks in solids. J. Appl. Phys. 72:2665-2611

Hiroe T, Matsuo H, Fujiwara K, Tanoue T, Yoshida M, Fujiwara S (1993) The production of cylindrical imploding shocks in solid by exploding wire rows. In: Proc. Joint AIRAPT/APS Conf. (in press)

Matsuo H, Fujiwara K (1990) Explosive-driven cylindrical imploding shocks. Phys. Fluids A 2:266-273

Nagayama K (1989) Solution of the high-pressure Riemann problem for solids including rigidity effects. J. Phys. Soc. Jpn. 58:1631-1638

Sod GA (1977) A numerical study of a converging cylindrical shock. J. Fluid Mech. 83:785-794

Model of Viscoelastic Unidirectional Composite with Coupled Thermal and Elastic Processes

A.D. Resnyansky* and E.I. Romensky†
*Lavrentyev Institute of Hydrodynamics, Russian Academy of Sciences, 630090 Novosibirsk, Russia
†Institute of Mathematics, Russian Academy of Sciences, 630090 Novosibirsk, Russia

Abstract. To analyze the shock wave processes in composite materials, a model is developed, based on the representation of the behaviour of matrix and reinforcement materials as viscoelastic media. For its construction, a method was formulated of phenomenological averaging of parameters not requiring an assumption of composite structure regularity. Hypotheses proposed for the closure of the model are described in detail. The interphase nonequilibrium and the structural parameters concerned are introduced. Two-phase composites of two types are considered: fibrous one-directional ones with parallel fibres, and laminated ones with parallel laminas. The resulting system of evolution and kinetic equations for the composite macroparameters is outlined. Numerical examples illustrating the influence of high-rate nonequilibrium are shown.

Key words: Composite materials, Dynamic problems, Microstructure

1. Introduction

Multicomponent media are of particular interest in dynamic problems due to peculiarities in their behaviour induced by the medium structure. Among these media are bubbles, gas-trickle, dust media, porous and composite materials. The use of equilibrium thermodynamical models is not successfull often even in homogeneous materials because, under loading, the parameters of such media vary with time and rather extended periods of time are needed, as in static and quasi-static problems, in order that an equilibrium description might be used; otherwise, only a few integral characteristics might be described.

For describing the behaviour of homogeneous isotropic solids we use a nonlinear viscoelastic medium model which is the generalization of the Maxwell model. The deformation $\dot{\varepsilon}$ of this medium is the sum of a reversible part $\dot{\varepsilon}_e$ and irreversible part $\dot{\varepsilon}_v$. The reversible strain is linked with the load σ in the material by either Hooke's law or its generalization for the case of finite nonlinear elastic strains. The irreversible part, not vanishing under unloading, is determined by the time of response to the load τ (relaxation time). In real media τ is a function of material state and is determined, for instance, by the behaviour of a defect structure in solids, by the type of reaction in chemically active media etc. In the simplest form, a viscoelastic Maxwell-type body is modeled by an elastic element in series with a viscous one. Then, the rate of total strain is defined as:

$$\dot{\varepsilon} = \dot{\varepsilon}_e + \dot{\varepsilon}_v \ , \text{ where } \ \dot{\varepsilon}_e = \dot{\sigma}/(\tau E) \tag{1}$$

where E is the elastic modulus. Thus, the irreversible part of the total strain in this model is determined by some kinetic equation for the structural parameters, of which, in the case of Eq.1, the role is played by the measure of irreversible strain ε_v.

In addition to the interaction of the elements of the homogeneous structure with each other (called internal friction or viscosity, induced by the interplay of dislocations, grains etc.), additional factors arise, caused by interaction of components with each other, giving rise to new structural parameters. The list of parameters increases with the possibilities of the components to interact between themselves. The inherent structure described by these parameters is an extra source of nonequlibrium and, for these structural parameters, corresponding kinetic equations must be obtained.

Shock Waves @ Marseille III
Editors: R. Brun, L. Z. Dumitrescu

Using unidirectional composites as an example, we can demonstrate a phenomenological approach, which lets us take into consideration the kinetic nonequilibrium due to structure as well as observe the composite material as a whole. Phases are considered to satisfy the equation of state linking elastic strains with stresses by Hooke's law.

2. Interaction of the components

For unidirectional structures two directions may be separated, components along them interact differently. The first direction $\boldsymbol{l}$ (the continuity direction) is aligned with stuck-up lines. Along the second direction $\boldsymbol{n}$ (the alternation direction), perpendicular to $\boldsymbol{l}$, components alternate. The direction $\boldsymbol{l}$ in fibrous composite is unique and coincides with the fibre direction. Any direction, normal to fibre, may be the alternation direction here. In contrast, the alternation direction in a laminated composite is unique and coincides with the direction of the normal to a plane of lamina. We can choose any direction lying in plane of the lamina as $\boldsymbol{l}$.

In order to formulate rules used for describing the interaction of components along the continuity direction, the Maxwell scheme (Eq.1) is applicable. For a realistic description of real processes in an element of solid, the dependence of relaxation time should be sharply varying. Therefore, to satisfy at least an ideal elastic- plastic $(\sigma - \varepsilon)$ diagram, τ is taken to be a function of the form: while ε changes up to a certain "elastic limit", τ is large enough so that only the elastic element is involved in the Maxwell series scheme. Next τ is diminished very rapidly and makes a "dumper" work and is a source of irreversible strain ε_v.

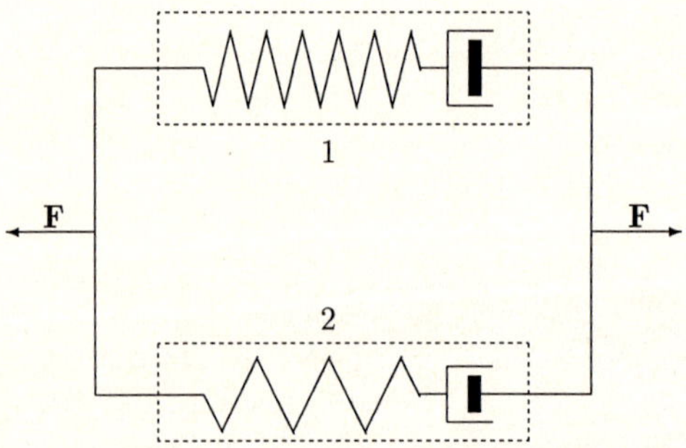

Fig. 1.

The connection between components is supposed to be perfect. The constituents of the composite are deformed in parallel when stretched along $\boldsymbol{l}$ according to the scheme on Fig.1. The elastic modulus of the first composite's component is E_1, of the second one is E_2, relaxation functions are τ_1 and τ_2, respectively. It is clear from the connection ideality that rates of total strain in Eq.1, denoted by $\dot{\varepsilon}_0$, are equal for both components. The elastic strain $\varepsilon^{(\alpha)}$ of the α -th component is related to the stress $\sigma^{(\alpha)}$ by Hooke's law $\sigma^{(\alpha)} = E_\alpha \varepsilon^{(\alpha)}$ and then corresponding evolution equations for $\varepsilon^{(\alpha)}$ result from Eq.1:

$$\dot{\varepsilon}^{(\alpha)} = \dot{\varepsilon}_0 - \varepsilon^{(\alpha)}/\tau_\alpha \tag{2}$$

In the parallel scheme of Fig.1 forces are summed up but total deformations are equal for both components. To be specific, we suppose that the first component has a higher modulus and strength than the second one. Let us analyze the following idealized $(\sigma - \varepsilon)$ diagrams of the first component, the second one and their mixture (curves 1-3 on Fig.2, respectively) at fixed

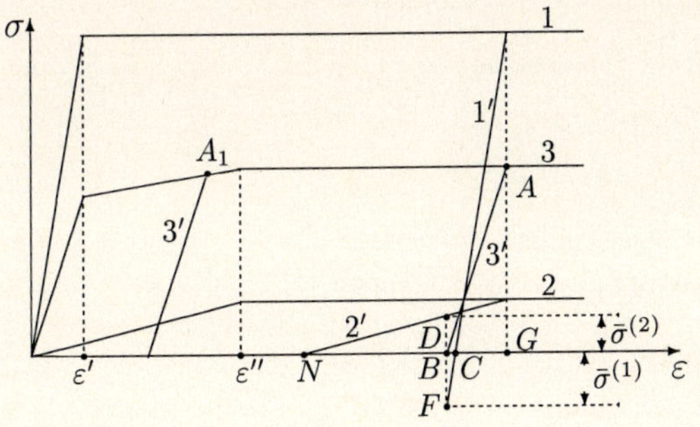

Fig. 2.

$\dot{\varepsilon}_0$. The formula for calculation of stresses in the mixture is the following (c is the first phase concentration):

$$\sigma = c\sigma^{(1)} + (1-c)\sigma^{(2)} = cE_1\,\varepsilon^{(1)} + (1-c)E_2\,\varepsilon^{(2)} \tag{3}$$

It is interesting that in the range of ε from ε' to ε'' in the first hard component irreversible deformation occurs, but the second soft component is deformed elastically. Let us consider an off-loading process, which can be performed in this scheme by reducing the deformation, for example, via change of the $\dot{\varepsilon}_0$ sign. The unloading from the point A (Fig.2) of the whole composite is performed along the path $3'$, simultaneously. The unloading of the constituents 1 and 2 is done along the paths $1'$ and $2'$. It is easily seen on Fig.2 that, at the instant the composite has been unloaded, its components are stressed. Here $BD = \bar{\sigma}^{(2)}$, $BF = \bar{\sigma}^{(1)}$, and from Eq.3 it follows that $c\bar{\sigma}^{(1)} + (1-c)\bar{\sigma}^{(2)} = 0$. These internal stresses (microstresses) will exist if at least one of the components is deformed irreversibly preceding the relief. At the end of the unloading, the residual elastic strains of the components are defined as $-\bar{\varepsilon}^{(1)} = BC = -\bar{\sigma}^{(1)}/E_1$, $\bar{\varepsilon}^{(2)} = NB = \bar{\sigma}^{(2)}/E_2$. Then from Fig.2 it can be seen that the elastic strain of the composite as a whole, $\varepsilon_e = BG$, may be compared to the elastic strains of the components $\varepsilon^{(1)} = CG$, $\varepsilon^{(2)} = NG$. Hence it gives

$$\varepsilon_e = \varepsilon^{(1)} - \bar{\varepsilon}^{(1)} = \varepsilon^{(2)} - \bar{\varepsilon}^{(2)} \tag{4}$$

i.e. the elastic strain of the composite along the $\boldsymbol{l}$ direction is equal to elastic strains of components reckoned from the relief state, where the elastic strains are $\bar{\varepsilon}^{(1)}$ and $\bar{\varepsilon}^{(2)}$. It is evident that the start of the unloading has no influence on the above reasoning (the unloading from the point A_1 in Fig. 2 results in the same except of a lower value of the difference $\bar{\varepsilon}^{(1)} - \bar{\varepsilon}^{(2)}$) and neither has the availability of microstresses before the loading (it produces only a shift of the starting points of curves 1 and 2 relative to the origin and changes the values of ε' and ε'', but doesn't affect the curves $1'$, $2'$, and $3'$). It is seen from Eq.4, that a quantity

$$\Delta = \varepsilon^{(1)} - \varepsilon^{(2)} = \bar{\varepsilon}^{(1)} - \bar{\varepsilon}^{(2)} \tag{5}$$

may be chosen as a measure of the internal stresses. Since Δ is a constant during unloading, it has no effect on the stresses of the composite as a whole but exerts influence on the internal microstresses. Thus, Δ may be interpreted as a structural parameter of the composite. Let us see how the kinetic equation for Δ and the required equation of evolution for ε_e can be obtained from the above. Residual elastic microstrains are found first from the definition of Δ and the condition of zero macrostresses in the relief point:

$$\bar{\varepsilon}^{(1)} - \bar{\varepsilon}^{(2)} = \Delta\ , \quad cE_1\bar{\varepsilon}^{(1)} + (1-c)E_2\bar{\varepsilon}^{(2)} = 0\ ,$$

then

$$\bar{\varepsilon}^{(1)} = (1-c)E_2\Delta/ < E > \ , \ \bar{\varepsilon}^{(2)} = -cE_1\Delta/ < E >$$

where $< E >= cE_1 + (1-c)E_2$. From that, the relations among the microstrains, macrostrains, and structure parameters are readily obtained:

$$\varepsilon^{(1)} = \varepsilon_e + (1-c)E_2\Delta/ < E > \ , \ \varepsilon^{(2)} = \varepsilon_e - cE_1\Delta/ < E > \tag{6}$$

Furthermore, using Eq.6 and subtracting the equation for $\varepsilon^{(2)}$ from the one for $\varepsilon^{(1)}$ of the system (Eqs.2) we can obtain the evolution equations for Δ and ε_e:

$$\dot{\Delta}=\varepsilon_e(\frac{1}{\tau_2}-\frac{1}{\tau_1})-\Delta(\frac{cE_1}{< E > \tau_2}+\frac{(1-c)E_2}{< E > \tau_1}) \ , \tag{7}$$

$$\dot{\varepsilon}_e=\dot{\varepsilon}_0-\varepsilon^{(1)}/\tau_1-(1-c)E_2\dot{\Delta}/ < E >$$

The composite behaviour along the alternation direction would be described reasonably in the series scheme, where two viscoelastic elements for the first and the second phases are involved. However it is necessary to take into consideration the complex stress state in a real composite body. In a compound bar with series elements the constituent elastic strains $\varepsilon^{(\alpha)}$ would vanish in unloading. In fact, they fall to $\bar{\varepsilon}^{(\alpha)}$, defined by the condition that all components of macrostresses are equal to zero. Here the residual elastic microstrains are the reference positions, too. Therefore, along with the condition of force equality for each phase in the series scheme, we should write the condition of elastic strain additivity:

$$\sigma = \sigma^{(1)} = \sigma^{(2)} \ , \ \varepsilon_e = c(\varepsilon^{(1)} - \bar{\varepsilon}^{(1)}) + (1-c)(\varepsilon^{(2)} - \bar{\varepsilon}^{(2)}) \tag{8}$$

Regarding the thermophysical parameters (temperature and entropy), various assumptions can be introduced. The rate of heat exchange in inclusions and their size are the cause whether or not the distribution of temperature in a composite element is uniform. We choose the simplest variant of uniform distribution assuming that inclusions are of small size (this doesn't limit the magnitude of their concentration c), and that the entropy is an additive function of mass. Thus,

$$T^{(1)} = T^{(2)} = T \ , \ \rho s = c\rho_1 s^{(1)} + (1-c)\rho_2 s^{(2)} \tag{9}$$

here $\rho_\alpha =$ const. is the component density, $\rho =< \rho >$ is the composite density.

3. The composite model

The above-developed system of hypotheses (Eqs.4, 5, 8, 9) applies for deriving the equations of a thermoviscoelastic composite model. In the case of linear relationships among stresses, elastic strains, and entropy, the dynamic equations of viscoelastic media, written for each component of the composite, are the following (Godunov 1978):

$$\frac{\partial \varepsilon_{ij}^{(\alpha)}}{\partial t}-\frac{1}{2}(\frac{\partial u_i^{(\alpha)}}{\partial x_j}+\frac{\partial u_j^{(\alpha)}}{\partial x_i})=-\frac{\sigma_{ij}^{(\alpha)}-\frac{1}{3}(\sigma_{11}^{(\alpha)}+\sigma_{22}^{(\alpha)}+\sigma_{33}^{(\alpha)})\delta_{ij}}{2\mu_\alpha\tau_\alpha} \ ,$$

$$\rho_\alpha\frac{\partial u_i^{(\alpha)}}{\partial t}-\frac{\partial \sigma_{ij}^{(\alpha)}}{\partial x_j}=0 \ ,$$

$$\rho_\alpha\frac{\partial U^{(\alpha)}}{\partial t}-\sigma_{ij}^{(\alpha)}\frac{\partial u_i^{(\alpha)}}{\partial x_j}=0$$

where $\varepsilon_{ij}^{(\alpha)}$ are the elastic strains, $u_i^{(\alpha)}$ are the velocity components, $\sigma_{ij}^{(\alpha)}$ are the stresses, μ_α are the shear moduli. The system is closed by the dependence of $U^{(\alpha)}$ on $\varepsilon_{ij}^{(\alpha)}$ and entropy $s^{(\alpha)}$, by formulae, linking $\sigma_{ij}^{(\alpha)}$ and temperature $T^{(\alpha)}$ with $\varepsilon_{ij}^{(\alpha)}$, $s^{(\alpha)}$:

$$\sigma_{ij}^{(\alpha)}=\rho_\alpha \frac{\partial U^{(\alpha)}}{\partial \varepsilon_{ij}^{(\alpha)}} , \; T^{(\alpha)}=\frac{\partial U^{(\alpha)}}{\partial s^{(\alpha)}}$$

and by functions $\tau_\alpha(\varepsilon_{ij}^{(\alpha)}, s^{(\alpha)})$ determined from indirect experiments. The assumption of the component-joining ideality gives the condition of the velocity equality in phases:

$$u_i^{(1)} = u_i^{(2)} = u_i \tag{10}$$

Denoting the direction along the composite anisotropy axis as x_1 we can write out systems of relations, to which the microparameters of the composite are satisfying. For fibrous composites the system is the following:

$$\sigma_{11} = c\sigma_{11}^{(1)} + (1-c)\sigma_{11}^{(2)} , \; \varepsilon_{11} = \varepsilon_{11}^{(1)} - \bar{\varepsilon}_{11}^{(1)} = \varepsilon_{11}^{(2)} - \bar{\varepsilon}_{11}^{(2)} ,$$
$$\sigma_{ij} = \sigma_{ij}^{(\alpha)} , \; (\alpha = 1,2) , \; (i,j) \neq (1,1) ,$$
$$\varepsilon_{ij} = c(\varepsilon_{ij}^{(1)} - \bar{\varepsilon}_{ij}^{(1)}) + (1-c)(\varepsilon_{ij}^{(2)} - \bar{\varepsilon}_{ij}^{(2)}) , \; (i,j) \neq (1,1) ,$$
$$T = T^{(\alpha)} , \; (\alpha = 1,2) , \; \rho s = c\rho_1 s^{(1)} + (1-c)\rho_2 s^{(2)} ,$$
$$\Delta = \varepsilon_{11}^{(1)} - \varepsilon_{11}^{(2)} = \bar{\varepsilon}_{11}^{(1)} - \bar{\varepsilon}_{11}^{(2)}$$

Correspondingly for the laminated composite we have:

$$\sigma_{ij} = c\sigma_{ij}^{(\alpha)} + (1-c)\sigma_{ij}^{(\alpha)} , \; (\alpha = 1,2), \; (i,j) = (2,2), \; (2,3), \; (3,3),$$
$$\varepsilon_{ij} = \varepsilon_{ij}^{(1)} - \bar{\varepsilon}_{ij}^{(1)} = \varepsilon_{ij}^{(2)} - \bar{\varepsilon}_{ij}^{(2)} , \; (i,j) = (2,2), \; (2,3), \; (3,3),$$
$$\sigma_{ij} = \sigma_{ij}^{(\alpha)} , \; (\alpha = 1,2), \; (i,j) = (1,1), \; (1,2), \; (1,3),$$
$$\varepsilon_{ij} = c(\varepsilon_{ij}^{(1)} - \bar{\varepsilon}_{ij}^{(1)}) + (1-c)(\varepsilon_{ij}^{(2)} - \bar{\varepsilon}_{ij}^{(2)}) , \; (i,j) = (1,1), \; (1,2), \; (1,3),$$
$$T = T^{(\alpha)} , \; (\alpha = 1,2) , \; \rho s = c\rho_1 s^{(1)} + (1-c)\rho_2 s^{(2)} ,$$
$$\Delta_{ij} = \varepsilon_{ij}^{(1)} - \varepsilon_{ij}^{(2)} = \bar{\varepsilon}_{ij}^{(1)} - \bar{\varepsilon}_{ij}^{(2)} , \; (i,j) = (2,2), \; (2,3), \; (3,3)$$

It follows herefrom that relationships among $\varepsilon_{ij}^{(\alpha)}$, $s^{(\alpha)}$ elastic macrostrains ε_{ij}, entropy of composite material s and structural parameters Δ_{ij} can be obtained as in Eq.6. Substituting these relations in Hooke's law a linear relation among σ_{ij}, ε_{ij}, and $s - \bar{s}$ is obtained; here the quantity $\bar{s}$ is linearly proportional to Δ and characterizes a nonremoving heat, remaining in the material after unloading. The yield of the corresponding relations for the fibrous composite can be found in Resnyansky and Romensky (1992). The coefficients, relating σ_{ij} to ε_{ij} for the laminated composite are identical to the effective elastic moduli available in the relevant literature (Christensen 1982). The system of evolution and kinetic equations for the unidirectional composite is obtained similarly to Eq.7 and complemented by the equations of momentum and energy conservation which hold because the component bond is ideal (Eq.10) and the energy is additive on mass:

$$\begin{aligned}
&\frac{\partial \varepsilon_{ij}}{\partial t} - \frac{1}{2}\left(\frac{\partial u_i}{\partial x_j} + \frac{\partial u_j}{\partial x_i}\right) = -\alpha_{ijmn}\sigma_{mn} - \beta_{ijkl}q_{kl} , \\
&\frac{\partial \Delta_{kl}}{\partial t} = -\beta_{klij}\sigma_{ij} - \gamma_{klps}q_{ps} , \\
&\rho\frac{\partial u_i}{\partial t} - \frac{\partial \sigma_{ij}}{\partial x_j} = 0 , \\
&\rho\frac{\partial U}{\partial t} - \sigma_{ij}\frac{\partial u_i}{\partial x_j} = 0
\end{aligned} \tag{11}$$

here $q_{ij} = \rho \frac{\partial U}{\partial \Delta_{ij}}$. The system is closed by the equation of state $U(\varepsilon_{ij}, s, \Delta_{ij})$ which is used for calculating σ_{ij} and T by the formulae: $\sigma_{ij} = \rho \frac{\partial U}{\partial \varepsilon_{ij}}$, $T = \frac{\partial U}{\partial s}$, and also by the functions $\tau_\alpha(\varepsilon_{ij}^{(\alpha)}, s^{(\alpha)})$ in which the coefficients α_{ijmn}, β_{klij}, and γ_{klps} are involved. It is interesting to note the symmetry of the right-hand side coefficients of the evolution equations for strains and structural parameters. It confirms the validity of the Onsager symmetry principle in the model. The evolution equation for entropy is a consequence of the energy equation of the system (Eqs.11), its right-hand part being nonnegative. This provides the nondecreasing of entropy in the model. It is easy to show, writing out the evolution equation for $\varepsilon_{11} + \varepsilon_{22} + \varepsilon_{33}$ from the obtained system, that the continuity equation has a right-hand part which vanishes at $\Delta_{kl} = 0$. This is readily explained because the densities of composite components do not remain as before for the residual elastic strains of the existing components.

4. Numerical example

For the system of Eqs.11 the traditional problem on uniform stretching of a thin bar with given strain rate can be easily formulated (Romensky et al. 1991). We modify somewhat the problem statement by changing the strain rate sign to opposite at some instant of the load. This enables us to get the hysteresis loop. We computed the stress-strain diagram for a test composite which contains copper fibres in an epoxy matrix (c =0.5) with three various strain rates $\dot{\varepsilon} = 10\ 1/s$; $10^3\ 1/s$; $10^5\ 1/s$. The respective values of residual stresses (the interval BD on Fig. 2) in the composite component are: 1.14; 1.6; 2.04 kbar. It is seen that the residual microstresses increase with increasing strain rate. This clearly illustrates the nonequilibrium effect on the phenomenological behaviour of the composite.

5. Conclusion

We have outlined a procedure which can be used for designing closed models using thermodynamical principles and general assumptions on phase-to-phase interaction. Among the main advantages of the approach are the following:

- regularity of the inclusion arrangement is not necessary;
- an uniform technique for obtaining the effective elastic moduli of various composites is used;
- derivation of composite dynamics equations in the terms of continuum mechanics is possible.

The above-mentioned example and computations of shock-wave propagation show the consistency of the results with the concepts fed into the model. Areas of model application are computing stress-strain diagrams and shock wave problems, as well as studying the influence of microstructure on the phenomenological behaviour of the composite.

Acknowledgement

This work has been supported by the International Science Foundation (ISF). One of the authors is also grateful to the ISF for the possibility to present this report at the Symposium according to Grant N 0192/3.

References

Christensen RM (1982) Mechanics of Composite Materials. Mir, Moscow

Godunov SK (1978) Elements of Continuum Mechanics (in Russian). Nauka, Moscow

Resnyansky AD, Romensky EI (1992) The dynamic deformation model of fibrous thermoviscoelastic composite. Fiz. Goreniya Vzryva 4: 430-436

Romensky EI, Resnyansky AD, Merzhievsky LA (1991) The model of viscoelastic composite. J. de Phys. IV, vol. 1, C3: 923-930

Underwater-Shock Consolidation of TiAl Powder

A. Chiba, M. Nishida, R. Tomoshige, K. Imamura and C. Moon
Faculty of Engineering, Kumamoto University, Kumamoto 860 Japan

Abstract. Explosive consolidation of PREPed TiAl alloy powder was performed by utilizing underwater shock waves with pressures of about 10 GPa. The as-consolidated specimen has high bonding strength and shows substantial interparticle bonding by melting of the powder surface. Relative density of the compact was 95% or more of the theoretical density. The compressive properties of the compacts were improved by subsequent annealing. Specimens annealed to 1623 K showed high compressive stress (1300 MPa) and high strain (32%) at room temperature. The microstructure of the annealed specimens showed a fine $(\gamma + \alpha_2)$ duplex structure.

Key words: Underwater shock wave, TiAl intermetallic compound, Shock consolidation

1. Introduction

Explosive powder consolidation has been highlighted as an effective process in producing high density compacts of advanced powder materials such as rapidly solidified alloy, amorphous and ceramics powders, because the required assembly is simple, processing time is short and a high pressure is easily obtained. Nevertheless, there are some fundamental problems which must be overcome, such as the difficulty in producing precision compacts, the formation of cracks and central melt holes, etc. We have developed a new powder consolidation technique by which the underwater-shock pressure transfers directly and homogeneously to the powders, as shown in the paper of Tomoshige et al. (1992).

The purpose of the present study is to clarify the microstructures and mechanical properties of TiAl compact produced by our newly developed consolidation technique.

The reason why we selected the TiAl powder is the following: The TiAl intermetallic compound has been a candidate material for high-temperature applications in aerospace and automobile industries, because of its excellent high-temperature strength and low density. The practical use of this compound is limited by poor ductility and formability at room temperature. Shock densification of Ti_3Al powder was hitherto reported by Shang and Meyers (1991), where Ti_3Al compact with high ductility can be produced by explosive shock pressure followed by hot isostatic pressing.

2. Experimental procedures

The material used in the present study was rapidly solidified TiAl alloy powder which was produced by the plasma rotating electrode process by Daido Special Steel Co. Ltd. The powder is about 150 μm in average diameter and of nearly spherical shape with a dendritic structure. The composition of the powder is 48 at% Al.

The experimental assembly consists of three parts, i.e. explosive container, water tank and powder container, as shown in Fig.1. The explosive used consists mainly of nitric ester, which is provided by Asahi Chemical Co. Ltd. Its detonation velocity is about 6900 m/s. The shock wave produced by the detonation of the explosive travels from top to bottom in the water tank. The energy of the underwater shock wave increases by convergence due to the reduction of the cross section area of the water tank. The increased shock pressure and shock duration act efficiently on the consolidation of powders. The pressure level obtained was estimated to be $4 \cdots 11$ GPa by Tomoshige et al. (1992).The as-consolidated specimens were subsequently annealed to different temperatures, ranging from 973 to 1623 K for 1 h to improve the mechanical properties. Archimedes

Shock Waves @ Marseille III
Editors: R. Brun, L. Z. Dumitrescu

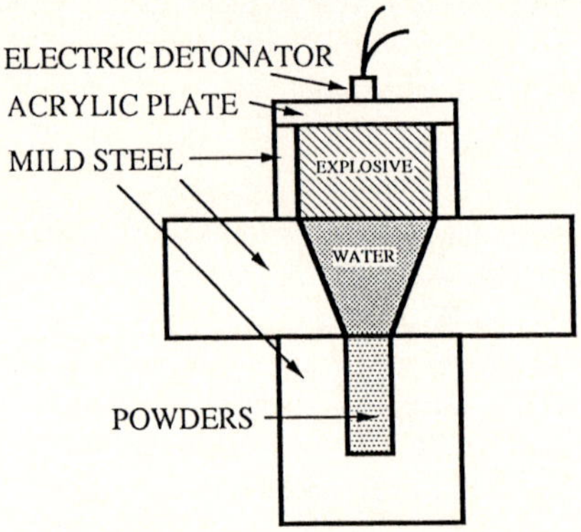

Fig. 1. Shock-consolidation assembly

method was used for density measurement. Microstructural observations were performed by optical, scanning, and transmisssion electron microscopy. The compression test was performed using an Instron testing machine at room temperature for the specimens as-consolidated and after annealing to various temperatures, and the initial strain rate was 2 to 3×10^{-3}/s.

3. Experimental results and discussion

3.1. Microstructure of the as-consolidated specimen

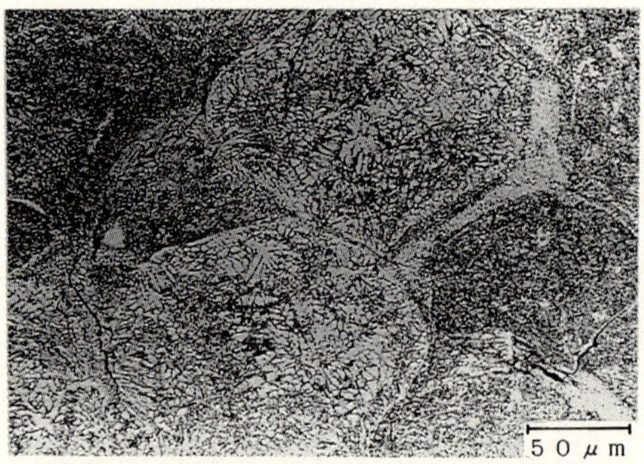

Fig. 2. A micrograph of the as-consolidated TiAl

Fig.2 shows an optical micrograph of the as-consolidated specimen. Each particle is deformed into a Japanese fan-like shape and stacked over each other to a well-consolidated microstructure. The dendritic structure is maintained in the powder. The melt layers were formed along the original powder surface. These features indicates that the consolidation is well performed without losing the characteristics of rapidly solidified powder by underwater-shock pressure. It is also confirmed by the fact that the density of the compact reached nearly theoretical density.

Fig.3(a) shows TEM micrographs of the melt zone and its adjacent region of powder particle. There are a few deformed structures around the melt zone because of a drastic temperature increment caused by the friction energy and adiabatic compressive deformation of the powder during the passage of the shock wave. However, the duration of the temperature increment is momentary, because the heavily deformed structures remained inside the powders. An enlarged electron micrograph of the melt zone is shown in Fig.3(b). There are equiaxed submicron grains with no deformation substructures in the melt zone, which are considered to be formed by rapid cooling from melt state. From the macroscopic and microscopic observations, it is clear that substantial interparticle bonding was achieved by the melting of the powder surface.

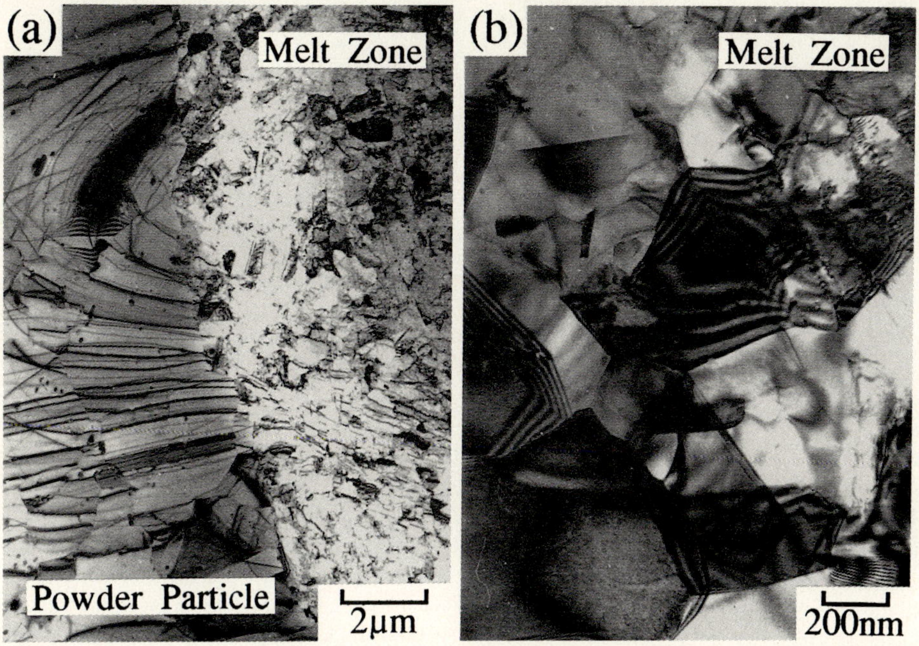

Fig. 3. TEM micrographs of (a) the melt zone and its adjacent region of powder particle and (b) high magnification of the melt zone in the TiAl compact

3.2. Effect of specimen annealing on microstructures and mechanical properties of the compact

The compressive fracture strength and strain of the as-consolidated specimens are about 1000 MPa and 14%, respectively. This shows that substantial interparticle bonding is achieved by the present underwater shock consolidation on account of not only the metallographic aspects mentioned above but also the mechanical properties. To further improve mechanical properties of the compacts, they were annealed to different temperatures, ranging from 973 to 1623 K for 1 h. The results are summarized in Fig.4. The density of the compacts reached the theoretical value, independent of annealing temperatures. The optimum annealing temperature for the compacts is considered to be 1623 K, since the compressive fracture strength and the strain are about 1300 MPa and 32%, respectively.

To identify the structures formed in the annealed specimens, X-ray diffraction experiments and optical microscopy observations were carried out. The raw powder used consisted mainly of $\alpha_2(Ti_3Al)$ phase with a few $\gamma(TiAl)$ phase, since the powder was brought into a metastable state by rapid solidification, as shown by Nishida et al. (1992). The compacts annealed to different temperatures, ranging from 973 to 1423 K almost transformed to $\gamma + \alpha_2$ lamellar structures, in which the volume fraction of γ was higher than that of α_2. By annealing to 1623 K, the volume ratio of γ and α_2 phases were nearly equivalent and there were no longer lamellar structures.

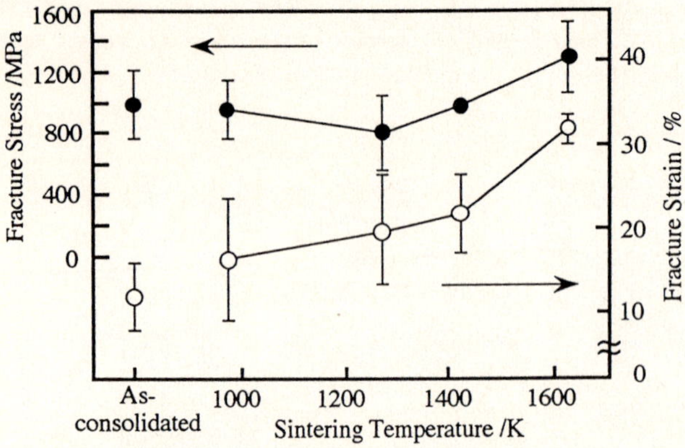

Fig. 4. Effect of specimen annealing to different temperatures on compressive fracture stress and strain

Fig.5(a) through (h) show SEM micrographs of the fracture surfaces of the as-consolidated specimen and the specimens annealed to 973, 1273 and 1623 K. The fracture mode shown in (a) and (b) is characterized by transparticle fracture although there is partially interparticle fracture as shown in (b). These also indicate that substantial interparticle bonding is achieved by melting at the powder surface. The fracture patterns in (c) and (d) are the evidence for the existence of the $\gamma+\alpha_2$ lamellar structure in the compact annealed to 973 K. The fracture patterns in (e) to (h) show that the grain refinement started at 1273 K, in which the average grain size is about 2 μm in diameter. It is likely that the improvement of the compressive fracture strain is related to the grain refinement. In the compact annealed to 1623 K, the average grain size increased to about 4 μm in diameter. However, the volume ratio of γ and α_2 were nearly equivalent as mentioned above. Therefore, the microduplex structure is formed in the compact annealed to 1623 K. The well-balanced strength and ductility of the compact may be caused by this duplex structure. From these results, it is notable that a sound TiAl compact can be produced by underwater shock consolidation without HIPing or hot-press treatments. In addition, a well-balanced material with high strength and ductility can be obtained by simple post-annealing , especially to 1623 K. The mechanism of the formation of the microduplex $\gamma+\alpha_2$ structure by annealing is now under study, but it may be related to the internal defects induced during the passage of shock wave.

4. Conclusions

Underwater-shock consolidation has been performed on PREPed TiAl powder. The microstructures and mechanical properties of TiAl compact were investigated in as-consolidated and subsequently annealed state. The following results have been drawn:

1. A sound TiAl compact can be produced by underwater-shock wave consolidation. Microstructural observation revealed that substantial interparticle bonding was achieved by the melting of the powder surfaces. The compressive fracture strength and strain were about 1000 MPa and 14%, respectively.
2. The mechanical properties of the compact were improved by post-annealing. The optimum annealing condition for the compact was determined to be 1623 K for 1 h. The compressive fracture strength and strain of the compact obtained were about 1300 MPa and 32%, respectively. The microstructure of the compact was $\gamma+\alpha_2$ microduplex structure.

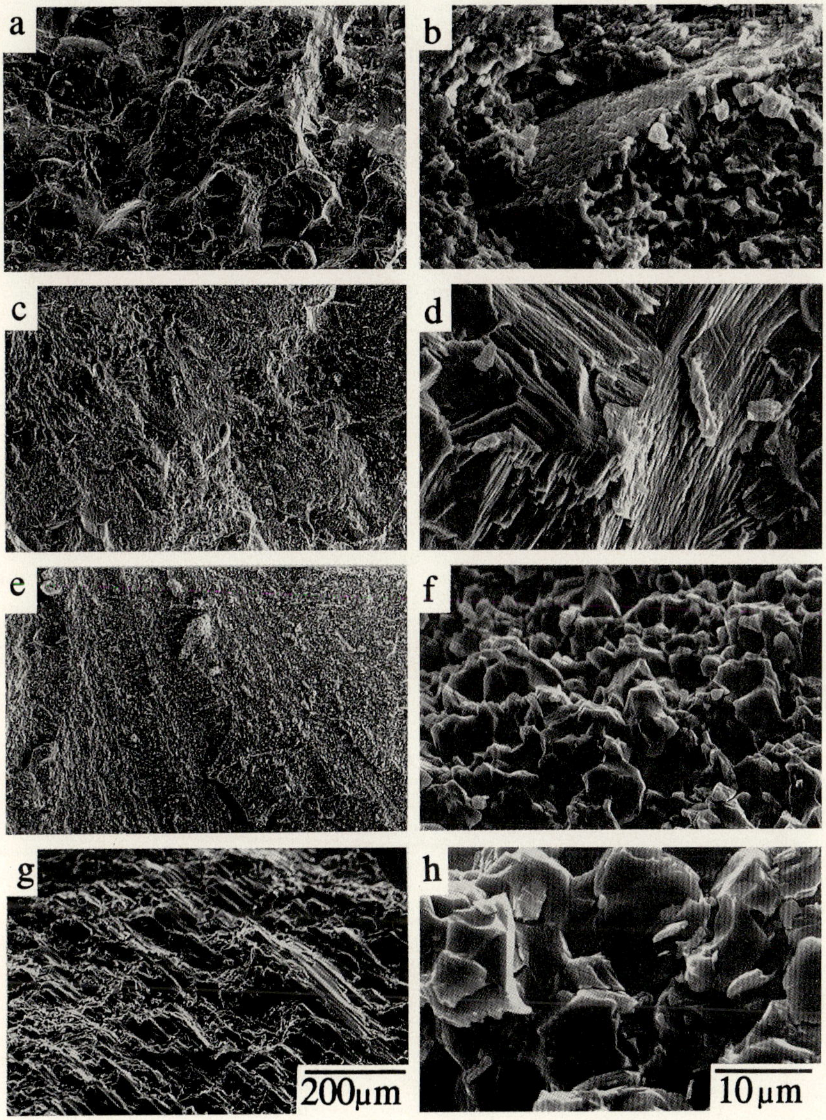

Fig. 5. SEM micrographs of fracture surfaces of TiAl compact, (a,b) as-consolidated, (c,d) annealed to 973 K, (e,f) to 1273 K and (g,h) to 1623 K

Acknowledgements

This work was partly supported by a Grant-in-Aid for Scientific Research on Priority Areas, Shock Wave Research, from the Ministry of Education, Science and Culture, Japan. We thank Mr. Y. Ishitani of Kumamoto University for assistance with the shock consolidation experiment. The explosive was kindly provided by the Asahi Chemical Industry Co., Ltd. Steel materials for making the consolidation assembly were kindly provided by Godo Steel Ltd.

References

Nishida M, Tateyama T, Tomoshige R, Morita K, Chiba A (1992) Electron microscopy studies of Ti-47at.% Al powder produced by plasma rotating electrode process. Scripta Metallurgica et Materialia 27:335-340

Shang SS, Meyers MA (1991) Shock densification/hot isostatic pressing of titanium aluminide. Metal. Trans. A22A:2667-2676

Tomoshige R, Chiba A, Nishida M, Imamura K, Fujita M. (1992) Explosive compaction of Si_3N_4 powder without additives using a water pressure medium. J. Cer. Soc. Japan, Intl. Ed.,100: 1209-1214

Tsujimoto T, Hishimoto K, Nobuki M, Suga H (1986) Structures and properties of an intermetallic compound TiAl-based alloys containing silver. Trans. Japan Institute of Metals 27:341-350

An Adaptive Mesh Method to Speed up the Numerical Simulation of an Underwater Explosion

Philippe Boyce* **and Thierry Hocquellet**[†]
*DCN-Centre Technique des Systèmes Navals-MDTC, Toulon, France
[†]Centre d'Etudes de Gramat, Gramat, France

Abstract. Due to the different time-scales involved, the numerical simulation of the effects of an underwater explosion is time consuming. In order to reduce the amount of CPU time required, an adaptive mesh technique has been added to a 1D spherical version of the hydrodynamic code HULL. Applied to an underwater explosion test, this method allows to save more than 80% of the computer time.

Key words: Underwater explosion, Numerical simulation, Adaptive mesh

1. Problem to be solved

To compute the effects induced by the detonation of an explosive charge fired underwater one has to take into account three major events :

- the propagation of the detonation wave through the high explosive,
- the build-up and propagation of the shock wave induced in free field water,
- the pulsating behaviour of the burt gases bubble during, at least, its first period.

2. Numerical difficulties

When using an hydrocode to compute the characteristic values of an underwater explosion (such as maximum pressure, time constant ... of the shock wave), the main difficulty comes from the different time scales that the three events exhibit. Typical values are: 0.001 ms for the detonation wave, 0.1 ms for the propagation of the shock in water, 100. ms for bubble pulsation. A typical pressure history profile recorded when firing a 2 kg charge is given on Fig.1.

Of course, the mesh size has to be suitable to reproduce the smallest time-scale. For instance, considering the following values:

- propagation velocity of the shock wave in water ~1500 m/s,
- thickness of the shock front (from hydrostatic pressure to maximum pressure) $\sim 1 \cdots 2$ mm,

the order of magnitude of the time step imposed by a classical finite-difference code would be around 1 microsecond.

This last point leads to tremendous CPU time when one wants to compute all the effects induced by the detonation of an underwater explosive charge. For instance, with the 2D eulerian code HULL (set in Gramat and used by CTSN-MDTC), computing the evolution of the bubble created by the detonation of a 2 kg charge with only 5 cells to discretize its radius (which is obviously not enough to study the shock wave) requires, on a CDC Cyber 962 computer, an order of 10 days CPU time !

3. Adaptive mesh principle

The method presented here is based on the observation of a typical spatial pressure profile induced in water (on Fig.2, an example is given for a 2 kg charge, 5 ms after firing). On such a curve, two zones of interest could be identified: the shock wave and the gas-water interface.

These two zones are located where the gradients (pressure, material velocity, density ...) present their highest values, that means they are the areas where the mesh has to be the finest. On the other hand, these physical phenomena are spread on a very narrow part of the total mesh;

Shock Waves @ Marseille III
Editors: R. Brun, L. Z. Dumitrescu

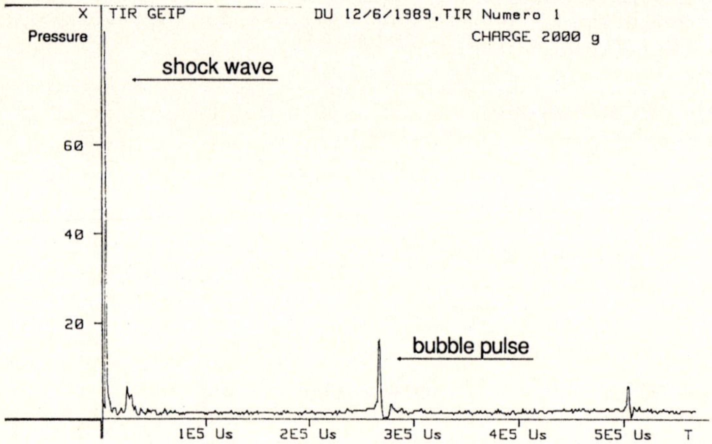

Fig. 1.

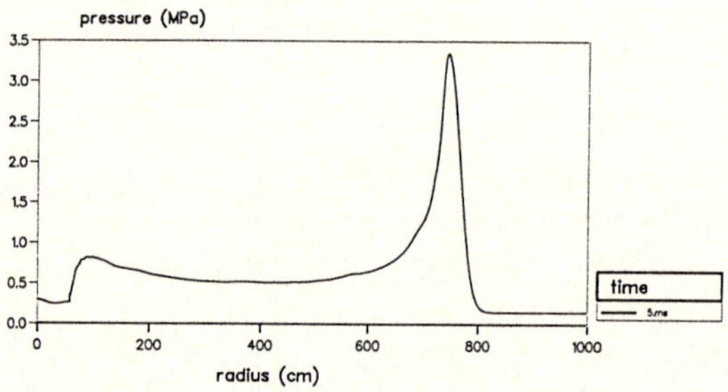

Fig. 2.

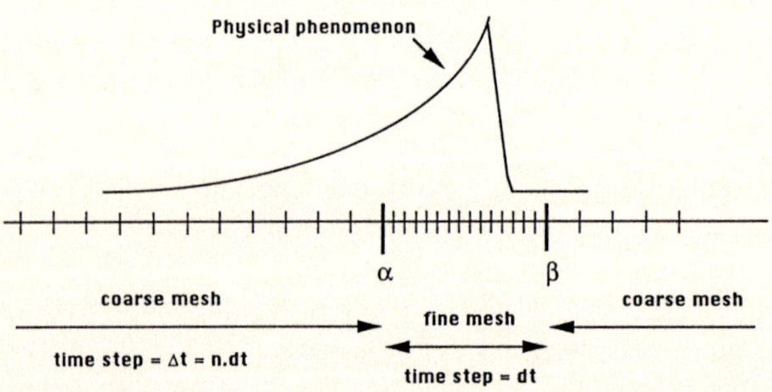

Fig. 3.

a typical underwater shock wave is modeled by a peak (rise time less than 1 microsecond) followed by an exponential decay.

So the method developed (Hocquellet 1990), written for a 1D version of the Eulerian HULL code, consists in bracketing these interesting zones by fictitious waves a and b which convey a small cell size mesh through a coarse one, as shown schematically in Fig.3.

Propagation of these fictitious waves is obtained by solving the two independent equations:

$$\begin{aligned} \alpha_t + (c_1\dot{\alpha})_x &= 0 \\ \beta_t + (c_2\dot{\beta})_x &= 0 \end{aligned} \tag{1}$$

c_1 and c_2 (and their signs) depend on which physical phenomenon is to be followed.

As the HULL time step is governed by the CFL condition $dt = k\dot{d}x/(c+u)$, dt is smaller in the fine mesh. So, in order to speed up the calculation, HULL computes N cycles in the fine mesh and only one cycle in the coarse one.

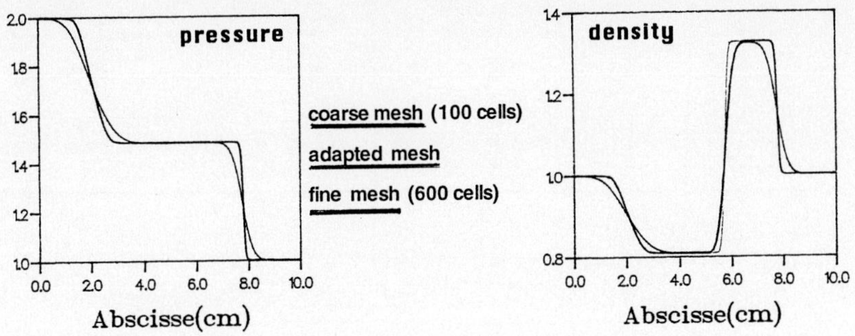

Fig. 4.

4. Numerical results

4.1. Reference test: shock tube flow

To test the feasibility and the accuracy of the modified code, the method has been applied to the shock tube problem. Pressure and density profiles can be seen on Fig.4. Three fine meshes were set in order to follow the two waves and the contact discontinuity. In each refined zone the cells are 5 times smaller.

The main advantage is to reduce significantly the smearing of shock fronts, characteristic of numerical diffusion. This is more sensitive for underwater applications, where it is usal to compute wave propagation on several meters (i.e during a large number of calculation cycles).

4.2. Calculation of an underwater explosion

The method has been applied to compute the effects induced by a 2 kg explosive charge detonating in free-field water. The computation is stopped after 10 milliseconds, so only the shock wave is studied. Three computations have been done:

- the reference one with a fine mesh all along the studied area;
- a computation with only a coarse mesh (the cell size is 5 times greater);
- the adapted mesh with fine mesh (cells being 5 times smaller) following the shock wave on the same coarse mesh as previously.

In Fig.5 and Fig.6, space and time history profiles of the pressure are reported. One can see that the adaptive procedure leads to the same accuracy as the one obtained with the fine mesh. However, the CPU time required is strongly decreased. From 0 to 10 milliseconds, the code needs:

- fine mesh ($\delta x = 0.3217$ cm): 1990 s;
- coarse mesh ($\delta x = 1.608$ cm): 102 s;
- adapted mesh ($\delta x_{\text{fine}} = 0.3217$ cm on $\delta x_{\text{coarse}} = 1.608$ cm): 366 s.

These CPU times are given for the 1D spherical HULL version, on a CONVEX C2 computer in Centre Technique des Systèmes Navals - Toulon. This represents a CPU time saving of 81%.

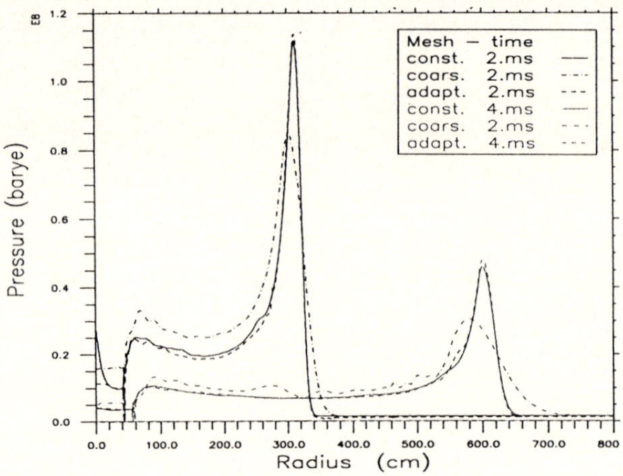

Fig. 5.

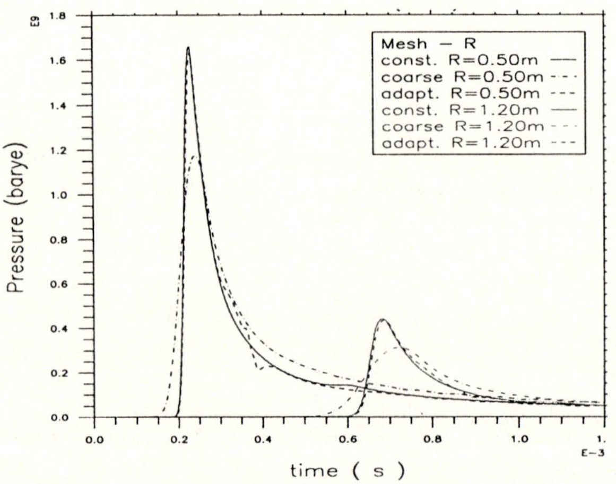

Fig. 6.

5. Further work

For two-dimensional cases, the feasibility of governing the fictive wave propagation has also been proved by Hocquellet (1990). Generalization of the system of equations (1) is made with the 2D Hamilton-Jacobi waves. With this set of equations, it is possible to follow, for example, an annular initial condition. Although the CPU time saving is supposed to be more interesting in multi-dimensional geometry, the adaptative mesh technique is not yet set in HULL 2D.

References

Hocquellet T (1990) Une technique eulérienne de maillage adaptatif contrôlé par des équations d'Hamilton-Jacobi. Thèse de Doctorat - Université de Bordeaux 1 (Studies made under the authority of Professor A-Y Leroux (Univ. Bordeaux 1 - France)

Converging Underwater Shock Waves for Metal Processing

S. Itoh, M. Fujita, S. Nagano and K. Kamohara
Department of Mechanical Engineering Faculty of Engineering, Kumamoto University 2-39-1 Kurokami, Kumamoto, Japan

Abstract. A new apparatus consisting of an ellipsoidal pressure vessel is proposed for punching holes on pipe walls from the inside. The phenomena taking place in the vessel were investigated experimentally and by a numerical method. When the aspect ratio (i.e. the ratio between the major and minor axes of the ellipsoid) was 2.0, the most suitable pressure distribution was obtained in the apparatus.

Key words: Converging shocks, Underwater shocks, Metal processing

1. Introduction

We have investigated the use of underwater shock waves in various metal processing applications, such as forming, cutting, welding or consolidation of powder etc. Now we will attempt to apply this technique to develop a new material, focusing on consolidation of difficult-to-consolidate powder such as CBN or difficult-to-work metal processing. In applications, the control of an underwater shock wave is most important and very difficult. An underwater shock wave can be generated by an explosion or by a sudden release of electrical energy in water. We have been researching explosions in water. The explosives have an extremely large amount of energy but a very short duration, so that it is very difficult to handle the energy efficiently. We have translated this energy to a high water pressure with a slightly longer duration. We have developed a suitable pressure vessel for generating and regulating an underwater shock wave to apply for metal processing.

The behaviour of underwater shock waves which were investigated by an optical measurement system is presented in this paper. This paper also discusses means to focus the underwater shock wave and to apply it for punching holes in pipe walls, as one of the examples. The profiles of the underwater shock waves and the pressure distributions in the pressure vessel, found in both experiments and numerical calculations, are also presented. We also would like to indicate the availability of converged underwater shock waves to apply for practical metal processing.

2. Experimental procedure and numerical calculation

We used detonating cord of 5.2 mm dia. and 280 mm length manufactured by Asahi Chemical Industry Co., Ltd., in experiments. We first investigated the behaviour of a shock wave occurring in underwater explosion by usual shadowgraph systems with a high-speed camera at 20 million frames per second. We also investigated the converging of an underwater shock wave in a pressure vessel as shown in Fig.1. This figure shows also the apparatus for punching holes in pipe walls and the pressure measurement ports. Two types of pressure vessels were investigated, one is an open type and the other is ellipsoid-shaped. In the latter case, the explosive is located at the first focal point of the ellipsoidal pressure vessel, in fact the shape of the upper part of this vessel is an ellipsoid.

When the explosion occurs, an underwater shock wave is generated from the site of the explosion which travels downstream and interacts with the wall of the pressure vessel to increase its strength. Finally, it enters into the pipe to make holes from the inside. The punching technique which makes a hole from the inside of the pipe is very useful because there is no after-treatment

Shock Waves @ Marseille III
Editors: R. Brun, L. Z. Dumitrescu

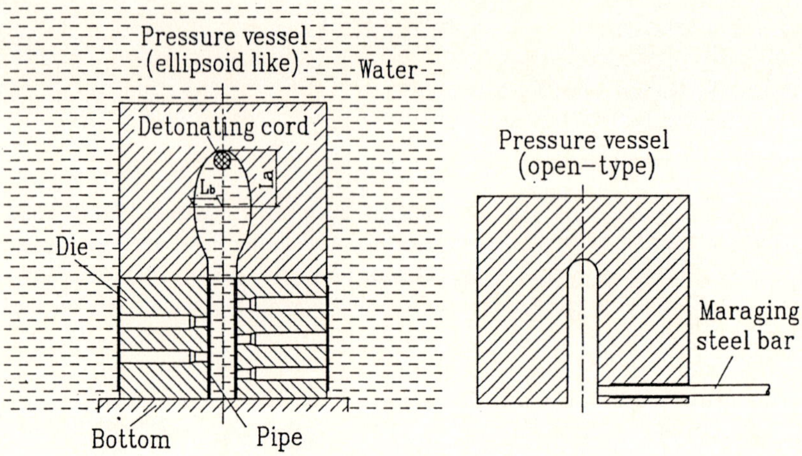

Fig. 1. Experimental apparatus for punching holes in a tube wall; (a) Ellipsoidal pressure vessel; (b) Open-type vessel with maraging steel bar for pressure measurement

that would reject a bar. We can use this technique for a relatively small inner diameter of pipe, less than 3.0 mm. On the other hand, if one wishes to use other mechanical tools such as a hydrostatic tool, there is a restriction of the inner diameter, due to the volume of the tool. The pressure was also measured by a maraging steel bar with a high conductive gauge, as also shown in this figure.

We also investigated the flow conditions of the underwater explosion in pressure vessels by numerical calculations. The basic equations are as follows:

$$\frac{\partial \boldsymbol{q}}{\partial t} + \frac{\partial \boldsymbol{f}}{\partial x} + \frac{\partial \boldsymbol{g}}{\partial y} = 0 \tag{1}$$

where $\boldsymbol{q}$, $\boldsymbol{f}$, $\boldsymbol{g}$ are vector functions as follows:

$$\boldsymbol{q} = \begin{pmatrix} \rho \\ \rho u \\ \rho v \end{pmatrix}, \quad \boldsymbol{f} = \begin{pmatrix} \rho u \\ \rho u^2 + P \\ \rho uv \end{pmatrix}, \quad \boldsymbol{g} = \begin{pmatrix} \rho v \\ \rho uv \\ \rho v^2 + P \end{pmatrix} \tag{2}$$

where u and v denote the velocity components in the x and y directions, respectively. ρ and P indicate the specific mass and the pressure, respectively. In calculation, we used the TVD finite difference scheme developed by Chakravarthy and Osher (1984). In calculation, we used Tait's equation of state for water because the maximum pressure is less than 3.5 GPa. We also used the pressure history of the detonating cord previously tested in experiment (Itoh et al. 1993). The pressure history is as follows:

$$P(t) = 3.0 \exp(10^{-5}\, t) \quad [\mathrm{GPa}]$$

3. Results and discussion

A sequential shadowgraph obtained in an underwater explosion is shown in Fig.2. The time is measured from the beginning of the ignition of the detonating cord. It is clear that the configurations of all shock waves are pseudo-stationary. The velocity of the detonating wave is 6,300 m/sec.

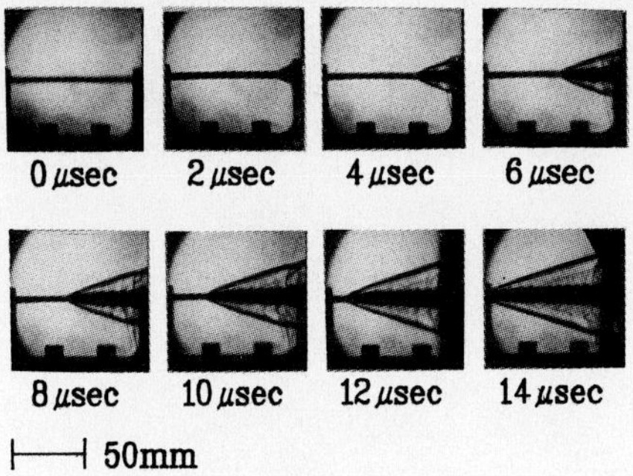

Fig. 2. High-speed framing photograph of an underwater explosion

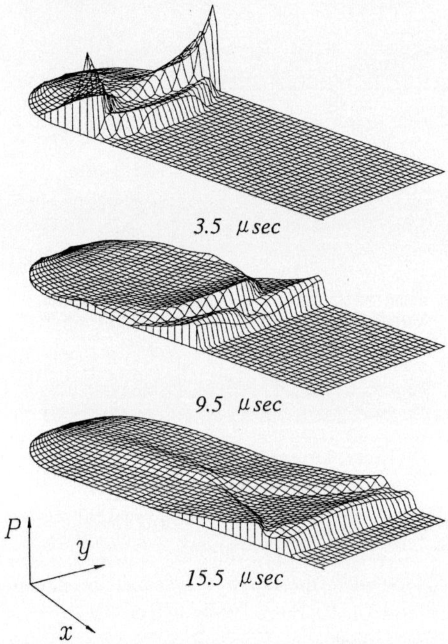

Fig. 3. Pressure distribution obtained in the open-type pressure vessel. The time is measured from the instant of explosion

Pressure distributions in the open-type pressure vessel obtained by the numerical calculations are indicated in Fig.3. We can see from these figures how the underwater shock wave converges. The pressure histories obtained by numerical calculation are shown in Fig.4. In the case of the open-type pressure vessel, we can recognize many pressure peaks as shown in this figure. d denotes the distance from the center toward the wall. About 21 μsec after the front of the shock wave arrived at the center of the pressure vessel, the pressure rises again, as denoted by symbol a. At the wall surface (Fig.3), it occurs 22.5 μsec after. The fact that there is a difference of time shows that the shock wave has not a plane configuration. The experimental pressure value obtained using our measurement system is also indicated in this figure. We can see that there is good agreement between the numerical and the experimental results. In our previous experiment, it was known that the pressure at the entrance of the pipe was required to be more than 350 MPa for punching the stainless pipe. Therefore we are searching for a shape of the pressure vessel which could produce over 350 MPa at the exit of the vessel. Although the peak pressure is over 350 MPa, if there are secondary or third-order peak pressures, the pipe will suffer sequential shock loadings. Once the first underwater shock wave goes through the pipe, it induces an elastic deformation of the pipe entrance by the pressure drop caused by the expansion fan. Sequential shock waves hit the pipe to produce collapse as shown in Figure 5 (a). The good result is shown in Fig.5 (b). The result (b) was obtained using a more suitable ellipsoidal pressure vessel as will be discussed later.

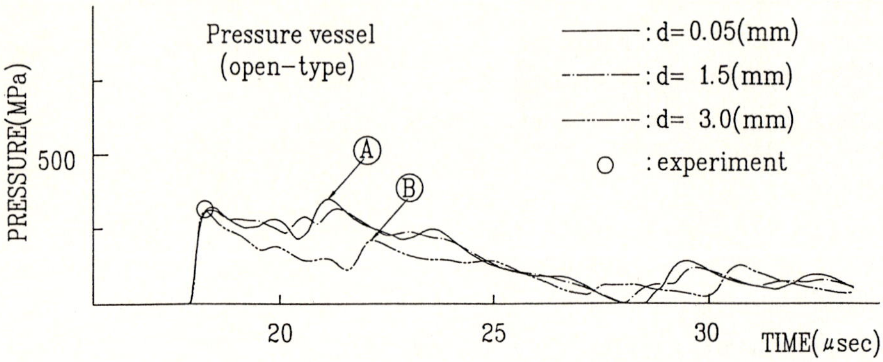

Fig.4. Pressure histories obtained in the open-type vessel. d denotes the distance from the vessel center

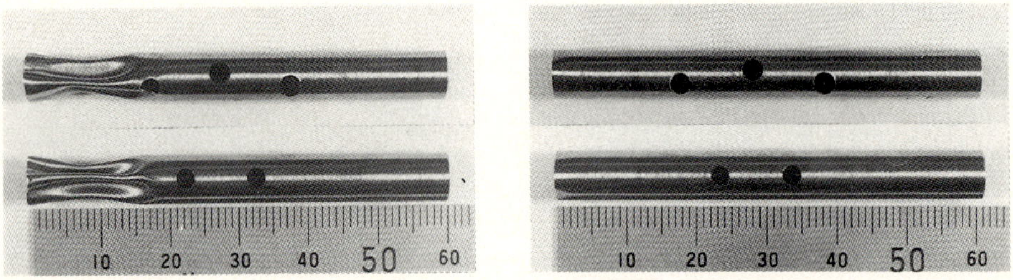

Fig.5. Photographs of a collapsed and intact pipe obtained in the open-type (a) and the ellipsoidal (b) pressure vessels, respectively

The collapse may be caused by buckling or other plastic deformation, but the effects of sequential shock loading are mostly responsible for the collapse. Thus there surely exists a restriction which caused both the pressure value and distribution to diverge.

The pressure distributions obtained in the ellipsoidal pressure vessel are shown in Fig.6. The vertical axis indicates the pressure values. Each time marked in this figure is measured from the instant of explosion. We can see the behaviour of the shock wave in the ellipsoidal pressure vessel. When the cord detonates at the first focal point of the ellipsoid, the shock wave propagates both backward and forward. After the backward progressing shock wave arrives at the wall, it immediately reflects from the wall. This reflected shock wave usually has a higher pressure and a higher velocity than those of the original one. Therefore the reflected shock wave can overtake easily the front of the shock wave propagated forward. When the reflected shock wave overtakes and collapses with the former, the strength of the shock wave increases.

The pressure contour maps and the velocity vectors are shown in Fig.7. The pressure contours are spaced at 30 MPa each. The concentrated lines indicate the shock wave. In the first stage of the explosion, the abrupt pressure difference between the front and the rear part of shock wave can be recognized. As the shock wave propagates, the pressure maps becomes broad. We can see the almost plane shock wave going into the downstream part of the pipe. This plane shock wave

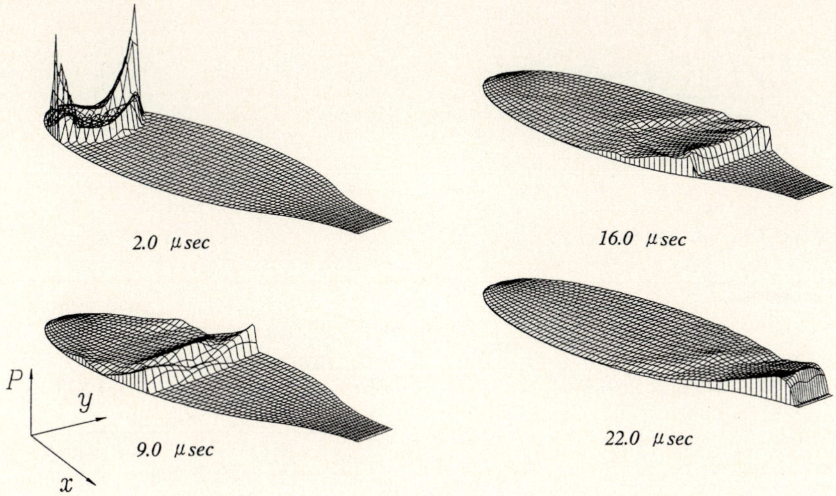

Fig. 6. Pressure distributions obtained in the ellipsoidal pressure vessel. Time is measured from the instant of the explosion

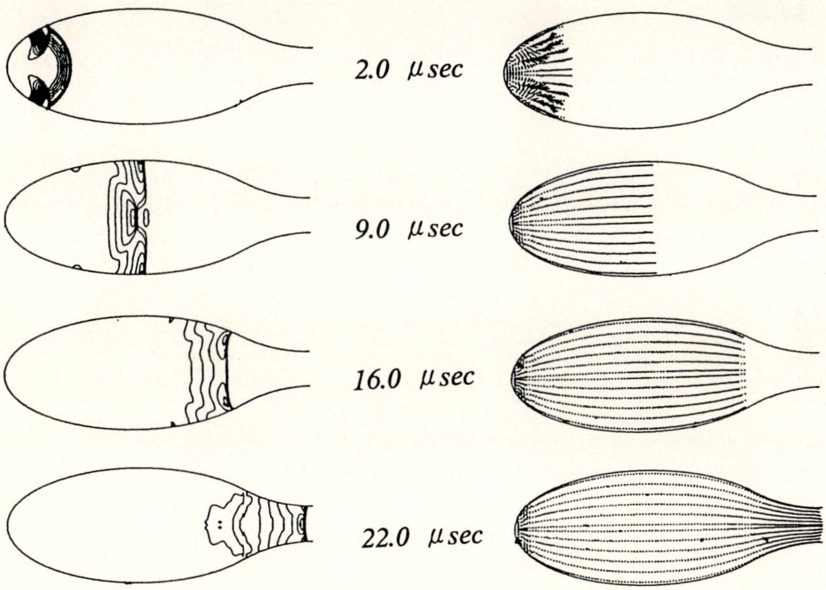

Fig. 7. Pressure contour maps (a) and velocity vectors (b) obtained in the ellipsoidal pressure vessel

can make holes on the pipe wall without any collapse. The shock configurations at the outlet of the ellipsoidal pressure vessel are shown in Fig.8. The shock wave is strongly deformed at the large corner radius (see Fig.8 (b)). In this case, we also could not get good results for making holes on the pipe wall.

Thus the configurations of the ellipsoidal pressure vessel are very important to get good results for punching holes in a pipe wall.

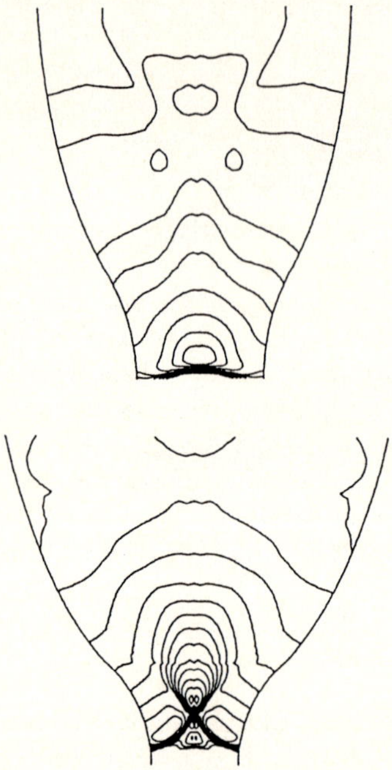

Fig. 8. Pressure contour maps obtained for different outlet corner radii and same major axes aspect ratio

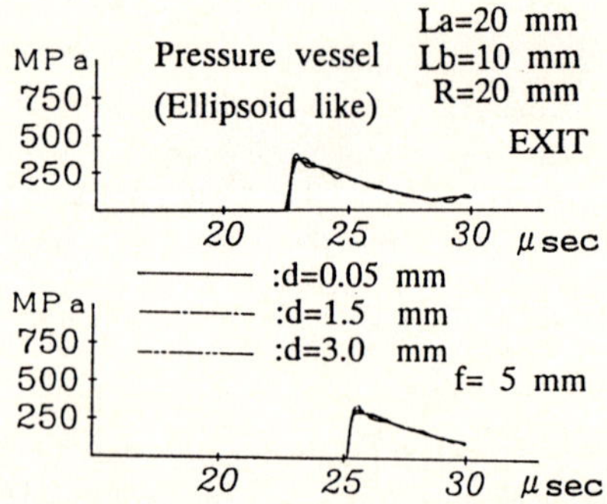

Fig. 9. Pressure histories obtained in the ellipsoidal pressure vessel. L_a and L_b denote the lengths of major and minor axes. R is the radius of the outlet corner

The pressure histories obtained at the exit of the pressure vessel are shown in Fig.9. The pressure histories obtained at a point 5 mm downstream from the outlet are also shown in this figure. d denotes the distance from the center of pipe. f denotes the distance from the exit of the pressure vessel. We never recognize the effects of the distance d on the pressure histories in all cases of various distances from the exit of the pressure vessel. We can see from this figure, that this ellipsoidal pressure vessel is quite suitable for punching holes in the pipe wall.

4. Conclusions

We proposed the ellipsoidal pressure vessel for generating and focusing underwater shock waves to make holes in pipe walls from the inside. First, the underwater shock wave produced by the underwater explosion of a detonating cord was investigated by an usual shadowgraph system. The pressure distributions were measured using semiconductor pressure gauges. The flow in the pressure vessels was investigated by a numerical method. We conclude as follows: this configuration of the pressure vessel is most effective with respect to the shock wave produced. When the aspect ratio (i.e. the ratio between the major and minor axes of the ellipsoid) is 2.0, the most suitable pressure distribution is obtained in this apparatus.

References

Chakravarthy SR, Osher S (1984) AIAA Paper 85-360

Itoh S, Kamohara K, Nagano S, Fujita M (1993) Explosion, Explosives and Pyrotechnics, No.54, Vol.30 (in Japanese)

Hugoniot Adiabate of Plasticized TATB-Based Explosive Compositions

E.V. Shorohov and B.V. Litvinov
Research Institute of Technical Physics, PO Box 245, Chelyabinsk-70, 454070, Russia

Abstract. Results of shock wave parameter measurements in a plasticized TATB-based explosive composition in the pressure range from 0.1 GPa to 40 GPa are given. Recording of shock and detonation waves was performed by means of the optical lever method. Under a shock compression stress lower than 0.7 GPa elastic and plastic waves are propagated in the matter. The amplitude of the elastic precursor having a prolonged front was estimated to be equal to 0.07 GPa. Experimental results were approximated in the form of a wave velocity - particle velocity relationship. Following the explosive transformation of the composition the compression parameters have been measured just inside the wave front. The average pressure at the edge of the front was equal to P_H= 38.5 $\pm$ 1.5 GPa. It belongs to the Hugoniot adiabate of the non-reacting explosive. The ratio of P_H to the pressure P_J at the Jouguet point is equal to 1.35.

Key words: Hugoniot adiabate, Explosive composition, Shock wave

1. Introduction

Plasticized TATB-based compositions are of interest due to their safe use in explosive technologies. These compositions, although being powerful explosives, have an uniquely high threshold of their initiation equal to 20 $\div$ 25 GPa (Jackson et al. 1976; Dick et al. 1988) as compared with other explosives of this class. This feature of explosive TATB-based compositions allows to investigate the predetonation part of the shock adiabate. In the present work the results of shock-wave loading of one of such compositions are presented. Its closest analogues are the known explosives LX 17 and PBX 9502 which contain 10% and 5% polymeric binder Kel F800 (polytrifluorochloroethylene), respectively. The composition being studied in the work includes 90% TATB - triaminotrinitrobenzene and 10% polymeric binder. The experimental results were obtained in the pressure range from 0.1 to 40 GPa. While choosing the method for carrying out the investigation, it was taken into account that, in the selected pressure range, many organic substances undergo transformations with volume change under shock compression. As a consequence, peculiarities can appear just in the wave-front structure. In this case a break can occur in the Hugoniot adiabate. In connection with this, methods based on the continuous recording of the processes appear to be preferable for recording the shock wave motion. For these purposes the known method of the optical lever was applied in this work (Fowles 1961; Gornovoy et al. 1989).

2. Experimental method

This method is based on the principle of following the motion of the free surface from the moment when the shock wave reaches it at a small angle. The sample under investigation has the form of a wedge (Fig. 1). Its free surface has the feature of mirror reflectivity. In our experiments we obtained this effect by applying a thin (20 and 50 μm thick) copper foil onto the explosives. Copper was chosen because it provides the required quality of the chronogram, although a material with less dynamic stiffness would have been preferable. The experimental set-up is shown in Figs.1, 2. Scheme 1 was used to obtain pressure values $P < 10$ GPa and scheme 2 for pressure values $P > 10$ GPa. When the shock wave reaches the free surface it makes it rotate. The free surface velocity W is determined by the angle of rotation, and the wave velocity D is defined via the motion velocity of the shock wave exit point onto the free surface along its surface:

Shock Waves @ Marseille III
Editors: R. Brun, L. Z. Dumitrescu

$$W = D_c \tan\alpha, \quad D = D_c \sin[\gamma(A, D^*)],$$

where

α is the angle of rotation of the free surface;

D_c- the velocity of the shock sliding along the free surface;

A- the angle of the wedge sample;

D^*- the detonation velocity of the explosive;

γ- the angle of shock wave incidence on the free surface.

To record the shock wave's angle of rotation and the sliding velocity, we placed slit light sources in front of the surface bearing the applied foil. The image of these sources in space and time is recorded by a rotating-mirror camera. The wave reaching the zone which reflects the respective source is marked by the change in its image. The change in size is proportional to the angle of rotation. This process is the same for every source. The wave sliding velocity is determined via the instants of successive displacements of the point source images.

The pressure in the wave front is determined by the following formula:

$$P = \rho D u,$$

where

u is the particle velocity of the shock wave, $u = W/2$;

ρ - the sample initial density.

The relative change in volume is found as

$$V/V_0 = 1 - u/D,$$

where

V_0 is the initial specific volume,

V - the final specific volume in the wave front.

3. Experimental results

Ten experiments of explosive loading of samples were performed. Four of them were run under the condition of sliding detonation (Fig.1), the rest were carried out under normal incidence of the shock wave onto the sample (Fig.2). In experiments according to Scheme 1, plastic explosive samples having a detonation rate of 7.8 km/s and thicknesses equal to 1, 2, 4, 6 mm were used. When loading the samples according to Scheme 2 the total thickness of the explosive was varied from 150 mm to 250 mm considering the generator of the plane detonation wave. For developing a pressure of the required value in the sample, shields were placed in front of the sample, at the side of the explosive. Copper, steel, lead and PMMA were used as shielding materials. The sample was a 30 mm thick disk of 120 mm diameter and bevelled at an angle of 15°. In experiments with sliding detonation the shock wave was attenuating, allowing to obtain up to 20 points of pressure in one experiment. Results of experiments are given in Figs.3, 4. A fair approximation of the dependence is of the form

$$D = 2.5101 + 2.4346\,u - 0.1983\,u^2, \quad 0.03\ \text{km/s} < u < 2.65\ \text{km/s}. \tag{1}$$

The most satisfactory description of the results was obtained by a piece-wise-linear correlation:

$$D = 2.50 + 2.35\,u, \qquad 0.03\ \text{km/s} < u < 0.7\ \text{km/s},$$

$$D = 2.88 + 1.82\,u, \qquad 0.7\ \text{km/s} < u < 2.65\ \text{km/s}. \tag{2}$$

Note that the wave velocity measurement error does not exceed 1-2% and it was no more than 5% for the mass velocity.

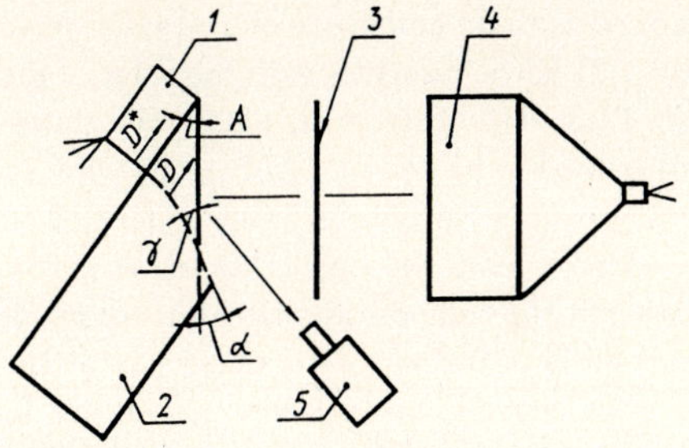

Fig. 1. Scheme of the experiment in the sliding detonation regime. 1 - plastic explosive; 2 - sample; 3 - raster; 4 - explosive argon light source; 5 - streak camera

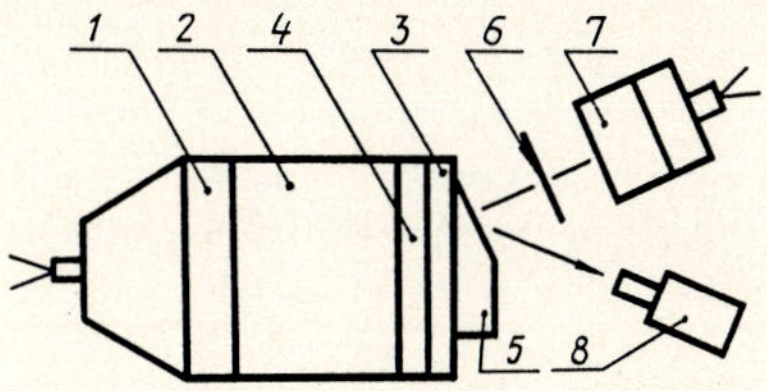

Fig. 2. Geometry of the experimental set-up with normal loading of the sample. 1 - plane-wave generator; 2 - explosive charge; 3 - copper, steel or lead; 4 - PMMA; 5 - sample; 6 - raster; 7 - explosive argon light source; 8 - streak camera

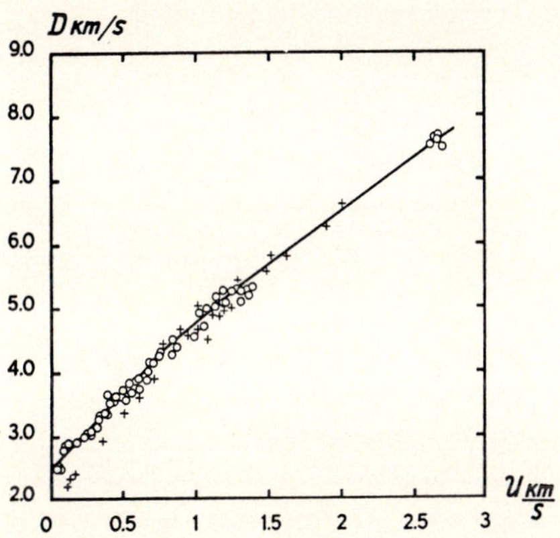

Fig. 3. Shock velocity vs. particle velocity data for the given explosive composition. o - data of present work; + - data of Dick et al (1988)

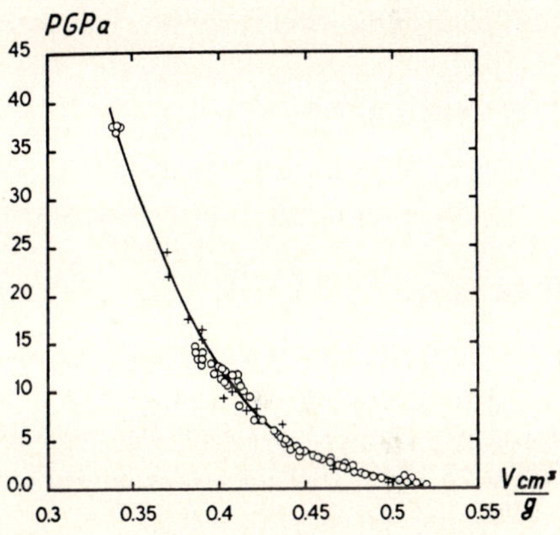

Fig. 4. Pressure/volume data for the explosive composition (V_0 = 0.5241 cm^3/g)

4. Discussion of the results

From the correlation (2) it follows that there is a break on the shock adiabate in the $D-u$ plane at $u = 0.7$ km/s and P =5.6 GPa. It might be supposed that here some structure change in the material takes place. However, the two-wave configuration of the shock waves which accompanies this polymorphous transformation was not realized by us. Its appearance was also not noted at $u = 1$ km/s ($P = 8.8$ GPa) where according to Dick et al.(1988) there is a break on the shock adiabate. Attention is paid to the existing difference in slopes in the $D-u$ plane between our data and data of the work by Dick et al.(1988) for $u < 0.5$ km/s and $P < 3.5$ GPa. This difference can be, obviously, explained by the sensitivity of the investigated material to such factors as the initial porosity and the sample manufacturing technology.

The elastic and plastic waves were recorded during experiments. The wave front of the elastic precursor is prolonged. The initial velocity of the precursor is equal to 2.91 km/s at the toe of the front but it is equal to 2.49 km/s ahead of the plastic wave with an amplitude of 0.14 GPa. Note that the obtained value of the elastic precursor velocity, 2.91 km/s, coincides with the value of the sound longitudinal velocity for the similar composition PBX 9502 (Dick et al. 1988). The elastic precursor amplitude before the plastic wave front is estimated to be equal to 0.07 ±0.01 GPa. In the plastic region extrapolation of Eqs.1 and 2 to $u = 0$ gives a wave velocity value equal to 2.50 km/s. It coincides with Olinger's calculation for the sound volume velocity (Dick et al. 1988). In a number of experiments the explosive transformation of the composition has been recorded. The wave and mass velocities have been measured and the pressure was determined just at the edge of the detonation wave front (~10 ns from the front arrival), this pressure turned out to be equal to $P_H = 38.5 \pm 1.5$ GPa. The pressure P_J at the Jouguet point is equal to 28.5 GPa according to Mader (1980), whence the relation $P_H/P_J = 1.35$. The detonation wave structure of this composition was studied by means of the photoelectric technique (Lubyatinsky and Vorobey 1991). The detonation front was shown to have a spike of pressure with a subsequent drop to the Jouguet pressure. Existence of an increased-pressure region in the detonation front has been noted by Evstigneev and Gatilov (1991) who applied the manganine transducer technique. As follows from Figs.3, 4, kinematic and amplitude parameters, being measured in the shock and detonation waves, are well described by Eqs.1, 2. They refer to the Hugoniot of the non-reacting material. Thus, one can draw the conclusion that first the precompression of the initial material takes place in the detonation wave front, and then the chemical reactions are initiated.

Acknowledgments

The authors thank their colleagues V.V. Murdasov, A.A. Denisenko, N.P. Baranova and T.N. Povasheva for their help in preparing experiments and performing calculations.

References

Dick JJ, Forest CA, Ramsay JB, Seitz WL (1988) The Hugoniot and shock sensitivity of a plastic-bonded TATB explosive PBX 9502. J. Appl. Phys. 63, 10: 4881

Evstigneev AA, Gatilov LA (1991) (unpublished data, private communication)

Fowles JR (1961) Shock wave compression of hardened and annealed 2024 aluminium. J. Appl. Phys. 32, 8: 77

Gornovoy AA, Kozlov EA, Muzyrya AK, Shorohov EV (1989) Investigation of elastic precursor relaxation kinetics in steel 3 and titanium. Sov. J. Phys. of Comb. and Expl. 1: 142

Jackson RK, Green LG, Barlett RH, Hofer WW, Kramer PE, Lee RS, Nidick EJ, Shaw JL, Weingart RC (1976) Initiation and detonation characteristics of TATB. Sixth Symposium (International) on Detonation, San Diego, California 24-27 August, pp 257-265

Lubyatinsky SN, Vorobey VA (1991) Study of detonation front structure in high explosives using a photoelectric technique. In Proceeding 5th Symposium (International) on Detonation, Krasnoyarsk, USSR 5-12 August, pp 369-373

Mader CL (1980) Los-Alamos Explosives Performance Data

Substance Compression Under Strong Shock Waves with Allowance for Nuclei Dissociation

B.V. Litvinov, N.P. Purygin, S.V. Samylov and I.V. Sanin
Institute of Technical Physics, P.O.Box 245, Chelyabinsk-70, 454070, Russia

Abstract. An evaluation is given of the influence of nuclei dissociation on substance compression in strong shock waves showing that, by taking dissociation into account, the compression behind the shock front increases.

Key words: Strong shocks, Solid matter, Nuclei dissociation

The ultimate compression of matter ($h = \rho/\rho_0$) in strong shock waves is determined by the effective adiabatic index γ :

$$h = (\gamma + 1)/(\gamma - 1)$$

In a real matter the value of γ is closely connected with, successively, molecule excitation and dissociation, atom ionization, radiation and, finally, with nuclei dissociation, when the shock wave intensity is increasing. These processes lead to decreasing values of γ, whereas the compression in the wave front is increasing but the temperature growth is slowing down. We shall estimate the extent of the influence of nuclei dissociation upon the compression ratio in a shock wave. Analysis is performed in a simplified form, in particular, the time for establishing the equilibrium state is not considered. In Table 1 temperatures calculated according to Zel'dovich and Novikov (1967) are given for 50% dissociation of some light nuclei. The approximate estimation of the influence of nuclei dissociation on the compression ratio in a shock wave can be made according to Zel'dovich and Raizer (1966),

$$h = h_0 + kE/E_0$$

where:

E - is the dissociation energy in the volume unit;
E_0 - the energy connected with pressure of particles and radiation;
h, h_0 - the ultimate compression, with or without the dissociation of the nuclei;
k - a coefficient depending on the adiabatic index γ: k=3 for γ=5/3, k=6 for γ=4/3.

Table 1.

Reaction	Total dissociation energy E_0 (MeV)	Temperature for 50% dissoc. $T_{1/2}$ (keV) ρ=1 g/cm^3	ρ= 10^3 g/cm^3
$D_1^2 \rightarrow H_1' + n$	2.25	86	105
$Li_3^6 \rightarrow He_2^4 + D_1^2$	1.49	57	67
$Be_4^9 \rightarrow 2He_2^4 + n$	1.58	30	36

We make an estimation for two limiting cases, i.e. for an optically transparent medium and an optically dense one, and consider the matter at the temperature corresponding to 50% nuclei dissociation.

Shock Waves @ Marseille III
Editors: R. Brun, L. Z. Dumitrescu

For an optically transparent medium, the pressure is determined by the particle pressure, $h_0 = 4$, and the formula for h is of the form:

$$h = 4 + E_0/T_{1/2}$$

where:

E - is the dissociation energy of one nucleus (keV),

$T_{1/2}$ - the temperature at 50% dissociation (keV).

For an optically dense medium, the pressure is mainly determined by radiation (so far as the temperature $T_{1/2}$ is sufficiently high), $h_0 = 7$ and the formula for h has the form:

$$h = 7 + 21\, \rho\, E_0/AT_{1/2}^4$$

where:

ρ is the density of matter in g/cm,

A - the atomic weight,

E, $T_{1/2}$ in keV.

In this case the nuclei dissociation effect becomes appreciable only at rather high densities ($> 10^3$ g/cm^3). The influence of the nuclei dissociation on deuterium compression is illustrated in Table 2.

Table 2. Ultimate compression for deuterium in a shock wave

Medium		$T = T_{1/2}$ (keV)	h_0	h
Optically transparent		86	4	30
Optically dense	$\rho = 10^3$ g/cm^3	105	7	7.2
	$\rho = 10^4$ g/cm^3	111	7	8.6

The dissociation of nuclei leads to the fact that, in the appropriate temperature range, compression is increasing and temperature growth is slowing down in strong shock waves. For matter in which nuclei consist of more than two nucleons the dissociation of nuclei is possible in several directions, therefore, the emergence of corresponding compression peaks should be expected in shock waves. The effect of slower temperature rise at the expense of the nuclei dissociation can exert an influence on the rate of thermonuclear reactions, especially under high temperatures, hence, its consideration in the involved processes,for example, in thermonuclear detonation (Fuller & Gross 1968) is desirable.

References

Fuller AL, Gross RA (1968) Thermonuclear detonation wave structure. Physics of Fluids 11, 3: 534

Zel'dovich YaB, Novikov ID (1967) Relativistic Astrophysics [In Russian]. Nauka, Moscow

Zel'dovich YaB, Raizer YuP (1966) Physics of Shock Waves and High-Temperature Hydrodynamic Phenomena [In Russian]. Nauka, Moscow, pp 181-182

Equations of State and Matter Transformation Parameters Under Shock Compression for Epoxy Resin

K.K. Krupnikov and V.P. Krupnikova
Research Institute of Technical Physics (VNIITF), PO Box 245, Chelyabinsk-70, 454070, Russia.

Abstract. Paper presents experimental results for hardened epoxy resin: principal shock adiabate, sound velocity along it, two-fold shock compression. On the principal shock adiabate a portion with elevated compressibility is observed. On the assumption that the matter is undergoing a transformation of the first-order phase transition type, equations of state for the phases were developed and thermodynamic parameters of the transformation were calculated. The endothermic variant of transformation satisfies the total set of experimental data. In this case, at the intermediate shock adiabate portion, which belongs to the phase mixture region, a temperature drop should accompany pressure increase. The nature of this transformation is not known for sure, though this classical variant of phase transition is quite possible, but the hypothesis that the transformation is connected with chemical reactions proceeding in the material is much more probable. The energy balance evaluation for one of these variants showed this explanation to be possible.

Key words: Shock compression of solids, Equation of state, Epoxy resin

The study of organic compounds under shock compression indicates the possibility of matter transformation behind the shock wave front (see for example Trunin et al. 1991). In the present work the following results are obtained for ED-5 epoxy resin hardened by polyethylene-polyamine, with the epoxy resin initial density being $\rho_{01} = 1/V_{01} = 1.19$ g/cm^3:

1. The principal shock adiabate within the 4-53 GPa pressures interval. The reflection method was used to find experimental points (Al'tschuler et al. 1958). In the intermediate (18-35 GPa) pressures range a peculiarity, i.e. an increased shock compressibility of the matter, is quite evident. A similar peculiarity is reported in papers by US investigators (Marsh 1980; McQueen et al. 1970).
2. The sound velocity C in the shock-compressed matter. This velocity was determined by the method of lateral load-release (Al'tschuler et al. 1960).
3. Two-fold shock compression. A two-fold shock compression was produced by the reflection of a first shock wave with 31.8 GPa amplitude from a copper barrier; the pressure in the second shock wave was 80 GPa.

Results of the experiment are shown by circle marks on the figures:

- Fig.1. Shock velocity D vs. particle velocity U for the principal shock adiabate;
- Fig.2. Shock pressure P vs. compression rate $\sigma = V_{01}/V$;
- Fig.3. Sound velocity C vs compression rate σ.

The error in determining D and U is $\pm(1-2)\%$, while for the velocity C, it is $\pm(5-7)\%$.

We failed to describe all experimental data by one common equation of state. An attempt was made to find an explanation of the peculiar matter behaviour in the thermodynamic equilibrium transformation of the first-order phase transition type. The matter is assumed to be an ideal compressed liquid; for the phases mixture the additivity is valid for specific volumes V, internal energies E, entropies S; particle velocities U are similar for both phases in the mixture. We believe that the lower and upper portions of the principal shock adiabate belong, respectively, to stability

Shock Waves @ Marseille III
Editors: R. Brun, L. Z. Dumitrescu

regions of the initial phase and of a more dense phase occuring behind the shock front, phase 2 of the matter, with the intermediate portion appertaining to the area of the equilibrium phase mixture.

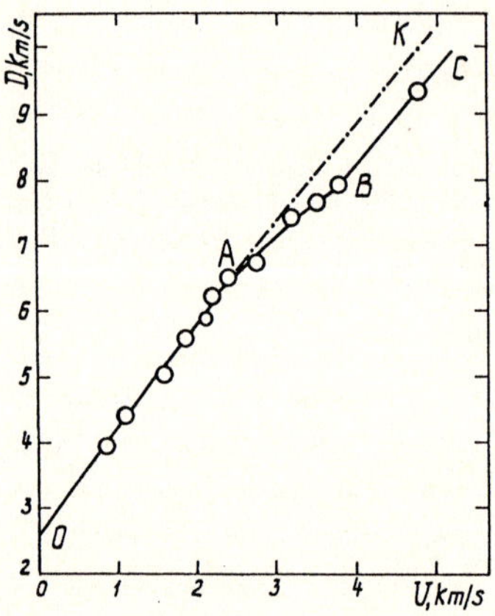

Fig. 1. Shock adiabate (principal). Shock velocity D vs. particle velocity U. Experiment: ○. Calculation: — OABC - equilibrium adiabate, —·— AK - "frozen" adiabate

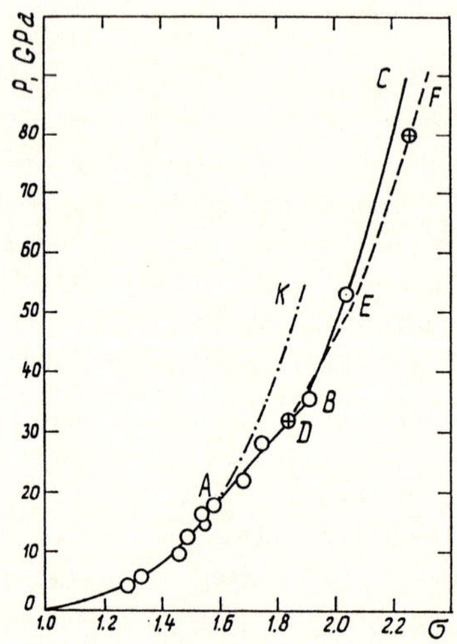

Fig. 2. Shock adiabates. Pressure P vs compression rate $\sigma = V_{01}/V$. Experiment: ○ - shock adiabate (principal), ⊕ - two-fold shock adiabate. Calculation: — OABC - equilibrium adiabate (principal), —·— AK - "frozen" adiabate (principal), -- DEF - two-fold shock adiabate (equilibrium). The portions AB and DE belong to the area of equilibrium phase mixture

To describe the total set of data for the experiment, it is necessary to determine equations of state for every phase as well as characteristics of the probable transformation, i.e. values of abrupt change in density, internal energy, entropy, interfaces position. For both phases the Mie-Grüneisen type equations of state (Zel'dovich and Reiser 1966) with variable Grüneisen coefficient $\Gamma = \Gamma_0(V/V_0)$ are taken:

$$\frac{V(P-P_c)}{E-E_c} = \Gamma,$$

where

$$P_C = \frac{C_0^2}{nV_0}\left[\left(\frac{V_0}{V}\right)^n - 1\right], \qquad E - E_C = \int_{T_0}^{T} C_V \, dT,$$

$$E_C = E_0 + \Delta E_C, \qquad \Delta E_C = -\int_{V_0}^{V} P_C \, dV, \qquad C_V = \text{Const}.$$

To solve the problem we use the known thermodynamic relations, phase equilibrium conditions (equality of pressures, temperatures and Gibbs potentials) (Landau and Lifshits 1976) as well as

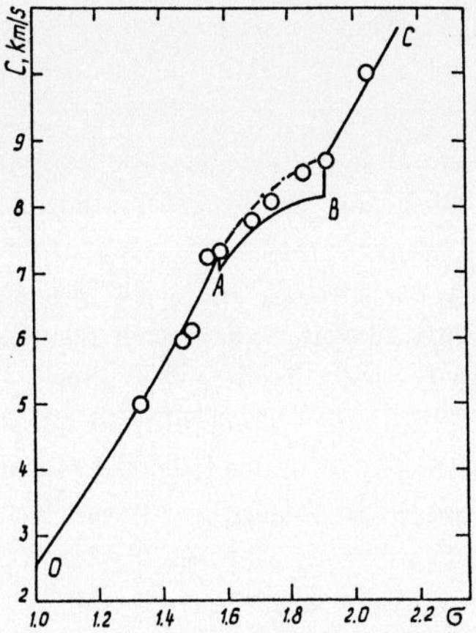

Fig. 3. Sound velocity along shock adiabate. Sound velocity C vs. compression rate $\sigma = V_{01}/V$. Experiment: ○. Calculation: — equilibrium sound velocity, —·— sound velocity in the "frozen" composition phase mixture

the additivity of extensive thermodynamic values; to compute the shock adiabates and sound velocities we use the Hugoniot equation and the isentrope equation in differential form. To fit the experiment, the parameters of the equations of state and the phase transition are chosen to satisfy all experimental data for the lower and upper pressure ranges as well as the shock compression data in the intermediate range (the sound velocity within this range is not taken into account). The required differences in initial internal energies and phase entropies are determined by:

$$E_{02} - E_{01} = \Delta E_{C1} - \Delta E_{C2} + \left(P - T\,\frac{dP}{dT}\right)(V_1 - V_2),$$

$$S_{02} - S_{01} = C_{V2}\Gamma_{02}\left(1 - \frac{V_2}{V_{02}}\right) - C_{V1}\Gamma_{01}\left(1 - \frac{V_1}{V_{01}}\right) - (V_1 - V_2)\frac{dP}{dT} - (C_{V2} - C_{V1})\ln\left(\frac{T}{T_0}\right)$$

where subscripts 1 and 2 relate to first and second phases, respectively.

Magnitudes in these formulae are taken at P (pressure) and T (temperature) - values that correspond to the "A" point, i.e. the point, where the shock adiabate's lower branch meets the phase equilibrium line. This point is determined on the basis of experimental data, as the point where the adiabate undergoes an inflexion. Specifying an approximate value of the shock adiabate inclination in the mixture range $(\partial V \,/\, \partial P)_H$ for this point one finds two corresponding inclinations dP/dT of the phase equilibrium line in this point:

$$\frac{dP}{dT} = \left(\frac{\partial P}{\partial T}\right)_{V1} \Theta^{-1},$$

$$\text{where } \Theta = A \mp \sqrt{A^2 + (\omega_H - 1)(k-1)k^{-1}}; \qquad A = 0.5\frac{k-1}{k}\left[2 + \tau(\ell - \omega_H)\right];$$

$$k - 1 = -T\left(\frac{\partial V_1}{\partial P}\right)_T \left(\frac{\partial P}{\partial T}\right)^2_{V1} C_V^{-1}; \qquad \tau = \frac{P}{2T}\left(\frac{\partial T}{\partial P}\right)_{V1};$$

$$\ell = \frac{V_1 - V_{01}}{P}\left(\frac{\partial P}{\partial V_1}\right)_T; \qquad \omega_H = \left(\frac{\partial P}{\partial V_1}\right)_T \left(\frac{\partial V}{\partial P}\right)_H$$

One of them is negative and corresponds to an endothermal transformation, the other is positive and corresponds to an exothermal one. After determination of the phase transition characteristics which correspond to these values of dP/dT the principal shock adiabate portion is calculated for the phase mixture range. In case of a considerable difference with the experiment a new value of $(\partial V/\partial P)_H$ at point A is chosen and the procedure is repeated until satisfactory agreement. The following values of constants for the equations of states were chosen for phases 1 and 2, respectively: initial density $\rho_0 = 1/V_0 = 1.19$ and 1.45 g/cm^3; initial sound velocity $C_0 = 2.58$ and 4.12 km/s; n=5.5 and 4.35. Heat capacity $C_V = 1.2$ J/gK and initial Grüneisen coefficient $\Gamma_0 = 0.8$ are taken identical for both phases.

The following phase transition parameters were chosen, which characterize the endothermal variant of transformation: shock adiabate inclination for the phase mixture range at point A: $\omega_H = 1.25$ and dP/dT =-68.8 MPa/K; coordinates of point A: pressure P =18 GPa, compression rate $V_{01}/V = 1.585$; differences in initial values of phases' internal energies $E_{02} - E_{01} = 3.31$ kJ/g and initial phases' entropies $S_{02} - S_{01} = 1.69$ J/gK. Figs.1-5 illustrate the calculation results. Portions OA and BC of the principal shock adiabate fall within the stability ranges of phases 1 and 2, respectively, and the intermediate portion AB belongs to the range of equilibrium phase mixture. The calculated adiabate of the two-fold shock DEF is shown. It starts from the initial point D (which lies in the phase mixture range at 31.8 GPa pressure) and passes the experimental point of the second shock at 80 GPa pressure.

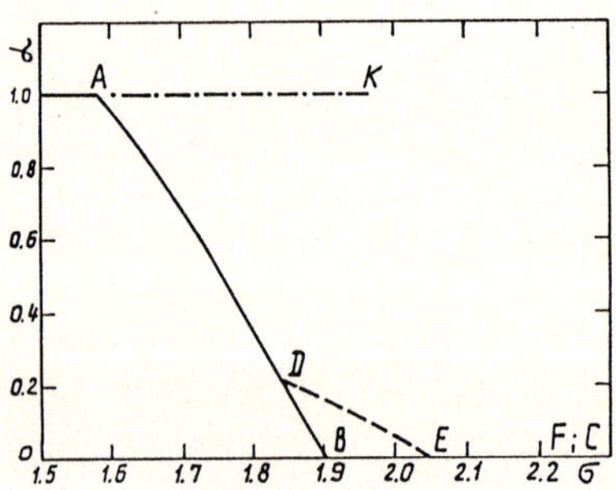

Fig. 4. Phase composition along the shock adiabates. Mass concentration of the first phase α vs. compression rate $\sigma = V_{01}/V$. Designations - see Fig. 2

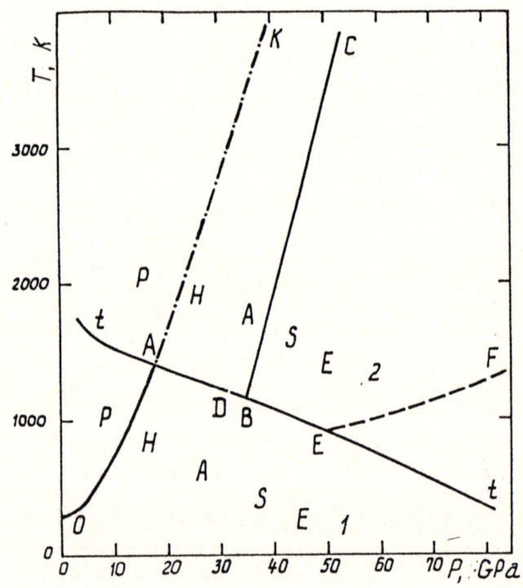

Fig. 5. Phase diagram. Pressure P vs. temperature T. $t-t$ - line of phase equilibrium. Other designations - see Fig. 2

From Fig.3 it is obvious that the calculated dependence for the sound velocity undergoes a jump in the range of equilibrium phase mixture. In the strict sense, the experimental data fail to reproduce this behaviour within this range. But with allowance for the possible errors of the experiment, they are still not inconsistent with calculation results. If it is granted that under isentropic expansion (in contrast to the shock compression) just two of the three phase equilibrium conditions are fulfilled (pressures and temperatures equalities), and the mixture composition remains as if "frozen" (holding on the value attained under shock compression), then the agreement between calculation and experiment is quite satisfactory. Fig.4 illustrates the change in phase composition along the related adiabate portions. Fig.5 gives the phase diagram in the (P,T) coordinates. The transformation of matter on the principal shock adiabate begins in point A at the pressure of 18 GPa and terminates at point B at 35.1 GPa; this transformation is accompanied by 5-6% matter density increase at fixed P; the heat of transformation is 2.5-2 kJ/g. The matter temperature on the AB adiabate portion decreases with the pressure growth. The same is observed for the two-fold shock adiabate portion in the phase mixture range. It should be noted that within this range the two-fold shock adiabate DE in (P, σ) coordinates is steeper than the principal adiabate of the single-shock compression AB. All these are as a result of energy absorption accompanying the matter transformation. In this case $dP/dT < 0$ and so the effective Grüneisen coefficient $\Gamma = V(\partial P/\partial E)_V$ in the range of phase mixture is negative. A similar pattern was observed for nitrogen by other authors (Radousky et al. 1986). Thus, the comparison of calculation results with the experiment illustrates that on the assumption of an endothermal transformation (with $dP/dT < 0$) we succeeded to satisfactorily approximate the total set of experimental data. The attempt to describe the experiment using the alternative, exothermic, transformation variant (with $dP/dT > 0$) was not successful. In general, it is not improbable that this transformation is connected with the generation of some unknown structures and appears to be a phase transition in its classical form. However it is likely to be conditioned by chemical reactions proceeding in the matter. The additional internal energy imparted to the matter under shock compression is rather large (for points A and B , 3 and 7 kJ/g, respectively, i.e. not less than 300-700 kcal/mol) and in magnitude is comparable with the energy of chemical bonds in a molecule. In the case of chemical reactions under the phase 2, the set of their final products should be implied. The epoxy resin molecule is rather complicated and a priori it is unlikely to reliably name the final reaction products. The molecule of hardened epoxy resin can be presented by the formula $C_{41}H_{52}O_7N_2$ where the following chemical bonds exist: 27 bonds (C-C), 12 (C=C), 47 (C-H), 11 (C-O), 3 (O-H), 4 (C-N), 2 (N-H). For the final products two variants will be considered. In the first variant, in accordance with the approximate reactions recommended in (Stanyukovich 1975) for explosives (whose molecules incorporate the same elements) we assume that the molecule can be decomposed either to 41 atom of C (here and further carbon is solid), 7 molecules of H_2O, 19 molecules of H_2 and one of N_2, or to 34 atoms of C, 26 molecules of H_2, 7 of CO and one of N_2. In the second variant we assume that aromatic rings incorporated in the resin molecule do not decay and it decomposes into two diphenyl molecules $C_6H_5C_6H_5$, 10 atoms of C, 16 molecules of H_2, 7 of CO and one of N_2. After applying the Gess law (the thermal effect of the reactions depends only on initial and final states) and evaluating the thermal effect for decomposition reactions the results are the following:

In both cases of the first variant we have the heat evolution (1.93 and 0.63 kJ/g, respectively). This result is inconsistent with the previous conclusion about the endothermicity of the shock transformation. In contrast, in the second variant thermochemical calculations showed that energy absorption takes place (0.88 kJ/g). This result coincides in sign and in order of magnitude with the above-given energetic characteristics of the shock transformation. In the thermochemical calculations performed the average values for the energies of breaking the chemical bonds in the initial molecule, energies of bond breaking in the molecules of final products, as well as the energy of carbon sublimation are taken from Glushko (1978, 1979) and Stanyukovich (1975). The result obtained merely indicates the possibility of generating reaction products, as done in the second variant. However, decomposition of the epoxy resin molecule into other complexes of atoms is quite possible. It is believed that the resulting energetic characteristics of matter transformation under shock compression, as well as the equations of state for the phases, can be factors whose allowance will enable to limit the freedom in choosing among the possible decomposition products of the initial matter molecule.

References

Al'tschuler LV, Krupnikov KK, Brazhnik MI (1958) Dynamical compressibility of metals under pressure from 400000 to 4 million atmospheres. JETP 34, 4: 886, in Russian

Al'tschuler LV, Kormer SB, Brazhnik MI, Vladimirov LA, Speranskaya P, Funtikov AI (1960) Isentropic compressibility of aluminium, copper, lead, and iron under high pressure. JETP, 38, 4: 1061, in Russian

Glushko VP (Ch. ed.) (1978, part I; 1979, part II) Thermodynamical Properties of Individual Matters, Reference book. Nauka, Moscow

Landau LD , Lifshits EM (1976) Statistical Physics. Nauka, Moscow

Marsh SP (ed) (1980) LASL Shock Hugoniot Data. University of California Press, Berkeley Los Angeles London

Mc Queen RG, Marsh SP, Taylor JW, Fritz JN, Carter WJ (1970) The equation of state of solids from shock wave studies. In: Kinslow R (ed) High - Velocity Impact Phenomena. Academic Press, New York London, pp 293-417, 526-527

Radousky HB, Nellis WJ, Ross M, Hamilton DC, Mitchell AC (1986) Molecular dissociation and shock-induced cooling in fluid nitrogen at high densities and temperatures. Phys. Rev. Lett. 57, 19: 2419

Stanyukovich KP (ed) (1975) Physics of Explosions. Nauka, Moscow

Trunin RF, Dorohin V, Zhernokletov MV, Kuznetsov NF, Sychevskaya NV, Shutov VV (1991) Shock compression of organic matters. Collection: 5 All - Union Conference on Detonation, v. 2, Krasnoyarsk

Zel'dovich JaB, Raiser JuP (1966) Physics of Shock Waves and High-Temperature Hydrodynamical Phenomena. Nauka, Moscow

Quasi-Spherical Pulsed Loading Impact on Solid Metal Balls

N.P. Purygin and V.I. Buzanov
Institute of Technical Physics, P.O. Box 245, Chelyabinsk-70, 454070, Russia

Abstract. Solid balls made of steel (Steel 3), brass (L63), and aluminium alloy (D16), 60 and 40 mm in diameter were subjected to the explosion of spherical explosive charges with external diameter of 80 mm, initiated from the surface at 12 points uniformly located over the sphere, with an asynchronism not exceeding 10^{-7} sec. Using massive casing restriction of the scattering of explosion products the balls survived, taking the form of a quasi-dodecahedron. In the absence of a casing, the balls were destroyed. A determination was made for the sizes of cavities formed in the center of surviving balls by cutting these balls in two parts; the sizes of the cavities before ball destruction were estimated by the sizes of surviving fragments.

Key words: Spherical implosion, Metal balls

Among the one-dimensional types of loading, the action of a spherical shock wave on solid balls of hard materials (Bakhrah et al. 1974; Buzanov and Purygin 1992; Buzanov and Purygin 1993; Kozlov 1992) is of great interest. Under certain conditions these balls remain intact (Bakhrah et al. 1974; Buzanov and Purygin 1992; Kozlov 1992). However, the generation of spherical flow is a complicated problem, therefore in many cases a quasi-spherical motion (Buzanov and Purygin 1992; Buzanov and Purygin 1993) is used. The present work generalizes the results of the works (Buzanov and Purygin 1992; Buzanov and Purygin 1993) in which a quasi-spherical impulse loading of metal balls was investigated. Solid balls of steel (Steel 3), brass (L63) and aluminium alloy (D16), 60 mm and 40 mm in diameter, were subjected to the explosion of a spherical charge having an outside diameter of 80 mm and being initiated in twelve surface points equally spaced over the sphere, with a non-synchronism not exceeding 10^{-7} s. For the 40 mm dia. balls an explosive of the PBX-9404 type was used whereas for the 60 mm dia. balls an explosive of the composition B-type was taken. The quasi- spherical detonation wave generated a quasi-spherical shock wave in the solid ball placed into the charge.

Two regimes for loading the balls have been set up. In the first regime, the scatter of the explosion products of the charge was restrained by a massive case around the charge. In our event, due to the simplicity of initiation, we succeeded in increasing the case rigidity and in providing a slower decay in the pressure impulse on the balls surface as compared with Kozlov (1992).

As a result of explosive action, the ball was deformed whereupon it became a quasi-dodecahedron in which the centers of faces coincided with the projections of the initiation points. On the surfaces of the balls (especially on the brass ones) one can see the collision lines of the detonation waves as well as the secondary shock waves in the explosive products in the form of thin lines (their widths and depths ranged from 0.2 to 0.4 mm depending on material and sizes of the ball). The 40 mm dia. ball of aluminium alloy has been destroyed, and this destruction took place along the collision planes of the shock waves propagating from the initiation points. The edges of fragments turning towards the center have melted, there were ejections of melted metal on the plane faces of the fragments. In this experiment one of twelve initiation points has accidentally not operated. The part of the ball around this point remained compact. After experiments the balls were cut in two. In the center of each ball (apart from the 60 mm dia. steel ball) there was a cavity. The edges of the cavity in the 40 mm dia. brass ball had melted, and

Shock Waves @ Marseille III
Editors: R. Brun, L. Z. Dumitrescu

1/3 of the cavity has been filled with melt flown down into it. The form of the cavity in the steel balls is close to spherical. In the balls there are hair cracks having mainly a radial direction and oriented mainly to the vertices and centers of the quasi- dodecahedron faces. In the 40 mm dia. brass ball, at the radius of 11 mm, there are spindle-wise flaws 4 mm in length and 2.5 mm in diameter which are oriented to the vertices of the quasi-dodecahedron. This fact demonstrates that there is some tangential flow alongside with the main radial motion.

Some geometric characteristics of the balls which have been loaded are presented in Table 1. It should be noted that in our case the sizes of the cavities which have formed in the central region of the balls are essentially smaller, compared with Kozlov (1992). This is, to some degree, explained by slightly less energy being transferred to the ball, in the case of the explosive charge initiation in 12 points, when compared with initiation over the surface or in a large number of points. However, the use of a more rigid case which restrains more effectively the scatter of the explosion products as well as the balls surface is in our opinion the decisive effect.

In the second loading regime for the 60 mm balls the heavy case was not applied, therefore one was not succesful in conserving the balls to be unbroken. Brass ball fragments being found have a pentahedral form whereas the destruction took place along the planes of shock wave collisions. In the zone of material continuity (this zone is adjacent to the ball surface) the planes of shock wave collisions have a slight roughness (0.1 mm) but in the zone of spalls (this zone is turned to the center) multiple spall separations of the material into layers are observed. The boundary of continuity zones and spalls is indistinct.

For balls of steel and aluminium it is difficult to identify destruction along the planes of shock wave collisions. These balls were crushed into some fragments of different sizes. The surface parts of all the fragments having been found is the ball surface. The emergence of columnar spall structures inclined at different angles to the balls surfaces is characteristic of spalls. It is possibly connected with the fibres orientation of the rod from which the ball was made. In experiments this orientation was not taken into consideration.

Table 1.

Material of the ball	R_0	R_1	R_2	R_3	R_4
Steel 3	29.9	29.0	30.0	0.0	0.0
Brass L63	29.9	28.0	30.0	11.0	3.9
Alum. alloy D16	29.9	26.1	30.4	15.0	4.1
Steel 3	19.9	19.0	19.9	–	1.9
Brass L63	19.9	17.8	20.3	9.0	5.1
Alum.alloy D16	19.9	broken down into fragments			7.9*
Steel 3	31.95				10.95**

In Table 1 the following notations were used:

R_0 - is the initial radius of the ball; dimensions in mm;

R_1 - the distance from the quasi-dodecahedron center to the center of its face;

R_2 - maximum distance from the center to the surface of the quasi-dodecahedron;

R_3 - maximum radius of the cracked zone;

R_4 - radius of the cavity having been formed;

* - estimation according to the sizes of the conserved fragments;

** - data from the work of Kozlov (1992).

From the analysis of the fragment sizes it follows that there is approximately a constant maximum thickness (h) of the fragments along the normal to the ball surface. This causes us to infer that the destruction took place as a result of all-round extension of the ball and, prior to destruction, a cavity had formed in the ball. The main data on the ball destruction are given in Table 2.

Table 2.

Material of the ball	Max. size of fragment l (mm)	Thickness of fragment along normal to ball surface h (mm)	Estimation of cavity radius prior to ball destruction r (mm)
Steel 3	54	18	12
Brass L63	38	17	13
Alum. alloy D16	23	14	16

By comparing the data of Tables 1 and 2 for 60 mm dia. balls one may reveal the influence of the casing on the size of the cavity being formed in the ball. One should remember that when the casing is not used the cavity size was evaluated according to the size of the fragments. The results are presented in Table 3. The outside radius of the balls with a cavity (R) was calculated on the assumption of a constant ball volume. For experiments without the casing this radius relates to the moment of ball destruction. From Table 3 it is obvious that in the presence of the massive casing the deformation of the balls in case of their outwards movement is small (up to $\sim 0.1\%$) and that they do not undergo destruction, and when the casing is absent the ball deformation is signifcant (up to $\sim 5\%$) and results in their destruction.

The experimental data on deformation and destruction of the balls under a quasi-spherical impulse loading may be useful for designing devices used for creating high pulse pressures, especially when it is necessary to conserve the matter under study.

Table 3.

Material of the ball	Presence of casing	Radius of the ball prior to destruction R_0 (mm)	Radius of cavity after loading r (mm)	Radius of the ball after loading R (mm)	Deformation $(R - R_0)/R_0$ %
Steel 3	+	29.9	0	29.9	0
	–	29.9	12	30.5	2.1
Brass L63	+	29.9	3.9	29.92	0.07
	–	29.9	13	30.7	2.7
Alum. alloy D16	+	29.9	4.1	29.93	0.09
	–	29.9	16	31.4	4.9

Acknowledgement

The authors express their gratitude to B.V. Litvinov and E.A. Kozlov for useful discussions, and to S.S. Kamenskih, A.F. Tarasenko, K.N. Bykov and N.P.Grachev for their assistance in the experimentations.

References

Bakhrah SM, Kovalev NP, Nadykto BA, Novikov SA, Chernyshov AK (1974) Doklady Akad. Nauk SSSR (DAN) 215, 5 : 1090

Buzanov VI, Purygin NP (1992) In: Theses of the reports at 10th Symposium on combustion and explosion, Chernogolovka, pp 131-132

Buzanov VI, Purygin NP (1993) Destruction of metal balls under quasi-spherical impulse loading. Chemical Physics [In Russian] 12, 2 : 180

Kozlov EA (1992) Experimental check of E.I. Zababakhin's hypothesis concerning limitation of energy cumulation in a spherically converging shock-wave front in a medium with phase transition. In: SC Schmidt et al. (eds) Shock Compression of Condensed Matter, Elsevier Science Publishers B.V., pp. 169-171

X-Ray Line Broadening Analysis of Dynamic Loaded Aluminium Metal Matrix Composite

C. Patuelli
Dipartimento di Fisica, Universita' di Bologna, Consorzio INFN, Via Irnerio 46, I-40126, Bologna, Italy

Abstract. The crystallite size, the residual lattice strain, the components of Gaussian and Cauchy functions of the line profiles obtained by Fourier analysis of X-diffraction lines, are correlated for a better knowledge of mechanical properties of infiltrated composites with ceramic reinforcements of different volume fractions (27, 19, 10%) and types (SiC whiskers, Al_2O_3 fibers and particles) after fast and slow deformation. The results are discussed taking into account the role of the reinforcing phase on the generation and relocation of dislocations along different crystallographic directions (111,200,220).

Key words: Dynamic load, Metal matrix composite, X-ray diffraction

1. Introduction

The mechanical properties of composites mainly depend on the distribution of the reinforcement, the grain size and the dislocation density. The dislocations generated at the matrix-fiber interface after various deformations can drastically improve the strength and the ductility of this advanced material. In a previous paper (Patuelli et al. 1992) the mechanical behaviour of an infiltrated 23% SiC whisker-Al composite after dynamic loading was examined. The aim of this work is to study the effects of the different volume fractions (27, 19, 10%), types (SiC whiskers, Al_2O_3 fibers and particles) of the reinforcing phase on the dislocation generation mechanisms of an aluminum metal-matrix composite after dynamic loading and, for comparison, after slow compression.

2. Experimental

The material used in this work is a metal-matrix composite (MMC) obtained by aluminum alloy (A356) infiltration into a ceramic preform manufactured by S.I.R. with proprietary technology. The infiltration is performed with gas pressure equipment under conditions reported in another paper (Barbieri et al. 1991). Fig.1,a,b,c shows SEM images of 27% SiC whiskers, of 19% Al_2O_3 fibers and 10% Al_2O_3 fibers and particles preforms, respectively.

The dynamic loading is obtained by an explosively accelerated flying plate which compresses samples with the same thickness (5×10^{-3}m) but different impact areas ($25, 49 \times 10^{-6}$m^2) against a quasi-rigid boundary. For the type of the explosive used calculations indicate peak loads of 0.5-1.5 GPa (Carli et al. 1981). An average compression rate of 1.5×10^{-2}m/s is used for slow conventional deformation. The XRD line profiles were examined by Fourier analysis; more details on the X-ray analysis are reported in a previous paper (Patuelli et al. 1992).

3. Results and discussion

Al/SiC 27% whiskers

Let us examine Tables 1,2 and Fig.2: the high deformability after dynamic loading with two different impact areas and the small increase in hardness of the composite material reinforced with SiC whiskers are associated with a drastic reduction of the crystallite dimensions together with an increase of the residual lattice strains of nearly half the value of the slow deformation one. As a consequence the structure of the dynamically-loaded Al/27% SiC whisker is softer than the one after slow conventional deformation.

Shock Waves @ Marseille III
Editors: R. Brun, L. Z. Dumitrescu

Table 1. Microstructural parameters obtained by Fourier analysis

Type of deformation	Deformability %	Hardness H_v (kg/mm^2)	Variance σ (kg/mm^2)	Al lattice plane hkl	Crystallite dimensions D (mm)	Lattice distorsion ϵ (x10^{-3})	Dislocation densities ρ_ϵ (x10^{14}m^{-2})	ρ_D
			SiC 27% whiskers					
not				111	56	2.0	6.09	3.19
deformed		143.1	0.4	200	61	1.5	3.47	2.72
				220	60	0.9	1.15	2.82
slow				111	45	2.3	7.48	4.95
impact area	69	185.7	3.5	200	45	2.2	7.17	5.05
25x10^{-6}m^2				220	46	1.2	2.10	4.80
fast				111	38	2.1	6.59	6.75
impact area	80	149.7	2.2	200	39	2.0	5.76	6.60
25x10^{-6}m^2				220	40	1.1	1.73	6.21
slow				111	41	2.3	7.85	6.06
impact area	65	165.3	3.4	200	44	2.3	7.45	5.20
49x10^{-6}m^2				220	45	1.1	1.79	4.94
fast				111	36	1.9	5.22	7.80
impact area	74	164.2	2.1	200	36	2.0	6.12	7.42
49x10^{-6}m^2				220	43	1.3	2.54	5.34
			Al/Al_2O_3 19% fibers					
not				111	47	2.1	6.42	4.44
deformed		105.4	1.0	200	49	1.6	3.94	4.16
				220	57	0.9	1.31	3.04
slow				111	46	1.7	4.01	4.81
impact area	77	117.1	3.1	200	47	1.8	4.49	4.60
25x10^{-6}m^2				220	53	1.2	2.01	3.54
fast				111	50	2.1	6.36	4.07
impact area	85	100.3	2.2	200	53	1.6	3.90	3.51
25x10^{-6}m^2				220	56	1.1	1.82	3.14
slow				111	36	1.9	5.57	7.91
impact area	70	112.8	3.0	200	47	1.7	4.26	4.47
49x10^{-6}m^2				220	50	1.1	1.63	4.02
fast				111	50	2.0	5.81	3.98
impact area	75	93.1	1.9	200	51	1.5	3.33	3.78
49x10^{-6}m^2				220	59	1.2	1.97	2.86
			Al/Al_2O_3 10% fibers+particles					
not				111	48	1.6	3.78	4.26
deformed		120.7	0.8	200	45	1.8	4.82	4.98
				220	54	1.2	2.08	3.38
slow				111	41	1.8	4.62	5.80
impact area	75	104.1	3.4	200	46	2.2	6.97	4.75
25x10^{-6}m^2				220	44	1.2	2.18	5.10
fast				111	40	2.0	5.82	6.12
impact area	82	115.9	2.3	200	44	1.9	5.35	5.10
25x10^{-6}m^2				220	53	1.1	1.78	3.59
slow				111	48	1.8	4.82	4.38
impact area	67	116.6	3.2	200	44	1.7	4.26	5.27
49x10^{-6}m^2				220	44	1.2	2.14	5.14
fast				111	47	2.0	5.73	4.47
impact area	75	122.3	2.4	200	50	1.7	4.29	4.06
49x10^{-6}m^2				220	55	1.3	2.34	3.25

Table 2. Microstructural parameters obtained by Fourier analysis

Type of deformation	Lattice plane hkl	β_c ($\times 10^{-4}$nm)	β_g ($\times 10^{-4}$nm)	k	Crystal size D (nm)	Lattice distorsion ϵ ($\times 10^{-3}$)
Al/AlSiC 27% whiskers						
not deformed	111	1.58	1.20	0.74	56	2.0
	200	1.33	0.80	0.93	61	1.5
	220	1.14	0.42	1.52	60	0.9
slow impact area 25×10^{-6}m^2	111	2.11	1.36	0.88	45	2.3
	200	2.32	1.26	1.04	45	2.2
	220	1.83	0.50	2.06	46	1.2
fast impact area 25×10^{-6}m^2	111	2.6	1.79	0.89	38	2.1
	200	2.57	1.54	0.94	39	2.0
	220	2.11	0.85	1.40	40	1.1
slow impact area 49×10^{-6}m^2	111	2.57	1.86	0.78	41	2.3
	200	2.35	1.28	1.04	44	2.3
	220	1.83	0.50	2.06	45	1.1
fast impact area 49×10^{-6}m^2	111	2.90	2.04	0.80	36	1.9
	200	2.65	1.31	1.14	37	2.0
	220	2.32	1.22	1.07	43	1.3
Al/Al_2O_3 19% (fibers)						
not deformed	111	1.94	1.30	0.84	47	2.1
	200	1.78	1.24	0.81	49	1.6
	220	1.15	3.83	1.70	57	0.9
slow impact area 25×10^{-6}m^2	111	1.84	1.31	0.79	46	1.7
	200	1.87	1.07	0.99	47	1.8
	220	1.62	1.05	0.87	53	1.2
fast impact area 25×10^{-6}m^2	111	1.78	1.04	0.96	50	2.1
	200	1.65	1.03	0.90	53	1.6
	220	0.001	0.90	0.91	56	1.1
slow impact area 49×10^{-6}m^2	111	3.07	2.3	0.75	36	1.9
	200	1.94	1.53	0.71	47	1.7
	220	1.74	1.68	0.59	50	1.1
fast impact area 49×10^{-6}m^2	111	1.69	0.87	1.08	50	2.0
	200	1.64	0.001	0.74	51	1.5
	220	1.41	1.04	0.77	59	1.2
Al/Al_2O_3 10% (fibers+particles)						
not deformed	111	1.70	1.24	0.77	48	1.6
	200	1.95	0.99	1.10	45	1.8
	220	1.51	0.94	0.91	54	1.2
slow impact area 25×10^{-6}m^2	111	2.16	1.54	0.79	41	1.8
	200	2.26	1.36	0.94	46	2.2
	220	2.03	0.86	1.33	44	1.2
fast impact area 25×10^{-6}m^2	111	2.31	1.56	0.94	40	2.0
	200	2.15	1.33	0.92	44	1.9
	220	1.65	1.49	0.62	53	1.1
slow impact area 49×10^{-6}m^2	111	1.80	1.29	0.79	48	1.8
	200	1.90	0.90	1.19	44	1.7
	220	1.92	0.95	1.14	44	1.2
fast impact area 49×10^{-6}m^2	111	1.85	1.06	0.98	47	2.0
	200	1.82	1.29	0.80	50	1.7
	220	1.69	1.18	0.81	55	1.3

Such enhanced ductility is particularly important in the continuing quest for improving the mechanical properties of composite materials characterized by low values of ductility. The uniform distribution of the dislocations (Fig.2) along (111,200) directions after both impact areas, as results from the data points falling near the 45° line, confirm the positive effects of dynamic loading deformation.

Al/Al_2O_3 19% fibers

The dynamically-loaded composite shows high deformability (up to 85%) with no variation of hardness and crystallite sizes, Table 1; there is only a small increase in the lattice distorsions. After slow deformation there is, instead, a reduction of the crystallite sizes, more relevant with the impact area of $49 \times 10^{-6}m^2$, which increases the overall dislocation density. The dislocation distribution is not uniform after fast deformation, whereas it is uniform after slow deformation with the impact area of $49 \times 10^{-6}m^2$.

Al/Al_2O_3 10% fibers + particles

Finally, let us examine the composite material with a low volume fraction of alumina fibers and particles. The behaviour of dynamic loaded samples is just the same as seen in the specimen reinforced with SiC whiskers: a high deformability, a nearly negligible variation of hardness, and a reduction of the crystallite sizes associated with an increase of the lattice distorsion. The corresponding dislocation density is lower and much more uniformly distributed than the one after slow deformation (Fig.2). Such effect shows the critical role of the alumina particles when mixed with the fibers in low volume fractions in the dislocations generation mechanisms of MMCs deformed by very high strain rates.

The uniform distribution of dislocations measured along the three planes (111,200,220) is in agreement with the low and near equal values of profile characteristic, $k = \beta_c/\beta_g\pi^{1/2}$, after fast deformation with the two different impact areas. As known (Warren 1969) the Cauchy component, β_c, depends on the crystallite sizes and twinning faults, whereas the Gauss component, β_g, on the lattice distorsions.

Therefore the values of $k < 1$, measured along the three different crystallographic directions, point to an equilibrium between the two components of the diffraction line profile, that is another indirect verification of the good deformation homogeneity due to the positve effects of the high strain rate as well as of the presence of alumina particles.

4. Conclusions

From the comparison after dynamic loading and slow conventional compression of an aluminium composite material with different volume fractions and types of reinforcing phases it was found that the dynamically-loaded composite reinforced with SiC whiskers or with a mixture of alumina fibers and particles showed a good deformation behaviour, in terms of high deformability, low hardness values as well as of a nearly negligible increase of the residual lattice strains together with a reduction of crystallite sizes.

Acknowledgements

The author wishes to thank Dr. L. Valdinoci for the experimental work and S.I.R. for supplying the composite material.

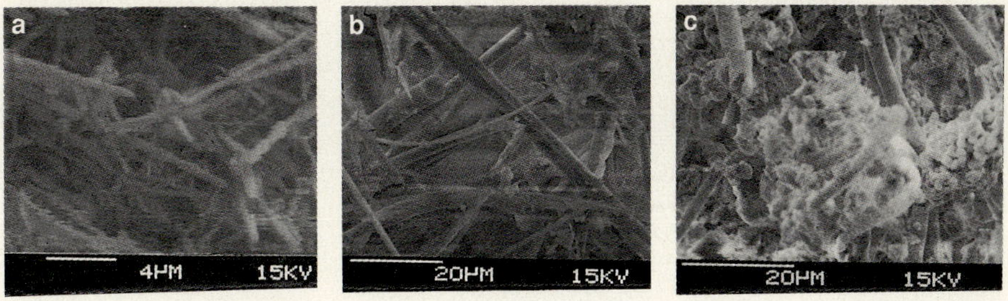

Fig. 1a,b,c. SEM images of ceramic preforms: a) 27% SiC whiskers; b) 19% Al_2O_3 fibers;c) 10% Al_2O_3 fibers and particles

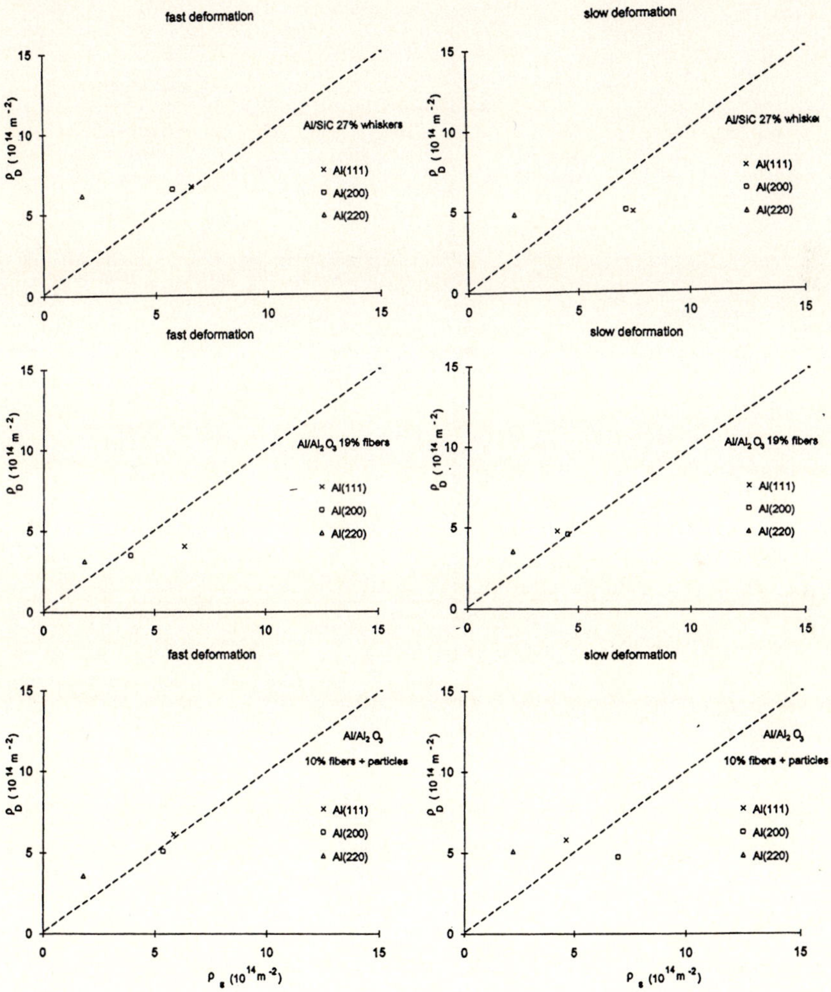

Fig. 2a. Correlation of dislocation density of deformed Al matrix after fast and slow deformation calculated from the crystallite size D ($\rho_D = 1/D^2$) and microstrain $< \epsilon >$ for a length parameter $L = 5$ nm ($\rho_\epsilon = 12 < \epsilon^2 > /b^2$); impact area 25 x 10^{-6}m^2

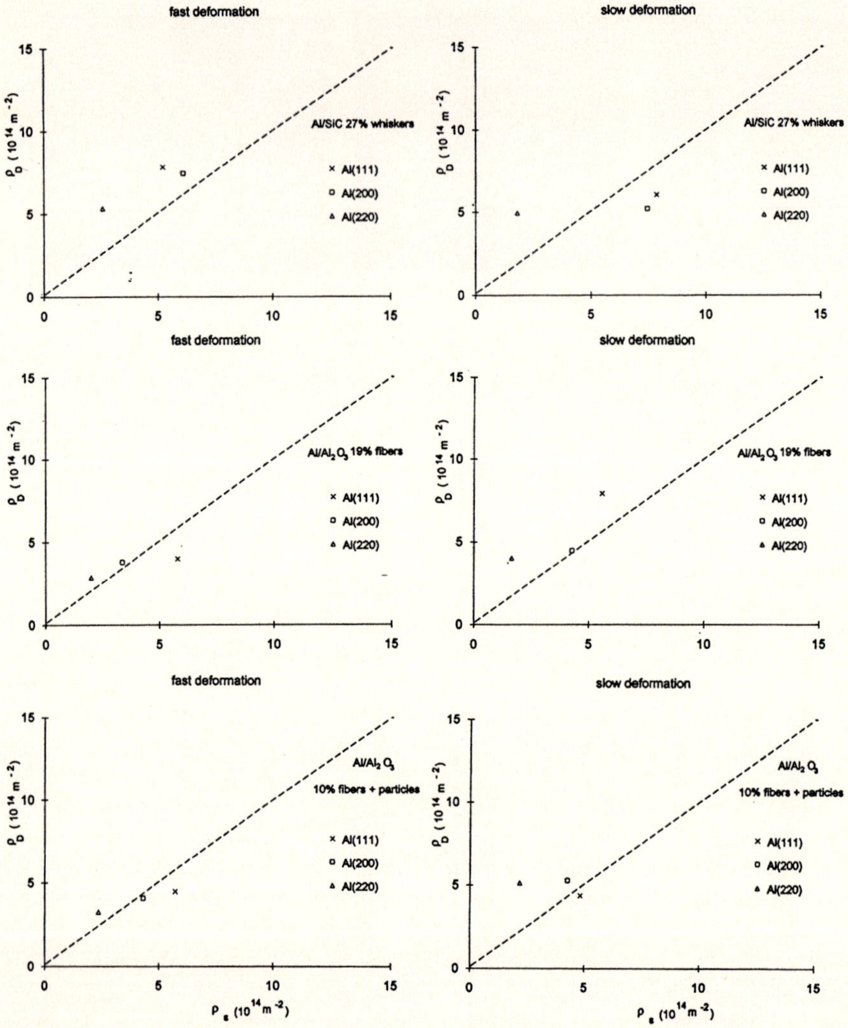

Fig. 2b. Correlation of dislocation density of deformed Al matrix after fast and slow deformation calculated from the crystallite size $D(\rho_D = 1/D^2)$ and microstrain $< \epsilon >$ for a length parameter L = 5nm($\rho_\epsilon = 12 < \epsilon^2 > /b^2$); impact area 49 x 10^{-6}m^2

References

Barbieri F, Giunchi G, Patuelli C, Zaffagnini M (1991) X-ray diffraction line broadening Fourier analysis of a metal matrix composite obtained by liquid infiltration. Mater. Sci and Eng. A135:199-202

Carli PS, Meyers MA (1981) Shock Waves and High Strain Rates Phenomena in Metals. Plenum Press, N.Y., pp 341-353

Patuelli C, Zaffagnini M(1992) X-ray diffraction line broadening Fourier analysis of a dynamic loaded metal matrix obtained by liquid infiltration. In: Takayama K (ed) Proc. 18th Intl. Symp. on Shock Waves. Springer, pp 387-392

Warren BE (1969) X-ray Diffraction. Addison Wesley, Reading Mass.

Application of Parallel Processors to Shock Wave Modelling

S.B. Bazarov, A.V. Shutov, O.Yu. Vorobiev, A.V. Zharov, V.E. Fortov and B.P. Krukov
High Energy Density Research Center, Izhorskaya 13/19, Moscow 127412, Russia

Abstract. In the present paper we demonstrate the application of parallel computation to high velocity impact problems. The base of the hydrocode is a 2D Godunov code on a moveable grid (Bushman et al. 1988). This code uses automatic subdivision of the computation domain into subdomains each time when it is required, for example, in case of great deformations of the matter. This feature of the code has not been used in the work, but such an organization of the computational algorithm is convenient also to the purposes of parallelization.

Key words: High-velocity impact, Moveable grid, Shock fitting, Shock waves

1. Introduction

The possibilities of numerical simulation in high energy density physics are limited firstly by difficulties in the theoretical description of thermodynamic and optical properties of matter in a wide range of pressures and temperatures and secondly by computational reasons. One such a reason is the nonlinear dependence of computation time on mesh size, which renders problematic accurate calculations of high energy density phenomena. While the difficulties associated with the complexibility of wide-range EOS can be overcome by using precomputed tables, computational difficulties require the development of efficient numerical methods and algorithms for parallel calculations.

2. Model and algorithm

2.1. Physical model

Condensed matter dynamics is treated by a hydrodynamic approach, as pressures considered substantially exceed the yield stress. Equations of motions are closed by wide-range semiempirical Equations of State (EOS), which describe the states of the medium both in condensed and gaseous phases (Bushman et al. 1989). To describe spallation in release waves a continuous kinetic model is used. This model is empirical and uses a few constants of the material obtained by comparison between experimental and numerical results (Fortov et al. 1991).

2.2. Numerical scheme

Equations of motion are integrated on a moveable grid according to Godunov's method (1959). The algorithm developed includes the following steps:

1. Riemann problem solution along the boundary;
2. Boundary displacement to the next time step;
3. Nodes distribution along the boundary curve approximated by arcs;
4. Correction of the location of the boundary nodes for better grid orthogonality;
5. Grid generation in subdomains and, if necessary, return to step 4;
6. Recalculation of conserved values to the next time step;
7. Mesh enrichment or coarsening, if necessary.

This algorithm allows to construct orthogonal grids adapted to the form of the domain. That is obtained by repeating steps 4-5 several times (usually 4 times is sufficient). One should not also worry about the change of the type of the boundary segments. A special searching algorithm provides automatically for that during the computation.

Shock Waves @ Marseille III
Editors: R. Brun, L. Z. Dumitrescu

2.3. Parallelization technique

Such an algorithm based on a moveable grid scheme can be efficiently implemented on a parallel processor by using known domain decomposition techniques. It corresponds to physical splitting of the hydrodynamic flow into a set of areas with different thermodynamic properties. The main cycle of computations - recalculation of conserved values (mass, momentum, and energy) - can be performed in parallel on a network of parallel processors (transputers) at every time step after determining boundary conditions for each subdomain. Then the procedure of matching boundary conditions should be executed by concurrently communicating processors through the transputers' interconnection links. Another important step - rebuilding and constructing curvilinear grids - can also be done in parallel, because the grid remapping algorithm is based upon the numerical solution of Poisson's equation. In our codes we achieve an efficiency of about 80% on parallel algorithms for the recalculation step and about 90% for the grid remapping step. This efficiency depends closely on the relation between the time of communications and calculation time. Consequently the size of the subdomain treated by one processor and the speed of link determines the efficiency of the algorithm.

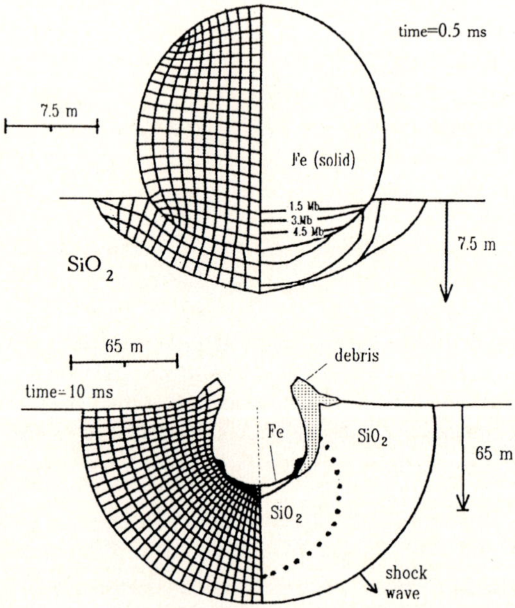

Fig. 1. Asteroid impact on rock, impact velocity 15 km/s

3. Simulation results

The first example concerns the problem of numerical simulation of the collision of an iron asteroid on rock. Mathematically the problem was formulated as a high velocity impact of a sphere onto a semi-infinite space. At the initial moment of collision an interaction takes place at the point of contact.

The surface of the contact increases in time. Shocks propagate into the earth and the asteroid until reaching the free surfaces. The motion law for the boundary inside the Earth should be described particularly. After consideration of the Riemann problem along the boundary one can compute the velocity of the front, which separates disturbed and undisturbed flow domains. This

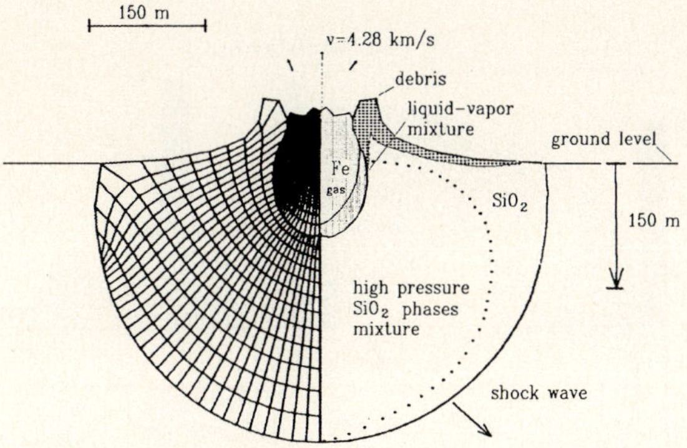

Fig. 2. Asteroid impact on rock, impact velocity 25 km/s, time 25 ms

velocity is equal to that of a corresponding shock or characteristic. Initially it coincides with the velocity of a characteristic, namely with the undisturbed sound speed. In time pressure waves overtake this boundary transforming it into a shock. Simulation results for an impact velocity of 15 km/s are presented in Fig.1 (a,b). The curvilinear grid is shown on the left side of the symmetry axes, and boundaries between the substances. Points designate the location of high pressure phases of SiO_2. The asteroid melts and fills the bottom of the crater. In case of a higher impact velocity - 25 km/s (Fig.2) vaporization of iron takes place.

One should note, that the problem considered above is characterized by the great difference in dimensions between the initial radius of the asteroid (10 m) and the size of the region where phase transformations occur (300 m). The procedure described allows to catch the shock and to perform calculations in the minimal region of influence.

The next example demonstrates other possibilities of the algorithm for the numerical simulations of penetration problems. In Fig.3(a,b,c,d) results of the collision of an iron projectile moving with a high velocity (20 km/s) into a two-layered spaced aluminum shield are presented. Initially the computational region consists of two subdomains. The first one corresponds to the projectile and the second belongs to the first layer of the shield (see Fig.3.a). Then these regions are divided into several subregions during the computation any time when they become deformed too heavily. For example, the region corresponding to the first aluminum screen is divided into subregions A, B, C and D at the moment $t = 4427$ ns (Fig.3.b). As the grid in every subregion is constructed independently, this allows to build rather uniform and orthogonal grids in spite of the strong deformations of the whole region.

When the jet generated after the collision of the projectile with the first layer reaches the position of the second layer, a new region (corresponding to the second layer) is inserted into the computational process. Note, that the projectile is highly porous at the moment of interaction with the second screen. Then the shock wave generated after collision with the screen compresses the matter of the projectile and porosity disappears (Fig.3.c,d). As the density in the projectile is greater at the edges, a nonuniform shock wave is generated in the second screen. It leads to nonuniform spallation of the second screen.

4. Conclusions

In conclusion one should note, that examples of calculations presented in this paper require not more than 2-4 hours of computer time for 24 transputers T800. Using an efficient moving grid code and a parallel algorithm allowed to obtain sufficient accuracy of calculations with rather moderate computer power.

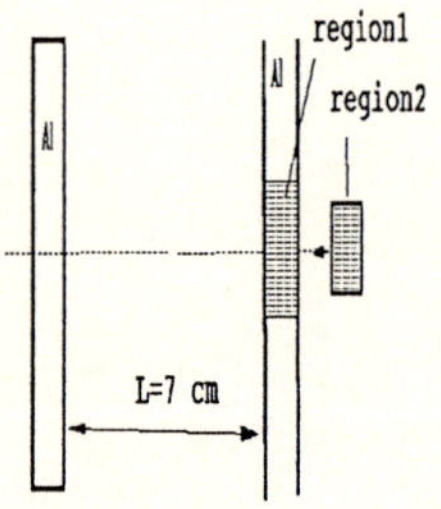

Fig.3a. Penetration of an iron projectile through a spaced aluminum target

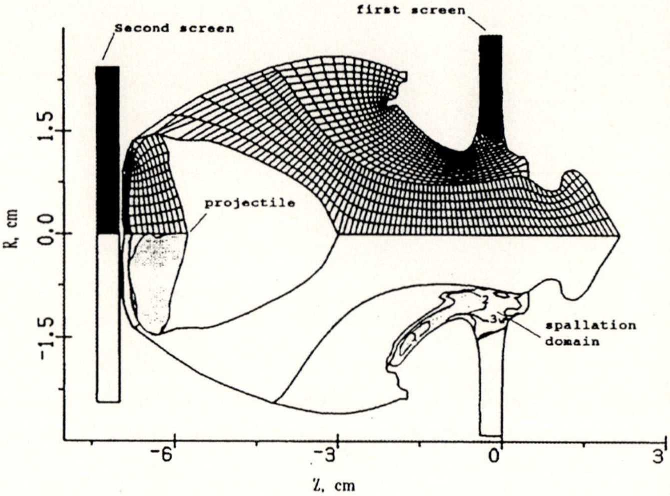

Fig.3b. Penetration of an iron projectile through a spaced aluminum target. Porosity levels:1-0.657, 2-0.438, 3-0.219

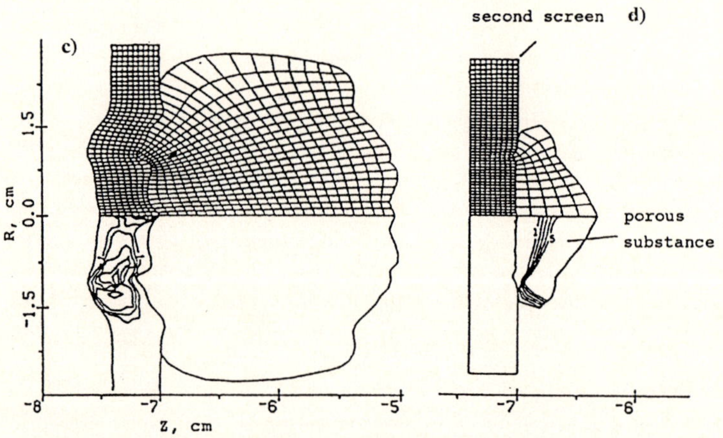

Fig. 3c,d. Projectile interaction with the second screen. c) time=4838 ns, porosity :1-0.06, 2-0.12, 3-0.18, 4-0.24, 5-0.3 d) time=6538 ns, porosity: 1-0.057, 2-0.114, 3-0.171, 4-0.228, 5-0.285

References

Bushman AV, Kanel GI, Ni AL, Fortov VE (1988) Thermal physics and dynamics of intense pulse action. Inst.of Chemical physics, Chernogolovka, Russia

Bushman AV, Fortov VE, Lomonosov IV (1989) Proc.Enrico Fermi School. Elsevier

Fortov VE, Kostin VV, Eliezer S (1991) J.Appl.Phys. 70, 8:4524-4531

Godunov SK (1959) Mat-Sb. 47:357

Part 4: Industrial Applications and Environment

Laser-Supported Absorption Waves in Radiation-Metal Interaction

C. Dumas, M. Ghrab, C. Prat and M. Autric
Institute of Fluid Mechanics - Laser-Matter Interaction Group. Parc Scientifique et Technologique de Luminy. 163 Avenue de Luminy, 13009 Marseille, France

Abstract. Many industrials applications require heat transfer studies between a CO_2 laser radiation and a solid material (machining, welding, cutting, surface treatment, shock hardening). With this type of interaction, at high intensity, a plasma occurs in front of the irradiated surface. This paper concerns the influence of laser beam intensity (intensity range : 3-120 MW/cm^2), the buffer gas (nitrogen, argon, helium at 10^5 Pa), and the ambient pressure (from 5.10^{-3} to 10^5 Pa in the case of nitrogen) on the hydrodynamic regimes governing the plasma expansion. Time-resolved argon laser interferometry was used to study the different dynamics of laser-produced plasma onto the sample surface (aluminium alloy (2017A), titanium alloy (TA6V) and steels (304L and 35NCD16)). The wave velocities were measured by means of a fast converter camera and a comparison was made to relate these data to different theoretical models.

Key words: Laser absorption wave, Plasma expansion, Aneurism, Blast wave, Detonation

1. Introduction

Laser-produced plasmas play an important role in the transfer of the laser energy in numerous applications such as machining, surface treatement, shock hardening and material removal. When a laser pulse irradiates a metallic target, we generally observe melting and vaporization of the solid target. At high irradiance, the vapour above the sample is ionized, leading to the formation of a plasma. With the type of laser pulse used in our experiments, the flowfield produced is divided into four regimes depending on the laser intensity and the ambient pressure. These regimes are discussed in the hydrodynamic study presented in this paper.

2. Experimental configuration

The experiments have been performed with a TEA CO_2 laser preionized by a cold-cathode electron gun functioning in gain-switched mode and able to deliver up to 1000 J in a 8 μs pulse duration. The laser power is constituted by a high-power spike 80 ns (FWHM) followed by a high-energy lower power plateau adjustable from 1.6 μs to 8 μs. The variation of the two ratios, spike intensity/average intensity and spike energy/total energy depends on the laser active medium pressure, pumping coefficient and the crowbar system. With our experimental configuration, we have $\Phi_{\text{spike}}/\Phi_{\text{average}} = 2.5$ and $E_{\text{spike}}/E_{\text{total}} = 0.10$ for 5 μs pulse duration. The delivered energy has been controlled by means of an absorption calorimeter and a photon drag. This allowed simultaneous observation of the temporal evolution and determination of the laser incident energy on the surface of the metallic sample. The total energy of the laser pulse has been varied with the aid of calibrated attenuators. The laser radiation has been focused with a spherical mirror of $f = 10$ m, into a vacuum chamber where the metallic samples have been placed. These samples have been irradiated with an average power density ranging from 3 to 120 MW/cm^2 (Spot area $\approx$0.4 cm^2). A Mach-Zehnder interferometer has been set up using a cw Argon laser at a wavelength of 5145 Å. The fringe shift induced by the variations of the refractive index related to the electron density and the wave propagation have been recorded by a fast converter camera (Imacon 790) allowing a framing rate up to 2.10^7 images/sec, and 1-100 ns/mm in the streak

Shock Waves @ Marseille III
Editors: R. Brun, L. Z. Dumitrescu

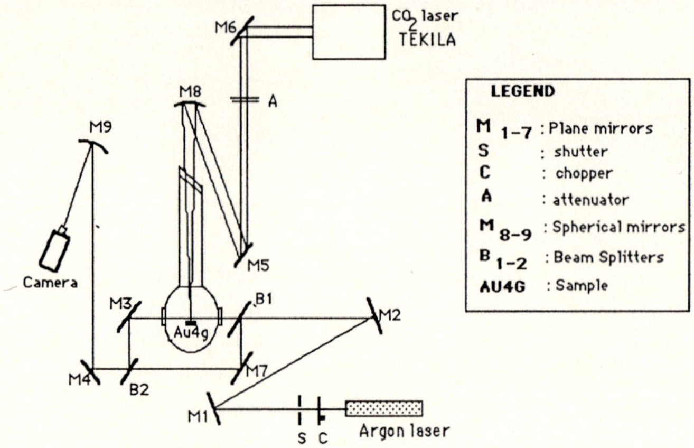

Fig. 1. Experimental set-up

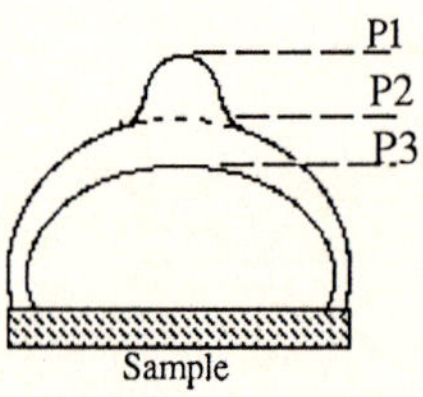

Fig. 2. Schematic of blast wave with aneurism formation

configuration. The plasma image has been formed onto the camera cell; in order to limit the cell exposition, both a shutter (2 ms) and a chopper (300 μs) have been used. The synchronization of these devices has been realized by means of a delay generator. The experimental arrangement is shown schematically in Fig.1.

3. Experimental results

Influence of the ambient pressure on the plasma expansion

For 5×10^{-3} Pa ambient pressure (nitrogen gas), the laser-produced plasma is close to the sample surface (aluminium sample). In this case, the absence of any aborption wave can be attributed to the reduced buffer gas density. This induces a permanent surface vaporization all along the duration of the interaction. For 10^2 Pa ambient pressure, we notice the ignition of two fronts, taking place in the target vapour. These fronts propagate with a velocity of 8 cm/μs. The first front expansion does not correspond to a spherical blast wave. Its evolution is proportional to (t^1) and not to $(t^{0.4})$ or $(t^{0.6})$ as predicted by Dabora (1972).

Complementary experiments were realized onto graphite 5890 samples, where we can easily observe regions of the shock front projecting well ahead of the neighboring spherical part. Stamper (1988) called them “aneurisms”. The axial expansion (P2) as well as the radial expansion (R) are proportional to time (t_1). This implies that the considered wave cannot be treated like a blast wave without taking into account aneurism-induced effects. In fact, this aneurism induces a hot air flow towards the target (Fig.5) involving a radial wave velocity increasing with respect to the classical blast wave velocity.

Influence of the ambient gas nature on the plasma expansion

During our experiments, we have observed a great influence of the ambient gas nature on the plasma expansion process. This influence can be explained by the breakdown threshold of the buffer gas ($I^*_{\mathrm{He}} = 110$ MW/cm^2, $I^*_{\mathrm{N}} = 19$ MW/cm^2, $I^*_{\mathrm{Ar}} = 7$ MW/cm^2).

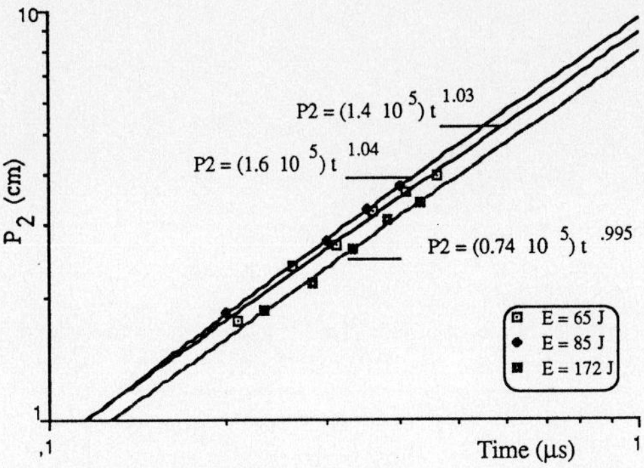

Fig. 3. Second front expansion at different incident energies $P = 10^2$ Pa (Graphite sample)

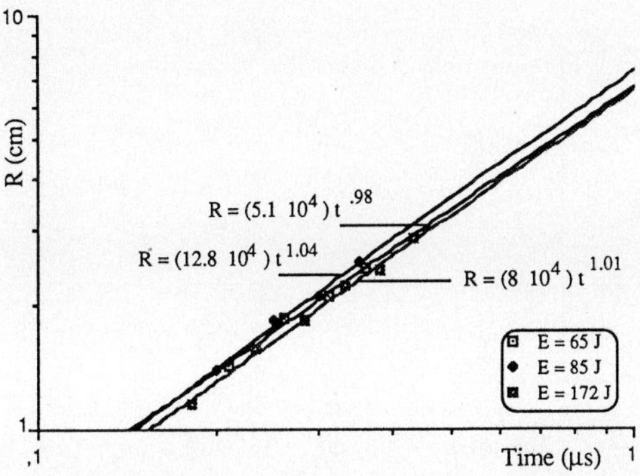

Fig. 4. Radial shock wave expansion at $P = 10^2$ Pa (Graphite sample)

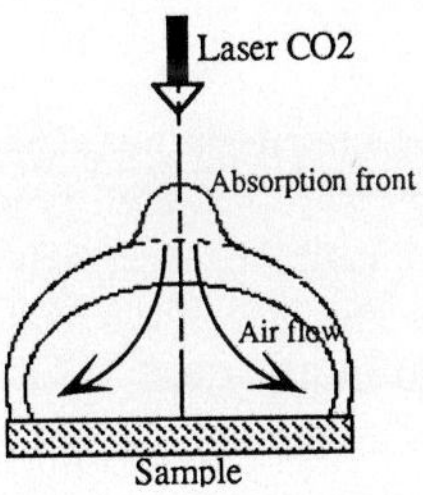

Fig. 5. Configuration of the air flow toward the target

In the case of irradiation in helium, the breakdown threshold is too high and the buffer gas cannot be ionized within our intensity range ($I \approx 3$ MW/cm^2). Then the plasma develops in the metal vapour. We have shown that the plasma ignition mechanism is associated with the presence of surface defects thermally uncoupled from the bulk material. So, the surface defects vaporize earlier and rough surfaces initiate dense and absorbing plasmas. The expansion of this kind of plasma can be described with a good agreement using the Boni-Su (1974) LSD model. On the contrary, polished surfaces induce a plasma partly absorbing, expanding like the LSC model developed by Pirri et al. (1978).

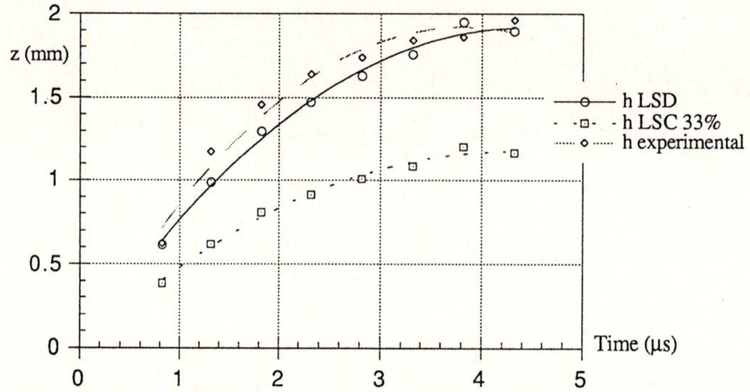

Fig. 6. Comparison between theoretical and experimental expansion (Stainless steel 304L sample with rough surface in helium $P = 10^5$ Pa)

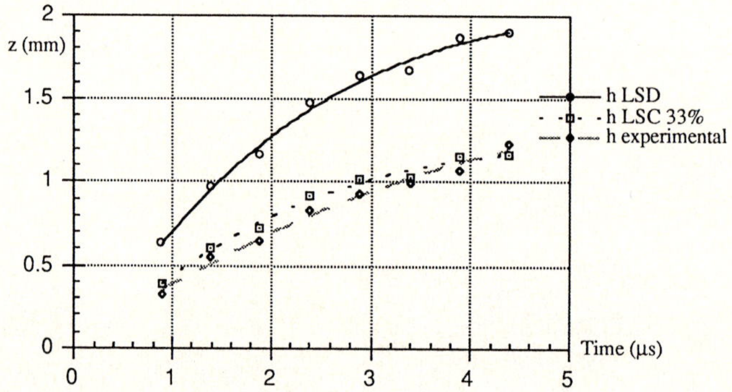

Fig. 7. Comparison between theoretical and experimental expansion (Stainless steel 304L sample with polished surface in helium $P = 10^5$ Pa)

In nitrogen, plasma ignition threshold is lowest and plasma develops in the vapour-gas mixture. This plasma is very dense and absorbing. In this case, the only absorption wave observed is a detonation one. On the first irradiations, we noticed an absorption front expanding more quickly than a detonation wave. Root (1989) has explained this process by a better efficiency of the heat conduction and/or radiation processes with respect to the heating induced by the shock wave. Hydrodynamic regimes observed in an argon atmosphere are almost the same than in nitrogen. Neverless, rough surfaces emit enough primary electrons to initiate breakdown in the ambient gas. The plasma appears to be separated in two parts: a semi-spherical one close to the target and on the top, along the laser beam axis, a conical one.

In our experiments, we have observed two changes of the hydrodynamic regime during the radial expansion of the plasma. The first one is a LSD-LSC transition, due to our spatial intensity repartition. The second one occurs when using helium as ambient gas (for which this expansion is larger). In this case, the plasma expands over the laser beam. Then, a part of the plasma is not supported by the laser energy and expands like a variable-energy blast wave (Dabora 1972). Furthermore, when the laser is off, all the waves expand as classical blast waves.

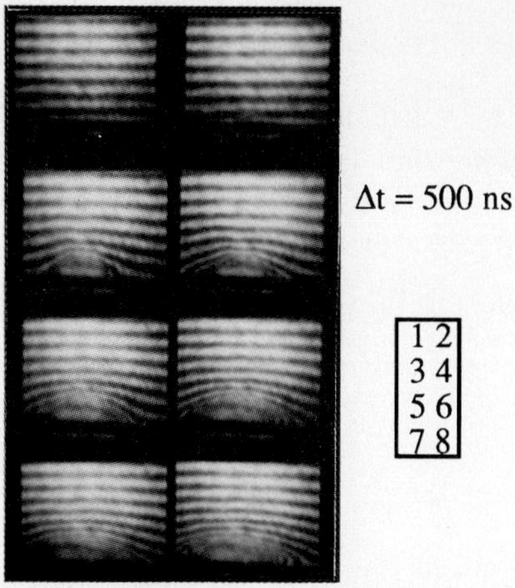

Fig. 8. Interferograms of plasma expansion (Aluminium sample $P = 5 \times 10^{-3}$ Pa)

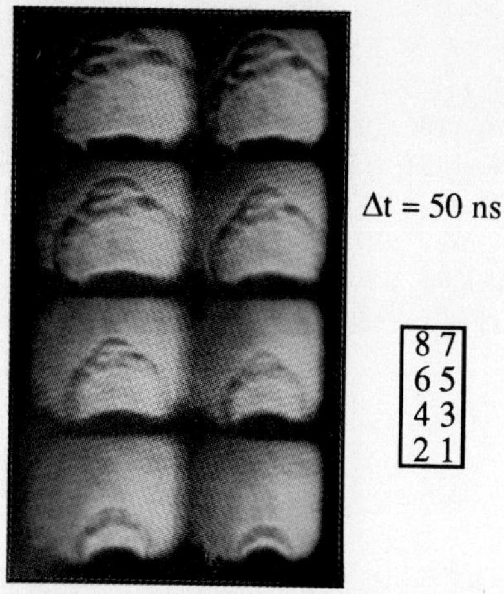

Fig. 9. Interferograms showing an axis aneurism growth (Graphite sample $P = 10^2$ Pa)

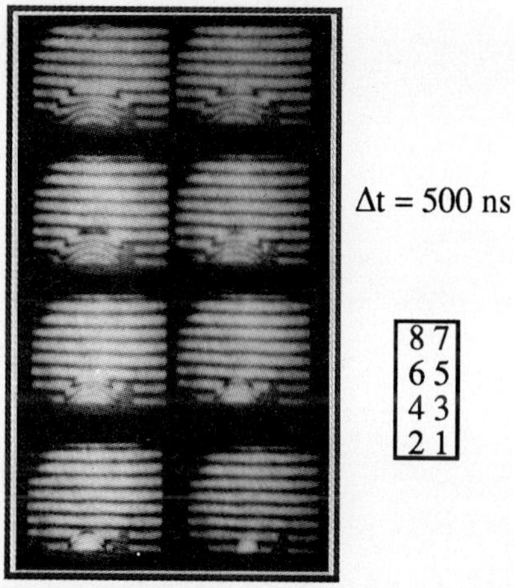

Fig. 10. Interferograms of laser supported detonation wave (stainless steel 304L with rough surface in helium $P = 10^5$ Pa)

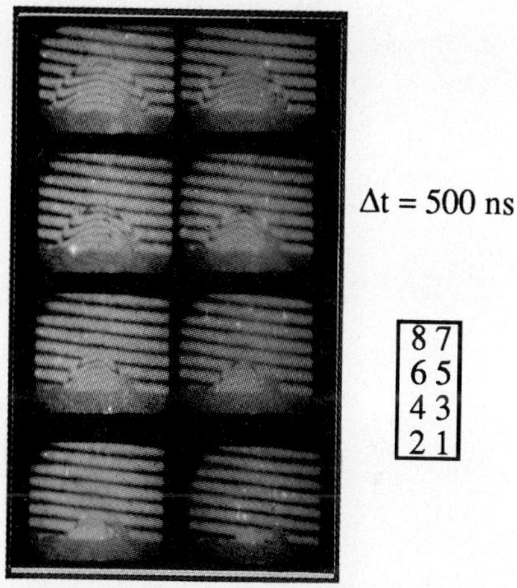

Fig. 11. Interferograms of laser supported combustion wave (stainless steel 304L with polished surface in helium $P = 10^5$ Pa)

Acknowledgement

This work was supported by the Eurolaser EU 194 Program "Powerful Lasers Industrial Applications", within the framework of the Groupement de Recherche GDR 911.

The authors thank Dr. G. Inglesakis for many helpful discussions, and J.P. Fragassi for his technical assisance throughout the course of the experiments.

References

Boni AA, Su FY (1974) An analytical technique for laser-driven shock waves. Acta Astronautica 1: 761-780

Dabora EK (1972) Variable energy blast waves. AIAA Journal 10, 10:1384-1386

Pirri AN, Root RG, Wu PKS (1978) Plasma energy transfer to metal surfaces irradiated by pulsed laser. AIAA Journal 16, 12: 1296-1304

Root RG (1989) In: Radziemski LJ, Cremers DA (eds) Laser induced plasma and applications, Dekker, NY, p 94

Stamper Ja (1988) Aneurisms in laser-driven blast waves. Phys. Fluids 31, 11: 3353-3361.

Shock Waves in the Cavity of a Xe-He Excimer Laser

S. Kosugi[†,‡], T. Ohishi[†], K. Maeno[‡] and H. Honma[‡]

[†]Heavy Apparatus Eng. Lab., Toshiba Corp., 20-1 Kansei-cho, Tsurumi-ku, Yokohama 230, Japan
[‡] Dept. of Mech. Eng., Fac. of Eng., Chiba Univ., 1-33 Yayoi-cho, Inage-ku, Chiba 263, Japan

Abstract. Shock waves, which are generated by pulse discharges in an excimer laser, cause arcing and nonhomogeneous excitation in the laser cavity. Experiments and numerical calculations are conducted to clarify the generation and propagation of shock waves in the cavity of an excimer laser. The shock waves are visualized by using a CCD color Schlieren method. Numerical calculations using Yee's symmetric TVD scheme are carried out for conditions which correspond to the experiments. The initial conditions of the calculations are determined from measured results with the laser Schlieren method, and from Schlieren photographs. Color Schlieren photographs are constructed from the numerical results. The numerical Schlieren photographs are found to be in good agreements with the experimental Schlieren photographs.

Key words: Shock wave, Flow visualization, Color Schlieren, Electric discharge, Excimer laser

1. Introduction

The repetition rate of existing high-repetitive excimer lasers is limited to 4 kHz (Goto et al. 1992). Shock or acoustic waves, which are caused by the periodic pulse discharges, limit the repetition rate (Baranov et al. 1983, Truong et al. 1992). Such waves break the homogeneous distribution of gas properties in the discharge region of the excimer laser. To clarify this problem, we have developed an excimer laser chamber with windows for flow visualization. Some results have been obtained by using this apparatus and a CCD Schlieren technique (Kosugi et al. 1991, 1992).

To simulate the flowfield of the discharge region, symmetric TVD calculations are conducted, under perfect gas assumptions. The essential problem for the numerical simulation of the flowfield in the laser cavity is characterized by the formation of narrow heated regions in the cavity, which act like the high pressure chambers in shock tubes, to generate shock waves. The main difference between the high pressure chamber of a shock tube and the heated region of a laser cavity is in their volumes. The volume of the latter region is much smaller than that of the former chamber. For example, the shock waves are generated from the surfaces of the main electrodes. The thickness of the heated layer is estimated to be about 0.2 mm from our Schlieren photographs. To test the reliability of the scheme, a series of one-dimensional TVD calculations were carried out changing the grid size (Kosugi et al. 1992). On the basis of the results of the test calculations, the two-dimensional symmetric TVD calculations are carried out with 300×100 nodes for the quarter physical region of the discharge cavity. The calculated results are exibited as color Schlieren photographs. A comparison between experimental and numerical results is discussed.

2. Experimental apparatus

Fig.1 indicates the cross section of the excimer laser cavity. The main electrodes consist of the upper anode and the lower cathode of about 300 mm length. The pin electrodes for preionization are located on both sides of the main electrodes. This discharge cavity is installed in a chamber with windows for visualization. Fig.2 shows the electric circuit consisting of a gap switch (G.S.), 20.7 nF main capacitor (Cs), 11.9 nF peaking capacitor (Cp), and 126 μH coil (L) to charge the main capacitor Cs. In our series of experiment, the DC supply voltage is fixed to be 20 kV. The energy E stored in the main capacitor Cs is 4.14 J.

Shock Waves @ Marseille III
Editors: R. Brun, L. Z. Dumitrescu

Fig.3 shows the experimental setup for Schlieren photography. A color striped filter and a stroboscope with xenon flash lamp are utilized for the color Schlieren technique. The xenon flash lamp is SUGAWARA NP-1A with a short flashing period of 75 ns. The Schlieren photograph is taken by a color CCD camera (TOSHIBA IK-T30). The stroboscope can be controlled to be flashed at any μs after the discharge. When the stroboscope is flashed, one Schlieren photograph is caught on the CCD camera and the image is recorded in the digital memory. Finally the image is preserved by using a VCR. The interval between the discharge and the flashing of the stroboscope is controlled by a digital delay unit (SUGAWARA RE-306). The interval can be changed from 1 μs to 1 second.

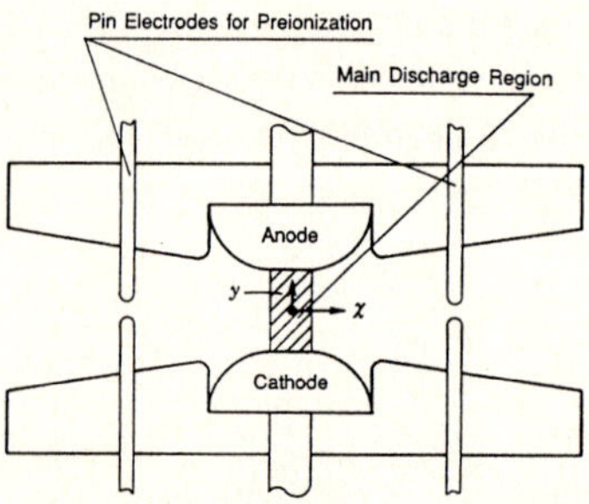

Fig. 1. Cross- section of discharge cavity of excimer laser

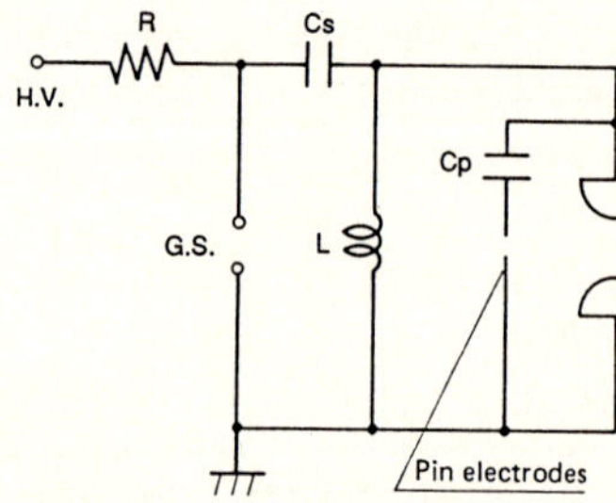

Fig. 2. Electric circuit diagram

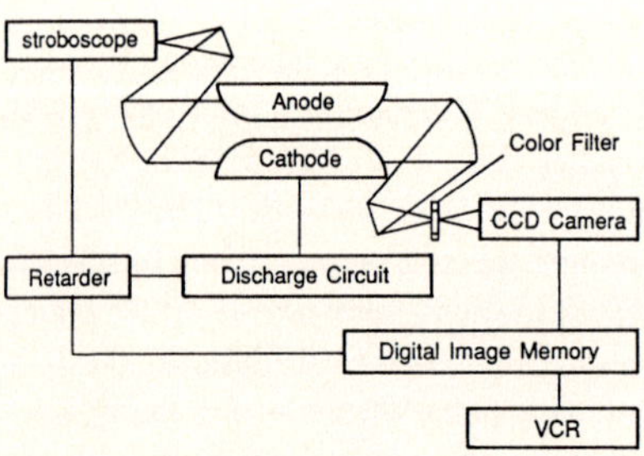

Fig. 3. Experimental set-up for Schlieren photography

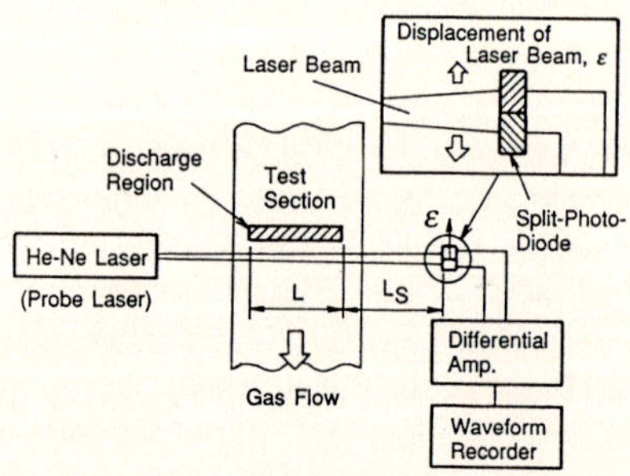

Fig. 4. Schematic diagram of the laser Schlieren method to measure the temperature distribution in the cavity of the excimer laser

3. Numerical calculation

Two-dimensional Euler equations in general coordinates are solved using Yee's symmetric TVD scheme (Yee 1986). To adapt the one-dimensional Yee's symmetric TVD method to the two-dimensional system of equations, the time-splitting method is used (Young and Yee 1987). The formulations given by Pulliam and Steger (1985) are used for the eigenvalues, the eigenvector matrices and their inverses.

In order to determine the initial condition, the temperature distributions in the main discharge region are measured by a laser Schlieren method. The method and the results will be discussed in another paper (Kosugi et al. 1993). Here we introduce the method briefly. Fig.4 shows the schematic diagram of the laser Schlieren method (Kiefer and Hajduk 1992, Koreeda et al. 1993). A He-Ne laser beam passes through the test section. The laser beam is set downstream nearby the discharge region. The gas flows at a low speed of about 10 m/s in the discharge region. The gas in the discharge region is heated by the pulse discharge instantaneously. Before the pulse discharge, the gas density is almost constant around the discharge region. Immediately after the discharge the heated gas starts to expand. The expansion of the heated gas continues only several tens of microseconds. When the gas in the discharge region reaches the He-Ne laser beam, the expansion has finished. The expanded gas crosses the He-Ne laser beam. The laser beam refracts according to the density change at the laser beam position. The refraction angle of the He-Ne laser beam is proportional to the gas density gradient which is perpendicular to the beam. In our experiment the refraction angle of the He-Ne laser beam is measured by using a split photo-diode. The index of refraction of the gas is proportional to the gas density as

$$n = 1 + K\rho, \tag{1}$$

where K is the Gladstone-Dale constant. When a density gradient is present, the laser beam is curved towards the higher density direction. The laser spot on the split photo-diode moves according to the density gradient. The density change of the gas is calculated from the change of the laser spot position by using the above equation and some geometrical relations. The density distribution is calculated from the time history of the density change, assuming constant gas flow speed. By the assumption of isentropic change in the expansion, the temperature distribution is determined from the density distribution.

By the laser Schlieren method, the temperature distribution in the x direction (Fig.1) is determined. Fig.5 indicates the temperature distribution in the case of the mixture of 1000 Torr helium and 5 Torr xenon. The voltage to maintain the glow discharge between the upper and lower main electrodes is almost constant regardless of discharge current. It is considered that the temperature distribution in the main discharge region is similar to the local current distribution. We assume that the temperature distribution is homogeneous in the y direction (Fig.1), because the current distribution is homogeneous in the y direction. The energy consumption in the main discharge region is estimated to be 2.9 J from the temperature distribution in Fig.5. The energy storage in the main capacitor is 4.14 J. The difference in energy can be attributed to the energy consumption in the sparking pin gaps and the main electrode surfaces. The energy consumed in the sparking pin gaps is estimated at about 0.5 J from the discharge characteristics. We assume that the remainder is consumed to heat the thin gas layers on the main electrodes. Because the voltage drops at the cathode and anode are constant regardless of the discharge current, the local temperature rise in the thin layers on the electrodes is proportional to the local current between the main electrodes. The temperature distributions of the thin layers are similar to the temperature distribution in the main discharge region in the x direction. The thickness of the thin layer is estimated to be 0.2 mm from the Schlieren photograph. In order to improve the efficiency of the calculation, the thickness of the layer is assumed to be 0.5 mm in the calculation. We assume that the same energy, 0.25 J, is consumed to heat the gas in the thin layers on the anode and the cathode. The flowfield is assumed to be symmetric with respect to the center of the main discharge region. The calculations are conducted for the quarter section of the discharge cavity. An orthogonal grid with 300×100 nodes is used for calculation.

The Schlieren photographs are obtained from the calculated density distributions by the color filter function corresponding to the filter in the experiments. Fig.6 is a schematic of the color striped filter used in the experiments. The filter is made of a color positive film. There are 5

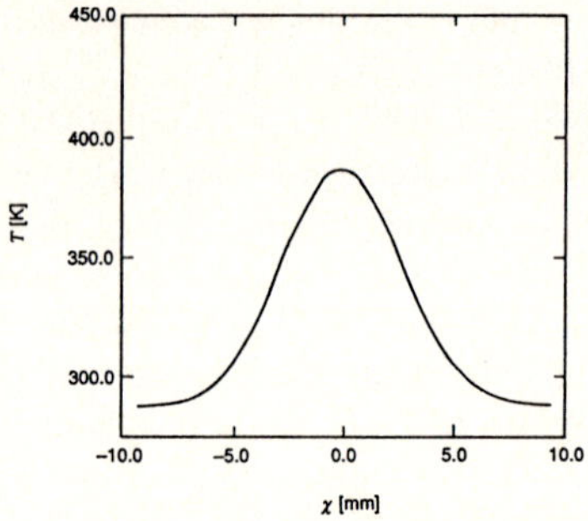

Fig. 5. Temperature distribution in the main discharge region in the case of a mixture of 1000 Torr helium and 5 Torr xenon

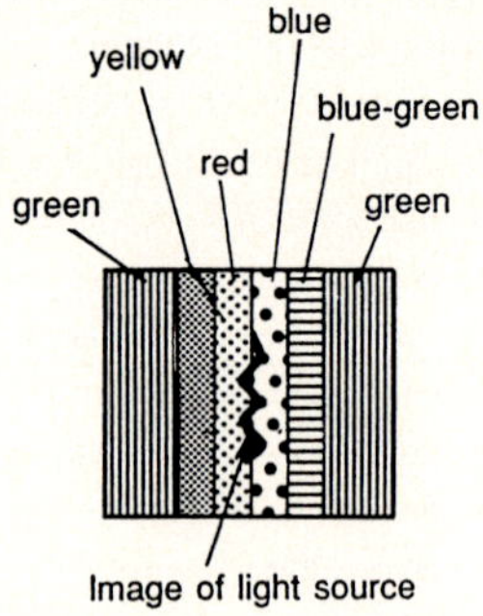

Fig. 6. Color striped filter

colors on the filter in rows. The image of the light source is focused on the filter. When no density disturbance exists, the image of the light source is focused on the boundary between the red and blue parts as shown in the Fig.6. In such cases, the Schlieren picture is violet. When a density gradient is present, the light is curved to the higher density direction and the image on the filter moves to the higher density direction. The light, which passes the point where the density gradient is present, changes the position of the image on the color Schlieren filter. The picture of the point is colored with the color where the image is focused. The displacement of the image on the color striped filter is calculated by using Eq.1 and some geometrical relations. The density gradient is assumed to be constant along the light path in the calculation.

4. Results and discussion

Figs.7 and 8 are typical Schlieren photographs taken at 17 μs after the discharge in the case of a mixture of 1000 Torr helium and 5 Torr xenon. Fig.7 is taken by the vertically striped filter, and Fig.8 is taken by the horizontally striped filter. In Fig.7, the horizontal density gradient $\partial\rho/\partial x$ is visualized by the vertically striped filter. Similarly, in Fig.8, the vertical density gradient $\partial\rho/\partial y$ is visualized by the horizontally striped filter. In these photographs, the horizontally propagating shock waves from the boundaries of the main discharge region, the shock waves from the spark pin gaps and from the surfaces of the main electrodes are observed. In Fig.7, the shock fronts of the shock waves from the boundaries are exibited as vertical and straight lines. The shock wave speed is nearly constant in the y direction (or in the discharge current direction). The temperature distribution in the main discharge region is also estimated to be nearly constant in the y direction.

Figs.9 and 10 show the calculated Schlieren photographs at 17 μs after the discharge for the same condition as those for Figs.7 and 8. Fig.9 shows the calculated result for the horizontal density gradient $d\rho/dx$, and it corresponds to the Schlieren photograph taken by the vertically striped filter. Fig.10 is calculated for the vertical density gradient $d\rho/dy$, and it corresponds to the Schlieren photograph taken by the horizontally striped filter. In Fig.9, the shock waves from the boundaries of the main discharge region are exibited by the same straight and vertical lines as shown in the experimental photographs. In both photographs by the calculation and the experiment, shock waves from the electrode surfaces are observed in the region between two shock waves caused by the main discharge. The calculated Schlieren photographs show a similar pattern to the experimental Schlieren photographs. This similarity also suggests that the initial conditions for the calculations are basically adequate to the present problem.

Hatanaka et al. (1989) measured the density fluctuation caused by a pulse discharge with a Michelson interferometer. They reported that the arcing was caused by density fluctuation for $\Delta\rho/\rho \geq 0.018$. Fig.11 shows the calculated density fluctuation at the center of the main discharge region. The solid line is the calculated result in the $x-y$ plane of Fig.1. The broken line is the calculated result in the $x-z$ plane (the z axis is perpendicular to the x and y axes). In the $x-z$ plane, it is assumed that the cavity length is 400 mm and the discharge length is 300 mm. This assumption corresponds to our experimental apparatus. The calculated density fluctuations in the $x-y$ plane and in the $x-z$ plane decrease to a level of $\Delta\rho/\rho \leq 0.018$ at 200 μs which corresponds to the period at 5 kHz operation. In these calculations, it is assumed that the gas pressure is constant on both sides of the laser cavity. So, if the gas pressure fluctuation is negligibly small outside the cavity, 5 kHz operation with a energy deposition higher than 70 J/L (0.27×10^6 J/Kg) could be achieved.

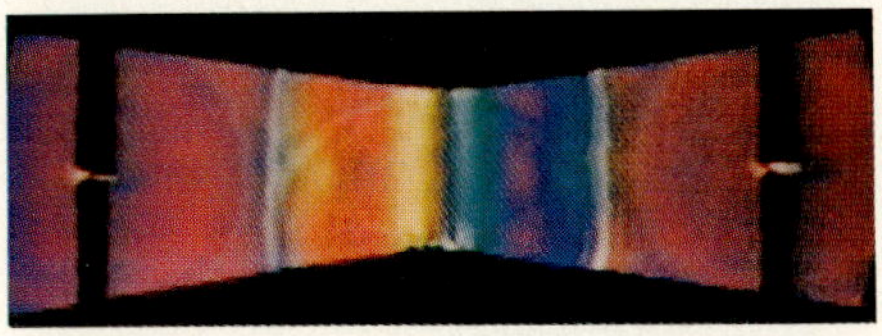

Fig. 7. Typical Schlieren photograph taken at 17 μs after the discharge with the vertically striped filter (mixture of 1000 Torr helium and 5 Torr xenon)

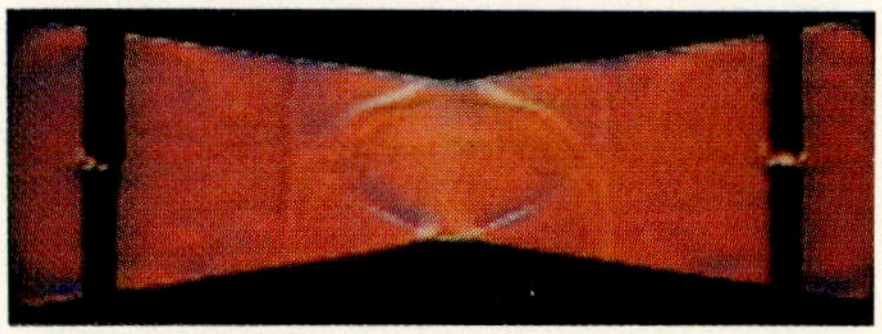

Fig. 8. Typical Schlieren photograph taken at 17μs after the discharge with the horizontally striped filter (mixture of 1000 Torr helium and 5 Torr xenon)

Fig. 9. Calculated Schlieren photograph at 17μs after discharge by the vertically striped filter (mixture of 1000 Torr helium and 5 Torr xenon)

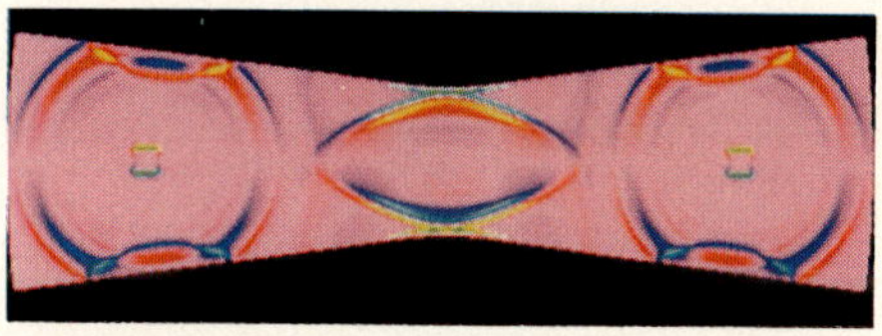

Fig. 10. Calculated Schlieren photograph at 17μs after discharge by the horizontally striped filter (mixture of 1000 Torr helium and 5 Torr xenon)

5. Concluding remarks

The shock waves are visualized by a CCD color Schlieren method with a color striped filter. Numerical calculations using Yee's symmetric TVD scheme are carried out for conditions which correspond to the experiments. The initial conditions of the calculations are determined from the measured results with the laser Schlieren method, and from the Schlieren photographs. On the basis of some assumptions, the calculations are conducted on an orthogonal grid with 300×100 nodes. Color Schlieren photographs are calculated from the numerical results. The numerical Schlieren photographs are found to be in good agreement with the experimental Schlieren photographs.

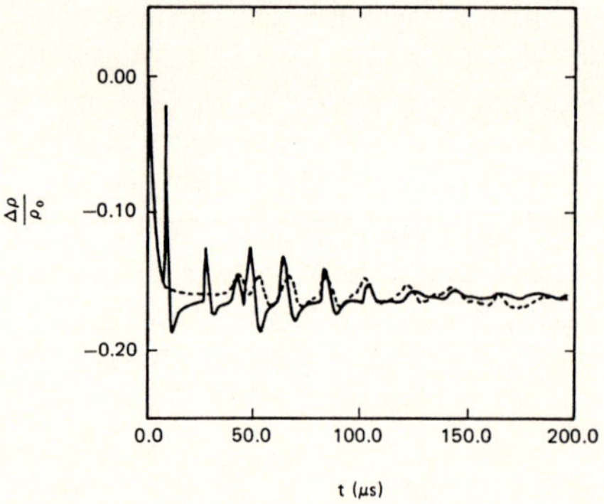

Fig. 11. Calculated density fluctuation at the center of the main discharge region (solid line: $x-y$ plane, dashed line: $x-z$ plane)

Acknowledgements

Part of this work was conducted within the program: 'Advanced Material-Processing and Machining System', consigned to the Advanced Material-Processing and Machining Technology Research Association from the New Energy and Industrial Technology Development Organization, which is carried out under the Industrial Science and Technology Frontier enforced by the Agency of Industrial Science and Technology, the Ministry of International Trade and Industry, Japan.

References

Baranov VYu et al. (1983) Characteristics of the pulse-periodic regime of excimer lasers. Sov. J. Quantum Electron. 13: 318-322

Goto T et al. (1992) Output characteristics of ultra-high repetition rate and high power XeCl excimer laser. In: Proc. 9th Intl. Symp. Gas Flow and Chemical Lasers, SPIE Vol. 1810, pp 376-381

Hatanaka H et al. (1989) Turbulence of the gas flow in the discharge region of a high repetition-rated TEA-CO_2 laser (in Japanese). Laser Science Progress Report of IPCR, Vol.11, pp 55-58

Kiefer JH, Hajduk JC (1980) Rate measurement in shock waves with the laser-Schlieren technique. In: Lifshitz A, Rom J (eds) Proc. 12th Intl. Symp. on Shock Tubes and Waves, Jerusalem, pp 97-107

Koreeda J et al. (1993) Front structures of strong shock waves in air. These Proceedings

Kosugi S et al. (1991) Experimental study on shock waves in the cavity of excimer laser. In: Takayama K (ed) Proc. 18th Intl. Symp. on Shock Waves, Vol.2, pp 1295-1299

Kosugi S et al. (1992) Effects of xenon gas on generation and propagation in the cavity of excimer laser. In: Proc. 9th Intl. Symp. Gas Flow and Chemical Lasers, SPIE Vol. 1810, pp 217-220

Kosugi S et al. (1993) Measurement of gas temperature profile in the discharge region of excimer laser with laser Schlieren method. Jap. J. Appl. Phys., to be published

Pulliam TH, Steger JL (1985) Recent improvement in efficiency accuracy, and convergence for implicit approximate factorization algorithms. AIAA Paper 85-0360

Truong JP et al. (1992) Efficient acoustic wave damping in a high pulse repetition rate XeCl laser. In: Proc. 9th Intl. Symp. Gas Flow and Chemical Lasers, SPIE Vol.1810, pp 430-434

Yee HC (1986) Numerical experiments with a symmetric high-resolution shock capturing scheme. NASA TM 88325

Aerothermodynamic and Thermal Phenomena in a High Pulse-Repetition Rate XeCl Laser

J.P. Truong, O. Uteza, M.L. Sentis, P. Delaporte, B. Forestier and B. Fontaine
Institut de Mécanique des Fluides de Marseille, UM 34 CNRS, 13003 Marseille, France

Abstract. Previous studies (Borisov et al. 1987; Sentis et al. 1989; Canarelli 1989) have shown that for high pulse repetition rate XeCl lasers ($\geq$ 300 Hz for an energy deposition of $\sim$ 50 J/l.b), the laser average power generally decreases. This decrease is mainly due to density fluctuations in the active medium, electrode material, composition of the gas mixture, flow characteristics, dust contamination level, preionization level, etc. (Baranov et al. 1984; Taylor 1986). In this paper, a new approach of the study of acoustic phenomena shows that one of the main processes of shock wave production is related to current over-densities and electric arcs occurring during the energy deposition in the laser gas mixture. A theoretical study of electrode temperature increase is also developed. It shows that thermal phenomena (electrode heating) leads to a drop of the output laser energy.

Key words: Excimer lasers, Shock wave, Electrical discharge

1. Main shock wave sources in excimer lasers

The aim of the present experimental study consists in establishing a correlation between the energy deposition by the electric discharge and recorded pressure histories which correspond to density fluctuations inside the laser cavity. In order to achieve this objective a time-resolved study of the discharge plasma has been done. This work was made on the test-bed named LUX (Sentis et al. 1991).

1.1. Discharge plasma evolution

Previous studies (Kosugi et al. 1991; Sentis et al. 1989) have shown the existence of three types of pressure waves generated by the input energy, which can be differentiated by the direction in which they propagate. Longitudinal waves propagate in the upstream and downstream directions, while transverse waves travel between the electrodes and between the optics. In the present work, our interest was focused on the influence of the discharge homogeneity on pressure wave generation. The evolution of the discharge plasma was visualised by taking series of nanosecond photographs (exposure time = 10 ns) with an ultra fast CCD camera (PCO model Dicam 2). Fig.1 shows four chronological transverse pictures of the discharge plasma, and the time evolution of discharge voltage and laser power pulse. The first photography (1) corresponds to the beginning of the discharge breakdown. The plasma is heating up progressively by the increase of the discharge current as the transverse discharge light emission is becoming more luminescent. At the time of the steady-state voltage (picture 2), the main discharge is homogeneous, and the maximum value of laser power is reached. But at the time of the end of the first voltage oscillation (picture 3) which corresponds to the beginning of the post-discharge, over-densities of current appear and predominantly persist for long time ($\sim$ 1 μs, pictures 3 and 4) at the ends of the electrodes. Such development of discharge instabilities has been studied in detail by Taylor (1986).

1.2. Correlation between the spatial distribution of the input energy and shock waves

In this study, the pressure fluctuations were recorded by a fast pressure probe (PCB 2222A) which allows to follow the propagation of shock waves reflected by the laser cavity walls. The spatial

Shock Waves @ Marseille III
Editors: R. Brun, L. Z. Dumitrescu

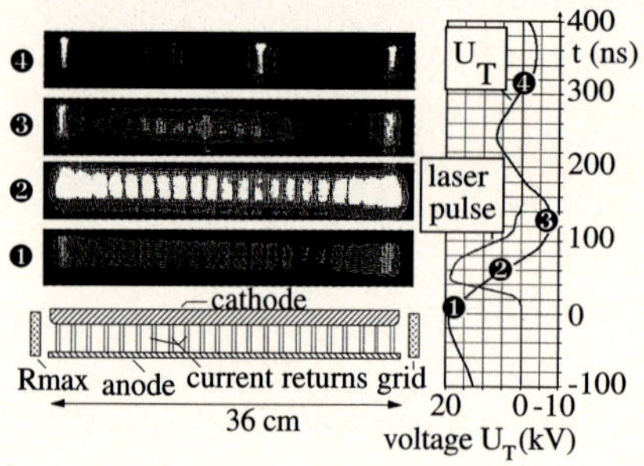

Fig. 1. Evolution of the plasma transverse emission for an energy deposition of 110 J/I in a gas mixture of Ne/Xe/HCI at a total pressure of 2.5 atm

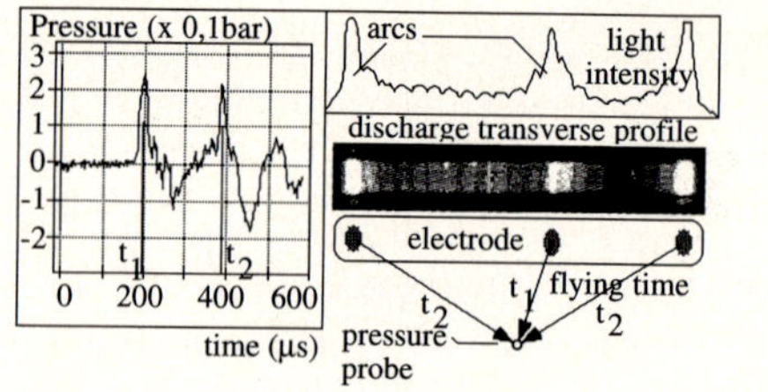

Fig. 2. Evolution of pressure fluctuations induced in the laser cavity by a discharge ending in three electric arcs

distribution of the total energy deposition was achieved with photographs integrated during the whole discharge (main and post-discharge included).

Fig.2 corresponds to a discharge with an energy deposition of 110 J/l at a total pressure of 2.5 atm. On the right-hand side, the upper figure shows an intensity profile of the discharge light which could be approximated in a first approach to the spatial distribution of the deposited energy while the lower figure is a schematic representation of electric arcs on the anode and their distances to the probe location. These two graphs are deduced from the integrated picture of the discharge (picture on the middle). On the left-hand side, the diagram corresponds to the time evolution of pressure fluctuations recorded with the pressure probe. It shows two overpressures at times t_1 and t_2. Elementary calculations show that t_1 and t_2 correspond to the time taken by a shock wave to travel from an arc position to the pressure probe location. The overpressure at t_1 is due to the arc located close to the middle of the electrodes, while the overpressure at t_2 is induced by both arcs located at the ends of the electrodes.

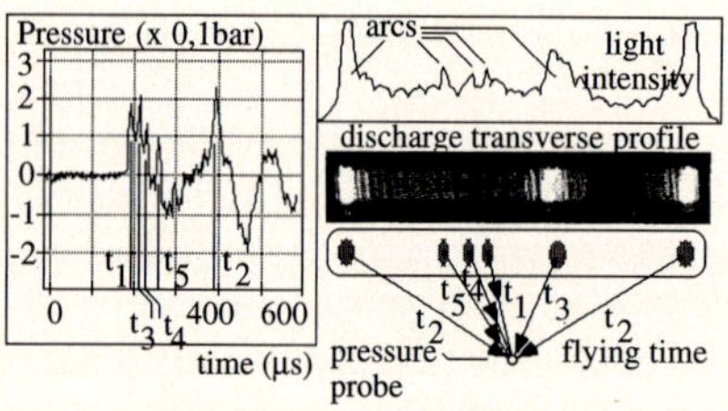

Fig. 3. Evolution of pressure fluctuations induced in the laser cavity by a discharge ended by several electric arcs

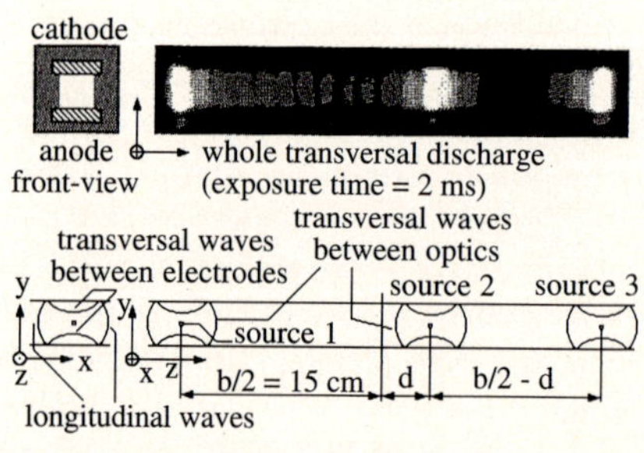

Fig. 4. Initial conditions for the 3D model. Each electric arc is a point source of a spherical acoustic or shock wave

When several light over-densities appear in the discharge volume, several overpressures are also recorded by the pressure probe as Fig.3 demonstrates. These shock waves may be approximated by cylindrical shock waves and be associated to longitudinal waves and transverse waves between optics.

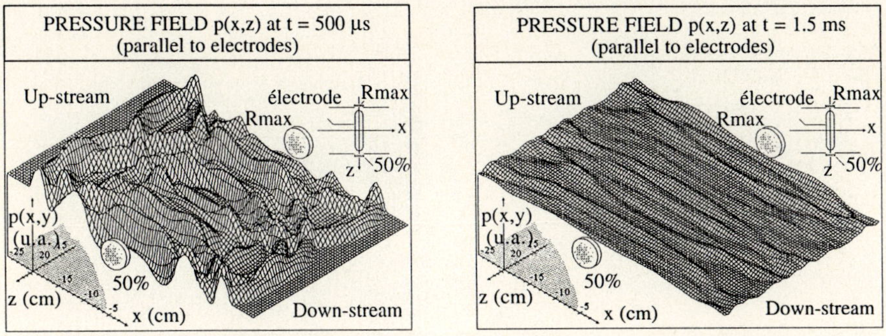

Fig. 5. Theoretical pressure distributions calculated with the 3D model along the xz direction in the laser cavity at $t = 0.5$ ms and $t = 1.5$ ms after the energy deposition (3 arcs)

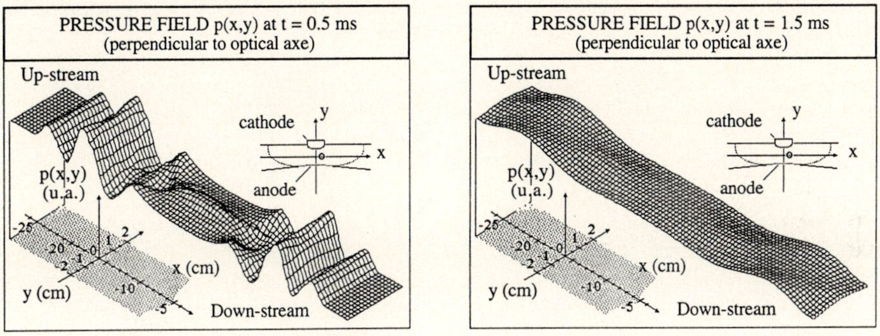

Fig. 6. Theoretical pressure distributions calculated with the 3D model along the xy direction in the laser cavity at $t = 0.5$ ms and $t = 1.5$ ms after the energy deposition (3 arcs)

2. Acoustic waves 3D model

We developed a 3D model of acoustic waves propagation based on geometrical optics which allows to describe the evolution of pressure fluctuations in the laser cavity during long time scales and to confirm the previous experimental results.

The initial conditions for the 3D model are shown in Fig.4. Acoustic waves are not supposed cylindrical but spherical. Although this hypothesis is severe, it allows to include in this interpretation a very simple model of both transverse waves between the electrodes and between the optics. However this hypothesis may be partly justified by taking into account that transverse waves between the electrodes are probably constituted by a great part of the spherical shock waves induced by the very intense anode and cathode zones of the electric arcs.

This 3D model only considers the contribution of electric arcs. For example, three current over-densities were taken into account in Fig.4, and each of them is considered as the source of a spherical wave whose location is taken in the middle of the electric arc. The laser cavity and the flow loop are modelled as a long box ($41 \times 480 \times 1800$ mm^3). The Mach number of these spherical acoustic waves is supposed to be constant and equal to unity. Another hypothesis is made on the wave width which is taken equal to the laser beam width (Zeitoun et al. 1992). Finally, all the reflections of these acoustic waves on the laser walls are considered as total, like in geometrical optics. The configuration of Fig.4 will be used later on (Figs.5 to 8).

Fig.5 represents the theoretical pressure distribution along the xz direction at two different times ($t = 0.5$ ms and $t = 1.5$ ms). Propagation of longitudinal waves and transverse waves between the optics are clearly visible in the upper view ($t = 0.5$ ms). At $t = 1.5$ ms, the transverse

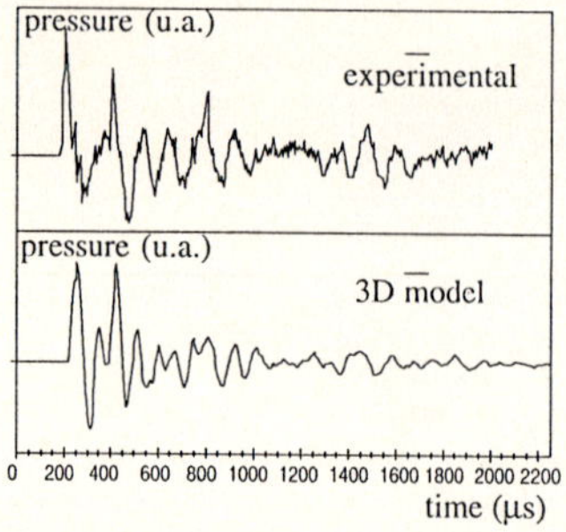

Fig. 7. Comparison of model and experimental pressure time evolution (probe location $x = -9$ cm upstream the discharge volume)

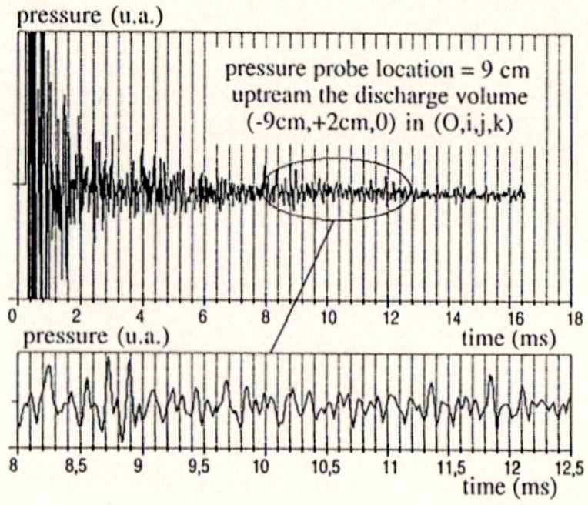

Fig. 8. Theoretical evolution of 3D model pressure fluctuations during 18 ms after the energy deposition (3 arcs)

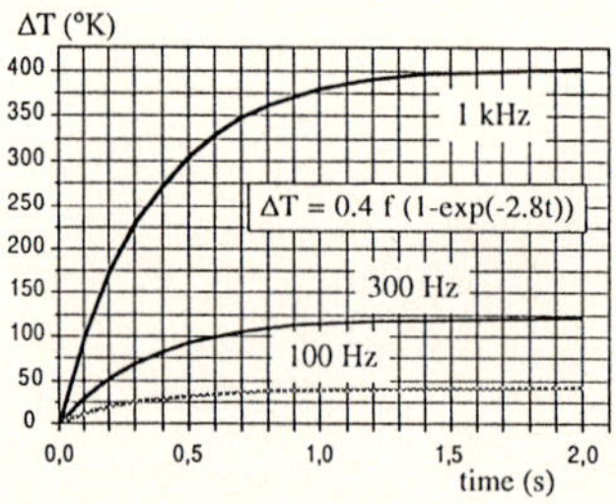

Fig. 9. Theoretical time evolution of electrode temperature during two seconds for PRF of 100 Hz to 1 kHz

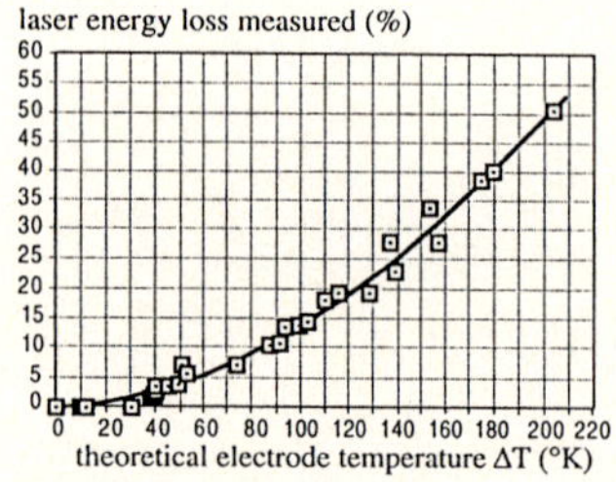

Fig. 10. Correlation between experimental laser energy loss and theoretical electrode temperature DT

waves between the optics still oscillate. They are confined by the walls of the laser cavity. The transverse waves between the electrodes are also observed along the xy direction in Fig.6. The geometrical diffusion is more intense for transverse waves between the electrodes than transverse waves travelling between the optics. In fact, these waves have been under-estimated because their intensity is probably three times more important than that of the longitudinal waves (Kosugi et al. 1991; Zeitoun et al. 1990).

A comparison of numerical and experimental pressure time evolutions is represented in Fig.7. The good agreement between experimental and model results demonstrates that most of acoustic and shock waves are induced by current over-densities or electric arcs.

For longer time scales (Fig.8), the time evolution of pressure fluctuations, deduced from the 3D model, follows an exponential decrease law: $a\exp(-bt)$. This behaviour is in very good agreement with experimental results (Sentis et al. 1989) and with previous theoretical studies (Knight 1986).

Such good agreement between experimental and theoretical results was never reached up to now. The majority of numerical models of aerothermodynamic phenomena does not really restore, up to now, the experimentally observed complexity of the pressure time evolution. These models are generally 2D models because they do not take such simple hypotheses as we did and so are much more complex. But 2D models cannot take into account the transverse waves between the optics and the three-dimensional nature of the energy deposition distribution. This approximation may explain the disagreement observed between 2D models results and experimental results. Although this 3D model could be improved by considering cylindrical longitudinal and transverse waves, the present work demonstrated that a 3D model is necessary to study such phenomena and that transverse waves between the optics must be considered in model development. Very recently, progress has been achieved in this way by Srivastava et al. (1992).

3. Thermal phenomena induced at high PRF

In high PRF excimer lasers, electrodes are strongly heated by the successive energy depositions. This electrode heating induces naturally an overheating of the gas mixture boundary layers near the electrodes. Strong density gradients with a distortion of the electrical field may appear, leading to a reduction of the energy deposition in the active medium, and finally to a decrease of the output laser energy. The electrode temperature increase results from the heating of the hot gas column (excimer lasers have low efficiency $\sim 1 \cdots 5\%$), and by different electric phenomena which occur between the discharge and the electrode surface (ionic and ohmic heating, etc.). This process exists on both electrodes, since the discharge voltage is unfortunately reversed several times during the same discharge.

A simple analytical resolution of the thermal phenomenon has been achieved with simplifying hypotheses. Considering the case of a thin flat electrode heating, the energy balance equation is the following:

$$mCd(\Delta T)/dt = f\,(Q_e - Q_c)$$

where ΔT is the temperature gradient between the electrode and the gas flow, f is the PRF, Q_e is the heat induced in the electrode by the discharge, $-Q_c$ is the heat convection term induced by the flow, and m is the mass of the material being heated. After some calculations and simplifications, we obtain:

$$\Delta T = \Delta T_\infty\,(1 - \exp(-2.8t))\,, \qquad \text{with } \Delta T_\infty = 0.4f$$

This result is in agreement with Baranov's data (1980) who has shown that the stationary temperature ΔT_∞ only depends on the PRF. Fig.9 shows the time evolution of electrode temperature for different PRF. At a PRF of 1 kHz, the theoretical electrode average temperature reaches a value of 400°C after one second of laser running.

Fig.10 represents on the same graph the evolution of the experimental laser energy loss (dots) and the theoretical electrode temperature for laser PRF of 100 Hz to 1 kHz. This graph has been built up by the following experimental procedure: For a fixed laser PRF (f) we were running the laser in a burst mode operation for 200 shots, recording the laser output energy for each shot. The first laser output energy of the burst was taken as reference to deduce the loss of any one of the other shots of the same burst. So for a shot number n ($2 \leq n \leq 200$) we knew the laser energy loss and we calculated the theoretical electrode temperature from:

$$\Delta T = 0.4f\,(1 - \exp(-2.8n/f))$$

This figure indicates a strong correlation between the decrease of laser output energy and the increase of electrode temperature. Further experimental studies of electrode temperature determination have to be undertaken to confirm this first result.

4. Conclusion

Sources and propagation modes of shock and acoustic waves have been specified. We demonstrated, in particular, that shock waves between the optics and longitudinal waves are strongly induced by current over-densities and electric arcs of both main discharge and post-discharge. Numerical simulation of aerothermodynamic phenomena induced by energy deposition in laser gas mixtures has to be done with 3D models. In the present work, we have developed and presented a very simple 3D model. It allows to simulate, in very good agreement with the experimental results, the propagation of acoustic waves in the LUX device during very long times (up to 18 ms). Such good agreement, never observed before, allows to confirm the very important role of current over-densities and electric arcs on the shock wave generation. A theoretical study of thermal phenomena has allowed to show a strong heating of the electrode surfaces (up to 400°C) due to different electrical phenomena that are occurring between the discharge and the electrodes (ohmic and ionic heating, etc.). This important heating produces thermal boundary layers on both electrodes, and could induce a strong decrease of extracted laser energy.

Acknowledgements

This work has been developed within the framework of the EUREKA-Eurolaser EU 205 program and supported by Laserdot, MESR and PACA Region.

References

Baranov VY, Malyuta DD, Mezhodov VS, Napartovich AP (1980) Influence of gas density perturbations on the ultimate characteristics of pulse periodic lasers with ultraviolet preionization. Sov. J. Quant. Electronics 10:12

Borisov VM, Vinokhodov AY, Kiryukhin YB (1987) Output energy evolution effects in a pulse-periodic XeCl excimer laser with an average power of ~ 400 W. Sov. J. Quant. Electronics 17:5

Knight CJ (1986) Sidewall muffler design for pulsed exciplex lasers. AIAA J. 24:11

Kosugi S, Ohishi T, Maeno K, Honma H (1991) Experimental study on shock waves in the cavity of excimer lasers. In: Takayama K (ed) Proc. 18th Intl. Symp. on Shock Waves, Sendai, pp 1295-1306

Sentis M, Delaporte P, Forestier B, Fontaine B (1989) Parametric studies of X-ray preionized discharge XeCl laser at single and at high pulse rate frequency. J. Appl. Phys. 66:8

Sentis M, Delaporte P, Forestier B, Fontaine B (1991) Design and characteristics of high pulse repetition rate and high average power excimer laser systems. IEEE J. Quant. Electronics 27:10

Srivastava B, Faria-e-maia F, Her J, Moran J (1992) High resolution computation of unsteady flows in pulsed lasers. AIAA J. 30:3

Taylor R (1986) Preionization and discharge stability study of long optical pulse duration UV-preionized XeCl lasers. Appl. Phys. B 41:1

Zeitoun D, Tarabelli D, Forestier B, Truong JP, Sentis M (1990) Effect of acoustic dampers on excimer laser flow. In: SPIE vol. 1397 GCL Madrid

Power Reduction of Excimer Lasers Caused by Electromagnetic Shock Waves due to Repetitively Pulsed Discharge

E. Hirai, N. Nagai and H. Yamakoshi
Mitsubishi Heavy Industries, Ltd., Japan

Abstract. It is well known that excimer lasers sometimes show output power reduction and instability at some pulse repetition frequencies. Although it is qualitatively pointed out that this phenomenon relates to the interaction between the repetitively pulsed glow discharges and the electromagnetic shock wave produced in the interelectrode region of the excimer laser, the detail mechanism of the phenomenon is not clearly explained. In this paper, the output power reduction corresponding to a certain exact shock position in the interelectrode region is investigated experimentally and a practical measure for shock dampening is proposed.

Key words: Excimer laser, Glow discharge, Electromagnetic shock wave, Shock damper

Nomenclature

R	: Gas constant	[J/kg K]	k	: Gas Specific heat ratio	[-]
T	: Gas temperature	[K]	C_p	: Gas specific heat	[J/kg K]
P	: Gas pressure	[Pa]	E	: Discharge input energy	[J/m^3]
ρ	: Gas density	[kg/m^3]	M_s	: Shock Mach number	[-]

1. Introduction

Excimer lasers are a promising light source for micromachining, semiconductor lithography, and others. For these industrial applications, it is necessary to develop excimer lasers with high power, high repetition rate and output power stability. However, the excimer lasers sometimes show output reduction (Fig.1) and instability at some pulse repetition frequencies (Kato et al. 1991).

The typical structure of an excimer laser is shown schematically in Fig.2. The laser gas recirculates through the interelectrodes in the chamber in order to sweep out the ionized gas. Discharges take place repetitively in the discharge gap and the rapid energy release during the discharge results in adiabatic expansion of the heated gas in the form of shock waves. According to a former investigation (Baranov et al. 1980), such accoustic and thermal disturbances due to the pulsed discharge generate fluctuation of the laser gas pressure and reduction of the laser output power.

In this paper, the output power reduction is investigated based upon the observation that the shock waves propagated both downstream and upstream reflect at the walls of the gas flow channel and go back to the interelectrode region, where the reflected shock waves interact strongly with the following pulse discharge. The output power reduction is demonstrated in relation to the reflected shock position and the discharge pattern.

2. Theory of the shock speed in excimer lasers

In order to estimate the shock speed, the Dzakowic's gas puffing model (Dzakowic and Wutzke 1973) was adopted. The duration of the discharge, approximately 20 ns, is so short that the discharge energy is injected into the gas in the glow discharge section without any thermal diffusion or convection, and that a plug of hot gas is produced. The Mach number of the shock wave immediately following glow discharge is calculated by the simple shock-tube theory and the Rankine-Hugoniot equations. As a result, the shock Mach number is expressed as follows:

Shock Waves @ Marseille III
Editors: R. Brun, L. Z. Dumitrescu

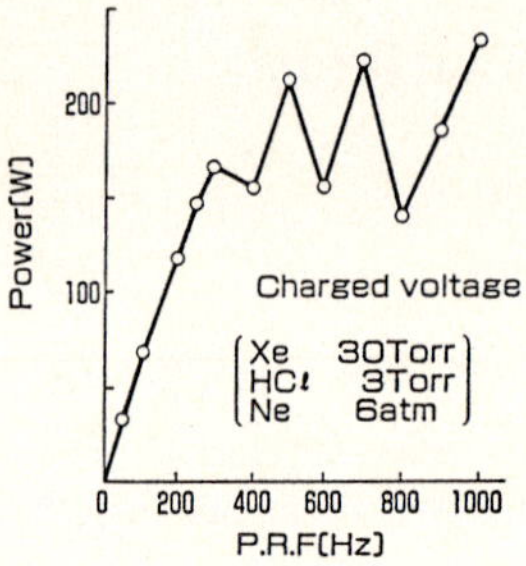

Fig. 1. Average output power of a high-repetition laser

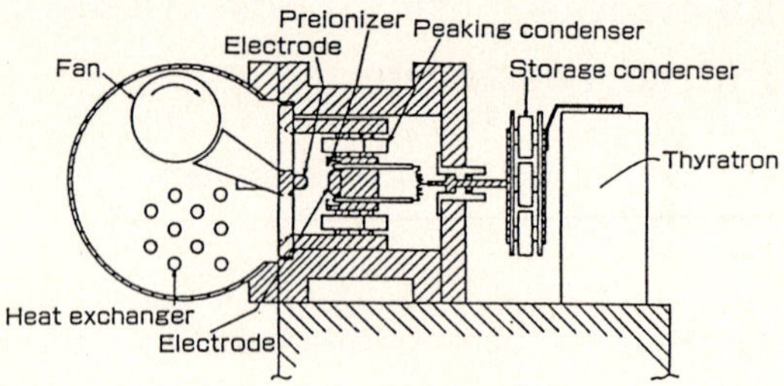

Fig. 2. Schematic diagram of the XeCl excimer laser

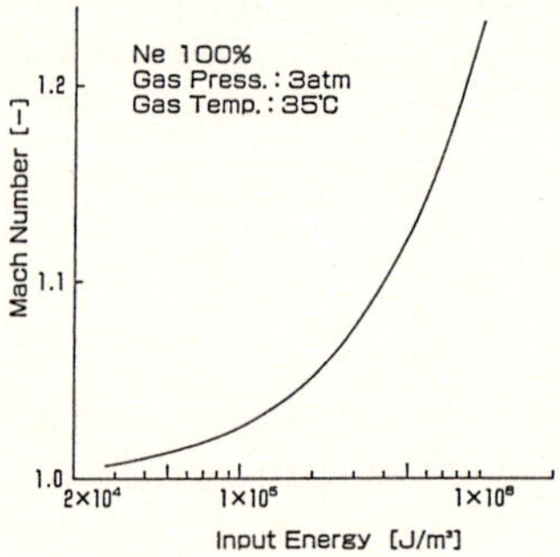

Fig. 3. Shock Mach number

$$\frac{R}{P}\left(T+\frac{E}{\rho C_p}\right)\rho=\frac{2\kappa M_s^2-(\kappa-1)}{\kappa+1}\left[1-\frac{\kappa-1}{\kappa+1}\sqrt{\frac{T}{T+\frac{E}{\rho C_p}}}\left(M_s-\frac{1}{M_s}\right)\right]^{-\frac{2\kappa}{\kappa-1}} \quad (1)$$

Fig.3 shows the relationship between the Mach number M_s and the injected energy E. The Mach number of the shock waves produced in excimer lasers is approximately one.

3. Shock wave visualization

The discharge test chamber used in our experiment is shown in Fig.4. In order to visualize the interaction between the reflected shock waves and the discharge pattern in the discharge gap, this test chamber consists of two excimer lasers linked to each other by a rectangular channel. One is called the pre-discharge section, which injects an electromagnetic shock wave into the other main discharge section (called the main excimer laser) through the channel.

The discharge duration of both excimer lasers is about 20 ns. The main excimer laser discharges a few milliseconds later after the discharge of the pre-discharge section. The delay is controlled by means of a pulse delay generator as shown in Fig.5. The laser output power is measured at the main discharge section using a Joule meter.

In order to determine the shock position in the discharge gap of the main discharge section, the shock wave passing through the interelectrode region was visualized by the Schlieren method using an IMACON 790 image converter with the plug-in 1/5T (Exposure: 2 μs) and a He-Ne laser as a light source. Fig.6 shows a representative shock wave visualization in the discharge gap. These photos were taken every 10 μs continuously after the pre-discharge. In this experiment T667 Polaroid ASA3000 film was used and the experimental conditions of the main discharge section are listed in Table 1.

Fig.7 shows the structure of the visualized shock waves corresponding to the picture at 1.171 ms delay (Fig.6). In Fig.7 the three observed shock waves are drawn schematically. Two of them,

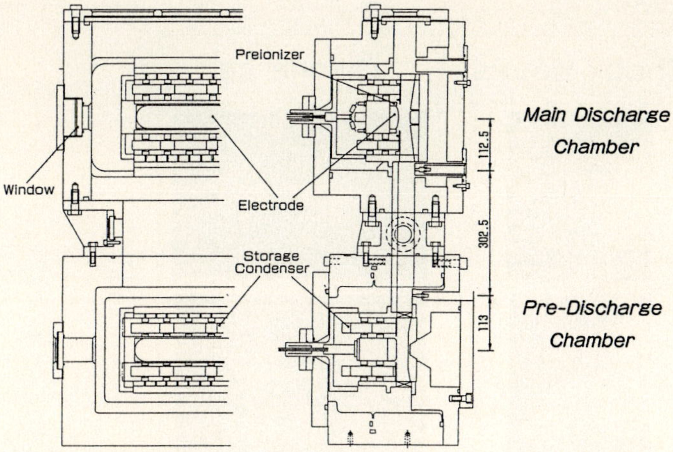

Fig. 4. Discharge chamber

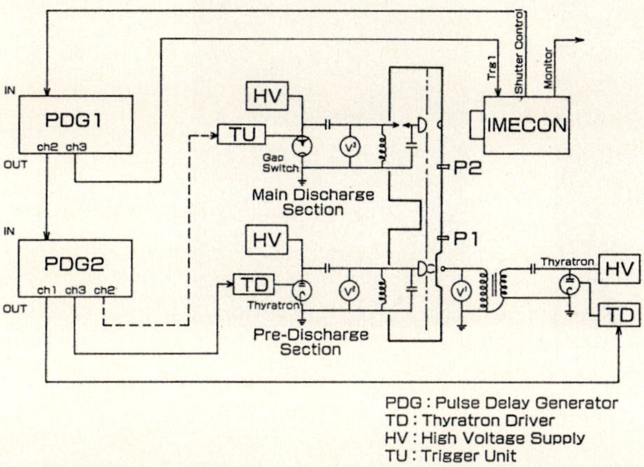

Fig. 5. Discharge operating circuit

which appear to be the shock front produced in the pre-discharge section, are straight and intensive. The reason why the shock front divides into two could be due to the cavity of the pre-ionization pins or due to the side effect of the electrodes. Another weak and arc-shaped shock wave appears to be the normal reflection shock wave which reflected off the connecting channel surface.

The gas pressure evolution of the connecting channel was also monitored by two pressure transducers (PCB Piezotronics 112A20) for measuring the shock speed. The measured shock speed was 447.5 m/s and it agreed well with the predicted value from Eq.1 with an accuracy of $\pm 1\%$.

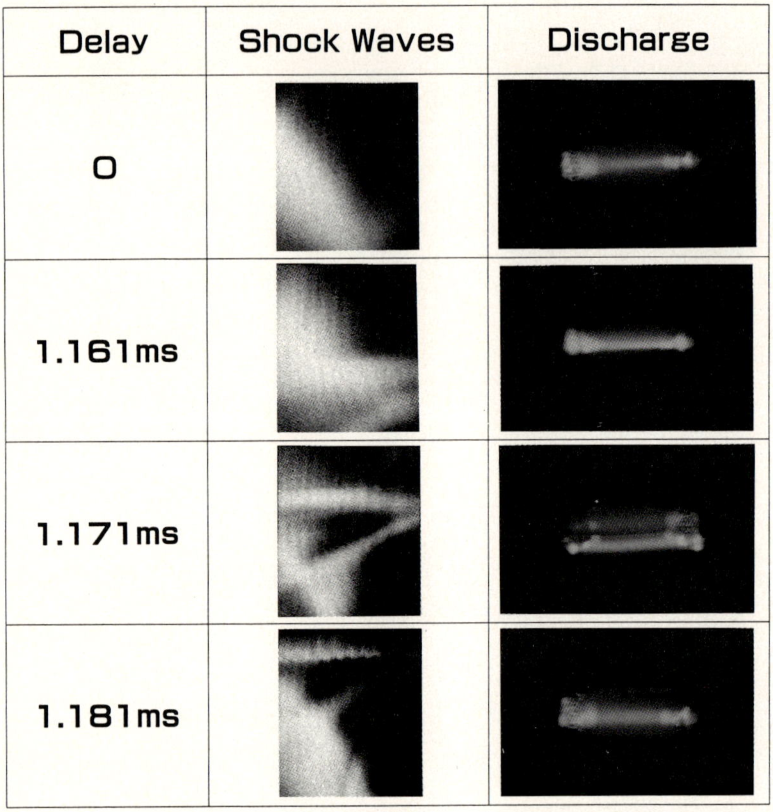

Fig. 6. Visualisations of shock and discharge

4. Power reduction measurements

The measured laser output power of the main excimer laser is shown in Fig.8. It is found that the laser output power falls around the 1.165 ms delay time, and, as visualized in Fig.6, around this delay time there exists a shock front in the center of the discharge gap. This fact suggests that the output power reduction is dependent upon the shock position. When the propagated shock position and the discharge timing coincide at the discharge gap, the laser output power is reduced drastically, because gas density purturbations due to the shock wave cause heterogeneous gas resistance between electrodes and result in arcing. In Fig.8, the reason why the output power after the delay of 1.18 ms does not completely recover to the initial output power level is that some pressure fluctuations (approximately 5%) remain for a long time in the laser gas flowfield even after the pulsed discharge has ended.

To investigate the output power reduction phenomenon in detail, the discharge patterns are observed, corresponding to the time when the laser output power falls, while a shock front exists in the discharge gap. In Fig.6, the observed discharge pattern is compared with a normal glow discharge pattern. A homogeneous glow discharge between the electrodes is generally desirable to get stable laser output power. However, at 1.171 ms delay time the discharge pattern is especially dissimilar and seems to split into two discharge paths. This discharge appears partially proceeding to an arc discharge.

Table 1. Experimental conditions

Gas components	HCl 4.4 torr Xe 30 torr Ne 3 atm
Gas temperature	292 K
Electrode	Gap 26 mm Length 195 mm
Discharge width	10 mm
Breakdown voltage	25 kV
Discharge energy	3.04 J/P (6.0×10^4 J/m^3)

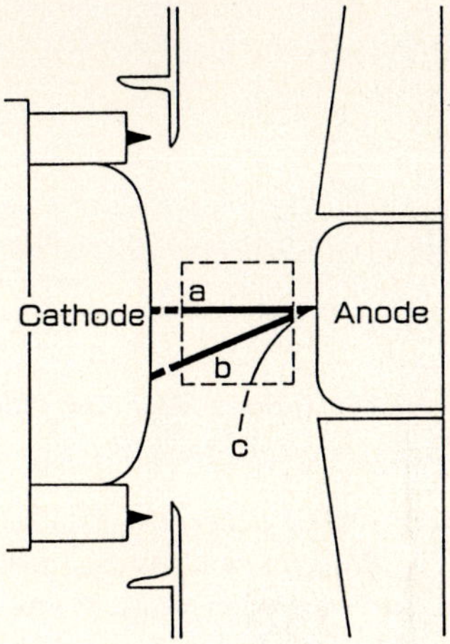

Fig. 7. Main discharge section

In this experiment the time characteristics of the laser output power were also measured at some delay times. The results are shown in Fig.9. While every curve of the output power has almost the same rise slope at the leading edge, the decay of each signal is different. This indicates that, although the discharge starts in the same way under any flowfield conditions, the duration of the discharge is affected by the homogeneity of the density in the discharge gap.

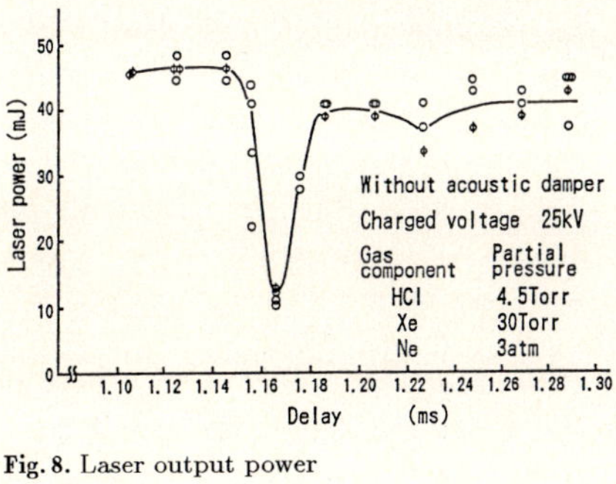

Fig. 8. Laser output power

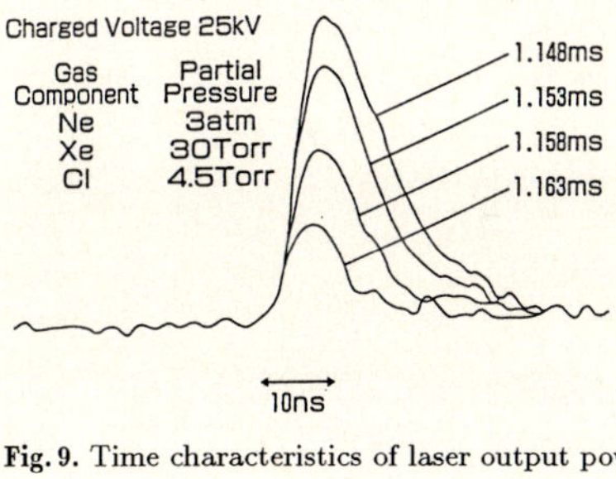

Fig. 9. Time characteristics of laser output power

As a conclusion, it is inferred that it is difficult to maintain a stable glow discharge when there exists a drastic pressure gradient like a shock front, and that the pressure gradient in the flowfield intensely affects the laser output power.

5. Shock wave damper in excimer lasers

In real excimer lasers, the electromagnetic shock wave should be attenuated and eliminated as soon as possible. For absorbing the produced shock wave, a porous aluminium plate (plate thickness: 10 mm, porous cavity diameter: 1 - 5 mm) was chosen instead of a conventional sidewall muffler (Knight 1985), because porous metal is commercially available and is easy to install to any wall shape of an excimer laser gas flow channel. In our experimental set-up, the porous plate was installed on the inside surface of the connecting channel. In our experiment no shock wave

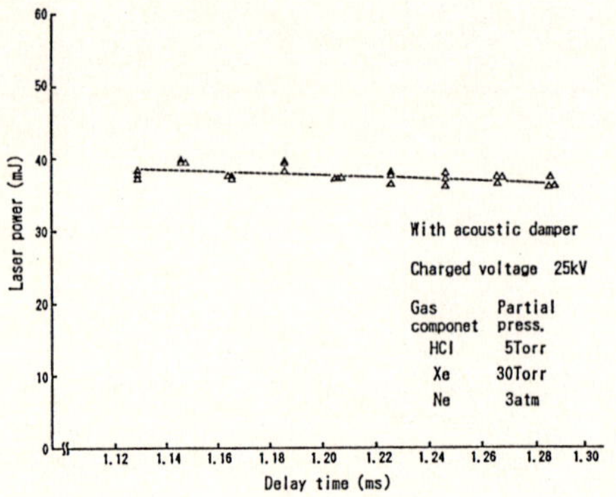

Fig. 10. Laser power vs. discharge delay time with acoustic damper

injected from the pre-discharge section reached the main discharge section. Therefore, no laser output power reduction was observed at any delay as shown in Fig 10. As a result, it was proven that porous aluminium plate is highly and practically effective as an electromagnetic shock wave attenuator of excimer lasers.

6. Conclusion

The power reduction of excimer lasers which takes place at certain pulse repetition frequencies was investigated experimentally. First, the structure and the position of the shock waves passing through the interelectrode region were visualized using the Schlieren method. Corresponding to the visualization, the power reduction of the excimer laser was measured. The shock wave existing at the interelectrode region strongly affected the power reduction of the excimer laser and the time when the power reduction takes place coincides with the time when the shock waves pass through the interelectrode region. The shock wave caused a drastic pressure gradient in the flowfield and accelerated the glow discharge to proceed on as an arc discharge. A very effective and practical measure to eliminate rapidly the electromagnetic shock wave produced was to install a porous aluminium plate on the inside wall of the gas channel; the effectiveness of the porous aluminium plate was demonstrated.

References

Baranov VYu et al. (1980) Average power limitations in high-repetition-rate pulsed gas lasers at 10.6 and 16 μm. Appl. Opt. 19, 6: 930-936

Dzakowic GS, Wutzke SA (1973) High-pulse-rate glow discharge stabilization by gas flow. J. Appl. Phys. 44, 1: 5061-5063

Kato M et al. (1991) Mitsubishi Heavy Industries Tech. Report, Vol.28, No.2

Knight CJ (1985) Sidewall muffler design for pulsed excimer lasers. AIAA 23rd Aerospace Science Meeting, AIAA Paper 85-0389

Attenuation and Distorsion of Propagating Compression Waves in a High-Speed Railway Model and in Real Tunnels

Toshiyuki Aoki*, Kazuyasu Matsuo*, Hideto Hidaka†, Yasuhiro Noguchi† and Syunji Morihara†

*Graduate School of Engineering Sciences, Kyushu University, Kasuga, Fukuoka, 816, Japan
†Technical Development Department, West Japan Railway Co., Kita-ku, Osaka, 530, Japan

Abstract. When a high-speed railway train enters a tunnel, a compression wave is generated ahead of the train and propagates along the tunnel. This wave subsequently emerges from the exit portal of the tunnel as an impulsive noise. In order to correctly estimate its magnitude, experimental and numerical studies are carried out on the attenuation and distortion of a compression wavefront in a high-speed railway model and in a real tunnel. The experimental pressure-time records in the model tunnel are compared with the calculated ones, where the unsteady friction is taken into account. The attenuation rate is affected by the strength of the compression wave and the unsteady boundary layer induced by the wavefront. In order to examine in which case the compression wave is steepened or spreads, an acoustic Reynolds number is defined and calculated for the compression wave propagating along the tunnel.

Key words: Compression wave, Railway tunnel, Attenuation, Distortion

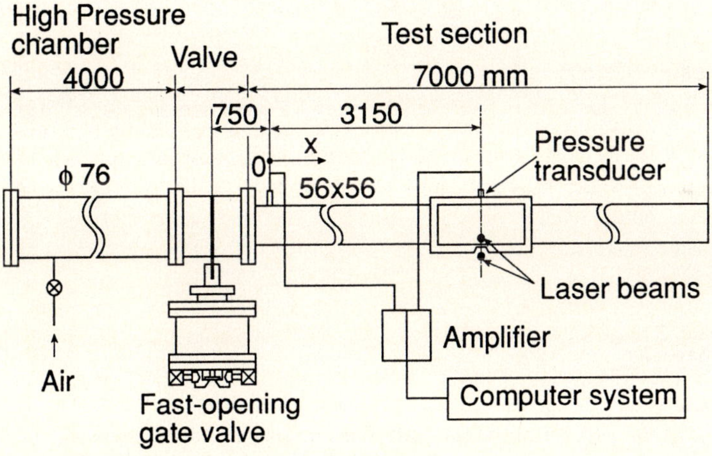

Fig. 1. Schematic diagram of experimental apparatus

1. Introduction

When a high-speed railway train enters a tunnel, a compression wave is generated ahead of the train and propagates along the tunnel, compressing and accelerating the air at rest in front of the wave. This wave subsequently emerges from the exit portal of the tunnel as an impulsive noise. With the arrival of higher-speed trains, such an impulsive noise could be large enough to present problems unless some attempts are made to alleviate the noise levels. According to the

Shock Waves @ Marseille III
Editors: R. Brun, L. Z. Dumitrescu

aeroacoustic theory, the magnitude of the impulsive noise at the tunnel exit is proportional to the rate of pressure change of the compression wavefront coming to the tunnel exit (Ozawa 1979). Therefore, in order to estimate the magnitude correctly, the mechanism of the attenuation and distortion of a compression wavefront propagating along a tunnel must be understood. There have been reported a number of experimental and theoretical studies on the shock attenuation in shock tubes (Mirels 1956). However, these results are not applicable because compression waves in railway tunnels are extremely weak and their thickness, i.e., width is much larger than that of a shock wave. In the present paper, experimental and numerical investigations are carried out on the attenuation and distortion of a compression wavefront using a high-speed railway model and a real tunnel. The experimental and numerical results are discussed from the viewpoint of an internal flow with the development of a wall boundary layer induced by compression waves. Finally, in order to examine in which case the compression wave is steepened or spreads, an acoustic Reynolds number, which is the ratio of the nonlinear term to the viscous dissipation term of Burgers equation, is defined and calculated for the compression wave propagating along the tunnel.

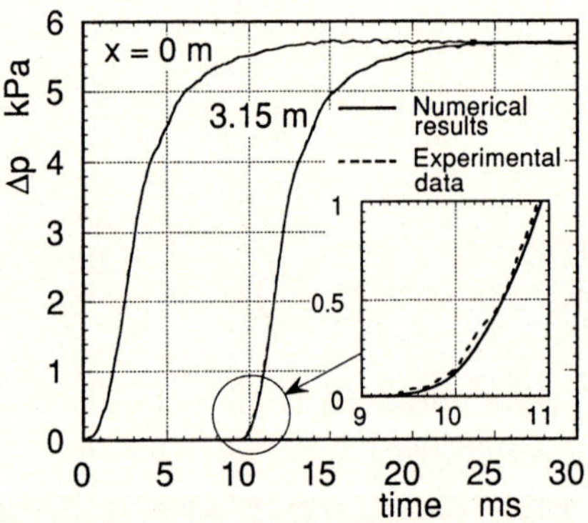

Fig. 2. Measured and calculated waveforms of compression waves in the model tunnel

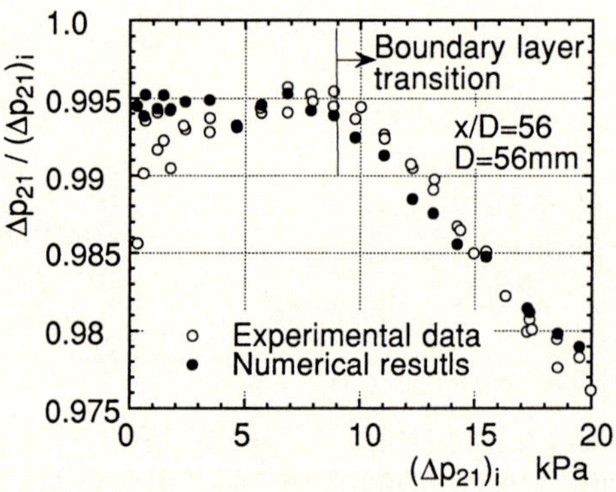

Fig. 3. Attenuation rate of compression wave propagating in the model tunnel

2. Experimental apparatus

The experimental apparatus used in the present work is shown schematically in Fig.1. This is a kind of open-ended shock tube with a fast opening gate valve instead of a diaphragm. The high pressure chamber has a circular cross section with an inner diameter of 76 mm and a length of 4 m. The test section has a rectangular cross section of 56×56 mm and a length of 7 m. The equivalent diameter D of the test section is 56 mm. The initial pressure p_1 in the test section is the atmospheric pressure. Since the opening time of the valve is finite, about 10 ms in the present one, some distance downstream from the valve is required to form a fully developed compression wave. The formation distance is about 750 mm from the gate valve. Accordingly, as shown in Fig.1, this point is regarded as the entrance of the model tunnel and the distance x will be taken from this point in the following discussions. The air in the high pressure chamber plays the role of a train entering a tunnel. The output signals of the pressure transducers at the stations $x = 0$ and 3.15 m are recorded. Laser differential interferometer techniques have been used to detect the transition region in the unsteady wall boundary layer induced by a compression wave (Aoki 1993). The laser beams enter the test section through the side wall as shown in Fig.1. Field tests with a high-speed railway train passing a tunnel have been carried out on the double track tunnels for the Sanyo Shinkansen line. The cross section of the real tunnel is 63.4 m^2 and its equivalent diameter $D = 9$ m. The train speeds in the tests were about 200 km/h.

3. Numerical calculation

The conservation laws for one-dimensional, unsteady, compressible flows with friction can be written in the form

$$\frac{\partial \boldsymbol{U}}{\partial t} + \frac{\partial \boldsymbol{F}}{\partial x} = 0, \quad \boldsymbol{U} = \begin{bmatrix} \rho \\ \rho u \\ e \end{bmatrix}, \quad \boldsymbol{F} = \begin{bmatrix} \rho u \\ p + \rho\, u^2 \\ (e+p)\, u \end{bmatrix}, \quad \boldsymbol{G} = \begin{bmatrix} 0 \\ f(t) \\ 0 \end{bmatrix}, \tag{1}$$

where ρ is the density, u is the velocity, p the pressure, e the sum of the internal energy and the kinetic energy per unit volume and $f(t)$ is the pressure drop per unit length, that is, the fluid friction term. We have assumed that the development of unsteady fluid friction behind the compression wave is taken into account by the method of Zielke (1968) and Trikha (1975) for laminar pipe flow, and Schultz and Sockel (1988) for turbulent pipe flow. These friction coefficients f are larger than values for steady flow at the wavefront and decrease, approaching the values f_0 of steady laminar or turbulent flow, asymptotically with the distance from the wavefront, respectively. The second-order symmetric total variation diminishing (TVD) scheme (Yee 1987) is employed together with the time-splitting technique introduced by Sod (1977). In the case of model tunnel simulation, since the fluid friction is considered to be related to the fact whether the boundary layer of flow induced by the compression wave is laminar or turbulent (Matsuo 1991), the position of the transition region from laminar to turbulent boundary layer is experimentally estimated (Aoki 1993). The friction coefficient is assumed to change gradually from the laminar flow value to the turbulent one in the transition region. In the case of real tunnel simulation, since the flow behind the compression wave in real railway tunnels is usually turbulent, the friction coefficient is assumed to be that of unsteady turbulent flow in rough pipes (Schultz and Sockel 1988). The initial waveform of the compression wave at the entrance ($x = 0$) is given by direct experimental data in the case of model tunnel simulation and as an empirical expression (Ozawa 1979) which fits the experimentally obtained compression wavefront in the case of real tunnel simulation.

4. Results and discussion

4.1. Model tunnel

Typical experimental pressure-time records at $x = 0$ and 3.15 m are shown in Fig.2. Let the pressure rise by a compression wave be denoted by Δp_{21} ($= p_2 - p_1$), where p_1 and p_2 represent the pressures before and after the compression wave, respectively, and the initial pressure rise at $x = 0$ be $(\Delta p_{21})_i$. The measured waveforms in Fig.2 are compared with the calculated ones by the TVD scheme, where the solid line represents the numerical results with unsteady friction for laminar flow (Zielke 1968; Trikha 1975). The calculated result is in good agreement with the experimental one. The attenuation rates $\Delta p_{21}/(\Delta p_{21})_i$ obtained from the measured pressure rise at $x/D = 56$ are plotted against the initial pressure rise $(\Delta p_{21})_i$ in Fig.3, together with the calculated values. It should be noted in this figure that the value of $\Delta p_{21}/(\Delta p_{21})_i$ initially increases with an increasing $(\Delta p_{21})_i$, becoming maximum when $(\Delta p_{21})_i$ is about 9 kPa, and then decreases. This decrease in $\Delta p_{21}/(\Delta p_{21})_i$ with $(\Delta p_{21})_i$ for $(\Delta p_{21})_i > 9$ kPa agrees well with the calculated value. This fact seems to mean that the transition from laminar to turbulent boundary layer may take place in the condition of $(\Delta p_{21})_i > 9$ kPa, which is confirmed by means of laser

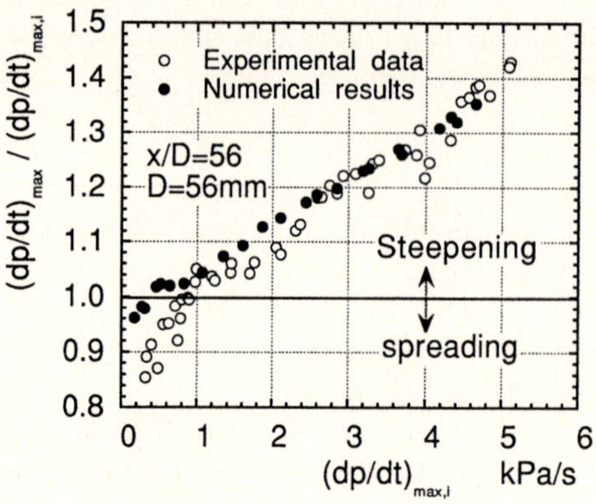

Fig. 4. Maximum rate of pressure change of the compression wave in model tunnel

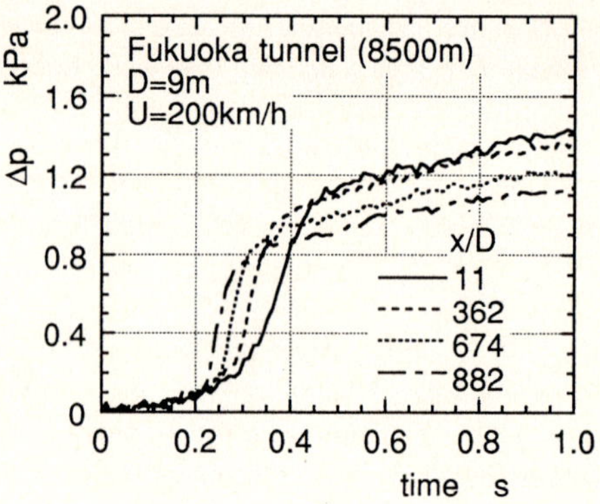

Fig. 5. Typical pressure-time records in real tunnel

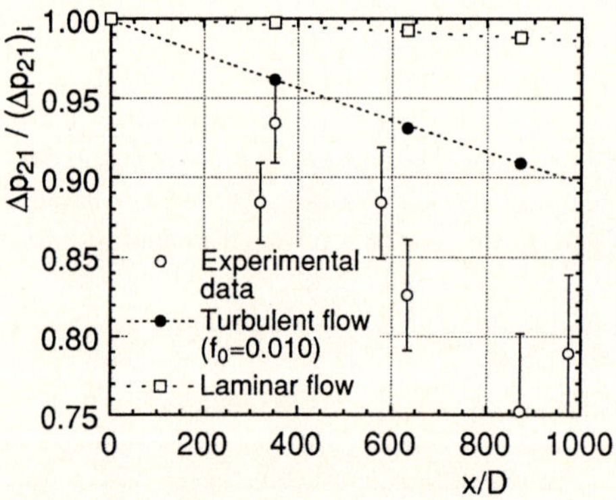

Fig. 6. Attenuation of the pressure rise of the compression wave with distance in real tunnel

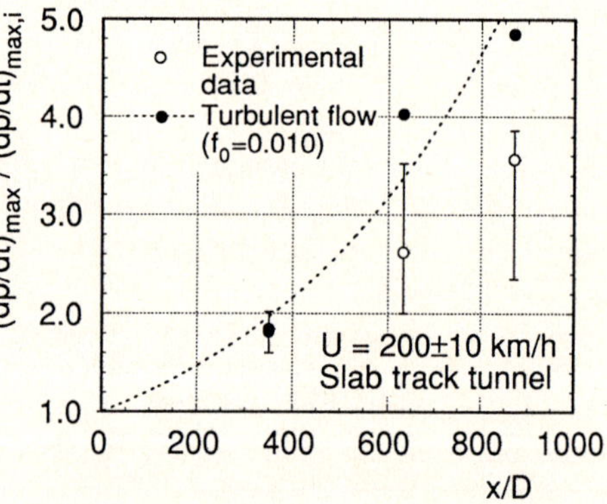

Fig. 7. Maximum rate of pressure change of the compression wave in real tunnel

interferometer detection (Aoki 1993). The reason why the attenuation rate slightly increases with $(\Delta p_{21})_i$ for $(\Delta p_{21})_i < 9$ kPa in the present experiment is understood, at present, to be related to unsteady laminar boundary layer development behind the propagating compression waves. In order to examine the steepness of the wave, let the maximum rate of pressure rise at a compression wavefront be $(dp/dt)_{max}$, and the ratio of the value at a measured point to the initial value at $x = 0$ be $(dp/dt)_{max} / (dp/dt)_{max,i}$. The values of $(dp/dt)_{max} / (dp/dt)_{max,i}$ measured and calculated are plotted against $(dp/dt)_{max,i}$ in Fig.4. From this figure, when a compression wave is weak, that is, $(dp/dt)_{max,i} < 0.8$ kPa/s, $(dp/dt)_{max} / (dp/dt)_{max,i} < 1$ because of the viscous effect of air. On the other hand, for a stronger initial maximum rate of pressure rise $(dp/dt)_{max,i} > 0.8$ kPa/s , $(dp/dt)_{max} / (dp/dt)_{max,i} > 1$ since the nonlinear effect of the wave overcomes the viscous effect. $(dp/dt)_{max} / (dp/dt)_{max,i}$ is overestimated in calculations in comparison with the measured data in region of $(dp/dt)_{max,i} < 0.8$ kPa/s. This is the subject for a future study.

4.2. Real tunnel

Fig.5 shows the attenuation and distortion of the wavefront of the compression wave in a long real tunnel with a concrete slab track (Fukuoka tunnel, length 8500 m). The horizontal axis represents the time measured from the head of the wavefront of the compression wave and the train speed is U = 200 km/h. It is evident from Fig.5 that the pressure rise Δp_{21} decreases and the maximum rate of pressure rise at the compression wavefront increases with distance x. The measured attenuation rates $\Delta p_{21}/\Delta p_{21})_i$ obtained from various slab track tunnels are compared with the calculated ones with unsteady turbulent flow (steady friction coefficient $f_0 =$ 0.01 for completely rough pipe) in Fig.6. Square points represent the calculated ones with unsteady laminar flow for reference. In this figure, the attenuation of the wave is underestimated in all calculations in comparison with the measured data. However, it should be noted that taking turbulent flow into account instead of laminar flow causes more attenuation of the wavefront and the calculated results are in better agreement with the experimental one. The values of $(dp/dt)_{max}$ $/(dp/dt)_{max,i}$ measured and calculated in the real tunnel are plotted against the nondimensional distance x/D in Fig.7. From this figure, $(dp/dt)_{max}$ $/(dp/dt)_{max,i}$ increases with x since the nonlinear effect of the wave overcomes the viscous effect. The numerical results agree well with the experimental data.

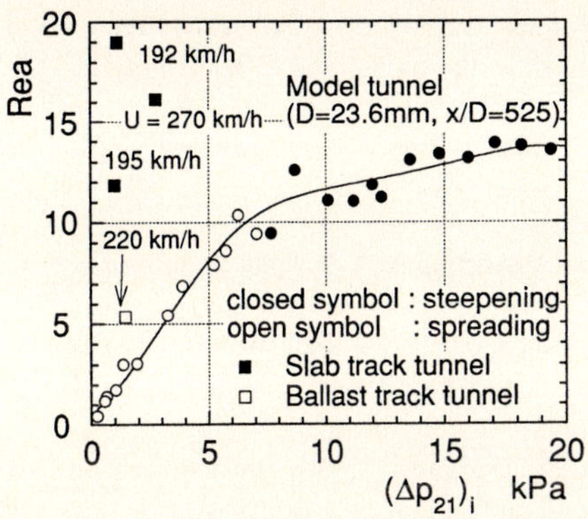

Fig. 8. Acoustic Reynolds number for compression wave propagating in the long model and real tunnels

4.3. Nonlinear effects of compression waves

According to the nonlinear acoustic theory (Cary 1968), the acoustic Reynolds number (Goldberg number) is a convenient parameter for assessing whether nonlinear effects dominate over attenuation, that is, the waveform steepens. In the present case, we assume that, first the nonlinear and attenuation effects of the compression wavefront are independent of each other and secondly, that the reference length is twice the length of the compression wavefront (Matsuo 1992). Consequently, in order to examine in which case the compression wave is steepened or spreads, an acoustic Reynolds number Re_a, which is the ratio of the nonlinear term to the viscous dissipation term of Burgers' equation, is defined by the following equation:

$$\mathrm{Re}_a = \frac{\pi\,(\gamma+1)}{2\,a_1}\,u_2\,\frac{1}{kL/D} \tag{2}$$

where γ is the ratio of specific heats, a_1 is the initial sound velocity and u_2 is the velocity of the main flow induced by the compression wave. k is an attenuation coefficient which is determined by experiments (Matsuo 1991).

The calculated acoustic Reynolds number is plotted against the initial pressure rise $(\Delta p_{21})_i$ of the compression wave in Fig.8. The circles and squares represent the data of the model tunnel (D =23.6 mm, L/D =525) (Matsuo 1991) and the real tunnel with a concrete slab and ballast tracks. The speed of the train entering the real tunnel is also shown in the figure. Further, the open and closed symbols denote the cases of spreading and steepening compression waves, respectively. It can be concluded from Fig.8 that the critical acoustic Reynolds number is about 10 in both cases of the model and real tunnels.

5. Concluding remarks

(1) In the case of the model tunnel, the waveform of the compression wave numerically calculated by the TVD scheme with an unsteady fluid friction is in good agreement with the measured data. The measured attenuation rate increases slightly with increasing initial pressure rise $(\Delta p_{21})_i$ at the entrance of the model tunnel, becoming maximum when $(\Delta p_{21})_i$ is about 9 kPa, and then decreases.

(2) In the case of the real tunnel, taking unsteady turbulent friction into account instead of unsteady laminar friction causes more attenuation of the wavefront and the calculated attenuation rate is in better agreement with the measured one.

(3) The acoustic Reynolds number is defined in Eq.2 and calculated for the compression wave propagating along the long model and the real tunnel. It can be concluded that the critical acoustic Reynolds number is about 10 in both cases: in the model and in real tunnels.

References

Aoki T, Matsuo K et al. (1993) to be published

Cary BB (1968) Prediction of finite-amplitude waveform distortion with dissipation and spreading. J. of Acoustical Society of America 43-6: 1364-1372

Matsuo K, Aoki T et al. (1991) Attenuation of compression waves in a high-speed railway tunnel simulator. In: Proc. 7th Intl. Symp. on Aerodynamic and Ventilation of Vehicle Tunnels, pp 239-252

Matsuo K, Aoki T et al. (1992) Numerical simulation of an impulsive wave emitted from an exit of high-speed railway tunnel. In: Proc. 3rd Intl. Conf. on Computers in Railways 2, pp 455-463

Mirels H (1956) Attenuation in a shock tube due to unsteady-boundary layer action. NACA Tech Note 3278: 1-60

Ozawa S (1979) Studies of micro-pressure wave radiated from a tunnel exit. Railway Tech. Res. Rept. of Japanese National Railways 1121 : 1-92 (in Japanese)

Schultz M, Sockel H (1988) The influence of unsteady friction on the propagation of pressure waves in tunnels. In: Proc. 6th Intl. Symp. on the Aerodynamics and Ventilation of Vehicle Tunnels, pp 123-135

Sod GA (1977) Numerical study of a converging cylindrical shock. J. Fluid Mech. 83-4: 785-794

Trikha AK (1975) An efficient method for simulating frequency-dependent friction in transient liquid flow. J. Fluids Engineering Trans. ASME 97: 97-105

Yee HC (1987) Upwind and symmetric shock-capturing schemes. NACA TM 89464, pp 1-127

Zielke W (1968) Frequency-dependent friction in transient pipe flow. J. Fluids Engineering Trans. ASME 90: 109-115

A Numerical and Experimental Study of Sonic Booms Generated in High-Speed Train Tunnels

A. Sasoh*, T. Saito†, S. Funabashi*, O. Onodera*, K. Takayama*, Y. Matsui‡ and R. Kaneko‡

*Shock Wave Research Center, Institute of Fluid Science, Tohoku University, 2-1-1 Katahira, Aoba-ku, Sendai 980, Japan

†Cray Research Japan Co. Ichiban Cho, Chiyoda Tokyo, Japan

‡East Japan Railway Co., Tokyo Kotsu Kaikan 7th Floor, 2-10-1 Yuraku-cho, Chiyoda-ku, Tokyo 100, Japan

Abstract. Entrance of a high-speed train (Shinkansen) into a long tunnel creates sonic booms which are sometimes very unpleasant, especially with the increase of the train speed. Paper describes a primary effort of understanding the mechanisms of sonic boom generation, both experimentally and numerically. Experiments were carried out with a scaled tunnel simulator in which a plastic rod was launched into a 40 mm dia. 30 m long inclined tube with speeds ranging from 60 to 120 m/s at the entrance. Sonic boom pressures were measured on site by using a PVDF piezofilm gauge. In order to interpret the experiments, a TVD finite-difference numerical simulation was carried out.

Key words: High speed train tunnel, Weak shock wave, Sonic boom

1. Introduction

Lukasiewicz (1976), comparing the cost of the Japanese high speed train network system (Shinkansen) with that of the Apollo project, wrote that these were indeed comparable. Similar to the French TGV high speed train network, the Japanese Shinkansen has boosted Japanese economy and stimulated nation-wide migrations of manpower. In Japan efforts have been made for increasing the train speed up to maximum 270 km/h in operation. Then Japanese railway companies seriously realized that sonic booms are actually generated at the tunnel exits. Since, unlike European countries, there are so many tunnels in Japan along the course of Shinkansen tracks, this prevents further increase of the train speed (Matsuo 1991). The goal of this paper is to clarify, from the point of view of shock wave dynamics, both experimentally and numerically, the mechanism of sonic boom generation in Shinkansen tunnels and the wave interaction in the train/tunnel system. The numerical simulation is briefly described in Section 2 and experiments using a scaled tunnel simulator and on-site pressure measurements are described in Sections 3 and 4, respectively.

2. Numerical simulation

In the Shinkansen tunnels, the ratio of the train cross-section to tunnel cross-section is approximately 0.2. Accordingly, in a scaled tunnel, the train/tunnel geometry is arranged to be identical to the real tunnel. In the present numerical simulation, it is assumed that an 18 mm dia. circular cross-sectional cylinder is moving along the center axis of a 40 mm dia. tube.

2.1. Basic equations and numerical scheme

The unsteady Euler equations in an axisymmetric flow are solved using the time-dependent explicit finite-difference scheme of Harten (1983) and Yee (1989). This is one of routinely used oscillation-free schemes known as TVD (total variation diminishing) algorithms, adopting flux limiters. The scheme has second-order accuracy in space for smooth regions and a Strang-type

Shock Waves @ Marseille III
Editors: R. Brun, L. Z. Dumitrescu

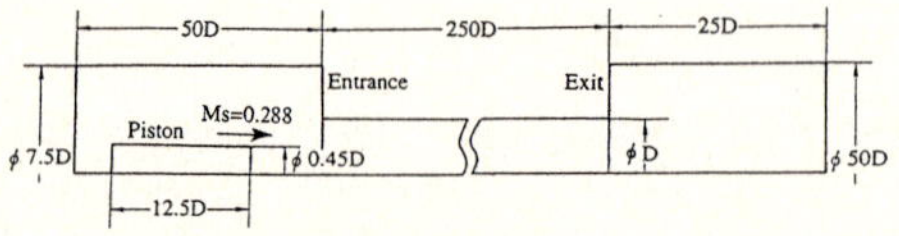

Fig. 1. Computational domain

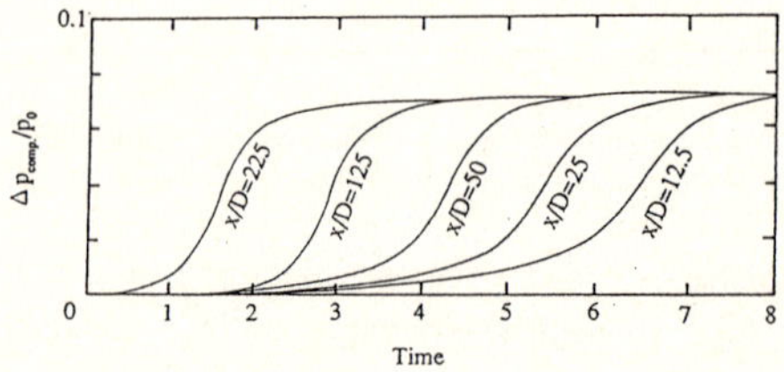

Fig. 2. Overpressure histories at different locations, numerical simulation

time-splitting method is used to keep the second-order time accuracy (Strang 1968). The flux limiters employed in the present computation are the so-called minimode limiter for the genuinely non-linear waves (for its robustness) and Superbee for linear waves.

2.2. Train model

The computational domain is shown in Fig.1 and is designed to compare the computational results with experiments conducted in a scaled tunnel simulator. The model train is 12.5 D long where D is the tunnel diameter, while the tunnel is 250 D long. A train model is accelerated in an entrance space and reaches constant speed before entering the tunnel. Relatively large entrance and exit spaces are provided in order, in addition to match with the scaled tunnel simulator geometry, to prevent waves created during the train acceleration from entering into the tunnel. The grid number is 2600 in the axial direction; 150 and 600 grid points are allocated for the tunnel and the entrance and exit spaces, respectively.

2.3. Boundary conditions

The location of the train model is traced and renewed at each time step when the train model traveled the distance of one grid interval. The boundary conditions at the front and rear faces of the moving train model are determined using a scheme similar to that of Groth and Gottlieb (1988).

2.4. Results and discussion

Compression waves are generated in front of the train and gradually build up a relatively steep rise. Fig.2 shows the sequential transition of the compression wave to a shock wave having a steep pressure rise. The train speed is 100 m/s and the Mach number 0.288. The ordinate is the ratio of overpressure Δp to ambient pressure p_0 and the abscissa is the dimensionless time ta/D, where a is the sound speed at room temperature. A steep pressure rise is built up at x/D =150. The parameter x/D is the dimensionless distance from the entrance of the tunnel.

For the train speed of 100 m/s, the $x-t$ diagram of the train entrance into the scaled tunnel simulator is shown in Fig.3. The compression wave becomes a weak shock wave as shown in Fig.2 and is running in front of the train, being followed by an expansion wave which is generated from the rear end. These waves are reflected at the exits of the tunnel and their reflection change their phase. These reflected shock and expansion waves successively encounter the train and create further reflection waves, as shown in Fig.3. This trend agrees well with our observations that

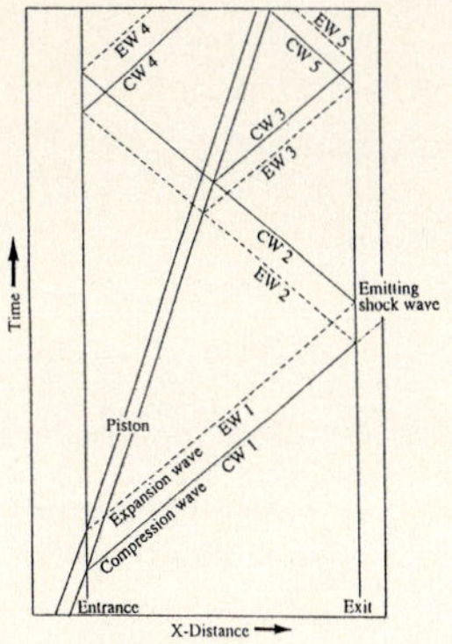

Fig. 3. Schematic $x - t$ diagram

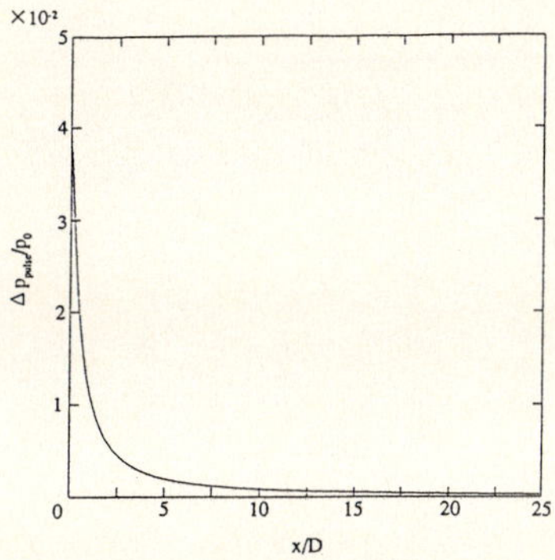

Fig. 4a. Peak overpressure along axis, M_{pst}=0.288

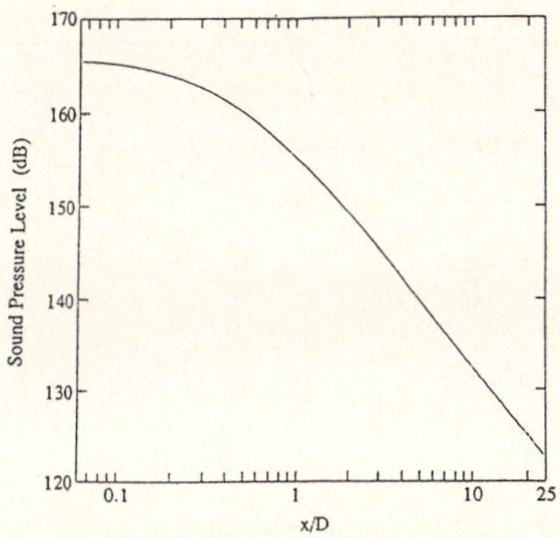

Fig. 4b. Peak sound pressure level, M_{pst}=0.288

when being on the Shinkansen train entering the tunnel, we feel the wall of the coach expanding suddenly but slightly outward. The pressure distribution along the Sinkansen train seems to be complicated even in the numerical results.

Being released from the exit of the tunnel, weak shock waves quickly attenuate and turn into sonic booms. Fig.4a shows the peak overpressure, and Fig.4b plots its peak sound pressure in terms of dB, where the ordinate represents the peak sound pressure in dB and the abscissa the dimensionless distance x/D from the tunnel exit. Although the shock wave is very weak, unlike in the acoustic treatment, a ring vortex appears at the tunnel exit and the particle velocity cannot be neglected. This implies that at least in the near field of the tunnel exit, non-linear shock wave dynamics is relevant in discussing the attenuation of the shock wave (Funabashi 1992).

3. Experiments: A scaled tunnel simulator

Prototype tunnels are too large to measure accurately various physical quantities on site. Therefore, a one-in-three-hundred scaled tunnel simulator was built in order to acquire accurately the various physical quantities representing train/tunnel interference. The measured data in the scaled tunnel simulator are also useful to validate the above-mentioned numerical scheme by comparing with numerical results. Fig.5a shows a schematic view of the scaled tunnel simulator, 40 mm in diameter and 35.6 m in length. A cylindrical plastic piston of 18 mm in diameter and approximately 200 mm in length was set in the driver chamber, as shown in Fig.5b. The 40 mm diameter tube and the 18 mm diameter plastic cylinder imitate the tunnel and the train, respectively. The

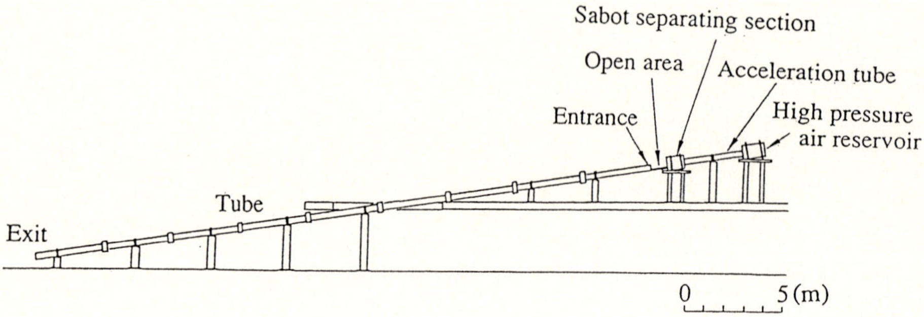

Fig. 5a. Scaled tunnel simulator

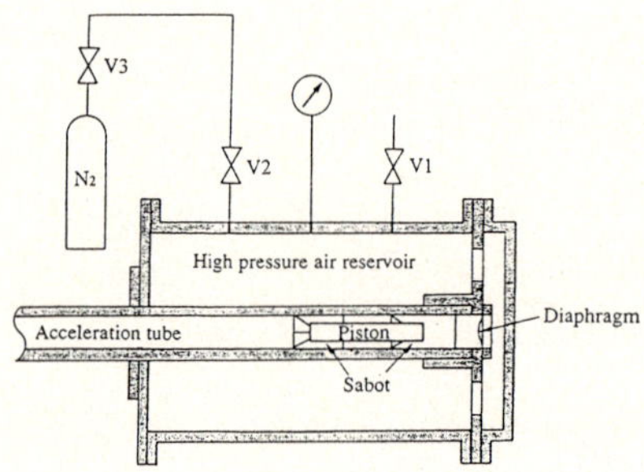

Fig. 5b. Driver chamber

plastic piston was supported, at its front surface, by a breakable sabot, and at its rear end by a second sabot. A thin Mylar diaphragm separated the rear surface of the second sabot and the high-pressure chamber. By rupturing the diaphragm, the projectile, supported on these sabots, was accelerated along the launch tube and at the entrance of the open section, being separated from the breakable sabot, moved into the tunnel simulator. The second sabot serves to stop the high-pressure gas flowing into the tunnel simulator. In order to prevent the cylinder speed from attenuating while moving along the tunnel simulator, the whole facility is inclined by 8 degrees. While the plastic cylinder is sliding down inside the steel tube, the gravitational force is expected to compensate the friction force between the plastic cylinder and the steel wall. Fig.6 shows the relationship between projectile speeds at the entrance and exit of the tunnel simulator, and the driver pressure in the high-pressure chamber. The speed of the projectile is calculated by a time-of-flight technique using laser beams passing across the tube. Open cycles and filled circles show the entrance and exit speed, respectively. The speed of the projectiles was slightly attenuated and the difference between the entrance and exit speeds is not more than 10 m/s against the entrance speed of 120 m/s.

Fig.7 shows the pressure history of the resulting shock wave in the tunnel simulator for projectile speeds at the exit of 60 m/s (220 km/h) and 75 m/s (270 km/h), respectively. The pressure was measured at 1.0 m from the tunnel simulator exit by using a Kistler model 603 B pick-

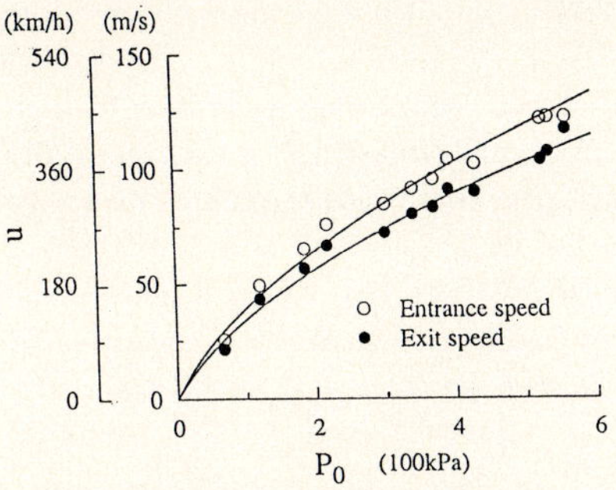

Fig. 6. Relationship between projectile speeds at the entrance and exit of the tunnel

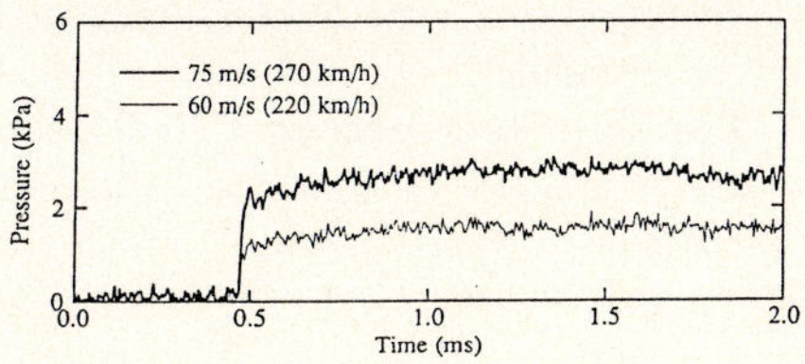

Fig. 7. Pressure history on the tube wall

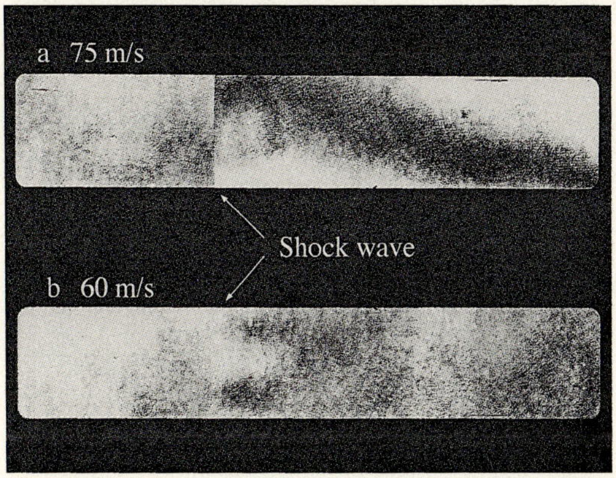

Fig. 8. Reconstructed holograms of a weak shock wave generated by a train model, **a**; 75 m/s, **b**; 60 m/s

up which was slightly recess-mounted on the tube wall. The measured overpressure behind the shock wave was approximately 2.4 kPa. Using the correlation $M_s = 1 + 0.43\Delta p/p_0$ between the shock Mach number M_s and the overpressure ratio of a very weak shock wave $\Delta p/p_0$ where p_0 is ambient pressure, the resulting shock Mach number was 1.01. Figs.8a and 8b show holographic interferometric photographs of weak shock waves generated by 75 and 60 m/s projectile speeds, respectively. These interferograms were taken by connecting a 30 mm × 40 mm rectangular cross sectional observation section at the end of the tunnel simulator.

4. Pressure measurement at a Shinkansen tunnel exit

Pressure measurements at a tunnel exit (Ichinoeki, Iwate, Japan, 10 m diam. and 10 km long) were also carried out using a home-made PVDF piezofilm pressure gauge (Fig.9). The gauge is placed on the tunnel wall at a distance of 15 m from the exit. The output from the gauge is shown in Fig. 10 for the train speed of 68 m/s, 245 km/h. The measured peak pressure is 570 Pa and corresponds to $M_s = 1.002$. This pressure wave is clearly heard as a sharp noise.

5. Summary

The process of weak shock wave generation when a high speed train passes through a tunnel is studied both numerically and experimentally. The following are obtained:

1. The process of forming a weak shock wave from compression waves inside the tube is numerically simulated. The numerical simulations also clarify detailed wave interactions in the whole process.
2. A test facility was built to study the current subject of weak shock wave generation by high speed trains in tunnels. The facility can well simulate an actual situation. It is experimentally confirmed that a shock wave, though weak, is actually generated inside the tunnel by a high speed train.
3. Pressure measurement at a Shinkansen tunnel exit was carried out. Using a home-made PVDF pressure gauge, it was confirmed that this sonic boom problem should be investigated based on treating the emitted wave not as an acoustic wave but as a weak shock wave.

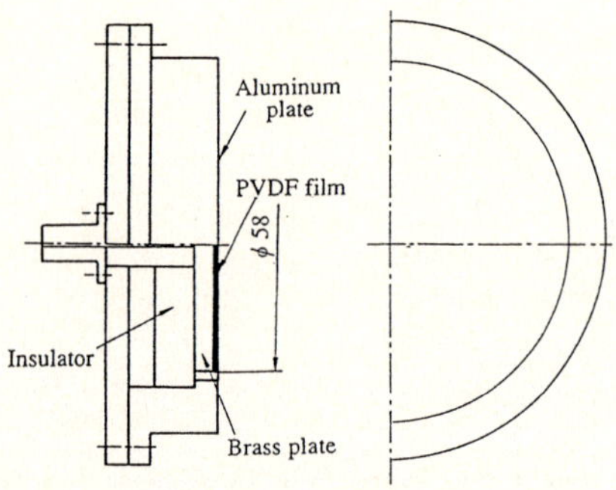

Fig. 9. PVDF pressure gauge

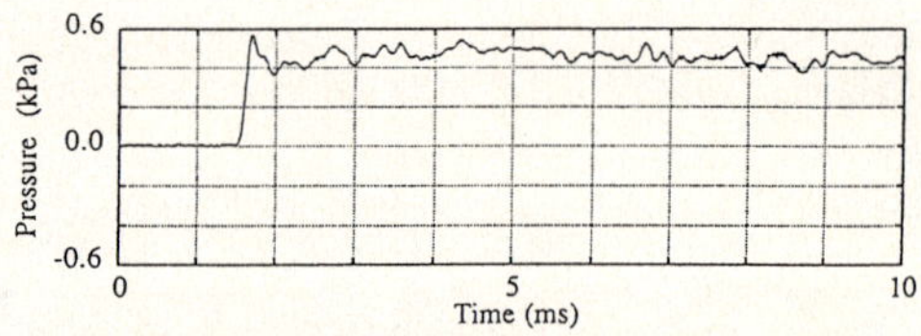

Fig. 10. Pressure history at the exit of the tunnel

References

Funabashi S (1992) Sonic booms generated by entrance of high speed train into tunnel. Master Thesis, Graduate School of Tohoku University

Groth CPT, Gottlieb JJ (1988) Numerical study of a two-stage light-gas hypervelocity projectile launcher. UTIAS Rep 327

Harten A (1983) High resolution schemes for hyperbolic conservation laws. J. Comp. Phys. 49:357-393

Lukasiewicz J (1976) The Railway Game. Carleton Contemoraries, MaCelland and Stewart Ltd

Matsuo K, Aoki T (1991) Wave problems in high-speed railway tunnels. In: Takayama K (ed) Proc. 18th Intl. Symp. on Shock Waves, Sendai, pp 95-102

Strang G (1968) On the construction and comparison of difference schemes. SIAM J. Num. An. 5:37

Yee HC (1989) A class of high-resolution explicit and implicit shock-capturing methods. NASA TN 101088

Effects of Shock Waves on Silencer Characteristics in the Exhaust Gas Flow of Automobile Engines

Noriaki Sekine*, Ichiro Kudo*, Osamu Onodera and Kazuyoshi Takayama****
*Fuji Heavy Industries, Co,Ltd.,3-9-6, Osawa, Mitaka, Tokyo, Japan
** Shock Wave Research Center, Institute of Fluid Sience, Tohoku University 2-1-1 Katahira, Aoba, Sendai, 980, Japan

Abstract. It is known that weak shock waves appearing in automobile gasoline engine exhaust pipes are responsible for the generation of exhaust gas noise. In conjunction with the reduction of this noise, an experimental investigation, using a shock tube, was made of the initiation and propagation of shock waves at $M_s = 1.1$, which are equivalent to those appearing in automobile exhaust gas pipes, and also the development of a favourable silencer configuration which may be effective to suppress the shock-induced noise. It was found that among various geometrical factors, the ratio of change of silencer cross-sectional area is the most important for attenuating effectively the shock wave and supresses the noise, that is, a double chamber with staggered pipe insertion. A numerical simulation was also carried out to interprete the experimental findings, which agreed well with the experiments.

Key Words: Automobile engine, Exhaust pipe, Silencer, Weak shock

1. Introduction

The non-linear transition of compression waves discharged from the exhaust ports of an automobile engine into shock waves generates unpleasant exhaust noises, which has been one of the urgent problems for automobile engine manufacturers to solve (Sekine et al. 1988, 1989). The goal of this paper is to clarify the characteristics of silencers which can effectively reduce this undesirable automobile engine exhaust pipe noise. A 60×150 mm cross-sectional shock tube is used for producing shock Mach numbers of approximately $M_s = 1.1$ in air; various passages representing basic configurations of silencers were inserted in the test section of the shock tube in order to visualize mutual interactions between the transmiting shock waves and the wall configuration. A numerical simulation was also carried out by a TVD finite-difference numerical scheme for a wide range of combinations of geometrical parameters of the silencers. It is confirmed that the motion of shock waves in silencers can be correctly simulated numerially. Both by experiments and numerical simulation, the most favourable configuration for the silencer is found to be the double chamber type having inserted inlet and outlet pipes.

2. Experiments

2.1. Shock tube

The shock tube consists of a 150 mm dia. high pressure chamber, 1550 mm long and a 60×150 mm cross-sectional low-pressure channel, 6750 mm long. Fig.1 shows the experimental set-up. As shown in Fig.2, the shock tube is conected to a sudden area expansion section, whose arrangement is analogous to the connection of an exhaust pipe to a silencer. The exit of this ficticious silencer is open to air. Pressures are measured, using transducers (KISTLER 603B type), at the entrance of this ficticious silencer and at 400 mm from its exit. This experimental set-up is so arranged as to simulate the interaction of weak shock waves with an automobile engine exhaust pipe and a silencer, both by pressure measurements and holographic interferometric flow visualization (Sekine et al. 1988, 1989). Silencers having various configurations are installed case by case in the test section. The geometry of the silencer models is shown in Table 1.

Shock Waves @ Marseille III
Editors: R. Brun, L. Z. Dumitrescu

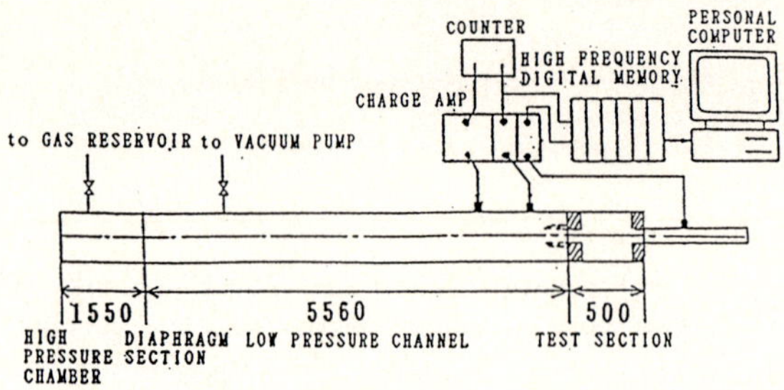

Fig. 1. Experimental set-up

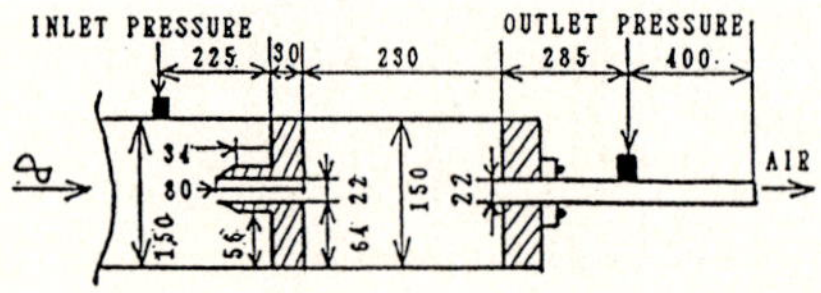

Fig. 2. Details of pressure-measurement set-up

Table 1. Geometry of silencer models

No	MODEL	D	d	L	l_1	a	b
1	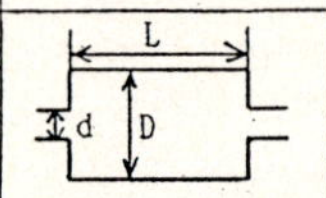	150.	22.	230.	0.	56.	56.
2	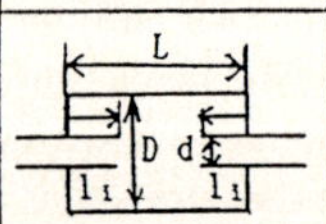	↑	↑	↑	50.	↑	↑
3	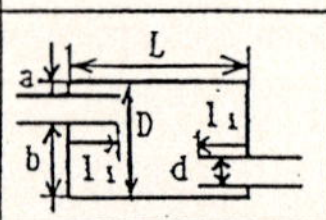	↑	↑	↑	↑	38.	89.
4	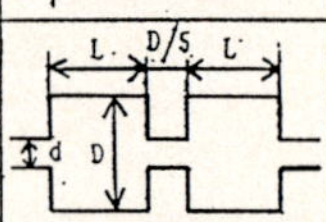	↑	↑	100.	0.	56.	56.
5	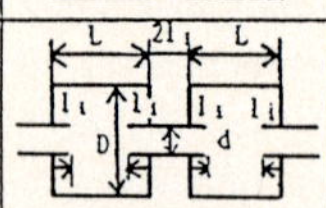	↑	↑	↑	15.	↑	↑
6	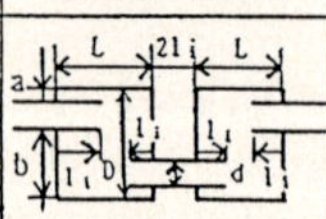	↑	↑	↑	↑	38.	89.

2.2. Method

The high-pressure chamber is filled with nitrogen at 0.131 MPa at room temperature. A 9 μm thick Mylar film separates the low-pressure channel at ambient conditions. Rupturing the diaphragm created a shock wave in the low-pressure test section at $M_s = 1.1\pm$ 1.0 %. Measured pressure signals were stored in a transient recorder with a sampling rate of 5 μs/word.

3. Results and discussion

3.1. Flow visualisation

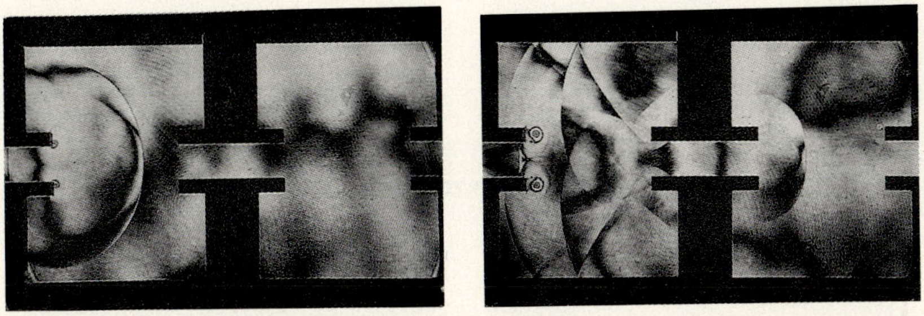

Fig. 3. Interferograms of shock propagation in a double chamber

Fig.3 shows sequential interferograms of shock propagation in a double chamber having inserted inlet and outlet pipes. It is found that while propagating in the double chamber, the shock wave attenuates very effectively due to limiting mass flow.

3.2. Pressure variations

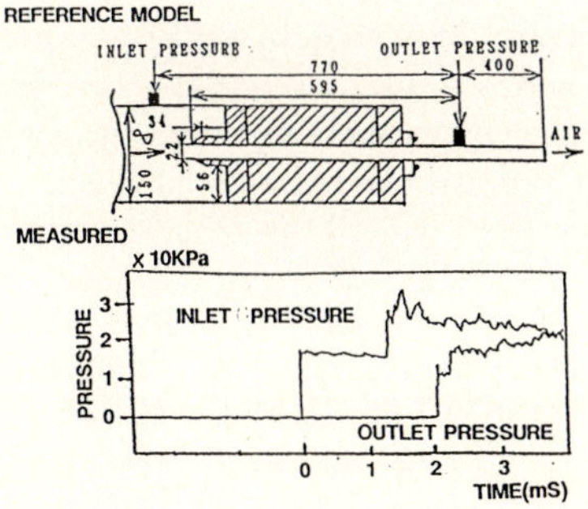

Fig. 4. Pressure variations at the entrance and exit

Fig.4 shows pressure variations at the entrance and exit as indicated in Fig.2. It is clearly observed that the shock pressure at the entrance is remarkably attenuated at the exit of the silencer. Although peak pressures are well attenuated, spike-shaped pressures, which represent the high-frequency components, still remain unattenuated. Fig.5 summarizes the pressure variations at the exit for various configurations of the silencer, as listed in Table 1 (Eriksson et al. 1985; Callow et al. 1988). For reference, the pressure variation at the entrance is shown in Fig.5(a). In the simplest case of inlet and outlet pipes inserted along the center axis in Fig.5(b), reflected shock waves are indeed attenuated and the pressure variation at the exit appears to be smooth.

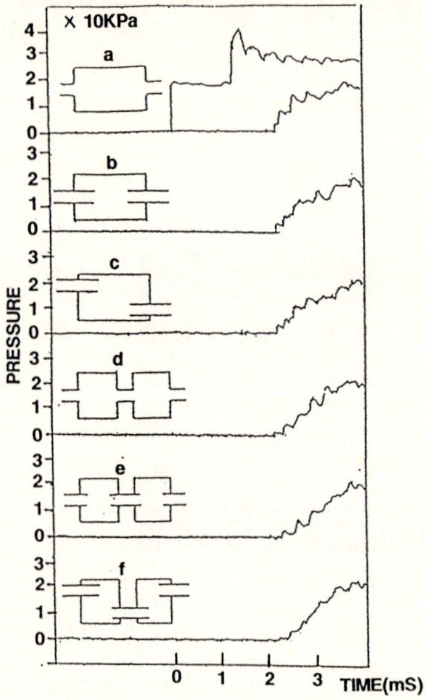

Fig. 5. Pressure variations at the exit of silencers

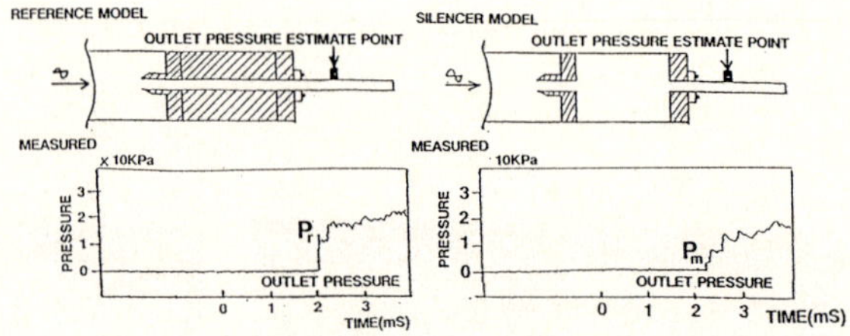

ATTENUATION OF OVERPRESSURE IN EXPERIMENT: **R** ARE DEFINED AS FOLLOWS

$$R = \frac{P_m}{P_r}$$

FACTOR	D/d	INSERTED	STAGGERED	NUMBER OF CHAMBER
R %	76%	3%	5%	16%

D/d :A RATIO OF EXHAUST PIPE DIA d TO SILENCER DIA D

Table 2. Order of effectiveness of various factors

In the case of the staggered insertion of inlet and outlet pipes, as in Fig.5(c), due to the shifted arrangement of the outlet pipe, the transmitted shock wave hits directly the end wall of the silencer so that its reflected shock wave cannot be attenuated as effectively as in Fig.5(a). A spike-shaped peak pressure remained, which indicates that the high-frequency component is not well suppressed. In a double-chamber type silencer, as in Figs.5(d) and (e), weak shock waves are attenuated more effectively than in single-chamber type silencers in Figs.5(a) and (b). Double-chamber type silencers having a staggered inlet and outlet pipe insertion are found to attenuate shock waves most effectively. However, the resulting pressure variation still contains pronounced high-frequency components.

3.3. Attenuation of overpressures

The ratio of exhaust pipe diameter d to silencer diameter $D : m = D/d$, determines the attenuation of overpressures. In the present series of experiments, four combinations of variables are examined. That is, 1) D/d, 2) whether or not pipes are inserted into silencers, 3) whether or not pipes are staggered, and 4) number of chambers, single or double. Table 2 shows the order of effectiveness of these factors. According to these silencer models, it is found that D/d is the dominant factor. In the design of a silencer, If D/d cannot be a large value, which is the case in many practical designs, the double chamber is a more important factor.

4. Numerical simulation

4.1. Set-up of the silencer

In this study, an upwind TVD finite-difference numerical scheme has been applied to the two-dimensional Navier-Stokes eqs. (Sekine et al. 1988, 1989; Yee et al. 1985). The silencer model and initial conditions are shown in Fig.6. In this Section, the effects of D/d and D/L on the attenuation of overpressure is examined. In order to clarify the attenuation of overpressures in silencers, a correlation between M_s and the overpressure attenuation is shown in Table 2.

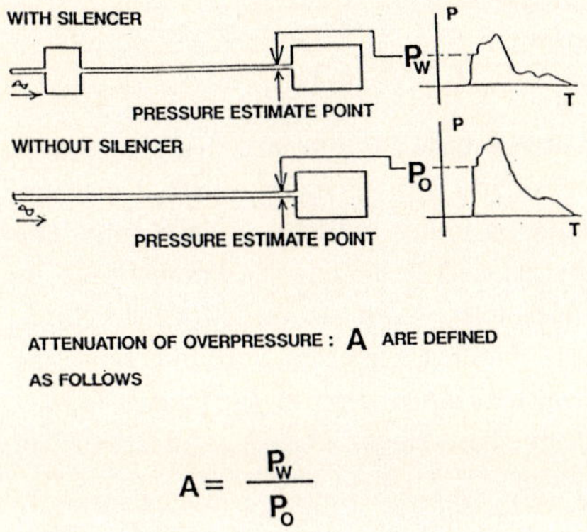

Fig. 6. Computation model and conditions

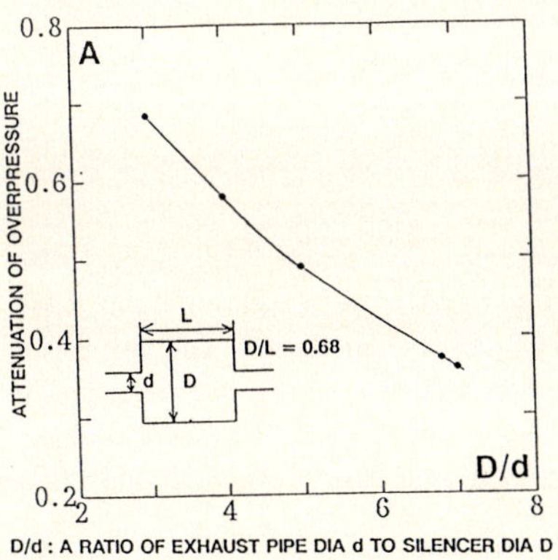

Fig. 7. Correlation of attenuation of overpressure and D/d

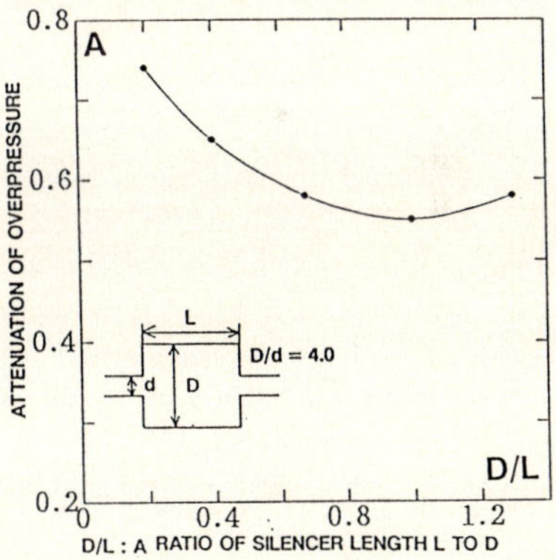

Fig. 8. Correlation of attenuation of overpressure and D/L

4.2. Results

Fig.7 shows the correlation of the attenuation of overpressure with D/d at $D/L = 0.68$. It is found that the smaller D/d is, the larger the attenuation of overpressure becomes. The D/d in practical automobile exhaust silencers cannot be over 7.0, so that in this numerical simulation the range of D/d is a maximum at 7.0. Fig.7 indicates that the attenuation of the overpressure is directly proportional to D/d.

Fig.8 shows the relationship between overpressure attenuation and D/L for $D/d = 4.0$. This value is the largest allowed for the automobile silencers of SUBARUs. Generally, the largest D/L allowed for a standard automobile silencer is approximately 1.3. Therefore, in the numerical simulation the largest value of D/L is 1.3. In the present computation D/L is 1.0 and the overpressure attenuation is a maximum. For a silencer of the single-chamber type, an optimum D/L will exist for given M_s and D/d. These results are different from design concepts of silencers proposed by acoustic theory (Davis et al. 1954) In the design of silencers of automobile engine exhaust pipe systems, the existence of an optimum D/L should be taken into account.

5. Measurements of pressures and sound pressures

5.1. Experimental conditions

The experiment was carried out on a water-cooled 0.66-litre 4-cycle gasoline engine having multi-point-fuel injection. The engine was connected to an electrical dynamometer through a transmission. The measurement was conducted in firing condition under full load at 4800 rpm. The pressure transducers (KISTLER 603B type) were set at stations 1 to 4 along the exhaust pipe, as shown in Figs.9(a) and (b). At the same time, the exhaust noise was measured at station No. 5. Silencers used in this pressure measurement were a double chamber expansion with insertion-pipe type and a single chamber expansion, where D/d, the length of inserted pipes, the staggered pipe length, and the volume of the chamber were identical for these cases. These specifications are what is practically used for the silencers.

5.2. Results

Figs.9(a) and (b) (lower trace), show the pressure variations measured along the exhaust pipes. It is found that the double-chamber type attenuates well the shock waves, but in the single-chamber type shock waves still remain at the outlet pipe and the unpleasant exhaust noise was not completely suppressed. It is found that a double-chamber type is more effective in attenuating shock waves than controlling the D/L.

Fig.10 shows a spectrum of sound pressure levels measured at the outlet of the silencer. The test conditions are the same as in the previous case. The frequency component of unpleasant exhaust noise caused by shock waves ranges from approximately 1/3 octave band 1 kHz to 3.15 kHz. Over the 1/3 octave band 3.15 kHz range, the sound intensity level in the double-chamber type is higher than that in the single-chamber type. Generally, the exhaust pipe noise in the range over 1/3 octave band 3.15 kHz is so-called "flow generated noise" (Lighthill 1952, 1954; Obata et al. 1978). High-frequency sound is to be reduced by adapting an usual sound absorption muffler. The double-chamber type is, therefore, less effective than the single-chamber type in suppressing high-frequency noise. In order to reduce the shock-induced noise, it is worth adopting a double-chamber type silencer combined with a sound-absorption muffler at the rear end of the double-chamber type silencer.

These results have been successfully applied to practical automobile silencers. Models IMPREZA and VIVIO produced by SUBARU in 1992 have already adopted a double-chamber silencer and the sound absorption muffler. For the decision of adopting the silencer model, the result of the present research has been taken into account.

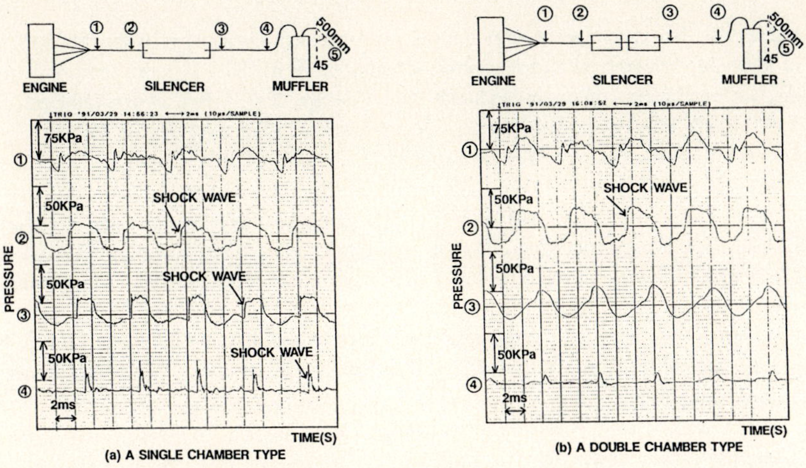

Fig. 9. Pressure variations in an automobile exhaust pipe

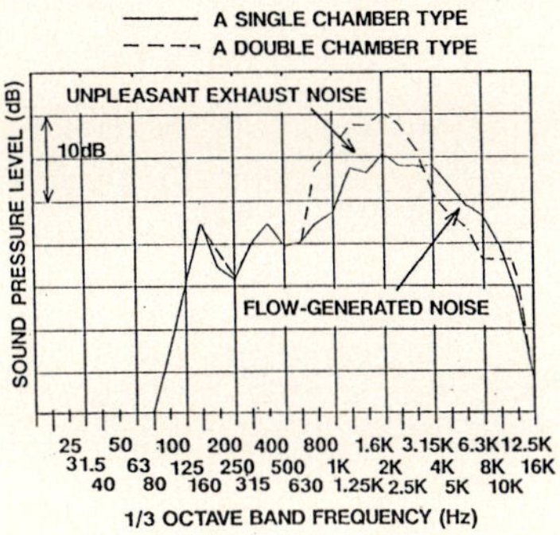

Fig. 10. The spectrum of sound pressure levels

6. Conclusions

The results obtained are summarized as follows:

(1) The ratio of exhaust pipe diameter d to silencer diameter $D, D/d$, is effective. Under conditions of less than 7.0, overpressure attenuation is directly proportional to D/d.

(2) For a silencer of the single-chamber type, an optimum ratio of silencer length L to D, D/L, will exist for given M_s and D/d. In the design of silencers of automobile engine exhaust pipe systems, the existence of the optimum D/L should be taken into account.

(3) The double-chamber type attenuates well the shock waves. However, this type is less effective than the single-chamber type in suppressing high frequency noise (so-called "flow-generated noise"). High frequency noise is to be reduced by adapting an usual sound-absorption muffler.

(4) Consequently, for practical automobile exhaust silencers, which cannot accept large values of D/d, in order to reduce the shock induced noise, it is worth adopting a double-chamber type silencer combined with a sound absorption muffler at the rear end of the double-chamber type silencer.

References

Callow GD et al. (1988) Insertion loss of engine intake and exhaust silencers. C19/99 IMechE

Davis Jr DD et al.(1954) Theoretical and measured attenuation of mufflers at room temperature without flow, with comments on engine-exhaust mufler design. NACA Report No.1192, 827

Eriksson LJ et al. (1985) Theory and practice in exhaust system design. SAE Paper 850989

Lighthill MJ (1952) On sound generated aerodynamically I. Proc. Royal Soc. London, A211, 564

Lighthill MJ (1954) On sound generated aerodynamically II. Proc. Royal Soc. London, A222, 1

Obata T et al. (1978) Estimation on acoustic power of flow-generated noise and on head losses of silencers. J. Acoustical Society of Japan. 34-9

Sekine N et al. (1988) Shock wave development and propagation in automobile exhaust systems. SAE Paper 880082

Sekine N et al. (1989) Generation and propagation of shock waves in the exhaust pipe of a 4 cycle automobile engine. In: Kim YW (ed) Proc. 17th Intl. Symp. on Shock Waves and Shock tubes, AIP Conference Procs. 208

Yee H C et al. (1985) Implicit total variation diminishing (TVD) schemes for steady-state calculations. J. Computational Physics 57

Noise Induced by Weak Shock Waves in Automobile Exhaust Systems (Effects of Viscosity and Back Pressure)

S. Matsumura*, **O. Onodera*** **and K. Takayama**†
*Subaru Research Center Ltd. Ota, Gunma, Japan
†Institute of Fluid science, Tohoku University, Sendai, Japan

Abstract. Paper reports on the effects of viscosity and back pressure on shock wave formation in an automobile exhaust pipe. In the present experiment, a periodic wave generator was designed and built for investigating the transition of compression waves into shock waves. Pipes with various diameters were tested to observe compression waves developing into shock waves under the influence of viscosity. A numerical simulation was also conducted using the TVD (Total Variation Diminishing) finite difference scheme applied to the Navier-Stokes equations. It was found that the viscosity hardly affects the transition of compression waves into shock waves as long as the pipe has a practical size suitable to automobile exhaust systems. The back pressure in the exhaust pipe is more important than the viscosity for the transition of compression waves into shock waves.

Subsequently, the mechanism of the noise generation at the exhaust outlet was discussed. From the aeroacoustic analysis, one infers that the magnitude of sound pressure emanated from the open end is proportional to the gradient of pressure variation in the exhaust pipe.

Key words: Noise, Weak shocks, Automobile exhaust

1. Introduction

Compression waves in an automobile exhaust pipe can develop into shock waves under certain conditions. Such shock waves generate loud and jarring exhaust noises. In our previous paper, the existence of shock waves in exhaust pipes was clearly observed using holographic interferometric flow visualization.

In order to reduce exhaust noise, the formation of shock waves in the exhaust pipe and the mechanism of the noise generation at the exhaust outlet have to be clarified. In the present study, the effect of viscosity on the transition of a compression wave into a shock wave was experimentally and numerically discussed. The analysis of noise generation at the exhaust outlet was conducted by aeroacoustic dynamics.

2. Experimental set-up

The experimental apparatus is shown in Fig.1. A high-pressure chamber (a part of the periodic wave generator) simulates an automobile engine whilst a straight pipe simulates the exhaust pipe. The straight pipe, being changeable, with different diameters, serves to check the existence of compression waves developing into shock waves under the influence of viscosity. The valve located in the high pressure chamber periodically opens and closes. High-pressure air is supplied by a compressor through a reservoir tank. Pressure transducers (Kistler 6001 type) are set along the straight pipe and pressure measurements are carried out.

Shock Waves @ Marseille III
Editors: R. Brun, L. Z. Dumitrescu

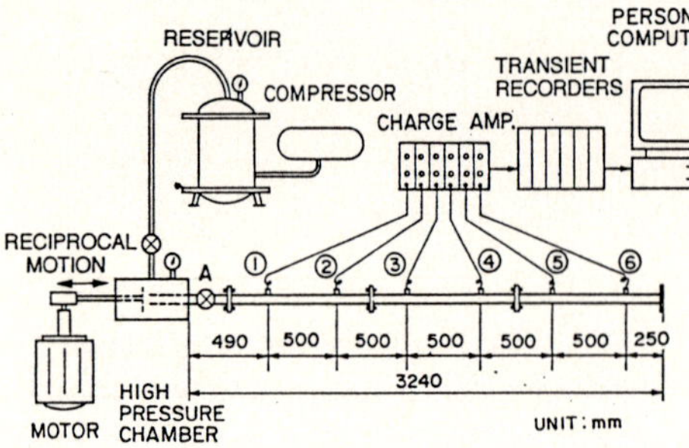

Fig. 1. Experimental apparatus

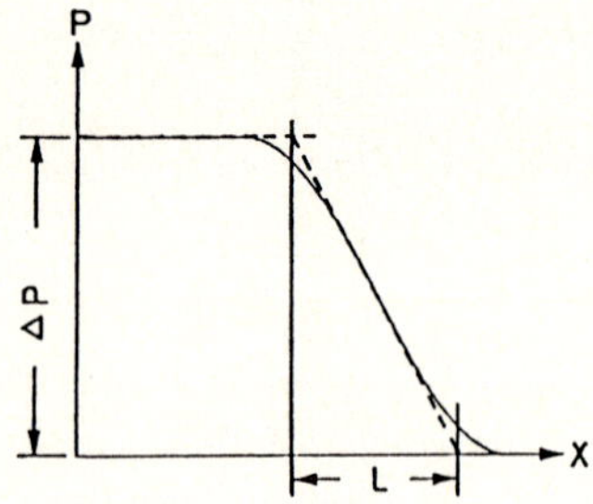

Fig. 2. Definition of thickness L

3. Numerical simulation

The TVD finite difference scheme (Harten 1983) was applied to the axisymmetrical two-dimentional unsteady Navier-Stokes equations (Reddy et al. 1991). Grid points in the exhaust pipe are 2500×40. Initial disturbances corresponding to the experimental pressure pulse were imposed at the pipe inlet. The computations were carried out on a Fujitu VP-2600 supercomputer of the Fuji Heavy Industries.

4. Results and discussion

4.1. Transition of a compression wave into a shock wave in the exhaust pipe

The compression wave propagating along the exhaust pipe is affected by counterbalancing phenomena. One is the nonlinearity which makes the waveform to become steeper. The other one works in the opposite sense, i.e. the dissipation of the compression wave. The dissipation is due to the viscosity of the exhaust gas. In the present experiment, pipes of 8 mm and 20 mm in diameter were tested. A numerical simulation was also conducted. Results are shown in Fig.3 for a 8 mm diameter pipe and in Fig.4 for a 20 mm diameter pipe.

Fig.3a shows the experimental result of the distortion of the compression wave propagating along the pipe. The abscissa is distance x and the ordinate is the overpressure of the compression wave. The numbers on the left hand side of Figs.3 and 4 indicate the locations of the pressure measurement stations. At location (6), a shock wave is formed. Fig.3b shows the result of the numerical simulation using the TVD finite difference scheme, which agrees well with the exprimental result. In order to compare the distortion of the compression wave, the thickness of the compression wave L was introduced and defined in Fig.2. Fig.3c shows the variation of a compression wave of thickness L along the pipe. The abscissa is distance x and the ordinate is the thickness of the compression wave. Fig.4 shows the results in the same manner as Fig.3. Comparing Figs.3 and 4, it is found that the viscosity does not play any important role in the shock wave formation since no clear distinction is seen between these two cases.

The numerical experiment was conducted to clarify the effect of the fluid viscosity on the transition of compression waves into shock waves. Fig.5 shows the variation of the thickness of the compression wave along the pipes for various pipe diameters. Though there are significant differences when D is less than 0.4 mm, very small differences can be seen when D is larger than 0.8 mm. Therefore, the viscosity of the gas hardly affects the transition of compression waves into shock waves as far as the pipe size is large enough to be used practically for automobile exhaust systems. However, it is known that for the smaller diameter it is hard to generate the exhaust noise caused by shock waves. In the exhaust pipe under running conditions, the gas flow generates back pressure. Of course the pipes of smaller diameter induce a larger back pressure. The efffect

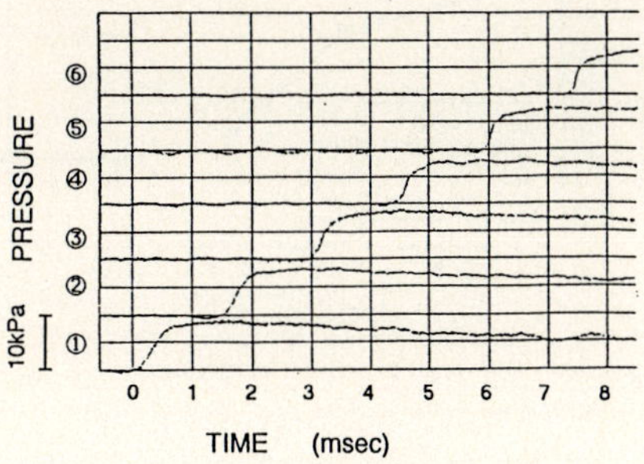

(a) Experiment

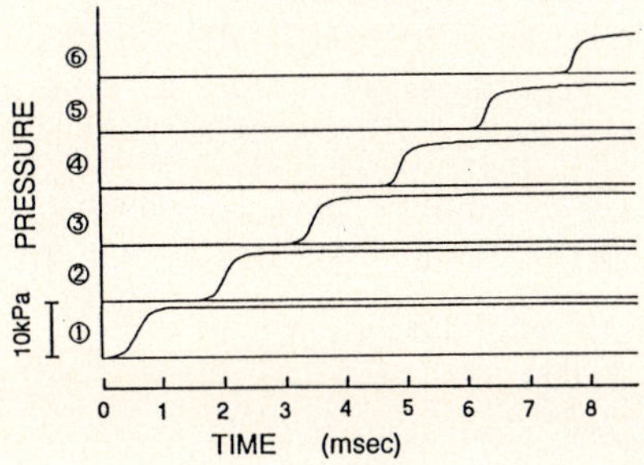

(b) Numerical simulation

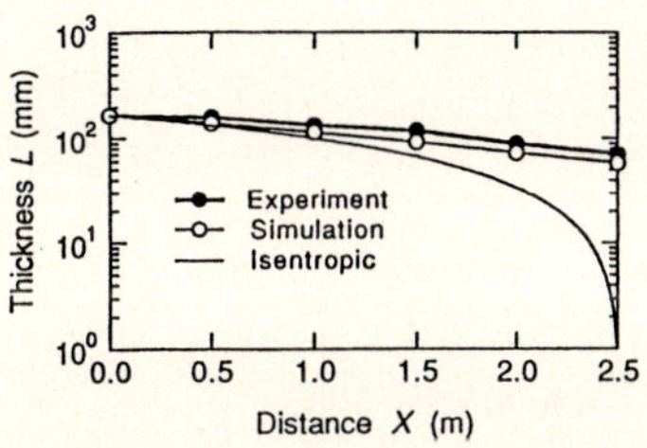

(c) Thickness

Fig. 3. Shock wave formation ($D = 8$ mm)

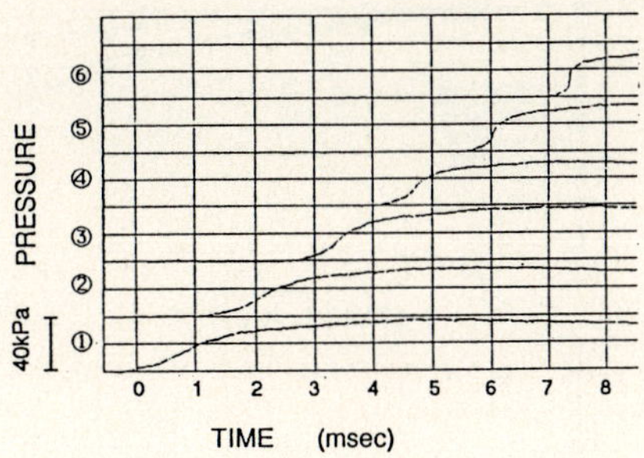

(a) Experiment

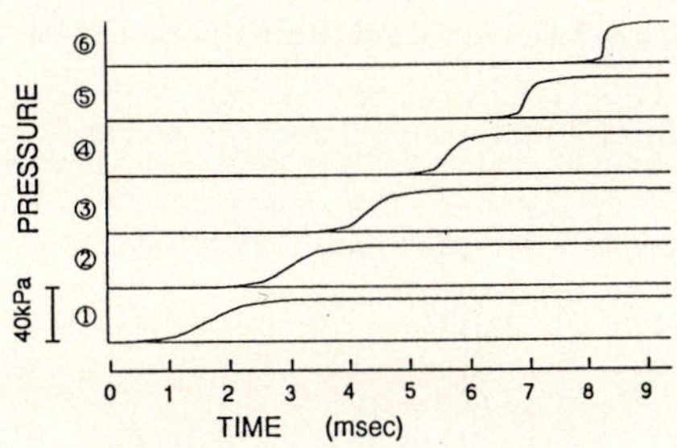

(b) Numerical simulation

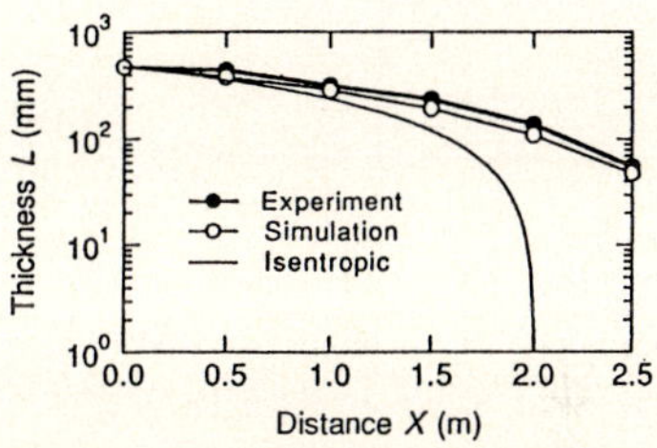

(c) Thickness

Fig. 4. Shock wave formation ($D = 20$ mm)

of back pressure is derived from the following equation (Liepmann et al. 1960). In the isentropic assumption, the local sound velocity U is expressed as

$$\frac{U}{a_0} = \frac{\gamma+1}{\gamma-1}\left[\left(\frac{p}{p_0}\right)^{(\gamma-1)/2\gamma} - 1\right] + 1 \tag{1}$$

where a_0 is the sound velocity at stagnation conditions, p and p_0 are the local pressure and back pressure, γ is the ratio of specific heats. Eq.1 indicates that the back pressure p_0 is closely related to the deformation of the compression wave.

Fig.6 shows the effect of the back pressure on the variation of thickness of the compression wave. The upper three curves show the effect of back pressure. The lower two curves show the effect of viscosity. Results indicate that the back pressure in the exhaust pipe is more influential than the viscosity in the transition of compression waves into shock waves.

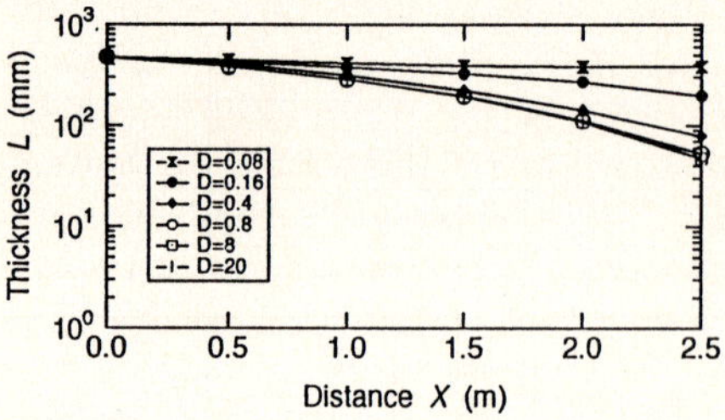

Fig. 5. Effect of viscosity on thickness of compression wave

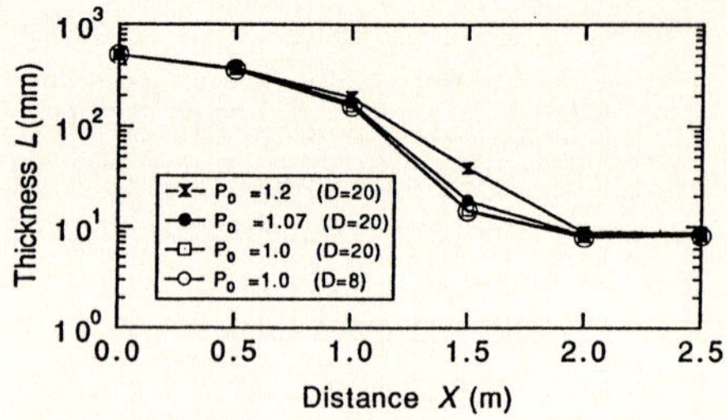

Fig. 6. Effect of back pressure on thickness of compression wave

4.2. Exhaust noise generation in the far-field

Assuming a pipe with an infinite baffle plate as shown in Fig.7, the sound pressure Δp which is radiated from the pipe exit is given (Lighthill 1952) by:

$$\Delta p(r,t) = \frac{1}{2\pi r}\frac{\partial m}{\partial t} \tag{2}$$

where r is the distance from the exit, t is time, and m is the mass flux.

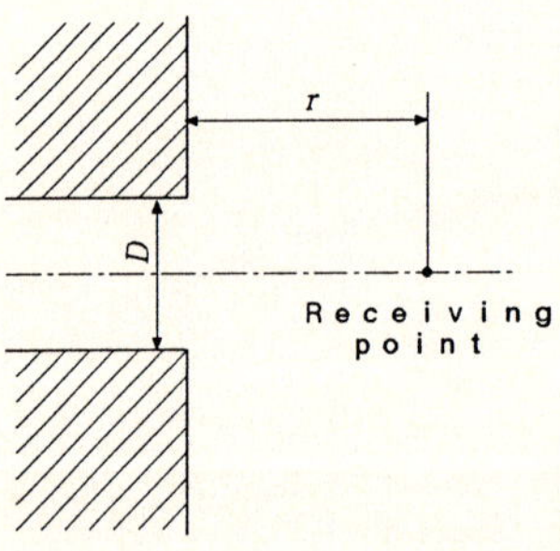

Fig. 7. Pipe with baffle plate

The overpressure p_i at the exit is

$$p_i = \rho\, a\, u \tag{3}$$

where r is the density, a the sound speed, u the particle velocity. The mass flux m is expressed as

$$m = (\pi/4)\, D^2\, \rho\, u \tag{4}$$

where D is the diameter of the pipe.

Substituting Eq.3 into Eq.4, and Eq.4 into Eq.2, we have

$$\Delta p(r,t) = \frac{D^2}{4ra}\frac{\partial p_i}{\partial t} \tag{5}$$

Applying the open end correction, the temporal pressure gradient is defined (Aoki et al. 1991) as:

$$\frac{\partial p_i}{\partial t} \equiv \frac{p_i(t+\Delta l/a) - p_i(t-\Delta l/a)}{2\Delta l/a} \tag{6}$$

where Δl is the correction length ($= 0.425\ D$). Eq.5 means that the magnitude of the radiated sound pressure is proportional to the gradient of the compression wave in the pipe (Ozawa 1980). Therefore it is found that the exhaust noise becomes large when shock waves are formed in the exhaust pipe (Takayama et al. 1981). Fig.8 shows the incident compression waveform and the radiated compression waveform when $D = 20$ mm and $r = 200$ mm. Each line in Fig.8a corresponds to a line of the same sort in Fig.8b. Fig.9 shows the relation between the maximum sound pressure and the incident pressure gradient. The abscissa is the pressure gradient, normalized to unity when the wave thickness L is equal to $2\Delta l$, and the ordinate is the maximum sound pressure.

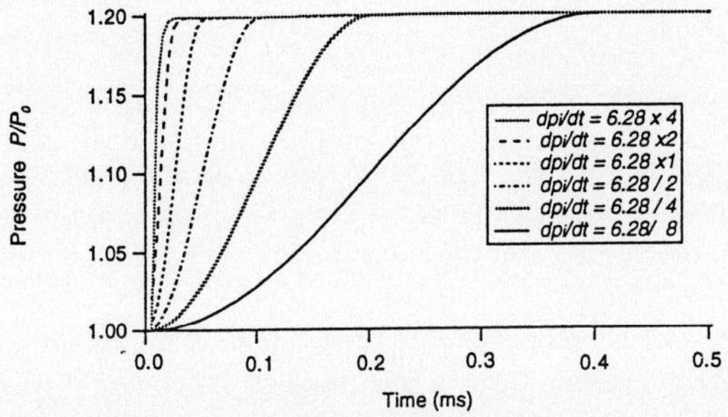

(a) Incident Compression Wave

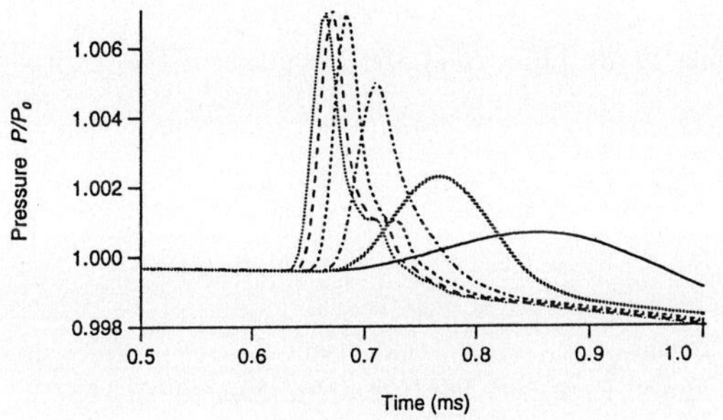

(b) Radiated Compression Wave

Fig. 8. Incident pressure with various gradients and radiated sound pressure

Results indicate :

- In the case $L > 2\Delta l$, the magnitude of the radiated sound pressure is proportional to the gradient of the incident compression wave.
- In the case $L < 2\Delta l$, the magnitude of the radiated sound pressure is proportional to that of the incident compression wave.

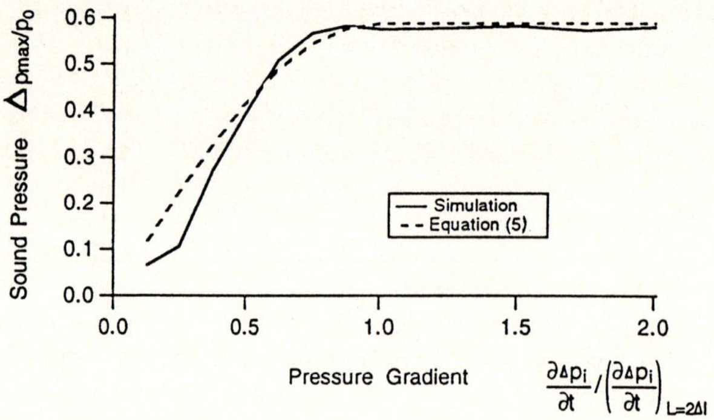

Fig. 9. Relation between maximum sound pressure and incident pressure gradient

5. Conclusions

Results obtained are summarized as follows ;

(1) The gas viscosity hardly affects the transition of compression waves into shock waves as long as the pipe has a practical size suitable to automobile exhaust systems. The back pressure in the exhaust pipe is more influential than the viscosity on the transition of compression waves into shock waves.

(2) The magnitude of the radiated sound pressure is proportional to the gradient of the compression wave in the pipe when the incident compression wave is thick enough. Therefore it was clarified that the exhaust noise becomes large when shock waves are formed in the exhaust pipe.

Acknowledgement

The authors would like to express many thanks to Mr. Ojima and Mr. Takahashi of the Institute of Fluid Science, Tohoku University.

References

Aoki T, Kashimura H, Nonaka Y, Matsuo K (1991) Discharge of compression wave from the open end of a tube. In: Takayama K (ed) 18th Intl. Symp. on Shock Waves, Sendai, pp 1331-1334

Harten A (1977) The artificial compression method for computation of shocks and contact discontinuities I. Single conservation laws. Comm. Pure and Appl. Mathematics 30: 611-637

Harten A (1983) High resolution schemes for hyperbolic conservation laws. J. Comp. Phys. 49: 357-393

Liepmann L, Roshko A (1960) Element of Gasdynamics. John Wiley & Sons, New York

Lighthill M (1952) On sound generated aerodynamically I. General theory. Proc. Roy. Soc. London A211: 564-587

Ozawa S (1980) Studies of micro-pressure wave radiated from a tunnel exit. Railway Tech. Res. Report of Japanese National Railways: 8-14 (in Japanese)

Reddy S, Klausmeyer S, Liu X, Papadakis M (1991) TVD formulations of the 2D Navier-Stokes equations for airfoil analysis. SAE Paper 910992

Takayama K, Sekiguchi H (1981) Formation and diffraction of spherical shock waves in a shock tube. Rep. Inst. High Speed Mech. Vol. 43 No 336, pp 89-119.

An Experimental Study of an Intelligent Muzzle Brake

K.C. Phan
Defence Research Agency (DRA) Fort Halstead, Sevenoaks, Kent, England TN14 7BP

Abstract. An investigation was made to study a novel aspect of muzzle brake design. It uses a concept which isolates and minimises the interaction of an expanding blast wave from the muzzle of a gun with a muzzle brake device, whilst at the same time maintains its function as a recoil impulse attenuator. The technique utilises the intrinsic energy of the exhaust gas flow to activate a set of active surfaces. During the gas ejection phase from a weapon launch, these surfaces become operative under the influence of the exhaust gas pressure, and function interactively with a muzzle brake surface to deflect some of the axial gas momentum. This effect produces a forward gas dynamic impulse by the brake and is subsequently contributing to the reduction of the overall recoil impulse exerted on a gun.

Key words: Blast wave, Muzzle brake, Recoil, Impulse

Nomenclature

AMB Adaptive (or Active) Muzzle Brake
P_a Ambient pressure
ΔP_B Peak overpressure with AMB
ΔP_R Peak overpressure without AMB
I_B Forward impulse exerted on AMB
I_T Total forward or gas ejection impulse
C Calibre of blast tube
β Peak overpressure ratio
η Impulse attenuation efficiency

1. Introduction

There are two factors that are commonly used to assess the performance of a muzzle brake. These are blast overpressures and a recoil impulse reduction efficiency parameter. When a muzzle device is configured for the conventional application on a gun, these factors function against each other and as a result, a compromise is often needed to obtain a balance between recoil suppression efficiency and acceptable level of blast overpressures. However such a compromise is not always favourable and a design is seeked to isolate these events to optimise the effectiveness of a muzzle brake.

In this paper, a novel approach to suppress recoil impulse without significantly enhancing the blast overpressures is described. The technique uses the intrinsic energy derived from the gas to operate a set of active surfaces. These surfaces become operative only during the gas ejection phase, omitting the interactions of the expanding blast wave with the muzzle device minimising the effect of blast reflection. This in turn, gives a brake design with little or no compromise between blast overpressures and recoil absorption efficiency. In this study, a number of configurations was considered and a presentation of the study is reported here.

2. A novel concept of muzzle brake design

The novel aspect of an intelligent muzzle brake is one that produces a large forward impulse when it is needed. It works by deflecting the exhaust gas flow after the main blast wave event has passed. A schematic of this design is shown in Fig.1a. The active component is a retractable flap which operates under the influence of the exhaust gas pressure, allowing the gas to escape and subsequently, to become and perform as an integral part of a muzzle brake device to yield a

Shock Waves @ Marseille III
Editors: R. Brun, L. Z. Dumitrescu

forward impulse by deflecting and decreasing the axial momentum of the gas. By the end of the gas ejection phase, the effect of gas pressure diminishes permitting the active surfaces to retract, thus completing a cycle of operation. The sequence is schematically shown in Fig.1b-f.

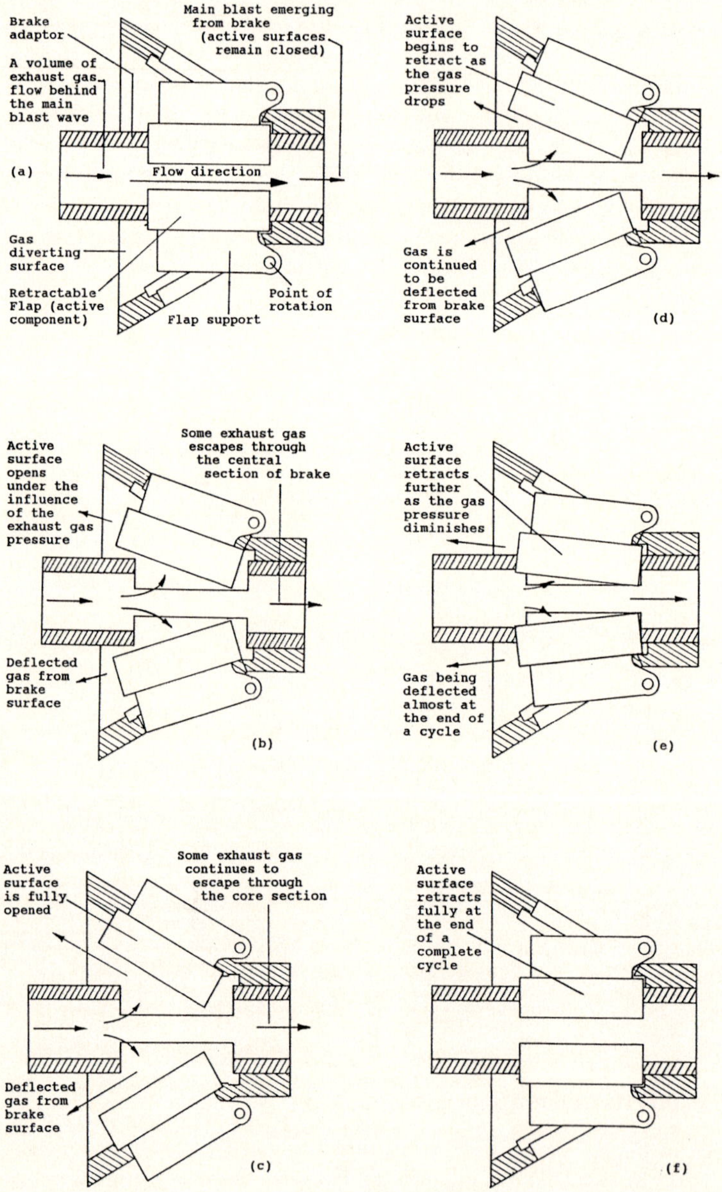

Fig. 1. Schematic of a novel muzzle brake design and illustration of one complete cycle of operation

3. Experimental details

The experimental investigation was conducted using the facility by Phan (1991). In the muzzle brake design, the active component was a pair of retractable surfaces. This, combined with a gas diverting surface and an adaptor, makes up the adaptive (or active) muzzle brake (AMB). The assessment of a muzzle brake performance is based on two parameters, the blast overpressure parameter (β) and the impulse attenuation efficiency (η). These are given as follows:

$$\beta = \frac{\Delta P_B/P_a}{\Delta P_R/P_a}; \qquad \eta = \frac{I_B}{I_T}$$

where P_a is the ambient pressure, P_B is the peak overpressure with an AMB and P_R is the reference peak overpressure without an AMB, I_B is the forward impulse exerted on an AMB and I_T is the total forward or gas ejection impulse.

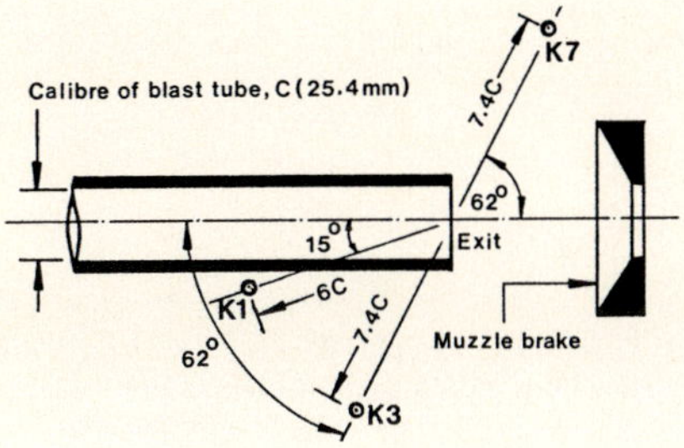

Fig. 2. Pressure measurement locations

The pressure measurements were taken at locations shown in Fig.2. The force measurement was obtained by the technique employed by Phan (1991). Tests were conducted on a number of configurations. An example of the discharging gas flow interactions with one of these configurations is shown in Fig.3.

Three test models are presented here and are shown as AMB1, AMB2 and AMB3 in Fig.4. These models had the same overall dimensions. The diameters of the gas diverting surface and centre passage hole were 0.1524 m (6") and 25.4 mm (1") respectively and the length of the cut-out segments on the brake adaptor was 76.2 mm (3"). A test model had two retractable closure flaps. Each flap was supported by a rectangular rib and depending on the configuration, the rib rotated either about a shaft T or with the shaft T about a pair of cylindrical sockets.

On AMB1, the spring loading action on each flap was provided by a pair of half S-shape strips of spring steel (about 0.01" thick). The rotating motion was about the shaft T. Tests had shown that though the flaps had returned to the fully closed position after each run, it was found that the spring loading action was not fully restored due to a partial distortion on the spring caused by the impulsive loading on the flaps. As a result, a corrective action was often necessary to restore the spring loading before another test can be made. This was unacceptable as AMB1 was unable to sustain repetitive cycles of operation. It was decided that a different spring arrangement was needed.

An improvement was made in the spring loading mechanism. This was tested on AMB2 and it had a spring arrangement in the form of a coil. It consisted of a strip of spring steel which coiled at each end of shaft T. One end of the spring steel was connected to the shaft while the other end was attached to the gas diverting surface. The shaft was spring loaded by winding it around the spring steel and its loading action transferred to each flap by securing rigidly the shaft to the flap's supporting rib. In this configuration, the flap and the shaft T rotated as an integral

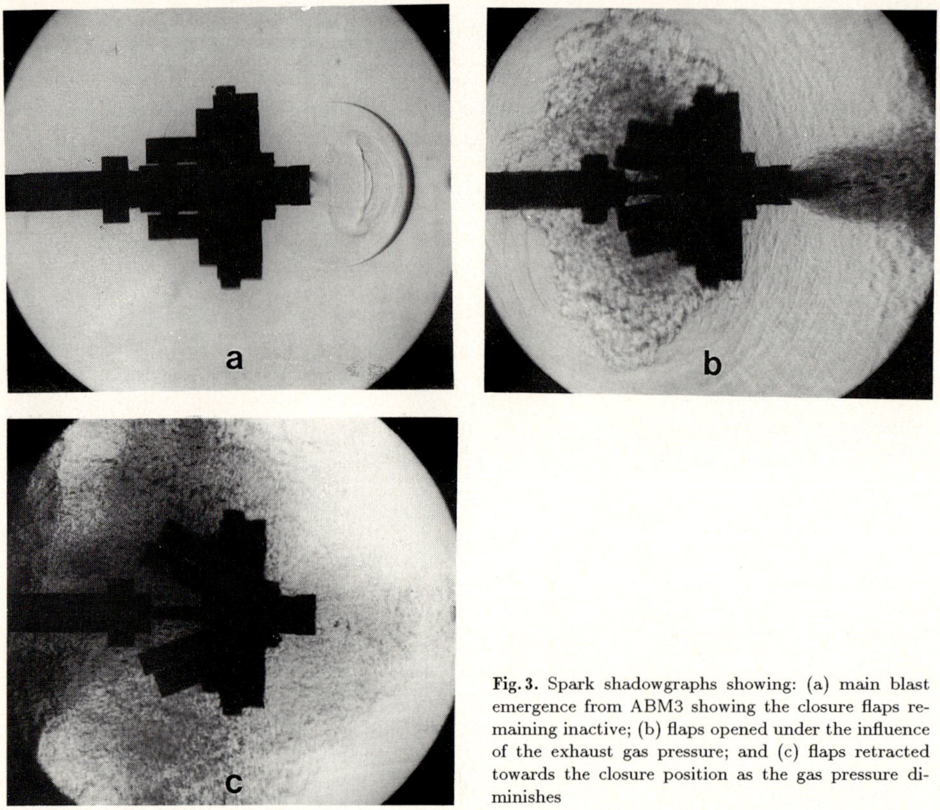

Fig. 3. Spark shadowgraphs showing: (a) main blast emergence from ABM3 showing the closure flaps remaining inactive; (b) flaps opened under the influence of the exhaust gas pressure; and (c) flaps retracted towards the closure position as the gas pressure diminishes

part about a set of cylindrical sockets as shown on AMB2. A series of repetitive tests was made. AMB2 was shown to endure successfully a successive cycles of operation without any corrective action. Later, this arrangement of control mechanism was extended to AMB3 where a conical gas diverting surface was deployed to achieve greater forward impulse. A summary of results is given in Table 1 (The reference values for $\Delta P_R/P_a$ at K1, K3 and K7 were 0.03292, 0.07623 and 0.3747 respectively. The total gas ejection impulse, I_T measured was 1.4746 Kgf.s). It can be observed that, when these were compared with results obtained on devices without the active components and of the same geometrical shape and dimensions, there was a significant difference in peak overpressures while at the same time still maintaining the impulse attenuation efficiency.

Table 1. A summary of results and a comparison with the conventional muzzle brakes of the same geometric shape and dimensions

Muzzle brake Configuration	Blast overpressure Parameter (β) at location			Impulse attenuation efficiency (η)
	K1	K3	K7	
AMB1 & AMB2	1.94	0.92	0.37	0.80
Flat Disc as gas diverting surface (compared with AMB1 & AMB2)	8.52	3.55	0.73	0.88
AMB3	2.86	1.57	0.41	0.86
35^0 Conical disc as gas diverting surface (compared with AMB3)	17.4	5.13	0.73	0.92

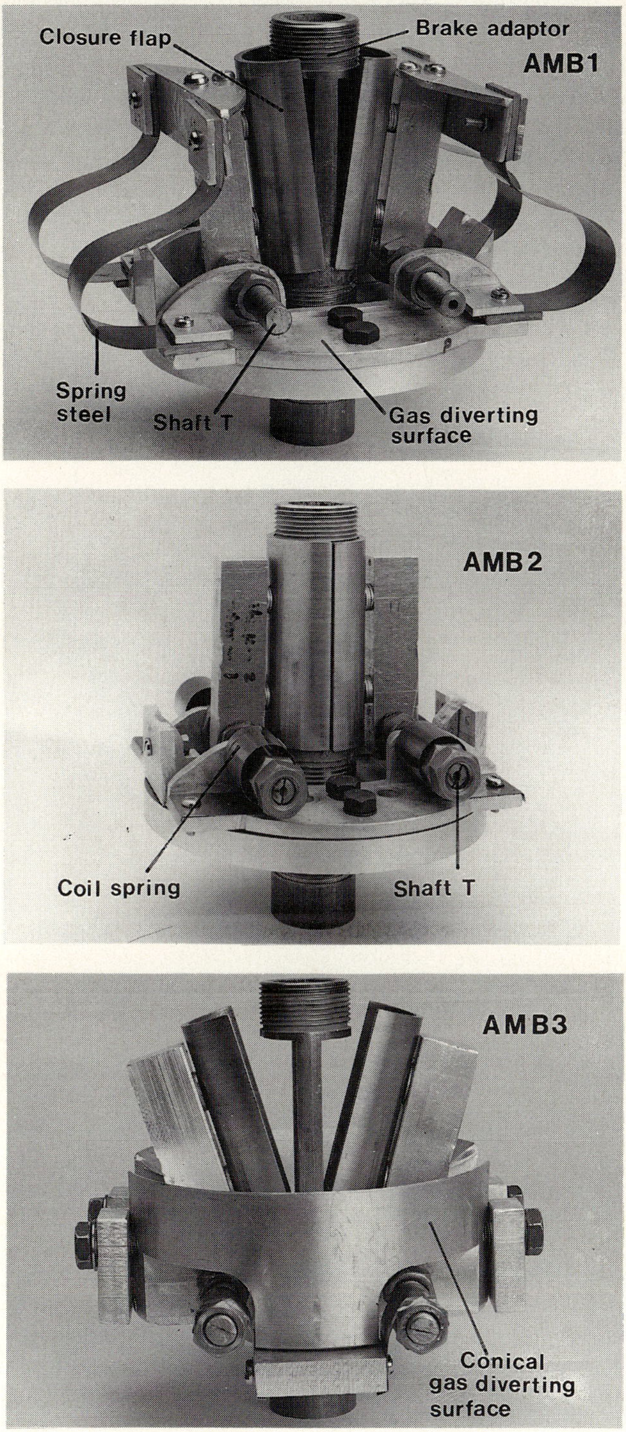

(a) ABM1 - half S-shape spring configuration (closure flaps partially opened)

(b) ABM2 - with coilspring configuration (closure flaps fully closed)

(c) ABM3 - with coil spring configuration & a conical gas diverting surface (closure flaps partially opened)

Fig.4. Test models of the adaptive muzzle brake (ABM) design

4. Conclusions

A novel concept of muzzle brake design was developed and tested using a suitable spring arrangement to regulate an active component. The study has shown that a muzzle brake can be designed to be operative only during the gas ejection phase, preventing the effect of blast wave reflection and as a result, averted an excessive level of hazard to be added to the existing blast field generated by the initial expanding blast wave. It was found that the flat coil spring technique employed on AMB2 and AMB3 was most successful. The spring arrangement was relatively simple to deploy. If needed, the torsional stiffness of the spring can be altered to provide the necessary spring loading transmitted to the flaps. This, in turn, can be used to regulate the pressure exerted onto the adaptor by the flaps, and may help to reduce blast wave pressure leaking between the flaps and the muzzle brake adaptor.

References

Phan KC (1991) On the performance of blast deflectors and impulse attenuators. In: Takayama K (ed) Shock Waves, Proc 18th Int Symp on Shock Waves, Sendai, Japan, pp 927-932

Effects of Shock Waves on Jet Mixing and Noise Generation

K. Kailasanath, J.P. Boris and A.M. Landsberg
Laboratory for Computational Physics and Fluid Dynamics,
Naval Research Laboratory, Washington, D.C. 20375-5344, USA

Abstract. Mixing and noise generation from underexpanded, supersonic rectangular jets have been studied using a recently developed simulation program, FAST3D, that uses the Virtual Cell Embedding (VCE) technique with the Flux-Corrected Transport Algorithm (FCT) to efficiently simulate flows in and around complex geometries. The simulations have been able to capture the shock-cell structure, the large-scale flow structures and their interaction with objects (paddles) in the flowfield. Detailed diagnostics have also been developed to characterize the flowfield and to provide a framework to understand the interactions between shocks, large-scale flow structures and objects in the flow field. For the cases reported, a flapping mode of flow instability is observed to be the dominant mechanism responsible for the mixing and noise generation. By introducing paddles into the flowfield, the mixing can be significantly enhanced but the screech amplitude also increases. Further simulations with more aerodynamic paddles and different amounts of underexpansion are needed to develop a better understanding of mixing and noise generation from these supersonic jets.

Key words: Mixing, Supersonic jets, Shock-cells

1. Introduction

There is empirical evidence suggesting that the presence of shock waves increases the amplitude of screech noise from supersonic jets. While it is possible, in principle, to design jet nozzles to minimize or eliminate shock waves, in practice, shock waves are always present during the off-design performance of such nozzles. Furthermore, one may not always want to eliminate shock waves because they can enhance the mixing and dispersal of jet exhausts into the environment. An ideal solution would be to take advantage of the beneficial effects of shock waves while minimizing the negative aspects. To do this, we need to understand the role of shock waves in noise generation and jet mixing. In this paper, we describe our efforts in using three-dimensional numerical simulations of the unsteady, compressible flow from underexpanded, supersonic jets as an effective tool to contribute to this understanding.

There is some evidence suggesting that the fundamental screech frequency is close to the most unstable shear layer mode for a given fully expanded Mach number (Hammitt 1961). However, this does not explain the scaling observed between the screech wavelength and the shock cell spacing (Powell 1953). It has also been suggested that the amplification of screech tones may be due to the interaction between large-scale coherent structures and shock waves (Seiner 1984). There is also substantial evidence indicating that feedback from flow or physical obstacles plays a significant role (Norum 1984). Recent experimental efforts (Rice and Raman 1992) indicate that the use of rods to increase jet mixing also results in an increase in screech noise.

The first step in using numerical simulations as an effective tool to isolate and understand the role of shock waves and their interactions with large-scale flow structures and obstacles in the flowfield is to develop an accurate and efficient approach to simulate the compressible, turbulent flow in and around complex objects. The Monotone Integrated Large-Eddy Simulation (MILES) approach (Boris et al. 1992) using the Flux-Corrected Transport (FCT) algorithm (Boris and Book 1973,1976) has been extensively used to simulate subsonic and supersonic turbulent flows. In particular, this approach has been used to isolate and study the interaction between acoustic waves and large-scale flow structures in confined axi-symmetric jets (Kailasanath et al. 1989).

Shock Waves @ Marseille III
Editors: R. Brun, L. Z. Dumitrescu

Recently, a new simulation program, FAST3D, has been developed that uses the FCT algorithm with a Virtual-Cell Embedding (VCE) technique to simulate compressible flows over arbitrary complex objects. A paper on this subject can be found elsewhere in these Proceedings (Landsberg et al. 1993).

In this paper, we report our progress in adapting and using the approach described above to solve problems in aeroacoustics. First, we briefly discuss the numerical techniques used. Then, we discuss two cases of underexpanded supersonic jets: one with paddles to interfere with the flow and the other without the paddles. By comparing these two simulations, we gain some understanding of the interactions between shock waves, paddles and large-scale structures and their role in the noise produced by supersonic jets.

2. The numerical simulation tool

The Virtual Cell Embedding (VCE) method is a new capability for resolving complex geometries on a structured, orthogonal grid with little sacrifice in speed or memory. This method has been incorporated into a versatile CFD simulation program, FAST3D. FAST3D solves the unsteady, three-dimensional, gasdynamic Euler equations using the Flux-Corrected Transport algorithm (Boris and Book 1973; 1976). An orthogonal, structured, rectilinear mesh is used but variable spacing between the grid planes, and even motion of the grid planes, is allowed. The FCT algorithm was originally designed to solve generalized continuity equations for space and time-varying coordinates. The most recent version of the algorithm, LCPFCT (Boris et.al. 1993), is fourth order accurate in space in smooth regions of the flow and second order accurate in time. The LCPFCT algorithm is well-tested, monotone (positivity-preserving), conservative, and relatively high-order. Monotonicity is achieved by introducing a diffusive flux and later correcting the calculated results with an antidiffusive flux modified by a flux limiter.

Typically, complex bodies would have to be very highly resolved to minimize the "staircasing" effects generated when using an orthogonal, rectilinear mesh. The staircasing effects are created since a smooth body must be aligned to the grid locally thus making the body computationally "rough". To cancel this staircasing effect, the VCE method was developed to allow the smooth surfaces of complex bodies to cut cleanly through the rectilinear grid. Thus, the VCE method greatly increases the accuracy of the solution around complex geometries by smoothing out the cell corners causing the "staircase" effect. With the VCE method, cells may be fully outside the body, fully inside the body, or partitioned by the body. It is these partitioned, boundary cells that require special treatment. The VCE method computes partial volumes and partial face areas for all of the cells which intersect the body. Thus, this approach only refines (subdivides) cells next to the body. CPU time is not sacrificed appreciably since only those cells next to the body require special treatment. The division into subcells can be made so fine that the body is essentially smooth without staircasing. The term " virtual" is used since the subcells embedded within a cell are not stored in memory and therefore are not integrated in the flow solution.

The special treatment required for the partitioned cells is a modification to the flux calculation in these cells. To compute the conserved fluxes into and out of each cell, the unobstructed "correct" face areas and volumes are used for the cells bordering the body as well as a flux coupling vector to correct for the apparent fluid compression in cells bordering the body. The details of this method are described in Landsberg et al. (1993). Little additional CPU time is needed since the geometric information from the subcells is only computed once and stored and these additional computations are only performed in the cells adjacent to the body.

FAST3D produced accurate results when tested on a series of two- and three-dimensional problems. This versatile program has been applied to shocked, supersonic reactive flow through a ram accelerator, to unsteady airflow over a Navy destroyer, and to vortex shedding from a number of complex shapes. Here it is being applied to simulate the mixing and near-field noise generated by a supersonic jet.

3. Results and discussion

The results from two numerical experiments on underexpanded jets will now be discussed. In the first case, there is natural screech due to the interaction between the large-scale flow structures and the shock cells inherent in such a jet. In the second case we use paddles or rods to interfere with the flow and thus modify the screech feedback loop. By comparing these two cases, we hope to gain some understanding of the relation between screech and jet mixing. For both cases, the simulated jet exits from a rectangular nozzle of aspect ratio 4.5 with the smaller dimension of 1.44 cm. The exit pressure is 1.7483×10^6 dynes/cm^2, the back pressure is atmospheric and the exit Mach number is unity, thus simulating an underexpanded jet issuing from a converging nozzle.

For each of the cases, detailed flow visualization at a sequence of times along with the Fourier analyses of the pressure and velocity fluctuations are used to obtain an overall understanding of the instabilities and noise produced by these jets. For flow visualization, in addition to velocity fields and density (to simulate Schlieren), static pressure and Pitot pressure distributions have also been obtained.

3.1. An underexpanded jet

The first case we discuss is that of a free jet issuing from an underexpanded, high aspect ratio rectangular nozzle. An instantaneous snapshot of the flowfield at 10.5 ms from the impulsive start of the jet is shown in Fig.1. Since the simulation is three-dimensional, various slices or planes have been shown. In the three longer frames, the flow is issuing from the left to the right and in the three smaller frames, the flow is coming out of the plane of the paper. The first frame is a side view of the density field with half the longer dimension of the nozzle being visible at the left. The plane chosen is close to the outer edge of the jet. The shock cells can be discerned from the differences in the gradations of the density field and the brighter regions in the downstream are due to an intense flapping motion of the jet, in and out of the plane of the paper. The shock cells are clearer in the static pressure distributions shown in the second frame. The point of view here is from above the jet, with the narrow dimension of the nozzle being visible to the left. In addition to the shock cells, brighter and larger blobs or islands are clearly visible. These are regions of high and low pressure and have been shown to be directly related to large-scale vortical structures. We note that several shock cells occur before the appearance of these vortical structures. In the bottom frame, the pressure that would be obtained by using a Pitot-probe to map out the flow is shown. That is, the correction for the pressure loss across the shock in front of the probe is taken into account in displaying this information. The Pitot pressure drop in the axial direction is often taken as a measure of the jet mixing in experimental studies (Rice and Raman 1992). This visualization provides a two-dimensional picture of the jet mixing, and flapping motion of the jet can be clearly deduced from this frame. A similar frame at a different time shows the change in the structure of the jet with time due to the flapping-mode instability. The three smaller frames to the right are also Pitot pressures but are slices of the flowfield at various stations in the axial direction of the jet. We note that even at 10 cm from the jet exit plane, there is very little spread or mixing of the jet with the surroundings. However, by 18 cm, there is significant mixing, though it is clearly not completely mixed into the surroundings. Similar flow visualizations have been done at a sequence of times to study the time-evolution of the flow. Space does not permit us to show the details here.

To obtain more quantitative information and to deduce some information on the noise produced by these jets, detailed time histories have been collected at various spatial locations in the flowfields. These data have then been Fourier analysed to bring out the dominant frequencies.

There are a number of fluctuation frequencies associated with various phenomena observed in the flow and a complete description of all these effects is beyond the scope of this paper. Since the dominant flow instability observed is the flapping-mode across the narrow dimension of the jet, the Fourier analysis of the velocity fluctuation in this direction (V_z) is particularly interesting. Even V_z has multiple frequencies and these change from one location to another. In the early part of the jet, a frequency of 2800 Hz is dominant and is probably associated with the shock-cells. There are two dominant frequencies further downstream where the flapping motion is intense and these are around 3200 and 4000 Hz. The pressure fluctuations are even more complex because they are sensitive to flow fluctuations in all directions. However, again the dominant frequencies are the same as above, suggesting that the flapping mode is the primary source of pressure fluctuations and hence of the noise produced from this jet.

Comparison of this simulation to one with paddles or rods in the flowfield will bring out the role of feedback from obstacles and this is the case discussed next.

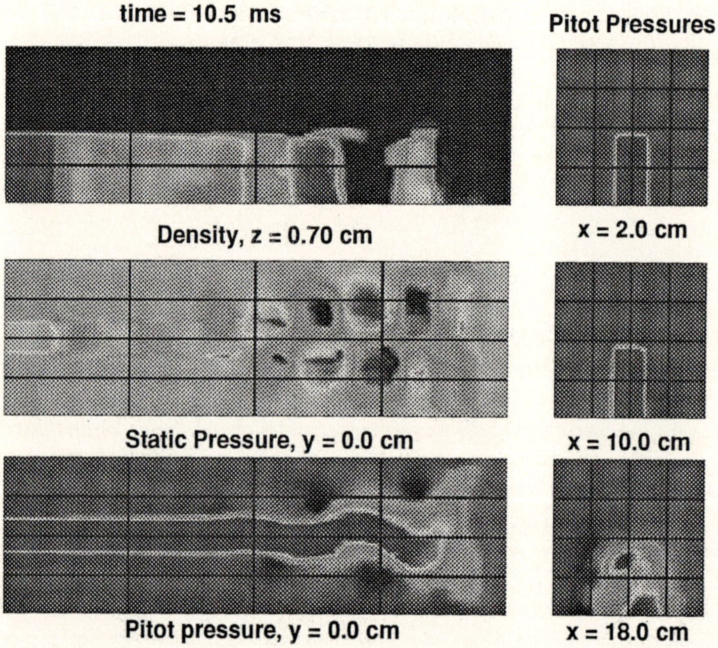

Fig. 1. Various flow diagnostics showing the instantaneous flowfield from the underexpanded jet without paddles.

3.2. Effect of paddles

Two, 3.4 cm long, square rods or paddles (.25 sq cm), were placed 10 cm downstream of the exit plane and 0.65 cm from the centerline of the nozzle. The simulations were started impulsively as before and the initial conditions were identical to those used in the simulations described above. The paddles interact with the outer edges of the jet during the flapping motion and we expect it to interfere with the feedback process. The instantaneous visualization of the flowfield at the same time (10.5 ms) as before is shown in Fig.2. The color scale used for the visualizations and the variables and views chosen are identical to those in Fig.1, so that a direct comparison can be made between the two cases.

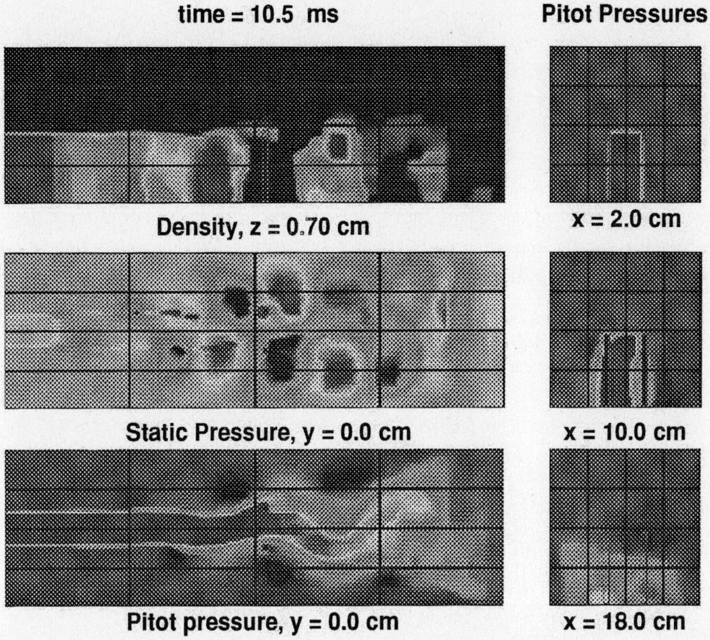

Fig. 2. Various flow diagnostics showing the instantaneous flowfield from the underexpanded jet with paddles located 10 cm from the nozzle exit plane.

The shock cells, the large-scale flow structures, and the flapping motion of the jet are all evident in this case also. However, we observe that the flapping motion starts closer to the exit plane in this case. The number of shock cells clearly visible before the appearance of the vortices in frame 2 are fewer and the Pitot pressure in frame three also goes unstable closer to the nozzle exit. The smaller frames to the right showing the simulated Pitot pressure measurements are particularly noteworthy. Unlike the previous case, there is significant spread even at 10 cm, where the actual paddles are also visible. Furthermore, by 18 cm the jet has essentially dissipated into the surroundings. From these observations, we conclude that the paddles have significantly altered the flow and enhanced the mixing of the jet with the surroundings for the particular configuration and conditions studied.

The Fourier analysis of the pressure and V_z fluctuations provide further information on the effect of the paddles. The dominant frequency in both the pressure and velocity fluctuations is now 3600 Hz at all downstream locations. Again, in the region close to the nozzle exit, a frequency of about 2800 Hz is observed as before suggesting that this frequency may be related to the shock structure. The amplitude of the fluctuations at 3600 Hz at the downstream locations is significantly larger (by as much as a factor of two) at certain locations compared to the case without the paddles. The paddles appear to have set up a resonance at 3600 Hz and the flow at the nozzle exit is being excited at this frequency. The energy previously distributed among multiple frequencies are now focussed at this frequency. Since the frequency is higher, we expect the instability to grow closer to the nozzle exit and the larger amplitude may also have some influence. However, further analysis is required to quantitatively explain the shift in the breakdown of the jet for this case. From this simulation, we conclude that by introducing paddles at selected locations in the flowfield, the mixing of the jet can be significantly increased but the screech amplitude is also increased.

3. Summary

Unsteady supersonic jets issuing from underexpanded, high aspect-ratio rectangular nozzles have been simulated using a recently developed code, FAST3D, which incorporates the Virtual-Cell Embedding (VCE) technique and the Flux-Corrected Transport (FCT) algorithm to efficiently simulate flows in and around complex geometries. Several new numerical diagnostics techniques have been developed and used to understand the mixing and noise generated from supersonic jets with shocks. For the cases simulated, the flapping mode is observed to be the dominant flow instability mode and responsible for the screech emanating from these jets. By introducing paddles downstream of the nozzle exit, the amplitude and frequency of the flapping mode was changed to enhance the mixing. However, the amplitude of the screech also increases and the square paddles used will introduce a drag penalty. More complex paddles with aerodynamic shapes and further studies on the effect of paddle location and amount of underexpansion need to be done to quantify the trade-off between mixing and noise generation. Simulations of fully expanded jets will also provide information on the absence of shock waves. The FAST3D code along with the detailed diagnostics being developed is a valuable tool to understand and alter the characteristics of supersonic jets.

Acknowledgements

The authors wish to acknowledge several useful discussions with Prof. Eugene Brown of VPI, Blacksburg, VA. and Drs. Ed Rice and Ganesh Raman of NASA-Lewis Research Center, Cleveland, OH. This work has been supported by ONR through NRL, ARPA, and the NASA-Lewis Research Center.

References

Boris JP, Book DL (1973) Flux-Corrected Transport 1. SHASTA, A fluid transport algorithm that works. J. Comput. Phys. 11: 38–69

Boris JP, Book DL (1976) Solution of the continuity equation by the method of Flux-Corrected Transport. Chapter 11 in: Methods in Computational Physics, Academic Press, New York, pp 85–129

Boris JP, Grinstein FF, Oran ES, Kolbe RJ (1992) New insights into large eddy simulation. Fluid Dynamics Research 10: 199–228

Boris JP, Landsberg AM, Oran ES, Gardner JH (1993) LCPFCT - A Flux-Corrected Transport algorithm for solving generalized continuity equations. NRL Memorandum Report 6410-93-7192

Hammitt AG (1961) The oscillation and noise of an overpressure sonic jet. J. Aero. Sci. 8:673-680

Kailasanath K, Gardner JH, Boris JP, Oran ES (1989) Acoustic-vortex interactions and low frequency oscillations in axisymmetric combustors. J. Prop. Power. 5: 165–17f

Landsberg AM, Boris JP, Young TR, Scott RJ (1993) Computing shocked flows through the Euler equations. These Proceedings

Norum TD, (1984) Control of jet shock associated noise by a reflector. AIAA Paper 84-2279

Powell A (1953) On the noise emanating from a two-dimensional jet above the critical pressure. Aeron. Quarterly 4:103-122

Rice EJ, Raman G (1992) The role of natural and induced screech in the enhanced mixing of a rectangular supersonic jet. 45th Annual Meeting of the Division of Fluid Dynamics, APS, Tallahassee, Florida

Seiner JM (1984) Advances in high speed jet aeroacoustics. AIAA Paper 84-2275

Numerical Simulations of Blast Wave Propagation Induced by Eruptions of Volcanoes

T. Saito*, T. Kitamura$, K. Takayama$, N. Fujii† and H. Taniguchi‡
*Cray Research Japan Ltd., Supercomputer Center, Institute of Fluid Science, Tohoku University, 2-1-1 Katahira, Aoba-ku, Sendai, 980 Japan
$Shock Wave Research Center, Institute of Fluid Science, Tohoku University, 2-1-1 Katahira, Aoba-ku, Sendai, 980 Japan
†Research Center for Seismology and Volcanology, School of Science, Nagoya University, Chikusa-ku, Nagoya, 464 Japan
‡Science Education Institute of Osaka Prefecture, 4-13-23 Karita, Sumiyoshi-ku, Osaka, 558 Japan

Abstract. Propagation of shock waves induced by two Japanese volcanoes is numerically simulated. The volcanoes are Fugendake at Unzen, Nagasaki prefecture and Sakurajima in Kagoshima prefecture. The results demonstrate the significant effect of local topography on the pressure distributions both on the ground and in the air. Simulations are carried out by using a second-order finite-difference scheme in general curvilinear coordinates.

Key words: Volcanic eruption, Blast wave, Air shock

1. Introduction

Investigations on how shock waves, created by volcano eruptions, propagate give useful information in establishing better countermeasures to volcanic disasters. The imprint which was left on a blast meter tells us that the eruption of Fugendake on June 8, 1991 generated a shock wave. The shock Mach number is estimated to be about 1.1 at the distance of 2.8 km from the crater where the blast meter was installed. Numerical simulations are carried out for the shock wave propagation by taking the real topography of the area into account. The results demonstrate the effect of local topography on the pressure distributions both on the ground and in the air.

The imprint left on the blast meter, however, provides only a rough estimate of the total energy imposed on the blast meter. It does not provide details such as pressure history at the site, etc. Consequently, it does not give enough quantitative data on the eruption for verifying the eruption models and initial conditions of the numerical simulations. Knowing how energy is released, not just how much, is also an important subject for understanding the mechanism of volcanic eruptions and their related phenomena. Then, in addition to Fugendake, numerical simulations on eruptions of Sakurajima were carried out. Sakurajima became active in 1955 and is one of the most active volcanoes in the world ever since (Ishikawa et al. 1992). It repeats hundreds of vulcanian type eruptions annually and regular observations are being carried out by the Disaster Prevention Research Institute at Sakurajima. Accordingly, much data are available on Sakurajima's volcanic activities. A set of observation data regarding the propagation of a shock wave induced by the eruption of Sakurajima is used for quantitative comparison with the numerical simulations.

The numerical work is done by using a second-order finite-difference TVD scheme in three-dimensional general curvilinear coordinates. Operator splitting of the Strang type is used to carry out multidimensional calculations by keeping the time accuracy to the second order.

In Section 2 the numerical code and mesh generation are described. The numerical results for eruptions of Fugendake are described in Section 3. The observed data of the shock propagation induced by the eruption of Sakurajima is described and compared with numerical simulations in Section 4. A summary and future plans are outlined in Section 5.

Shock Waves @ Marseille III
Editors: R. Brun, L. Z. Dumitrescu

2. Numerical simulation

2.1. Basic equations

Three-dimensional Euler equations in a general curvilinear coordinate system of the following form are solved together with the equation of state for ideal gases:

$$\boldsymbol{Q}_\tau + \boldsymbol{E}_\xi + \boldsymbol{F}_\eta + \boldsymbol{G}_\zeta = 0$$

where $\boldsymbol{Q}$, $\boldsymbol{E}$, $\boldsymbol{F}$ and $\boldsymbol{G}$ are, respectively, the vectors of conserved variables and fluxes in the computational space, which are expressed as follows:

$$\boldsymbol{Q} = \frac{1}{J}\begin{pmatrix} \rho \\ \rho u \\ \rho v \\ \rho w \\ e \end{pmatrix} \quad \boldsymbol{E} = \frac{1}{J}\begin{pmatrix} \rho U \\ \rho u U + \xi_x p \\ \rho v U + \xi_y p \\ \rho w U + \xi_z p \\ (e+p)U \end{pmatrix} \quad \boldsymbol{F} = \frac{1}{J}\begin{pmatrix} \rho V \\ \rho u V + \eta_x p \\ \rho v V + \eta_y p \\ \rho w V + \eta_z p \\ (e+p)V \end{pmatrix} \quad \boldsymbol{G} = \frac{1}{J}\begin{pmatrix} \rho W \\ \rho u W + \zeta_x p \\ \rho v W + \zeta_y p \\ \rho w W + \zeta_z p \\ (e+p)W \end{pmatrix}$$

The symbol J represents the Jacobian of transformation from Cartesian coordinates (x, y, z) to the curvilinear coordinates (ξ, η, ζ), i.e. $J = \partial(\xi, \eta, \zeta)/\partial(x, y, z)$. Symbols p, ρ, u, v, and w are the pressure, density and Cartesian velocity components, respectively. The contravariant velocity components U, V and W are expressed as:

$$U = \xi_x u + \xi_y v + \xi_z w, \quad V = \eta_x u + \eta_y v + \eta_z w, \quad W = \zeta_x u + \zeta_y v + \zeta_z w$$

The total energy per unit volume e is expressed as $e = \rho\epsilon + \frac{1}{2}\rho(u^2 + v^2 + w^2)$; where ϵ is the specific internal energy of the gas which, for ideal gases, is expressed with the specific heat ratio γ as $\epsilon = p/(\gamma - 1)\rho$.

2.2. Numerical scheme

The numerical scheme used in the present work is the time-dependent explicit finite-difference scheme of Harten (1983) and Yee (1989). It is one of the oscillation-free schemes devised with a TVD (Total Variation Diminishing) stability algorithm using flux limiters. The scheme is second-order accurate in space for smooth regions and the Strang-type time splitting method is adopted to keep the second-order time accuracy for multidimensional calculations (Strang 1968). The flux limiters employed here are the so-called minmod function limiter for the genuinely nonlinear waves for its robustness and Superbee for linear waves (contact surfaces), for preventing excess smearing. Although the code requires relatively complex programming, just as in any modern shock capturing methods, it is well vectorized and parallelized. A maximum performance of 200 Mflops is achieved by using a single CPU of the Cray Y-MP8 system of the Supercomputer Center of the Institute of Fluid Science, Tohoku University. The degree of parallelism is also high and a speedup of 3.9 by using four CPUs is achieved. This makes the total maximum speed of the code to about 1.5 Gflops when all eight CPUs are used. Considering the fact that the ratio of the number of multiplications to the number of additions in the current coding is about 5 to 3, this performance is quite satisfactory. Memory arrangement is also important for a large CFD simulation and this code needs about 34 storage locations per grid point. The memory size can be reduced further by sacrificing the computation time.

2.3. Grid generation

The computational domains consist of 200×200×40 grid points corresponding to areas of approximately 10×10 km and height of 4 km in the case of Unzen and 10×7 km and height of 4 km in the case of Sakurajima. The bottom of the computation volume corresponds to the ground, reproducing the topography of either Unzen area, where Fugendake is located, or Sakurajima area. Both Fugendake and Sakurajima are located in Kyushu, one of the four major Japanese islands. The digital elevation map of the Geographical Survey Institute of Japan was used for producing the ground topography of the Unzen area. The ground surface of Sakurajima was created by extracting elevation data out of an ordinary map with a digitizer since the digital elevation map of the area is not available. The grid points between the bottom and the top planes of the computation space are obtained by solving an elliptic PDE for smooth grid distributions. The grids of a vertical plane including the volcanoes and the contour maps of the area are shown in Fig.1. The increase in altitude for each contour line is 25 m.

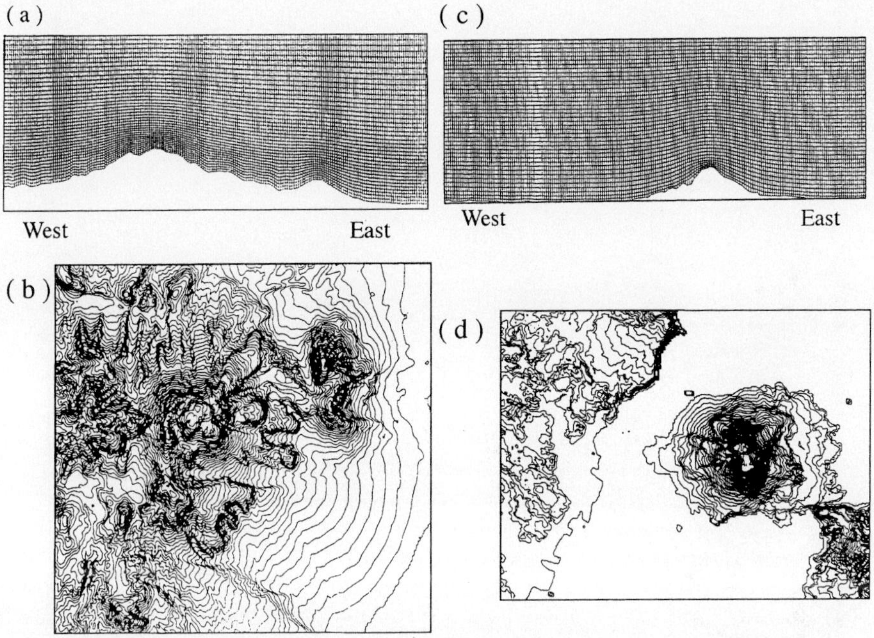

Fig. 1. Computational grids and contour maps : (a,b) Unzen-Fugendake and (c,d) Sakurajima

2.4. Eruption model

Two eruption models which are employed in the current work are shown in Fig.2. One of them is the point source explosion model, in which high pressure air is released instantaneously from a cubic container located at the crater of the volcanoes. Although this model rigorously should be called "Cubic shock tube model " or "Three dimensional shock tube model" since the container has a finite volume, it is called "point source (explosion) model" in this paper. The other model is the "(open-end) shock tube model" in which a vertical shock tube is connected to the crater. The conditions at the crater are provided by actually calculating the flow inside the open-end shock tube. It is expected that the shock tube model approximates the real volcano better than the point source model. Comparisons between the two models are done in the following sections.

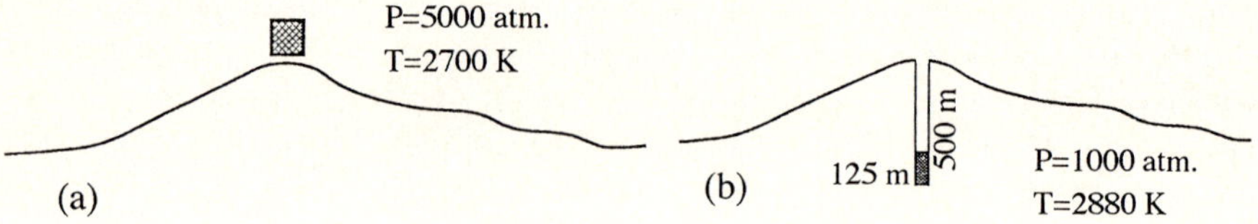

Fig. 2. Initial conditions and models of volcanic eruption : (a) Point source explosion model and (b) Shock tube model

3. Numerical results for the eruption of Fugendake

A series of numerical simulations was carried out firstly with the point source explosion model. The initial pressure and the temperature are assumed to be 5000 atm and 2700 K. This initial condition is chosen rather arbitrarily simply because we didn't have enough information.

Contour plots of the pressure distribution on the ground at two different times are shown in Fig.3. The interval of contour lines is 5% of the maximum pressure difference. We can see the complex structure due to the topography.

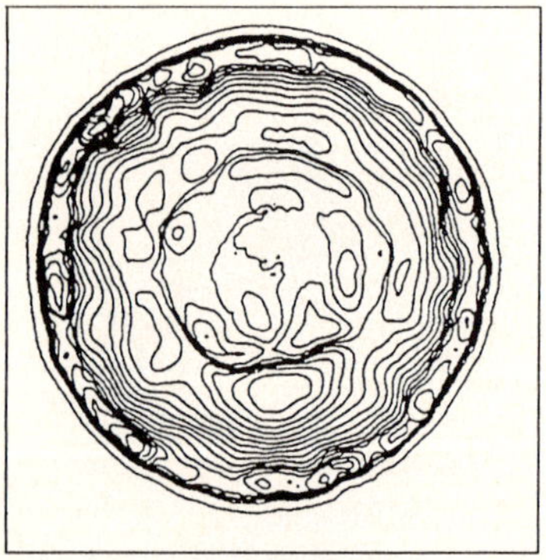

Fig. 3. Pressure contours on the ground (point source explosion model)

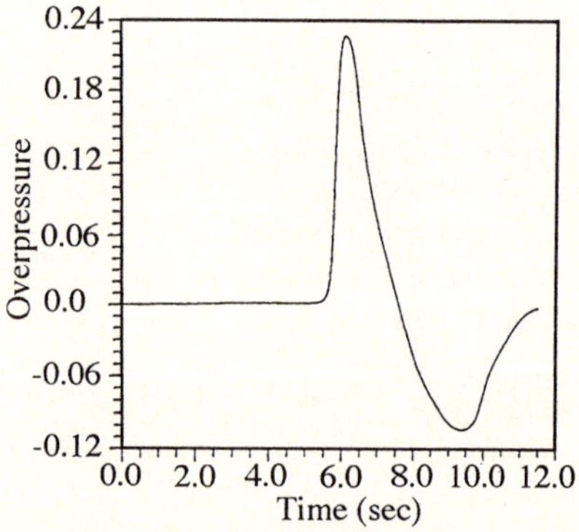

Fig. 4. Overpressure history at a point 3.5 km Southwest (point source explosion model)

Fig.4 shows the overpressure history at a point 3.5 km Southeast from the crater where a city is located. The calculated peak overpressure (0.23 bar) is slightly higher than the estimated value (0.2 bar) from the blast meter imprint. Since a blast meter gives very rough quantitative data, no further attempts were done to match the two data.

Figs.5 and 6 are pressure distributions in the vertical plane obtained by the point source explosion model and the open-end shock tube model, respectively. Again the topographical effect is seen as the complex pressure distributions behind the expanding shock wave. It is also seen that the shock wave propagates faster and stronger in the vertical direction in Fig.6 while the shock wave propagates more uniformly in all directions in Fig.5.

The cubic volume of the initial high pressure/temperature region has side lengths of approximately 100 meter. The eruption energy then is estimated to be about 400 kilotons of TNT

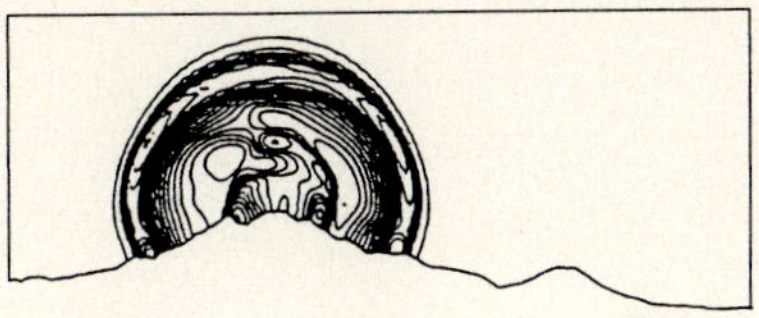

Fig. 5. Pressure contours on the vertical section for the point source explosion model at time = 2.2 sec

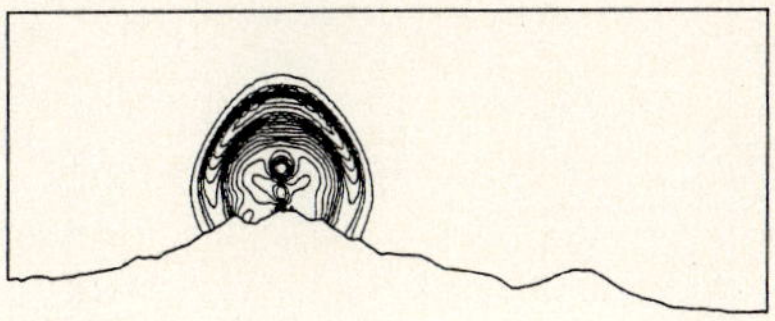

Fig. 6. Pressure contours on the vertical section for the shock tube model at time = 1.82 sec

equivalent. Typical CPU time is about one and a half hour for calculating 500 time steps which is enough for the shock wave to propagate out of the computational area.

4. Observed data and numerical simulations of the eruption of Sakurajima

4.1. Observed data

It is often observed that a cloud disappears and reappears due to the compression and expansion phases of a spherically expanding shock wave at the time of volcanic eruption. Such phenomena were recorded with a video camera at Sakurajima on May 25th 1982. The video film shows the phenomena for the two layers of cloud at the altitudes of 1200 m and 1600 m with a time mark in each frame. The data were analyzed by Ishihara et al. (Ishihara et al. 1985; Kamo et al. 1980) and the $x-t$ diagram and the pressure histories at the two altitudes were obtained. According to those results, the maximum overpressures were estimated to be 0.2 bar and 1.0 bar at the altitudes of 1600 m and 1200 m, respectively. These data induced from the observation are compared with the results of numerical simulations.

4.2 Numerical simulation

Numerical simulations are carried out for eruptions of Sakurajima by changing the initial pressures for both eruption models, i.e. the point source explosion model and the open-end shock tube model. Comparisons of the simulated data with the previously described observed data are carried out. It is found that the shock wave decelerates faster in the point source explosion model for the same initial pressure between the two eruption models. The observed shock wave, however, decelerates even faster than those that are obtained numerically. This character of slower rate in deceleration in the numerical simulations is seen in Fig.7 which shows pressure histories at the altitudes of 1200 m and 1600 m for both eruption models.

It is found that the ratio of peak pressures at the two altitudes are higher for the observed data compared to that of either one of the numerical results with different eruption models. As a consequence, if the initial pressure is increased in order to match the calculated peak pressure to the observed value at the altitude of 1200 m, then the calculated peak pressure at 1600 m becomes too high. On the contrary, if the initial pressure is reduced to match the peak pressures at the elevation of 1600 m, then the numerically obtained value at 1200 m becomes too low. It is impossible to match the numerically obtained pressure histories to the observed data at the two altitudes simultaneously. This tendency is common to both eruption models and no superiority between the two eruption models is seen. One of the reasons for that, perhaps, is the lack of modeling of atmospheric conditions. The atmospheric pressure and temperature decrease with the altitude with the rates of about 1.2 mb/10 m and .64 K/100 m, respectively. These changes in pressure and temperature are not taken into account in the current simulations. Poor grid resolution is also a candidate for the discrepancies between the observed and the simulated data. In the current simulations, the shape of the actual crater is not modeled well enough. Since the diameter of the crater and the distance from the crater to the two layers of cloud are comparable,

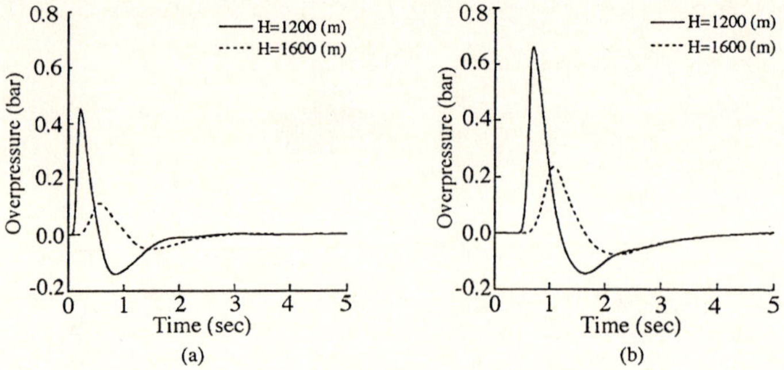

Fig. 7. Duration of the pressure: (a) point source explosion model and (b) shock tube model

the shape of the crater is expected to have an influence on the pressure fields in the region of interest. The eruption energy was estimated to be about 4 kilotons of TNT equivalent, by assuming an initial pressure of 100 atm for the point source explosion model.

5. Summary and future work

It is demonstrated that the local topography affects the pressure distributions both on the ground and in the air from a series of numerical simulations carried out for the eruption of Fugendake. Dfferences in the numerical results due to two different eruption models are also demonstrated. Some quantitative comparisons are done on the eruption of Sakurajima. However, the superiority of either one of the eruption models over the other is not determined. This is due probably to the limitations of the mesh resolution and to the neglection of changes in atmospheric conditions with altitude.

The eruption energies are estimated to be 400 kiloton TNT equivalent for the eruption of Fugendake on June 8, 1991 and about 4 kiloton TNT equivalent for the eruption of Sakurajima on May 25, 1982.

It is planned to do numerical simulations with finer grids so that the shape of the crater will be better resolved. It is also planned to do simulations by considering the effect of the pressure and temperature changes with the altitude. Direct pressure measurements are also planned for the eruptions of Sakurajima.

References

Harten A (1983) High resolution schemes for hyperbolic conservation laws. J. Comp. Phys. 49: 357-393

Ishikawa H (1992) Sakurajima. 7, Kyouritsu (in Japanese)

Ishihara K (1985) Dynamical analysis of volcanic explosion. J. Geodynamics 3: 327-349

Kamo K, Nishi K, Takayama T, Ueki S (1980) Seismicity in the region south of Sakurajima volcano and the region of abnormal propagation of seismic waves. 5th Geoscientific Observation of Sakurajima Volcano, pp 131-138 (in Japanese)

Strang G (1968) On the construction and comparison of difference schemes. SIAM J. Num. Anal. 5: 37

Yee HC (1989) A class of high-resolution explicit and implicit shock-capturing methods. NASA YN 101088

Experiments and Analysis of Orbital Debris Shock-Induced Penetration of Pressure Vessels in Space

Robert M. Schmidt, Kevin R. Housen and Harlow G. Ahlstrom
Boeing Defense & Space Group, M/S 8H-05, Seattle, WA, USA 98124

Abstract. An experimental technique developed by Schmidt et al. (1994), for simulating the response of single-bumper aluminium debris shields at impact velocities up to 18 km/s, is extended here to the case of double-bumper shields. The aluminium impactor and bumper(s) of interest are replaced with cadmium surrogates that have the same dimensions as the prototype, but have melt and vaporization specific energies an order of magnitude lower. Dimensional analysis is used to show that these lower specific phase transformation energies are the key to simulating impact velocities that are a factor of 3.1 higher than previously attained. The particular uniqueness of cadmium is that, by coincidence, its density is a factor of 3.1 greater than aluminium, thereby reproducing the initial momentum of an aluminium projectile and the impulse distribution delivered by the debris cloud to the pressure hull wall. The tests reported here investigate the behavior of a double bumper shield at both normal and 45° angles of incidence.

Key words: Space debris, Debris shields, Double bumper, Cadmium simulation, Perforation

1. Introduction

Orbital debris pose an increasing threat to spacecraft. Therefore, efforts are being directed toward developing effective light-weight debris shields. The role of a shield is to break up and distribute a colliding particle over the largest possible area on the spacecraft's pressure hull. In its simplest form, a single thin bumper is spaced as far from the pressure hull as practical to allow a large dispersion of the debris cloud. However because standoff distance is limited by overall envelope considerations and support structure weight, some attention has been directed at multiple bumper designs.

Debris impacts involve a number of complex and often poorly understood mechanisms, such as fragmentation, melting, vaporization and structural failure. Because of this, theoretical and numerical analyses of shield response have met with limited success, and so a shield's effectiveness is usually evaluated through experimentation. However, experimental studies have been limited by the fact that centimeter-size objects other than disks (Chhabildas et al. 1992) or near-molten shaped charges (Mullin 1993, personal communication) cannot be launched faster than about 7 km/s. This is in contrast to debris impacts on spacecraft in low Earth orbit, 85% of which occur at velocities above 7 km/s, with a median velocity of 10 to 11 km/s.

Present design approaches assume that shield response at high velocities can be determined by extrapolating results at 7 km/s, using arguments based on energy (Christiansen 1993) or momentum (Wilkinson 1969) scaling. However penetration at 7 km/s is dominated by solid and partially melted fragments. At higher velocities, the completion of melting and the onset of vaporization create different physics and loading conditions for the pressure hull. Most notable is the reduction of fragment or melt droplet size and the expansion of the vapor cloud leading to more of a blast loading, as opposed to the point loads presented by the collection of small particle impacts at 7 km/s.

To address the physics relevant to high velocities, Schmidt et al. (1994) used surrogate materials to simulate the phenomena. Dimensional analysis was used to derive similitude requirements for the surrogate impactor and bumper to reproduce the impulsive loading against a common

Shock Waves @ Marseille III
Editors: R. Brun, L. Z. Dumitrescu

back wall of structural aluminium. They demonstrated that an impact at velocity U of a cadmium sphere onto a cadmium bumper delivers the same impulse as an impact of an aluminium sphere on an aluminium bumper at a velocity of 3.1 U. Concurrent work by Holsapple (1993) based upon a continuum similarity analysis, along with finite difference calculations for shock loading and release, supported the value of this velocity scale factor.

The previous study (Schmidt et al. 1993) considered normal impact on a single metal bumper protecting an aluminium wall. The present work incorporates an additional "layer" consisting of a metal sheet and MLI * placed midway between the bumper and the back wall. As discussed in Section 3, if material strength effects are important, the scaling arguments are not so obvious as to whether the intermediate layer should consist of cadmium, or aluminium as in the prototype. To test assumptions regarding the composition of the layer and the role of material strength, both configurations were tested and compared with an all aluminium prototype at 7 km/s and normal impact angle.

The scaling arguments of Schmidt et al. (1993) were based on the observation that at high velocities (> 7 km/s for aluminium), wall failure is due to large plastic deformation and tensile rupture resulting in tears or cracks in the material. This is in contrast to particle penetration and spall noted at 7 km/s and below. In the shield considered here, the role of the intermediate bumper and MLI is to further smooth the collection of particles, droplets and vapor, producing an axially symmetric smoothly varying impulse to the rear wall.

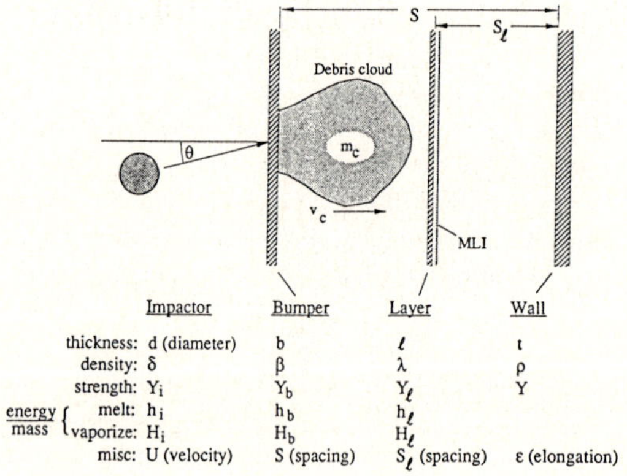

Fig. 1. Definition of variables used in scaling analysis

2. Description of experiments

The impact experiments were conducted by A. J. Piekutowski and K. L. Poorman at the University of Dayton Research Institute. A description of the facility and experimental techniques is given by Piekutowski (1987).

Thirty-two experiments were performed and are summarized in Table 1. A schematic illustration of the shield geometry is given in Fig.1. The wall, or pressure hull, is protected by an outer metal bumper, and an intermediate layer consisting of a metal sheet and MLI. "Baseline" tests were performed at roughly 7 km/s using aluminium for the projectile and shield components. Cadmium simulations were performed at actual velocities ranging from 2.2 to 5.24 km/s, which

* Multiple layer insulation: used for thermal protection of spacecraft. The MLI used here consisted of 14 layers of 6 μ-thick aluminized Kapton

simulate aluminium impacts at velocities from 6.8 to 16.2 km/s. In the simulations, the projectile and outer bumper were composed of cadmium. The intermediate layer was composed of either cadmium or 6061-T6 aluminium, as noted in Table 1. In all of the tests, the impactor was spherical, the wall was 3.2 mm thick 2219-T87 aluminium plate, and the distance, S, from the wall to the bumper was 73.4 mm. The intermediate layer was located midway between the wall and the bumper. The lateral dimensions of the wall were 254 × 254 mm. The intermediate layer was typically 100 × 150 mm, and the bumper was 100 × 100 mm. The bumper, layer and wall were supported at each of their four corners, and the MLI was clamped on to the layer.

If the wall was perforated, the extent of the perforation was noted; otherwise the extent of any bulge produced in the wall was measured. Time-resolved radiographs were obtained to provide average wall deformation rates, which can be used to bound the failure times and indicate the degree of damage. Example radiographs are shown in Figs.2 and 3.

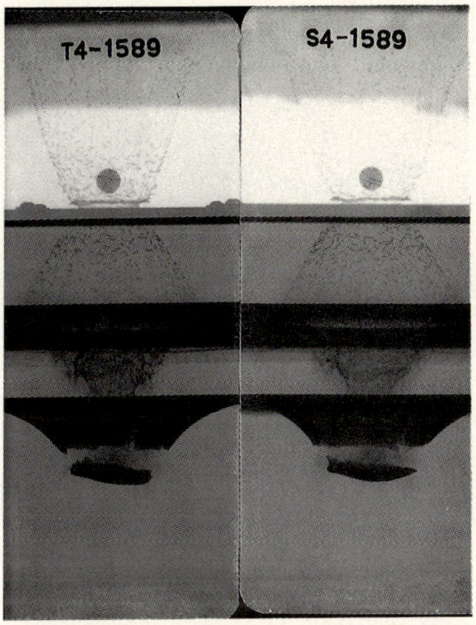

Fig. 2. Flash X-ray of shot 1589, an impact of an aluminium projectile onto an aluminium double-bumper shield at 7.17 km/s. The top and bottom parts show perpendicular views. Each part is a superposition of three images taken at different times. The first shows the projectile just before impact, the second shows penetration of the second layer and the third shows a disk-shaped section being removed from the wall

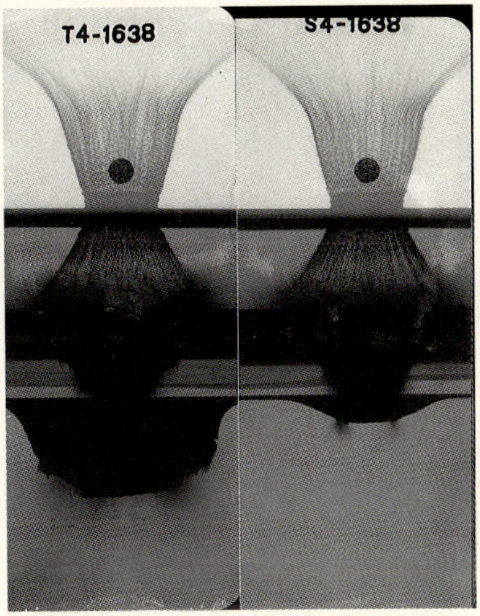

Fig. 3. Flash X-ray of shot 1638, a cadmium simulation of an aluminium impact at a scaled velocity of 16.2 km/s. As in Fig.2, the top and bottom parts show perpendicular views, each containing three superimposed images. However, the third image in the lower view was taken 45.2 μs later than the third image in the top view. The delay between these images is used to determine the deformation velocity of the wall

3. Similarity requirements

The use of surrogate materials to model the response of an aluminium debris shield at high velocities is based on a dimensional analysis of the problem. Schmidt et al. (1994) noted that when the debris cloud consists largely of melted and vaporized material, and not solid fragments, failure of the wall occurs when the specific impulse (impulse/area) exceeds a threshold value. The similarity analysis identifies the conditions that must be met for a cadmium simulation to provide the same impulsive loading on the hull as that provided by a prototype event of interest, in which an aluminium projectile strikes an aluminium shield.

The objective here is to find the ballistic limit curve for a shield, i.e. the diameter, d of the smallest projectile which is able to perforate the pressure hull, for a given impact velocity. To accomplish this, d is expressed as a function of nondimensional ratios of the variables shown in Fig.1. By imposing the three following conditions, the functional relationship is simplified because many of the nondimensional ratios are constant: (1) the impactor and bumper are composed of the same material; (2) all linear dimensions in the simulation are the same as in the prototype, and (3) the wall material in the simulation is the same as in the prototype. As such, d in the simulation is the same as in the prototype so long as the following functional relationship is satisfied:

$$f\left[\frac{\lambda}{\delta}, \frac{Y_\ell}{\lambda H_\ell}, \frac{h_\ell}{H_\ell}, \frac{H_\ell}{H_i}\right] = \text{Const.} \tag{1}$$

One way to satisfy Eq.1 is to require that all four terms on the left side be individually constant. Although this cannot be accomplished exactly, there are two ways to approximately satisfy this requirement. First, if a cadmium intermediate layer is used in the simulation, the first and fourth terms on the right side of Eq.1 exactly match those of the aluminium event (because in both events the impactor and intermediate layer are composed of like materials). Furthermore, the ratio of the energy for melt to that for vaporization is, to within 10%, the same for cadmium and aluminium (Holsapple 1993), so that the third term is also nearly constant. However, the second term is not exactly matched by the cadmium layer. Aluminium (6061-T6) has an ultimate tensile strength of 3.1 Kbar and a density of 2.70 g/cm^3. Handbook values for the strength of annealed cadmium range from 1 to 2 Kbar, with a density of 8.65 g/cm^3. The heat of vaporization for aluminium is a factor of 3.12 higher than for cadmium. Therefore, the ratio of the value of $Y_\ell/\lambda H_\ell$ for aluminium to that of cadmium is in the range of 0.5 to 1. In other words, the strength parameter for the cadmium intermediate layer is probably higher than in the prototype aluminium event. If the strength of the intermediate layer is important, this suggests that the cadmium simulation would deliver less impulse to the wall than the aluminium prototype it attempts to simulate, because the scaled strength of the layer in the simulation would be too large.

The alternative is to use aluminium for the intermediate layer. This has the advantage that the second and third terms on the right side of Eq.1 are exactly matched in the simulation. However, in this case the density ratio is a factor of 3.2 smaller in the simulation, and the fourth term is a factor of $3.1^2 = 9.6$ too large. It is hard to assess the effect of the mismatch in the density ratio. The mismatch in the heat of vaporization (the fourth term), however, means that the intermediate layer will not melt and vaporize as easily as in the prototype. Therefore, it is more likely that the contribution of the layer to the debris cloud will be in the form of solid fragments, rather than melt or vapor. Because solid fragments are efficient penetrators, the simulation will deliver a more lethal cloud than the prototype.

It is therefore difficult to devise an exact simulation. The analysis indicates that a cadmium intermediate bumper provides a slightly less severe loading than the aluminium prototype, and an aluminium intermediate layer provides a slightly more severe loading. The magnitude of these differences depends on the dominant mechanisms involved in the penetration of the intermediate layer. If the layer's strength is not too important, e.g. if the dominant mechanism is melting and vaporization, then the similarity violation incurred by using cadmium as the intermediate layer will be small. On the other hand, if the penetration of the layer is controlled primarily by it's strength, with melting and vaporization playing only a small role, then it is preferable to use an aluminium layer in the simulation. In any event, separate experiments with cadmium and aluminium as the intermediate bumper should provide upper and lower bounds for the response of the aluminium prototype shield. As discussed in the next section, this does appear to be the case at 7 km/s (the largest velocity at which a prototype and a simulation can be compared).

4. Results

The initial conditions and results of the impact tests are summarized in Table 1. The result of each test was either the formation of a bulge (whose depth is given in the table) in the wall, or perforation of the wall. For scaled velocities near 7 km/s, the results shown in Table 1 can be used to compare the minimum projectile diameter for perforation of the prototype (aluminium projectile striking an aluminium shield) with those of cadmium simulations using either a cadmium layer or an aluminium layer. Six shots are summarized below giving this comparison:

			Bulged only		Perforated	
Projectile	Bumper	Layer	shot	*d* (mm)	shot	*d* (mm)
Cd	Cd	Cd	1615	8.77	1617	8.94
Al	Al	Al	1591	7.90	1589	8.49
Cd	Cd	Al	1618	7.65	1627	7.89

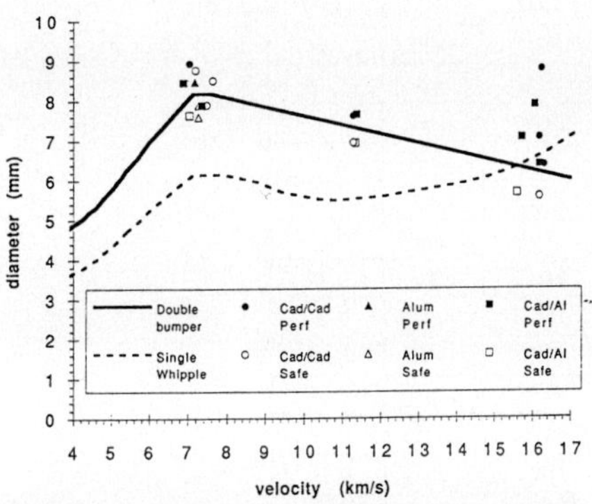

Fig. 4. Results of tests with all-aluminium double-bumper shields with simulations employing both cadmium and aluminium intermediate bumpers. The dotted curve is for a single-bumper shield from Schmidt et al. (1993)

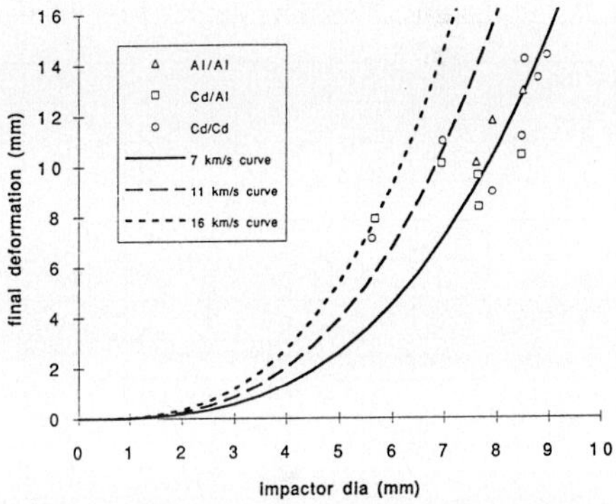

Fig. 5. Measurements of the maximum deformation (bulging) of the rear wall for impacts which did not perforate the wall. For a constant impactor size, the deformation increases as impact velocity increases, as expected. Similarly, for a constant velocity, larger impactors produce greater deformation. The curves show the cubic dependence predicted by the impulsive wall failure model of Schmidt et al. (1993)

For the aluminium prototype, the minimum diameter for perforation is between 7.90 and 8.49 mm. The cadmium simulations with a cadmium layer gave a somewhat higher estimate: the midpoint of the range shown above is 8% larger than the midpoint of the range shown for the aluminium prototype. On the other hand, the cadmium simulations with an aluminium layer slightly underestimated the threshold diameter: the midpoint of shots 1618 and 1627 is 5% below the midpoint for the prototype. This behavior is consistent with the discussion in Section 3, where it was noted that cadmium simulations using a cadmium layer slightly underestimate the loading presented to the wall, while an aluminium layer probably overestimates the loading. Having noted these differences, the cadmium experiments provide good bounds for the aluminium prototype for scaled velocities near 7 km/s.

Table 1. Summary of test conditions and results

			m	d	U	U		wall
Shot number	Shield (mm/mm)	angle (°)	mass (gm)	diam (mm)	vel (km/s)	scaled (km/s)	Bulge (mm)	vel (km/s)
4-1565a	1.96/0.95	45	0.7327	7.94	6.09	6.09	perf	-
4-1581c	2.11/1.07	45	2.2542	7.92	3.55	11.01	6.4	-
4-1582c	2.11/1.07	45	2.7708	8.49	3.77	11.69	perf	-
4-1584c	2.11/1.07	45	2.7670	8.48	2.90	8.99	perf	-
4-1585c	2.11/107	45	2.7893	8.51	2.35	7.29	6.4	-
4-1586c	2.11/1.07	45	2.8035	8.52	5.01	15.53	perf	-
4-1587c	2.11/1.07	45	2.0091	7.63	5.04	15.62	perf	-
4-1589d	2.21/0.94	0	0.8679	8.49	7.17	7.17	perf	-
4-1590d	2.21/0.94	0	0.6235	7.60	7.26	7.26	10.2	-
4-1591d	2.21/0.94	0	0.7009	7.90	7.26	7.26	11.8	-
4-1592c	2.11/0.94	0	2.2316	7.90	2.41	7.47	9.0	-
4-1593c	2.11/0.94	0	2.7932	8.51	2.46	7.63	14.2	-
4-1613c	2.11/0.94	0	2.7476	8.46	2.24	6.94	11.2	-
4-1614x	2.11/0.94	0	2.7436	8.46	2.22	6.88	perf	-
4-1615c	2.11/0.94	0	3.0567	8.77	2.32	7.19	13.5	-
4-1616x	2.11/0.94	0	2.0074	7.62	2.27	7.04	9.7	-
4-1617c	2.21/0.94	0	3.2400	8.94	2.27	7.04	perf	-
4-1618x	2.21/0.94	0	2.0260	7.65	2.27	7.04	8.4	-
41620c	2.11/1.07	0	2.0250	7.65	3.63	11.25	perf	-
4-1627x	2.21/0.94	0	2.2275	7.89	2.37	7.35	perf	-
4-1634c	2.21/0.94	0	2.0045	7.62	3.64	11.28	perf	-
4-1635x	2.21/0.94	0	1.5091	6.93	366	11.35	8.0	0.11
4-1636c	2.21/0.94	0	1.5220	6.95	3.64	11.28	10.3	0.12
4-163x	2.21/0.94	0	2.0288	7.65	3.67	11.38	perf	0.21
4-1638c	2.21/0.94	0	3.057	8.77	5.24	16.24	perf	0.55
4-1639x	2.21/0.94	0	2.2161	7.88	5.18	16.06	perf	0.36
4-1641x	2.21/0.94	0	1.6002	7.07	5.07	15.72	perf	0.24
4-1642x	2.21/0.94	0	1.1807	6.39	5.23	1621	perf	0.18
4-1643c	2.21/0.94	0	1.5979	7.07	5.22	16.18	perf	0.14
4-1644x	2.21/0.94	0	0.8212	5.66	5.03	15.59	7.9	0.12
4-1646c	2.21/0.94	0	0.7808	5.56	4.99	15.47	7.1	0.05
4-1647c	2.21/0.94	0	1.1719	6.37	5.26	16.31	14.4	0.18

Shield column is bumper thickness/layer thickness. Shot number suffixes denote:

a: 2017-T4 impactor and 6061-T6 bumper & layer. c: Cd impactor, Cd bumper & Cd layer.

d: 1100F impactor and 6061-T6 bumper & layer. x: Cd impactor, Cd bumper & 6061-T6 layer.

Cadmium simulations were performed to estimate the minimum diameter for perforation at velocities above 7 km/s, where aluminium tests cannot be conducted. To bound the actual value for aluminium, simulations were performed using both a cadmium layer and an aluminium layer. Fig.3 summarizes the results as a function of scaled impact velocity (i.e., 3.1 U) for normal impacts onto double-bumper shields. Shots that perforated the wall are shown with filled symbols; those that did not perforate are shown with open symbols. Fig.3 shows that, at scaled velocities near 11 km/s, the results for cadmium and aluminium layers agreed to within less than 10% in the projectile diameter. Similar results are shown for a scaled velocity near 16 km/s. Fig.4 shows that the threshold diameter for perforation decreases as the impact velocity increases. This behavior is in contrast to that observed by Schmidt et al. (1994) for a single-bumper shield (shown as a dashed line in Fig.4). This demonstrates that the velocity-dependent response of a debris shield can vary markedly with the shield geometry and that simple extrapolations based on results at 7 km/s and an assumed curve shape can be grossly in error. For tests in which the wall was not perforated, the maximum deformation (bulge) of the wall was measured and is shown in Fig.5. The results for 11 km/s and 16 km/s were obtained from careful measurements of the deformed wall with a mechanical profilometer, which has a precision of ±0.1 mm. Because of practical constraints, the results for 7 km/s were estimated (without the use of a profilometer), and are known to within ±1 mm. The scatter in the 7 km/s data prevents detailed comparisons between the two types of cadmium simulations. Even so, the simulations show general agreement with aluminium tests at 7 km/s. For a given impactor size, the higher velocity results show larger deformations, as expected. Similarly, for a given velocity, the deformation increases as the impactor size increases. The dependence on impactor size is in general agreement with the model of wall failure developed by Schmidt et al. (1994), which predicts that the deformation should increase as the cube of the impactor diameter. The curves shown in Fig.5 are of the form $k\,d^3$, where the proportionality constant, k, for each curve was chosen to make it pass through the set of data for a given velocity.

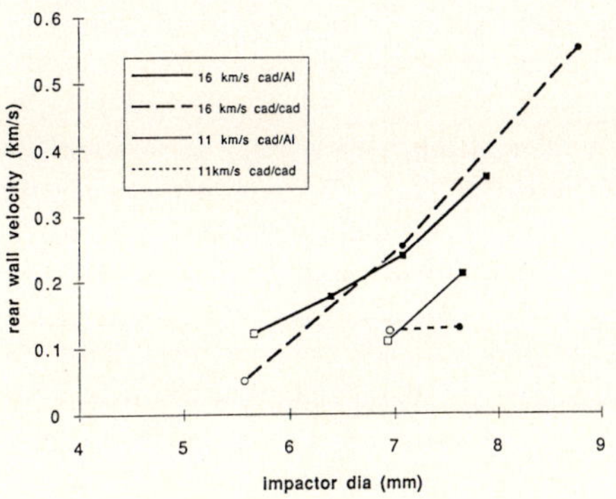

Fig. 6. Measured (lower bounds) of rear wall velocity for cadmium simulations of impacts at normal incidence

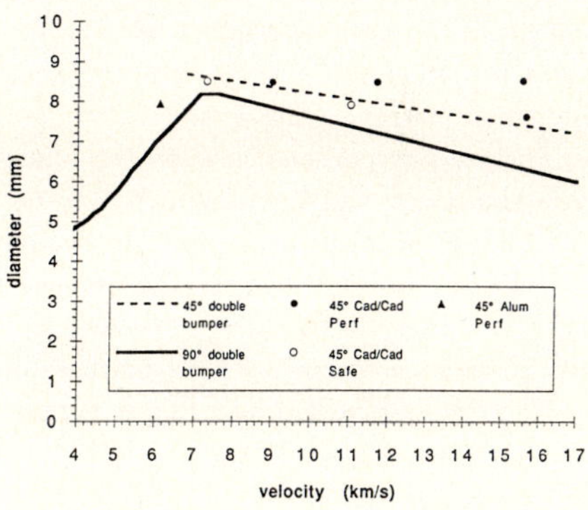

Fig. 7. Results of cadmium simulations of impacts on a double bumper shield for an impact angle of 45°

Time-resolved flash X-rays were used to measure the deformation velocity of the wall. Fig.6 shows a summary of these measurements for normal impacts at 11 and 16 km/s. For a given impact velocity, the velocity imparted to the wall increases as the projectile size increases. Furthermore, perforation occurs when the wall velocity exceeds roughly 0.1 to 0.2 km/s. These results will provide useful comparisons for numerical (hydrocode) simulations and analytical models of wall failure, such as that presented by Schmidt et al. (1994).

Fig.7 shows results of cadmium simulations of impacts at an angle of 45°. For comparison, the curve from Fig.4 is also shown. The 45° results show the same general trend as the normal impacts. That is, the threshold diameter for perforation slowly decreases as the impact velocity increases. If the scaled impact velocities for the 45° tests are multiplied by $cos(45°)$, the data points are in general agreement with the curve for normal impacts shown in Fig.7. This suggests that the impact angle can be accounted for by using the normal component of the impact velocity. This conclusion is based on a rather small data set, and so should be considered as preliminary. Moreover, as noted above, it is unwise to extrapolate the results for one shield to shields of differing geometries. Therefore, a cosine dependence may not apply to other shield types (e.g. single bumper shields).

5. Conclusion

A simulation technique for investigating the response of aluminium single-bumper debris shields at velocities above 7 km/s was extended and applied to the case of double-bumper shields. This technique is designed to allow low velocity (< 7 km/s) impacts to simulate the effects of melting and vaporization that occur for aluminium at high velocities. In a simulation, the actual structural aluminium wall of interest is used, and the impactor and bumper(s) are replaced with surrogates that have the same dimensions as the aluminium counterparts, but an order of magnitude lower specific energies for melt and vaporization. As a result, experiments can be conducted with surrogate materials to study impacts at much higher velocities than those that can otherwise be attained in the lab.

Cadmium was used for the surrogate material because it most closely satisfies the required conditions for similarity. In addition, cadmium provides a velocity scale factor of 3.1, which allows conventional light-gas guns to be used to simulate impacts on aluminium shields at velocities up to roughly 21 km/s.

Minor similarity violations occur for the case of double bumper shields. Arguments were presented, which concurred with experimental results, that the response of an aluminium shield is bounded by two cases in which the intermediate bumper is composed of either the surrogate cadmium, or aluminium. Impacts on aluminium shields at 7 km/s compared well with cadmium simulations at scaled velocities of 7 km/s. In particular, the minimum projectile diameter for wall perforation in the simulations was found to be within 10% of the value determined from the aluminium prototype impacts.

For impacts perpendicular to the bumper, the minimum projectile size for perforation decreased as the scaled impact velocity was increased from 7 to 16 km/s. This is in marked contrast to earlier simulations using a single bumper shield, in which the minimum diameter increased for velocities above 10 km/s. This underscores the errors that can occur in using the velocity dependent response of one type of shield to predict that of another. Impacts at 45° showed trends similar to those observed for the normal impacts.

The results reported here and in earlier studies support the use of cadmium as a means for simulating the response of aluminium debris shields at high velocities. This technique is quite useful because, at present, it is the only means to experimentally validate the effectiveness of debris shields over the entire range of velocities pertinent to the space debris environment.

References

Chhabildas LC, Hertel ES, Hill SA (1992) Whipple bumper shield tests at over 10 km/s. In: Schmidt SC, Forbes JW, Dick R (eds) Shock Waves in Condensed Matter-1991. Elsevier Science Publishers

Christiansen EL (1993) Design and performance equations for advanced meteoroid and debris shields. In: Proc. 1992 Hypervelocity Impact Symposium, Intl. J. Impact Engng. 14: 145-156

Holsapple KA (1993) Hypervelocity impacts: Testing in surrogate materials. In: Proc. 1992 Hypervelocity Impact Symposium, Intl. J. Impact Engng. 14

Hopkins AK, Lee TW, Swift HF (1972) Material phase transformation effects upon the performance of spaced bumper systems. J. Spacecraft 9: 342-345

Piekutowski AJ (1987) Debris clouds generated by hypervelocity impact of cylindrical projectiles with thin aluminum plates. Intl. J. Impact Engng. 5: 509-518

Schmidt RM, Housen KR, Piekutowski AJ, Poormon KL (1994) Cadmium simulations of orbital debris shield performance to 18 km/s. J. Spacecraft and Rockets (in press)

Wilkinson JPD (1969) A penetration criterion for double-walled structures subject to meteoroid impact. AIAA J. 7: 1937-1943

Shocked Quartz, Silica and Carbon Materials in Meteorites and Impact Craters

Y. Miura*, K. Takayama†, T. Kato*, N. Kawashima§, M. Imai*, G. Iancu* and M. Okamoto‡

*Faculty of Science, Yamaguchi University, Yoshida, Yamaguchi 753, Japan
†Shock Wave Research Center, Institute of Fluid Science, Tohoku University, Katahira, Aoba, Sendai 980, Japan
§Institute of Space and Astronautical Science, Yoshinodai, Samagihara, Kanagawa 164, Japan
‡Institute of Earth and Planetary Science, Okayama University, Okayama 730, Japan

Abstract. Characterization of shocked phases has been carried out by X-ray and electron devices on selected grains. Shocked phases are considered to be not a single phase of mineral but aggregates of various crystalline and amorphous phases. X-ray calculated density-deviation reveals itself to be an useful indicator of shock metamorphism at impact based upon selected data from the shocked aggregates. Main cause of the high density of shocked quartz is shrinkage of atomic distances mainly due to relic crystalline structure by quenching under high pressure and temperature conditions. Shocked glasses give important information on impact target rocks. Shocked graphites and chaoite-like carbon with high-density are also found in shocked graphite nodule aggregates of the Barringer crater.

Key words: Shocked aggregates, Shocked graphite, Shocked quartz

1. Introduction

Shocked quartz silica and carbon minerals have been found from meteorites, impact craters and Cretaceous-Tertiary (K/T) geological boundary; shocked quartz has been discussed by optical directions, mean optical refractive index and X-ray data (French 1968; Short 1968; Stöffler 1971; Bohor et al. 1987).

The purpose of the present study is to discuss the selected and analytical data of shocked quartz, graphite and chaoite-like carbon phases found in the Barringer crater and some impact craters (Miura 1991; Miura and Kato 1992; Miura et al. 1992; Miura et al. 1993).

2. Experimental procedure

Shocked quartz aggregate (SQA) samples were used from natural and artificial impact craters, the Cretaceous and Tertiary (K/T) geological boundaries. Shocked graphite aggregate (SGA) samples were first used from the Barringer crater, U.S.A. (Miura et al. 1993b).

Shocked grains from the SQA and SGA were selected and analyzed in composition and structure by X-ray diffractometry (XD) and analytical scanning electron microscopy (ASEM) at Yamaguchi University. The detailed measurement procedure of shocked samples is the same as reported previously by Miura (1991) and Miura et al. (1992).

3. Results and discussions

3.1. Refractive index data

The discrepancy between optical refractive index and X-ray calculated density with high-precision lies in the fact that the mean refractive index $R(\mathrm{m})$ is produced from two major mixed micrograins of shocked quartz crystal $R(\mathrm{c})$ and shocked glass $R(\mathrm{g})$ as follows:

$$\text{Mean refractive index}: R(\mathrm{m}) = R(\mathrm{c}) + R(\mathrm{g}).$$

Shock Waves @ Marseille III
Editors: R. Brun, L. Z. Dumitrescu

Thus, the quenched crystal part of the SQA or SGA shows high X-ray density and some significant information on the shock metamorphic process, whereas the mean bulk refractive index $R(\mathrm{m})$ of shocked aggregates reveals mixing value with crystal (c) and glass (g) parts without any significant information.

3.2. X-ray powder diffraction pattern

The X-ray powder diffraction (XD) patterns are characteristic of selected shocked data from mixed crystalline and glassy aggregates. In fact, the XD data of the SQA reveal that all X-ray peaks are split into major "three peaks" composed of low Bragg angle 2θ with low-density (LQ), normal Bragg angle with normal density (NQ) of standard sample and high Bragg angle with high-density (SQ) as follows:

$$\text{Bulk XD peaks}: P(\mathrm{SQA}) = P(\mathrm{LQ}) + P(\mathrm{NQ}) + P(\mathrm{SQ}).$$

On the SQA sample from the K/T geological boundary, the microscopic split XD data of (110), (200), (201), (202) and (211) in measured hexagonal cells are actually observed even in shocked samples, whereas these crystalline planes cannot be observed in macroscopic optical refraction data reported on the shock metamorphic samples.

The abundance of amorphous shocked aggregates can be shown by the relative X-ray intensity. In fact, the X-ray intensity values among the LQ, NQ and SQ phases indicate that the SQ phases of various aggregates are present in 36 to 53 % by volume. This suggests that the SQA contain major parts of amorphous phases (SG1, SG2) by impact.

3.3. Silica aggregates by impact experiments

The various silica minerals in powdered aggregates can be obtained in artificial impact crater experiments of various type target rocks as follows:

SQA (in powder state) = Low density quartz (LQ1) + Normal quartz (rock crystal : NQ)
+ High density shocked quartz (SQ1) + Amorphous silica glass (SG1).

The powdered SQA can be found as diffuse scattered X-ray diffraction patterns from low to high 2θ of the Bragg angles around the strongest peak.

A new interpretation of these X-ray diffraction spots is not "broken crystal structures from one single quartz structure under solid-solid reaction" which is the previous interpretation, but "quenched aggregates of shocked silica phases depressed from various solid-liquid-vapour reaction processes under high temperature-and-pressure conditions". This is mainly because the SQA in fine ejecta consists of quenched dendritic texture as discussed later.

In fact, among four phases of the SQA, volcanic ashes from Fugen-dake of Mt. Unzen contain silica aggregates of LQ1+NQ+SG1 without the SQ1 phase. However, all shocked aggregates of LQ1+NQ+SQ1+SG1 can be found in 20 terrestrial impact craters and artificial impact craters, as well as in the K/T boundary samples. Thus it is very important to detect only the SQ1 from various SQA by using X-ray diffraction (XD) to check the degree of shock metamorphism (cf. Miura 1991).

On a single grain, each silica data can be obtained by the XD method. Miura (1991) indicates that there are at least three combinations of single silica grain as follows:

Shocked quartz aggregates (SQA) (in single grains):

(a) Low-density quartz (LQ2) + silica glass (SG2),
(b) Normal-quartz of standard rock-crystal (NQ) + silica glass (SG2),
(c) High-density quartz (SQ2) + silica glass (SG2).

When we select only higher X-ray intensity peaks of the SQ2 by computer, we can obtain the characteristic high X-ray density of the SQ2 phases about 1 % higher than normal rock-crystal which is really direct evidence for shocked quartz (SQ) of impact reported by Miura (1991) and Miura et al. (1992).

3.4. X-ray density of shocked quartz

Higher values of density-deviation obtained from diaplectic SQ grains are summarized as follows:

(a) $+0.7 \pm 0.2\,\%$ in the Barringer meteorite crater,

(b) $+0.8 \pm 0.2\,\%$ in the K/T boundary sample from Clear Creek North, Colorado, U.S.A.

(c) $+0.3 \pm 0.2\,\%$ in artificial sandstone crater, and

(d) $+1.7 \pm 0.5\,\%$ in the Zagami achondritic meteorite.

Experimental error of the density-deviation is $\pm 0.11\,\%$ at 15 times the measurements of standard rock-crystal or the CCN K/T boundary shocked quartz. Thus the density-deviation exceeding $0.1\,\%$ has significant meaning for the shock metamorphic effect. There is no shocked quartz in primordial chondritic meteorites formed under high temperature of magmatic evolution, although small chondrite or achondrite fragments having collided in cosmic space before entering to Earth may have fine grains of shocked quartz.

3.5. Atomic structure and chemical composition of shocked quartz SQ

X-ray structure analysis of shocked quartz grains SQ has been carried out on the terrestrial impact craters and the CCN K/T boundary samples by using a full matrix-Fourier transfer computer program at Yamaguchi University. Although the peak numbers are limited due to its diaplectic feature, the following results have been obtained:

(a) Strongly shocked quartz (density-deviation $+0.7\,\%$) reveals both largely-deviated O-O and Si-O atomic distances of $-0.61\,\%$ and $-0.62\,\%$, respectively.

(b) The major cause of high density of the SQ is considered to be shrinkage of the atomic structure.

Chemical composition of the SQ by using electron and ion probe microanalyzers shows almost pure silica without Al element, though amorphous silica glass (G2) contains Al contamination which has been checked by analytical high-resolution electron microscopy.

3.6. Characterization of the SQ materials

The four major parts of the SQA from the K/T boundary and impact crater consist in silica glass (G1), low-density quartz (LQ), normal quartz (NQ) and high-density shocked quartz (SQ) which have been formed by artificial impact experiments in powder samples (Miura 1991; Miura et al. 1992): that is, SQA=SQ+NQ+LQ+SG1. Single quartz grain SQG is furthermore divided into crystalline SQ (or LQ, NQ) and diaplectic glasses SQ2 as follows (cf. Table 1):

$$\text{Shocked quartz grain (SQG)} = \text{SQ} + \text{SG2};\ \ \text{NQ} + \text{SG2};\ \ \text{LQ} + \text{SG2}$$

Shocked quartz grains are mixtures even in grains. Such a diaplectic silica glass (SG2) is considered to be the best indicator of target materials. In fact, the composition of the glass is silica at artificial impact craters of sandstone target rock as silica, K-feldspar at granitic target rock, and pure silica or mixture of Si+Ca+Fe in the Barringer crater.

Table 1. Different values of density-deviation of shocked quartz and co-existing shocked silica minerals (LQ, NQ, SQ) in the impact craters and the K/T boundary. X-ray calculated density deviation is obtained only in cristalline parts

No.	Sample name	Aggregate name	Density deviation (%)	Others
1	Barringer meteorite crater (Arizona, USA)	1. Low-density quartz (LQ)	-0.1 ± 0.1	SQG
		2. Normal quartz (NQ)	0.0 ± 0.1	
		3. High-density quartz (SQ)	$+0.7 \pm 0.1$	
2	Artificial impact crater on sandstone target rock	1. LQ	$+0.3 \pm 0.2$	SQG
		2. NQ	0.0 ± 0.1	
		3.SQ	-0.4 ± 0.1	
3	K/T boundary sample (CCN, Colorado, USA)	1. LQ	$+0.8 \pm 0.2$	SQG
		2. NQ	0.0 ± 0.1	
		3. SQ	-0.1 ± 0.1	

Anomalous experimental data of shocked quartz with high density and shrinkages of atomic arrangements are observed only in the SQ crystallline phase. Thus the term of "shocked quartz" means only crystalline SQ phase in this study, whereas mixed shocked silica materials including glasses should be called as "shocked quartz aggregate (SQA)" or "shocked quartz grain (SQG)". The following results are characteristics of shocked quartz materials in this study:

(a) the bulk SQG or SQA is a product of shock wave impact as a whole.

(b) the fine SQ phase with high density is found only at shock wave impact.

(c) the complex phases of LQ+NQ+SG are also formed under magmatic high- temperature which are not an useful indicator of shock impact.

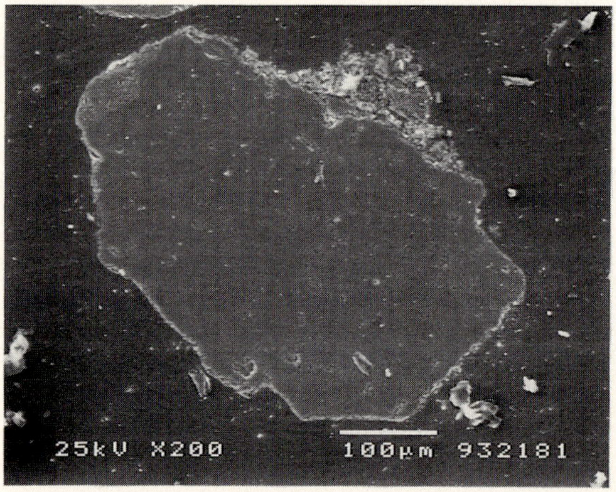

Fig. 1. Scanning electron micrograph of shocked quartz grain (SQG) from artificial impact experiments

3.7. Shocked silica phases

Shocked silica phases with high density are also found in coesite and stishovite phases, although shocked quartz is mainly found with larger density-deviation. For example, fine-grained shocked coesite from the Barringer crater has +0.5 % of density-deviation, though shocked stishovite from

Fig. 2. Scanning electron micrograph of shocked carbon with morphology of synthetic CVD diamond of (100) and (111) planes, although its X-ray intensity peaks reveal shocked mixture of amorphous carbon, shocked graphite and shocked chaoite-like carbon

the Barringer crater cannot be obtained in this study due to its poor crystalline state. The only data for shocked stishovite are obtained from the K/T boundary as +0.2 % of density-deviation, although these data are close to normal stishovite.

3.8. Shocked carbon phases

The similar relic crystal phases with high-density crystal and glass phases formed by impact processes can be found in carbon phases of shocked graphite and shocked chaoite-like carbon from shocked graphite blocks collected at the western outside (ca. 2 km) of the Barringer crater.

Shocked graphite is defined as low X-ray intensity (i.e. low crystalline state), high Bragg angle-shift (i.e. high density) and fine-grained states. For example, the density-deviation of shocked graphite from the Barringer crater is 0.9 %. Shocked carbon of hexagonal chaoite-like phase from the Barringer crater, which has been considered to be diamond from meteorite parent body, shows a high density-deviation of +0.6 %. The main reasons of its formation by impact are considered to be (a) main graphite nodule block with mixed composition of target limestone and sandstone with iron meteorite, (b) that showing mixed texture with tiny iron and carbon in shocked graphite in matrix, and (c) carbon supplied from vaporized target rock of limestone. Thus the shocked carbon as shown in Fig. 2 is considered to be (a) clear morphology of synthetic CVD diamond with (111) and (100) planes in shocked mixed vein surrounding with mixed kamacite iron, and (b) mixture of shocked graphite and shocked chaoite estimated by X-ray diffraction peaks. Such anomalous shocked carbon data suggest that the first formation of CVD diamond under shock impact condition was rapidly transformed and stopped at lower pressure and temperature condition after the major impact at high temperature and pressure.

4. Summary

The experimental results of this study are summarized as follows:

(1) Shocked phases are characterized by X-ray and electron devices at selected grains.
(2) Shocked phases are not single crystalline phase but aggregates of various crystalline and amorphous glasses.

(3) X-ray calculated density-deviation of the shocked phase is considered to be an useful indicator of shock metamorphism at the impact process based upon selected crystalline data from shocked aggregates.
(4) The main origin of the high density of shocked quartz is considered to be the shrinkage of atomic distances mainly due to relic structure by abrupt quenching from high shock condition.
(5) Shocked glasses show significant information of major sources of impact target rocks.
(6) Shocked graphites and shocked hexagonal chaoite-like carbon with high density are also found in this study.

Acknowledgement

The present study is supported partly by the Grant-in-Aid for Scientific Researches of the Japanese Ministry of Education and Science, 1990FY-1992FY of the senior author.

References

Bohor BF, Foord EE, Modreski PJ, Triplehorn DM (1987) Mineralogic evidence for an impact event at the Cretaceous-Tertiary boundary clays. Science 224:867–869

French B (1968) Shock metamorphism as a geological process. Shock Metamorphism of Natural Materials, pp 1–17

Miura Y (1991) Evidence for shock wave effect of meteoritic impact. Shock Waves 1:35–41

Miura Y, Kato T (1992) Shock metamorphism on the Moon. Celestial Mechanics and Dynamical Astronomy 54:245–248

Miura Y, Takayama K, Kato T, Kawashima N, Yamori A (1992) Shock metamorphism in artificial impact craters. In: Takayama K (ed) Proc. 18th Intl. Symp. on Shock Waves, Sendai, pp 403–408

Miura Y, Noma Y, Iancu G (1993) New occurence of shocked graphite aggregates at Barringer crater. Meteroritics 28: 402

Short N (1968) Experimental microdeformation of rock materials by shock pressure from laboratory-scale impacts and explosion. Shock Metamorphism of Natural Materials, pp 219–242

Stöffler D (1971) Coesite and stishovite in shocked crystalline rocks. J. Geophysical Research 76:5464–5488

Shock Metamorphic Evolution of Meteorites and Impact Craters

Y. Miura*, K. Takayama†, N. Kawashima‡ and T. Kato*
*Faculty of Science, Yamaguchi University, Yoshida, Yamaguchi 753, Japan
†Shock Wave Research Center, Institute of Fluid Science, Tohoku University, Katahira, Sendai 980, Japan
‡Institute of Space and Astronautical Science, Yoshinodai, Sagamihara, Kanagawa 164, Japan

Abstract. Shock metamorphic evolution which is followed by crystallization after major shock impact metamorphism has been investigated by impact experiments. Shocked phases of quartz, anorthite (An) -poor plagioclase, Fe-rich feldspar and Fe-rich pyroxene can be obtained by shock metamorphic evolution. Shocked quartz aggregates with "dendritic texture" are considered to be formed by rapid cooling process under "solid-melt-vapour" reaction at overall shock-impact. Two major types of shocked (diaplectic) plagioclases are obtained as (a) large-coarse-grained feldspar with partial anomalous composition, and (b) fine-grained plagioclases (-like) grains with decrease of An-content. The experimental results in this study are applied to the evolution process of initial impact materials following crystallization on the primordial and evolved surfaces of the Moon and Earth-type planetary bodies.

Key words: Earth-type planetary bodies, Shock metamorphic evolution

1. Introduction

It has been considered that quartz minerals and anorthite (An)-poor and Fe-rich plagioclases are formed by evolved magmatic crystallization on the Earth or Earth-type planets under high-temperature condition of the magma. However, Miura et al. (1992a) reported that when similar high-temperature condition can be obtained at impact processes, various shocked phases of silica minerals, An-poor plagioclase and Fe-rich feldspars (or pyroxene) which will be found at the surfaces of the evolved Earth-type planetary bodies should be formed even under impact condition (Miura 1991; Miura et al. 1991; Miura et al. 1992b; Miura and Kato 1992).

The purposes of the present study are to obtain direct evidence for solid-liquid-vapour condition at artificial impact experiments based on compositional, structural and textural changes of shocked phase aggregates (compared with original target rocks), and to apply the experimental impact results to formations of silica and plagioclases phases on meteorite parent body, primordial lunar and Earth-type planetary bodies following crystallization by major shock metamorphism (i.e. shock metamorphic evolution; Miura et al. 1992a).

2. Experimental procedure

Target rocks of impact experiments are gabbroic anorthosite from Mt. Kohyama, Yamaguchi (Japan), granite from Tokuyama (Japan), basalt from Hagi, Yamaguchi (Japan), and sandstone from India. The achondritic meteorite sample is Zagami (probably from Mars). Lunar samples are all from Apollo missions made available by NASA, U.S.A. to the senior author. Experimental procedures of impact are the same as the previous method by Miura et al. (1992a; 1992b). Shocked grains from impact experiments were selected and analyzed in composition and structure by X-ray diffractometry (XD) and analytical scanning electron microscopy (ASEM) at Yamaguchi University. The detailed measurement procedure of shocked samples is the same as reported previously by Miura (1991) and Miura et al. (1992).

Shock Waves @ Marseille III
Editors: R. Brun, L. Z. Dumitrescu

3. Results and discussion

3.1. Silica phases by impact experiments

Various shocked silica minerals (low-, high- and normal-density types of quartz etc.) can be obtained in artificial impact experiments on various types of target rocks.

Fine-grained shocked quartz aggregates crystallized from vaporization of feldspars (F) or quartz (Q) compositions are shown by increased abundance of shocked quartz (SQ) and feldspar (SF) at the "fine ejecta"; that is, SQ/SF=3.0 and 5.6 in granite and gabbroic anorthosite, respectively (Miura e al. 1992a, 1992b) (cf. Fig.1). This type of shock metamorphism is a solid-melt-vapour reaction under the largest shock loading (cf. Table 1).

Table 1. Silica data and ratio of shocked quartz (SQ) to feldspar (SF) of artificial impact craters of three target rocks

Sample No.	Target rock	Density ρ (g/cm^3)	Density dev. ρ/ρ_0 (%)	Ratio SQ/SF	Remarks
1G	**Granite**				
	Original (GP)	2.640(1)	-0.19(5)	1.6	Q/F value
	Fine ejecta (GEF)	2.645(0)	+0.00	3.0	High SQ/SF
2A	**Babbroic anorthosite**				
	Original (BP)	2.642(8)	-0.11(7)	0.02	Q/F value
	Fine ejecta (BEF)	2.651(2)	+0.22(6)	5.60	High SQ/SF
3S	**Sandstone**				
	Original (SP)	2.645(1)	+0.00	-	
	Fine ejecta (SEF)	2.656(5)	+0.3(2)	-	High density

* Numbers in parentheses are standard deviations to the last decimal place

Impact effects of density change and shocked quartz formation (SQ/SF) are larger in the fine-grained target rock of gabbroic anorthosite than in the hard coarse-grained target rock of granite.

Table 2. Shock metamorphic materials formed by impact experiments

Target rocks	Shock metamorphic materials
Anorthosite:	
Plagioclase	a) Shocked metamorphic quartz (SQ)
($An_{50}Or_3$; $An_{54}Or_3$)	b) Albite plagioclase($An_{14}Or_8$)
(Q/F=0.02)	c)*Andesine plagioclase ($An_{33}Or_{18}$)+Fe(4 %)
Pyroxene	a) Fe-rich pyroxene ($Fs_{32}Wo_2$; $Fs_{47}Wo_4$)
($Fs_{14}Wo_{33}$; $Fs_{23}Wo_7$)	b)*Px ($Fs_{52}Wo_{21}$) + Plg ($An_{31}Or_{18}$?)
Granite:	
Plagioclase	a) Shocked metamorphic quartz (SQ)
(An_9Or_8; $An_{12}Or_2$)	b) Albite plagioclase ($An_{18}Or_7$)
(An_7Or_{64})	c)*Fe(projectile) + albite feldspar
Alkali feldspar	a) Alkali feldspar ($Or_{82}An_1$)
($Or_{98}An_2$; $Or_{21}An_{19}$)	
Quartz	a)*Fe(+Cu)+feldpsar
Pyroxene ($En_{52}Fs_{47}$)	

* Mixed amorphous phase

Shocked stishovite and shocked coesite crystals could not be observed in the small space of laboratory experiments (French 1968; Short 1968; Stöffler 1971). The highest density of shocked quartz crystal (SQ) in the fine-grained ejecta can be obtained in quartz-rich target-rock of sandstone, which is similar implication to the Barringer impact crater with sandstone (Miura et al. 1991).

3.2. Compositional change of plagioclase by impact experiment

There are two types of shocked (diaplectic) plagioclase (cf. Tables 2 and 3) as follows:

(1) Small compositional change type:
Chemical compositions of large coarse-grained fragments broken by impact processes reveal partly anomalous diaplectic feldspar grains with irregularly wavy (undulate) extinction and nonstoichiometric composition. The similar anomalous composition of diaplectic plagioclase can be found in the wall rocks of artificial and natural impact craters. This type of diaplectic plagioclase shows small change of composition and atomic arrangement within solid-solid (-liquid) reaction by impact.

(2) Large compositional change type:
Fine-grained plagioclase (-like) composition found in fine ejecta of artificial impact crater experiments reveals decrease of An-content and mixing with the projectile to form Fe-rich feldspars or pyroxenes. For example, plagioclase from Kohyama gabbroic anorthosite (bytownite-labradorite) changes to albite (i.e. decrease of An_{40} mol. %).There are variuous mixture phases among plagioclase, pyroxene and projectile element (i.e. Fe). The type of diaplectic plagioclase shows intermediate to maximum degree of shock loading (i.e. reaction under melt condition), as shown in Table 2.

If it includes solid-vapour reaction of impact, there are three types of vapour, melt and solid-solid reactions in the shocked metamorphic materials, as shown in Table 3.

Fig. 1. Fine-grained quartz silica aggregates with ca. 1 to 10 μm in width and large parts of X-ray amorphous states found in the fine-ejecta from target-rock of gabbroic anorthosite formed by the solid-melt-vapour reaction. BEI image. Steel projectile.

3.3. Direct evidence for solid-solid-vapour reaction of shock metamorphism

The shocked materials formed by large impact with solid-liquid-vapour reactions can be found in shocked quartz aggregates and projectile mixture, as follows:

(1) Very fine-grained shocked quartz (SQ):
Analytical data of SEM with EDX indicate that shocked quartz formed by the largest shock loading shows fine-grained aggregate of silica composition (about 1 micrometers) with dendritic texture (i.e. quenched texture from vapour-melt condition) and large parts of amor-

phous state (cf. Fig. 1). Abrupt crystallization from high temperature-and-pressure condition makes mixture with amorphous (large) and crystalline (small) states.

(2) Mixture of Fe-projectile and feldspar:
Projectile of Fe (mainly in steel projectile) and Cu (mainly in plastic projectile) has melted to produce mixtures of Cu/Fe and plagioclase, and various types of droplets and mantle-core concentric structure (cf. Fig. 2). The shocked concentric texture in the impact experiment is well similar to the planetesimal structure, where impact materials consist in a Fe (and Co)-rich region in core of the concentric texture, and a Cu (and Al, Mg, Ca, K, Na, Si)-rich region in the mantle of the concentric texture.

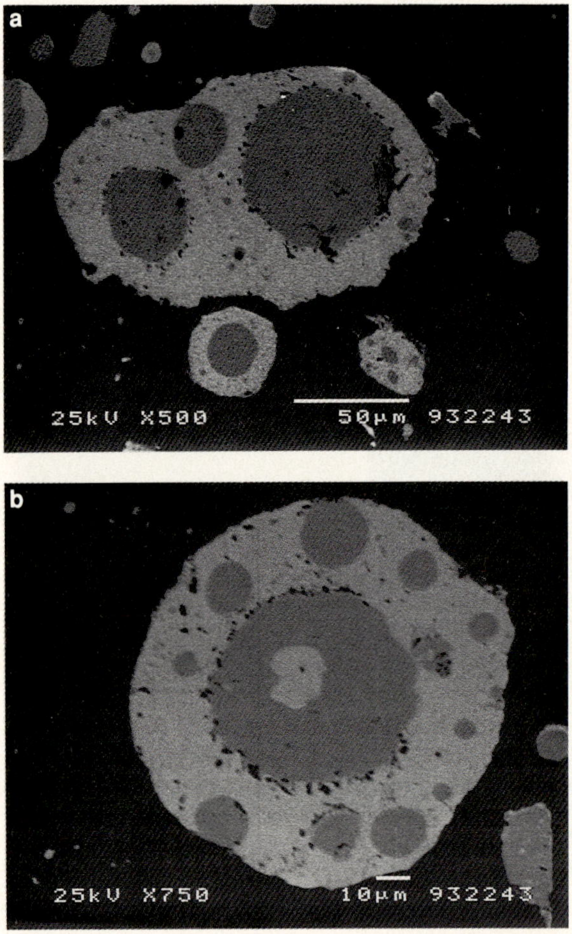

Fig. 2. Evidence of solid-melt reaction of shock metamorphism. Various types of (a) droplets, and (b) mantle-core concentric structure can be observed in fine-ejecta of granitic target-rocks. Plastic-Cu/Fe projectile. BEI images.

3.4. Shock metamorphic evolution of planetary materials

Based upon the experimental results of shock metamorphism, it is inferred that plagioclase minerals of anorthositic rock formed by magmatic ocean processes on the primordial planetary and lunar highland surfaces, can be changed to shocked silica minerals, An-poor plagioclase, diaplectic plagioclase (with mixture of iron elements), and Fe-rich feldspars (or pyroxenes) by impact shock effects, resulting in the formation of shocked quartz (SQ) by rapid crystallization. The SQ phases formed by impact can be changed to normal quartz (Q) by long-time crystallization or magmatic evolution process under high temperature, as shown in Table 4. The overall process is called a "shock metamorphic evolution". The present results of "shock metamorphic evolution" will be applied to lunar impact and recrystallized rocks (cf. the lunar rocks of KREEPy and lunar rocks with major feldspar and tiny quartz aggregates) as shown in Fig. 3.

Table 3. Various types of shock metamorphism in silica-plagioclase-projectile aggregates

Target rock, plagioclase	Shocked materials	Reaction	Shock wave
1) Gabbroic anorthosite			
Labradorite-bytownite	a) Shocked quartz Ag.	Solid-vapour	The largest
	b) Albite	Melt (-vapour)	Intermediate
	Andesine+Fe	Melt (-vapour)	Intermediate
	c) Nonstoichiometric	Solid-melt	Intermediate
	labradorite	Solid-solid	Small
2) Granite			
Albite (+Quartz)	a) Shocked quartz A.	Solid-vapour	The largest
	b) Albite+Fe	Melt (-vapour)	Intermediate
	Projectile mixture	Melt (-vapour)	Intermediate
	c) Nonstoichiometric	Solid-melt	Intermediate
	albite	Solid-solid	Small
3) Sandstone			
Quartz	a) Shocked quartz A.	Solid-vapour	The largest

Table 4. Shock metamorphic evolution on primordial and evolved surfaces of Earth-type planetary bodies and the Moon

1) On primordial lunar and Earth-type planetary surfaces:		
Feldspar (F) → (impact shock)		
a) → (melt condition)	→	An-poor plagioclase (+Fe)*
b) → (solid-solid reaction)	→	Diaplectic plagioclase*
c) → (vaporization; rapid depression)	→	Shocked quartz (fine-gr.: SQ)
→ (slow cooling; recrystallization)	→	Quartz (coarse-grained: Q)
2) On evolved Earth and Martian surfaces:		
Quartz (Q) → (impact shock) → High-pressure type silica		
a) → (rapid depression)	→	Shocked quartz (fine-grained: SQ)
b) → (rapid cooling)	→	Cristobalite or Tridymite
c) → (slow cooling; recrystallization)	→	Quartz (coarse-grained: Q)

* Maskelinite (found in meteorites). Similar phases can be found in lunar rocks, and wall rocks of impact craters

These shocked materials can be found in lunar meteorites, Martian SNC meteorites, interplanetary dust particles (IDP's), and artificial and terrestrial impact craters (Miura and Kato 1992).

Primordial materials ⟶ (Impact shock) ⟶ Shock metamorphic materials
1) ⟶ (Slow cooling) ⟶ Large coarse-grained crystals
2) ⟶ (Magnetic/tectonic process) ⟶ Large crystals

Fig.3. The process of shock metamorphic evolution on Moon and Earth-type planetary bodies

4. Summary

Experimental impact results in this study can be summarized as follows:

(1) Shocked phases of quartz, An-poor plagioclase, Fe-rich feldspar and Fe-rich pyroxene can be formed by shock metamorphic evolution.

(2) Shocked quartz aggregates with "dendritic texture" are formed by rapid cooling process under "solid-melt-vapor" reaction.

(3) Shocked quartz aggregates can be formed in all types of target rocks (anorthosite, granite, sandstone, and basalt).

(4) Shock metamorphic evolution can be discussed on primordial and evolved surfaces of Earth-type planetary bodies and the Moon which can be indicated by artificial impact experiments.

(5) The initial stage of impact process can be obtained in artificial impact craters, though almost all shock metamorphic phases (except small dendritic phases) can be found in well-developed terrestrial impact craters.

Acknowledgements

The authors thank the cooperative programs of Tohoku University and the ISAS for impact experiments. The present work is supported partly by the Japanese Monbusho Scientific grants from 1990 FY to 1992 FY.

References

French B (1968) Shock metamorphism as a geological process. Shock Metamorphism of Natural Materials:1–17

Miura Y (1991) Evidence for shock wave effect of meteoritic impact. Shock Waves 1:35–41

Miura Y, Kato T, Kawashima N, Yamori A, Imai M (1991) New shocked materials in artificial impact craters. Lunar and Planetary Science (ISAS, Tokyo) 24:46–52

Miura Y, Kato T (1992) Shock metamorphism on the Moon. Celestial Mechanics and Dynamical Astronomy 54:245–248

Miura Y, Kato T, Imai M (1992a) Shock metamorphic evolution on the moon and impact craters applied by shock impact experiments. Meteoritics 27:261–262

Miura Y, Takayama K, Kato T, Kawashima N, Yamori A (1992b) Shock metamorphism in artificial impact craters. In: Takayama K (ed) Proc. 18th Intl. Symp. on Shock Waves, Sendai, pp 403–408

Short N (1968) Experimental microdeformation of rock materials by shock pressure from laboratory-scale impacts and explosion. Shock Metamorphism of Natural Materials, 1st edt., pp 219–242

Stöffler D (1971) Coesite and stishovite in shocked crystalline rocks. J. Geophysical Res. 76:5464–5488

Interaction of Shock Waves with Tree Crowns and the Front of Crown Forest Fires

A.M. Grishin
Department of Physical Mechanics, Tomsk State University, Tomsk, Russia

Abstract. Fires are known to cause great damages to forests all over the world. Traditional ways of extinguishing town fires by water or by various chemical compositions are not effective in localizing and extinguishing forest fires, as the form of the fire front changes with time and its thickness is considerably less than its length. Besides, forest fires as a rule take place in almost inaccessible places, and supplying large masses of water to the front of a forest fire by ground transport is hampered, while throwing out water masses from safe flight altitudes is not efficient.

In this connection, a new concept of fighting forest fires by means of relatively small energetic effects on the fire front is proposed (Grishin 1992). Specifically, it is proposed to use relatively weak shock waves, more or less harmless to the trees themselves.

In the present paper a review of experimental and theoretical results is presented on the interaction of shock waves with tree crowns and the forest fire front, which have been obtained by the research workers of the Tomsk State University during a period of five years.

Key words: Explosive wave, Forest fire front, Continuous reactive medium

1. Methods and results of experimental investigations

The method of conducting experiments was as follows: On an equal area young pine trees freshly cut down were dug in and two forest bands of size $3 \times 10 \times 3$ m^3 were created (breadth by length by height). The fist (control) band was not ignited; on the forest bed of the second band a layer of lichen Cladonia, 15 cm high, was laid out for the purpose of full reproduction of the natural structure of pine undergrowth. At the edge of this band an ignition zone was formed, made of dry wood whose combustion initiated a crown forest fire at a wind speed $V > 2.5$ m/s. While the fire spread, needles and thin twigs (of diameter < 7 mm) which will be called in future forest combustible materials (FCM) were burned down. Fuse charges of explosives, which are usually used for suppressing forest bed fires, were placed in the crowns of the trees at a definite height. The explosive charges were exploded by means of an electric detonator. After the explosion in the first band, the pressure field was determined with transducers of various designs (Gavrilenko and Nikolayev 1982; Grishin 1992), and the degree of falling off of FCM was ascertained:

$$\overline{E}_i = (m_1 - m_3)/(m_1 - m_2), \qquad (1)$$

where m_1 is the initial mass of trees, m_2 is their mass after the explosion, m_3 is the mass of trees after combustion of FCM in crowns.

Table 1. Experimental data for the degree of falling off of FCM

	E_1	E_2	E_3	E_4	E_5	E_6	E_7
Control band	0.70	0.71	0.39	0.30	0.45	0.24	0.19
Burning band	0.81	0.79	0.60	0.35	0.41	0.30	0.29

In Table 1 the average degrees of falling off of needles and thin twigs in the control and burning forest bands are presented.

Shock Waves @ Marseille III
Editors: R. Brun, L. Z. Dumitrescu

In the above, $\overline{E}_i$ is the average degree of falling off of thin twigs and needles for the ith row of trees in each of which there were four young pine-trees of height $H = 3$ m. The distance between the trees in the row and between the rows was 0.7 m. The charge PZhV-20, usually employed by PO "Avialesookhrana" in laying out mineralized bands, was used. In the given case a hose charge of 3 kg was placed in the tree crowns at a height of 1.5 m.

In experiments the mean values were obtained of absolute pressures in atmospheres, measured by means of a transducer AIDA-M (Grishin 1992), at stations identified by indices i and j where $i, j = 1 \ldots 5$.

The scheme of disposition is presented in Grishin (1992).

On the whole more than 26 experiments on the action of shock waves (SW) on FCM in burning and control forest experiments were carried out.

Table 2. Experimental data for absolute pressure in control and burning bands

	P_{14}	P_{23}	P_5	P_{12}	P_{34}
Control band	10.28	3.26	10.80	10.83	4.10
Burning band	11.26	4.75	12.10	12.14	5.30

As a result of data analysis it was established that with sufficient mass of explosive, flame-out and cessation of spreading of the crown forest fire along the forest band takes place. In both bands falling off of the FCM takes place, with the degree of falling off of the FCM being 15–30 % higher in the burning band than in the control one. Besides, the pressure at similar points and at the same instants of time is 9–13 % higher on the underlying surface of the burning forest band than in the control one. As the analysis of experimental data shows, the shock is attenuated as a result of medium resistance and backpressure. It should be mentioned that the effect of the latter on SW attenuation in air is analyzed in detail in the work of Korobeinikov (1985). These results, however, cannot be used for analysis of SW attenuation in the forest phytocenosis which, from the point of view of continuous media, represents a specific multiphase dissipative medium. Therefore a theoretical analysis of this problem is of interest.

2. The main mathematical model and methods of solving the problem

For the theoretical investigation of the problem of interaction of shock waves with FCM and forest fire fronts, a special mathematical model was developed, within the frame of which the forest was modeled as a porous medium susceptible to undergo chemical reactions. For simplicity of analysis a number of assumptions were made and a special set of gasdynamic equations was used (Grishin 1992).

At the explosion of a cylindrical elongated charge, located in the cupola of the forest, perpendicular to the direction of spread of the forest fire, a cylindrical wave is formed, which later, at time $t = t_1$, is reflected from the underlying surface. Besides this, at $t = t_2$, the shock wave goes into the ground layer of the atmosphere through the upper boundary of the cupola. Therefore, for $t < \min(t_1, t_2)$ in a system of coordinates related to the centre of the cylindrical charge, the problem may be considered to be one-dimensional, while for $t > \min(t_1, t_2)$ it is two-dimensional. In this connection, within the limits of the above given model a one-dimensional problem has to be solved by Godunov's method (Godunov et al. 1976), for the interaction of the shock wave with the front of a forest fire; this latter, according to Grishin (1992) has a complicated structure, consisting of zones of heating, drying FCM, pyrolysis of FCM, combustion of gaseous products of pyrolysis of FCM. Also, a two-dimensional problem is to be studied, about the breaking-off of FCM elements by the shock wave.

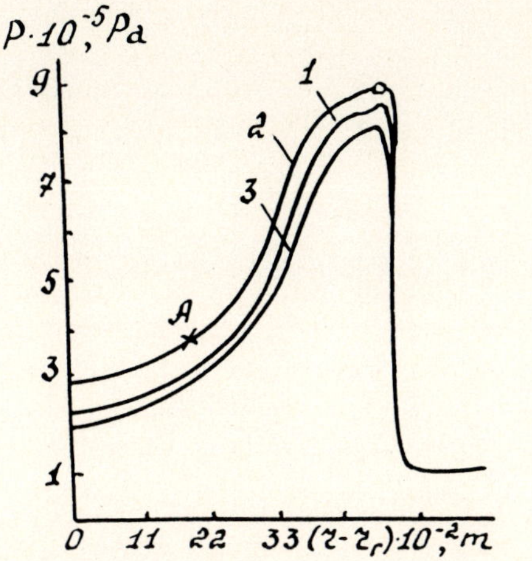

Fig. 1. Pressure profiles after the explosion of a cylindrical charge PZhV-20
1 – in air, 2 – in pyrolysis products with air, taking into account chemical transport, 3 – the same without account being taken; r_0 corresponds to the position of interface products of explosion - pyrolysis products; the experimental value of pressure is denoted by a circle, the position of the back frontier of the zone of chemical transformation is denoted by a cross

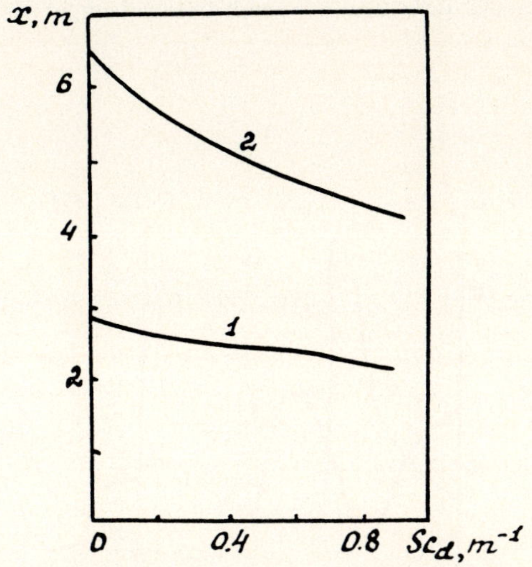

Fig. 2. Dependencies of the range of the explosion wave on the parameter SC_d (S is the specific surface of vegetation, C_d is the aerodynamic drag coefficient) for the values of the critical pressure differential $(\Delta P)_*$ equal to $1.2 \cdot 10^5$ (1) and $0.4 \cdot 10^5$ Pa (2).

3. Statement and solution of the one-dimensional problem

For a numerical solution of the one-dimensional problem, profiles of temperature and concentration of components were used, as obtained in the work of Grishin (1992) in investigatng the structure of forest fire fronts. In order to simply the thermochemical model of the process of interaction of the shock wave with the front of a forest fire it was considered that the gas phase consists of five components ($N = 5$): $^{(1)}N_2$, $^{(2)}O_2$, $^{(3)}CO$, $^{(4)}CO$, $^{(5)}H_2O$, (where the upper left index denotes the number of a component) and that an effective reaction of oxidation of carbon monoxide takes place. The concentration of CO was chosen so that the time of thermal explosion of a real mixture of air with the products of FCM pyrolysis and a hypothetical mixture of air with carbonic oxide, coincided. Thermokinetic chemical reactions were taken from the work of Grishin (1992). As initial conditions in the mixture of air and combustible gas the condition of zero flow rate and known values T, ρ and c_α for the front of the forest fire were used. Besides this it was considered that detonation of the condensed explosive PZhV-20 takes place instantaneously but, as initial conditions for the detonation products, profiles obtained from the solution of the self-similar problem were used. A cylindrical system of coordinates connected with the centre of the cylindrical charge was used.

Fig.1 presents pressure profiles after the explosion of a cylindrical charge in various media. Curve 1 corresponds to the explosion of the charge in pure air, 2 – in a mixture of products of FCM pyrolysis taking into account chemical transformations, and 3 – the same without taking into account chemical transformation; r_0 corresponds to the position of the interface of the explosion product with a mixture "products of pyrolysis of FCM and air". The experimental value of the pressure is designated by a circle and the position of the back boundary of the zone of chemical transformation at the interaction of the front of the forest fire with the shock wave is designated by a cross.

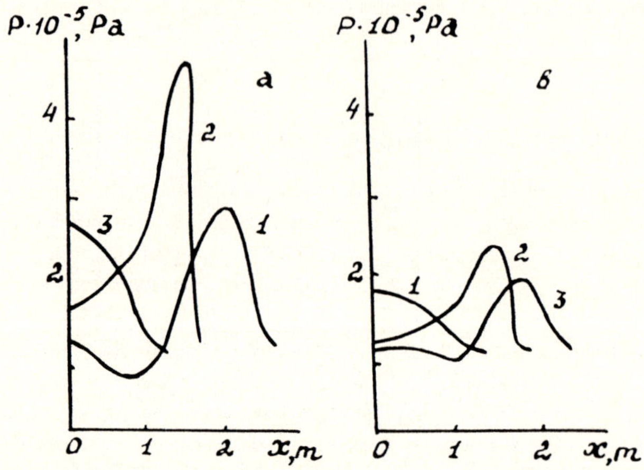

Fig. 3. Pressure profiles on the underlying surface (a) and the upper boundary of the forest cupola (b) for $t = 2.5$ ms
Here and in Fig.4-5 the values of h_1, m are: 0.75 (1), 1.5 (2), 2.25 (3)

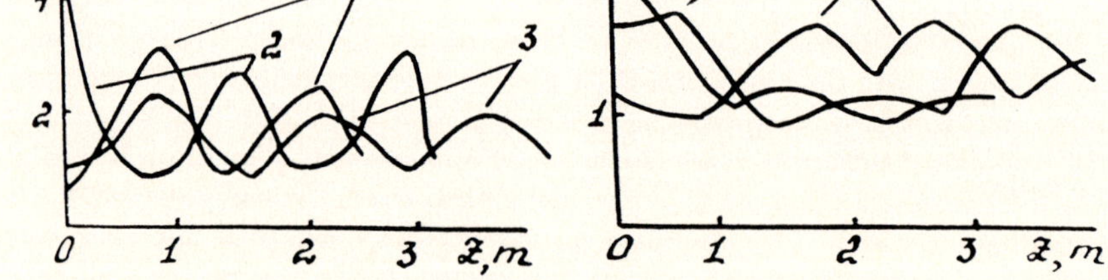

Fig. 4. Pressure profiles in the symmetry plane for t equal to 1.4 (a) and 2.5 ms (b).

As a result of the analysis of theoretical and experimental results, one can state that the main reason of the strengthening of the shock wave striking the front of a forest fire is the explosion of the mixture of gaseous products of pyrolysis with air at their interaction with the SW.

Besides that, there is the problem of determining the range of the shock wave. Define the range r_* of a blast wave in a certain direction e as the distance travelled by the shock wave from the centre of the charge up to a point in space lying on the ray e, where the overpressure (ΔP) at the front of the SW becomes less than a preassigned critical value $(\Delta P)_*$ of the pressure differential at which the degree of tearing off$E = 0.75$ for FCM takes place. The value $(\Delta P)_*$ depends on the type and age of vegetation, type of explosive and location of the charge. The dependence of r_* on SC_α with $(\Delta P)_* = 1.2 \cdot 10^5$ Pa and $(\Delta P)_* = 0.4 \cdot 10^5$ Pa is presented in Fig.2. It is easy to see that the "range" of the explosion, by which one understands the maximum area of the zone free from FCM, decreases according to a hyperbolic law with the growth of the value SC_α: the greater $(\Delta P)_*$, the faster x decreases.

4. Starting and some results of the solution of the two-dimensional problem

The functions which were obtained as a result of the solution of the one-dimensional problem were assumed as initial conditions for the pressure, density of internal energy and components of velocity. It is obvious that the problem of the evolution of a shock wave in a control volume of phytocenosis has a plane of symmetry $x = 0$. Therefore the following boundary conditions were used (Grishin 1992):

In the mathematical modeling of the evolution of a shock wave in a forest phytocenosis it was supposed that the standard blasting fuse charge FCH-IP was used as a cylindrical charge. The main problem of calculation consisted in determining isobars at various moments of time. In Fig.3 profiles of pressure on the underlying surface (a) and on the upper boundary of the forest cupola (b) with $H = 3$ m, $t = 2.5$ ms and various heights of location of the charge 1 – 0.75 m, 2 – 1.5 m, 3 – 2.25 m are presented.

In Fig.4 profiles of pressure are given, in the plane of symmetry, for t equal to 1.4 (a) and 2.5 ms (b) for the same value h, as in Fig.3.

The analysis of the pressure profiles at various moments in time shows that after the explosion of a charge of explosives in a forest cupola a complicated pattern of incident and reflected shock waves and expansion waves arises, which changes in time. Assigning a critical pressure differential $(\Delta P)_*$ an optimum location of the charge at which the greatest quantity of FCM is broken off with the same mass of explosive charge was determined.

As a result of solving the two-dimensional problem, the dimensions of the zone in the forest band where the break off of 75% of FCM took place were obtained.

In Fig.5 the graphs of the value of volume V cleared of phytomass per unit length of charge are presented for the same heights of location of the charge as in Figs.3 and 4 . From the analysis of the curves of Figs. 3 - 5 it follows that the optimal height of the location of the charge is $h = 1.5$ m as, placing the charge at this height, leads to raising the pressure and increasing the volume V.

Comparison of the results of mathematical modelling of the shock wave evolution and FCM break-off in the cupola of a forest with the experimental data shows that they agree with each other satisfactorily.

5. Conclusion

1. An experimental procedure was developed and experiments on shock wave - vegetation and forest fire front interaction were conducted.
2. It was established from experiments and theory that a shock wave intensified when interacting with a forest fire front. It is due to an explosion of gas - fuel products of pyrolysis - air and to interaction with the shock wave.
3. A mathematical model of shock wave - forest vegetation interaction was devised; and it may predict the space volume cleaned from forest burning material.

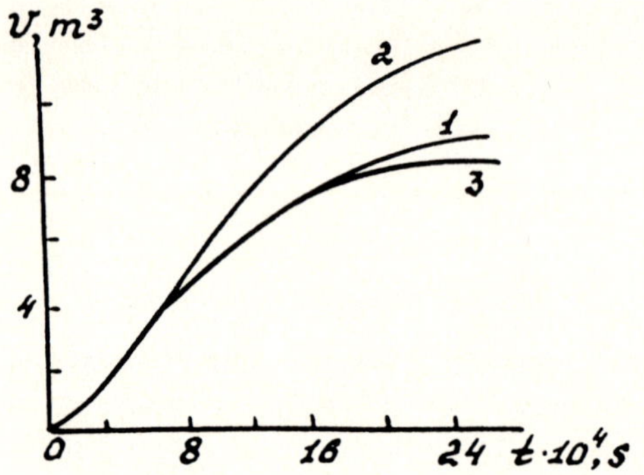

Fig. 5. Graphs of the volume cleared from phytomass per unit of the charge length for various locations of the charge

4. The results obtained offer a new concept for fighting forest fires using relatively low-intensity shock waves. The intensity of the shock waves should be sufficient to burn the mixture of gas - fuel products of pyrolysis - air in the zone of forest fire front pyrolysis. At the same time, the shock wave intensity should be lower than the one which would cause trees to fall out in front of the fire.

Acknowledgement

This paper has been prepared with the financial help of the Russian Fund for Fundamental Investigations.

References

Gavrilenko TP, Nikolayev YuA (1982) Piezoelectric transducer of pressure. Fizika goreniya i vzryva 3:127

Godunov K, Zabrodin AV, Ivanov MYu et al. (1976) Numerical solution of multidimensional problems of gas dynamics. Nauka, Moscow

Grishin AM (1992) Mathematical modelling of forest fires and new methods of fighting them. Nauka, Novosibirsk

Korobeinikov VP (1985) Problems of the theory of point explosion. Nauka, Moscow

Shock Tube Pyrolysis Research of Chinese Coals

Xi-Yu Chang, Ding-An Liu, Miu-Zhi Li and Hui Li

Department of Astronautics Technology, National University of Defense Technology, 410073, Changsha, China

Abstract. The results of pyrolysis investigation on three chinese coals in a shock tube are presented. The shock tube was designed especially for pyrolysis process research. Overall length is 7.0 m. Internal diameter is 70 mm. He/Ar was used as driver/driven gases. According to the different driver/driven length ratios, the pyrolysis time ranged from 0.39 ms to 2.0 ms. The infrared emission signal from the heated coal particles recorded by a photomultiplier tube was considered as the temperature-time history of the coal particle cloud. The principle results of the investigation are: 1. A relationship was established between the concentration of the products CO, CO_2, H_2, CH_4, C_2H_2, C_2H_4, C_2H_6, C_3H_6, C_4H_8 and $n-C_6^0$ and the particle heating time. 2. The effect of the pyrolysis pressure on hydrocarbon yield was assessed. On the basis of the experimental results, it appears that by regulating the coal dust fineness, that is the heating rate, the shock tube can be used for studying pyrolysis processes occuring in an industrial boiler. Moreover, because the pyrolysis pressure, as well as the temperature, is also an important parameter that strongly influences the hydrocarbon yield, to get an universal kinetic model is unrealistic.

Key words: Coal pyrolysis, Shock tube

1. Introduction

The extension of shock tube techniques to research related to coal technology was discussed in a series of papers (Lowenstein and von Rosenberg 1977; Frieske et al. 1981; Doolan et al. 1981, 1982). In the preceding studies, an important aspect is the study of the pyrolysis phenomenon. Their fundamental work not only have provided some knowledge on the coal pyrolysis process, but also have demonstrated that, compared with other used apparatus, the shock tube has a lot of unique advantages. Among these, the most spectacular is the complete collection of all thermal decomposition components and solid residues, allowing thus a quantitative analysis. It is the purpose of this paper: 1. To describe the specially-designed shock tube and experimental method; 2. To present the observation results on the pyrolysis of three kinds of chinese coal and to derive a pyrolysis kinetics model.

It has been found that the pyrolysis process of coal is affected by the heating rate, the pyrolysis temperature, the pressure, the coal particle fineness and the residence time, etc. (Lowenstein and von Rosenberg 1977; Frieske et al. 1981; Doolan et al. 1981, 1982; Szydlowski et al. 1981). The heating rate mainly affects the CO yield which is sharply raised at a high heating rate. The total yield of gaseous hydrocarbons increases with the pyrolysis temperature. Under the same conditions the yield of volatile matter of small fineness coal is higher than for coarser material. The pressure mainly affects the CO yield. These are general conclusions, whereas there are exceptional cases because of the different kinds and origins of the coal. Therefore it is necessary that the kinetic research of the designated coal be carried out.

Furthermore, there are also some limitations of the shock tube capacity to simulate the pyrolysis process of coal dust occuring in an industrial boiler. The heating rate is rather high. The pyrolysis conditions (region "5") should be regulated to the appropriate conditions of the chamber of a furnace. A third purpose of the paper is to discuss the possibility to use the shock tube for industrial coal dust combustion.

Shock Waves @ Marseille III
Editors: R. Brun, L. Z. Dumitrescu

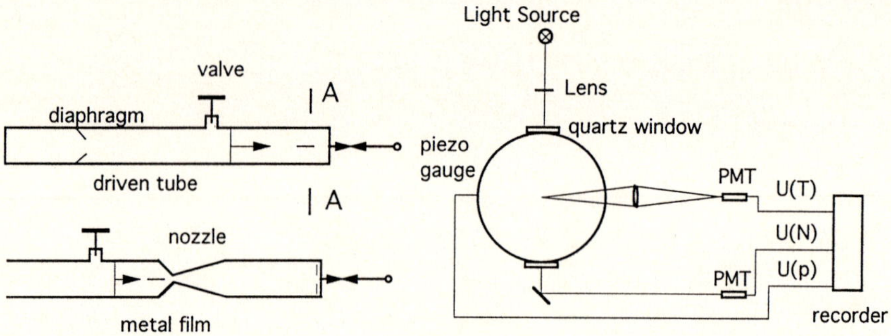

Fig. 1. Experimental set-up

2. Shock tube and experimental method

The shock tube and associated equipment used in this investigation are described below. The shock tube is made of stainless steel tube of 70 mm inside diameter, total length is 7.0 m. The driver section length may be set at 1 m, 2 m and 3 m. The variable driver/driven length ratios correspond to different coal particle residence times in high temperature Ar (0.40-2.00 ms). He/Ar was used as driver/driven gases in all experiments. Driver pressure was $P = 2.0$ MPa. The shock wave velocity is determined by measuring the transit time of the shock wave between two piezoelectric transducers. The pyrolysis pressure at the test location (region "5") is measured by a piezoelectric transducer. The pyrolysis temperature is determined by the shock Mach number and the pyrolysis pressure from the standard one-dimensional perfect-gas shock relation. The residence time of coal particles is determined by the plateau length of the pressure wave record. The shock tube and associated equipment are shown diagrammatically in Fig.1.

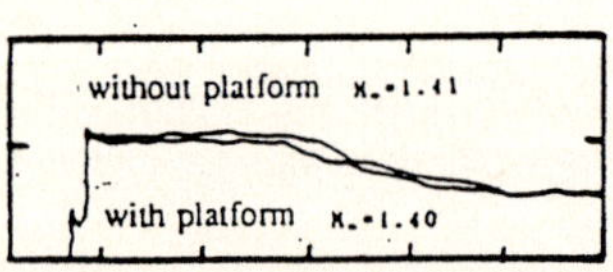

Fig. 2. Comparison of pressure p_5 with platform, and without platform (Air/Air)

At first, the coal sample is placed on a platform, which is a thin steel piece supported by the tube wall. Fig.2 shows that the condition in region "5" is not influenceed by the platform.

Two methods can be adopted to collect the pyrolysis product. The first one is setting a thin metal film near the nozzle; immediately after the metal film rupture, the heated gas and solid slag particles flooded in the collecting section, the isolating valve was closed isolating the collection section from the shock tube. The quenched gas in the collection section is analysed by gas chromatography. The second one is incorporated in the driver tube and forms a part of the driven tube (see Fig.1). The first one is used only to observe (distinguish) what are the pyrolysis concentrations when the coal is heated and decomposed.

In order to investigate the time-history of species evolution and to determine the kinetic model, operational conditions were set so that the duration of the heat pulse could be varied independently of other parameters, and the pyrolysis pressure remained basically constant. The range covered by these parameters is given in Table 1.

Table 1. Experimental conditions

	L_4/L_1 (m/m)	P_4 (MPa)	P_1 (MPa)	M_s	P_5 (MPa)	T_5 (K)	t (ms)
1	2/4	0.34	0.015	2.70	0.504	1700	1.68
2	2/4	0.37	0.019	2.52	0.560	1500	1.47
3	1/3	0.36	0.025	2.32	0.560	1300	0.87
4	1/3	0.34	0.031	2.13	0.565	1100	0.58
5	1/4	0.34	0.045	1.89	0.570	900	0.39

Three different local coals named CC-01, CC-02 and CC-03 were used in the experiments. The test sample was 40 mg for all runs. Coal particle size were 20 and 70 μm. The approximate and ultimate analysis results are given in Table 2.

Table 2. Coal sample data

	Wt% daf					Wt%			
	C	H	N	S	O	Ash	Moist.	V_0	FC
CC-01	80.09	6.04	1.86	0.88	11.13	19.37	4.61	46.10	29.92
CC-02	81.55	5.07	1.24	0.27	11.87	4.09	10.57	37.69	47.65
CC-03	76.52	3.89	1.01	0.37	18.23	6.81	21.59	32.04	39.56

The rate of evolution of each volatile species (denoted by i) is assumed to be first order in the amount of product yet to be produced

$$dV_i/dt = K_{\nu i}\,(V_{i\infty} - V_i) \tag{1}$$

where $V_{i\infty}$ is the maximum yield of the ith volatile, $K_{\nu i}$ are the Arrhenius rate constants

$$K_{\nu i} = A_i \exp\left(-E_i/(R \cdot T_p)\right) \tag{2}$$

Here E_i is the activation energy, A_i the frequency factor, T_p the coal particle temperature, R the gas constant.

Eqs.1-2 were least-squares fitted to experimental yield data obtained in the above-cited experimental conditions. So E_i and $K_{\nu i}$ were determined.

The principle observations of the investigation are:

1. The concentrations of ten products: CO, CO_2, H_2, CH_4, C_2H_2, C_2H_4, C_2H_6, C_3H_6, C_4H_8 and $n-C_6^0$ varying with the particle residence time have been measured;
2. The effect of the atmospheric pressure and fineness on hydrocarbon yield in previous measurements was considered.

3. Results and discussion

Figs.3, 4 and 5 give the time-history of ten kinds of volatile species of CC-01 coal. Utilizing the least-squares method and fitting the experimental results with a first order evolution model, Table 3 presents some characteristic constants obtained with the simple model, where $V_{i\infty}$ is the maximum yield (when $t \to \infty$) of the ith volatile. $K_{\nu i}$ is the Arrhenius rate constant, E_i the Arrhenius parameter.

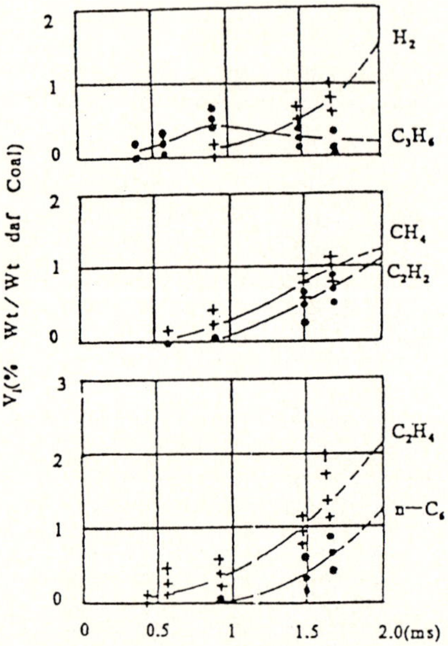

Fig. 3. Time history of species volatilisation (1)

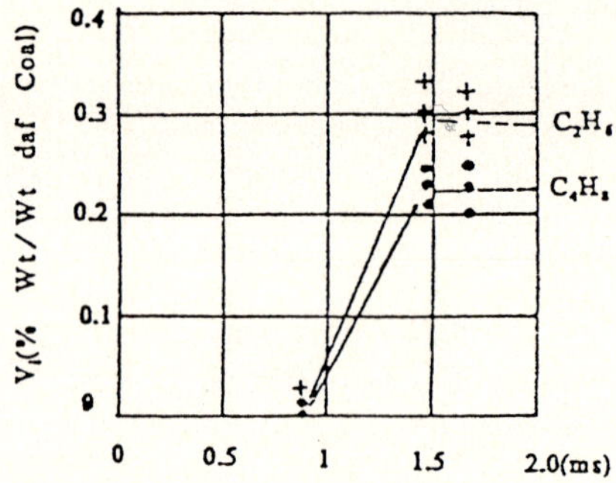

Fig. 4. Time history of species volatilisation (2)

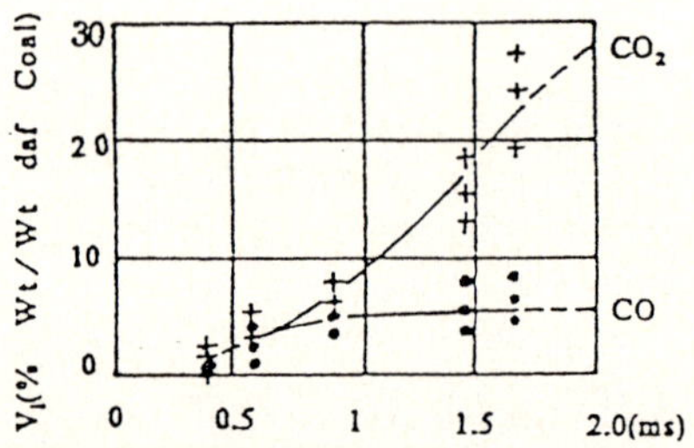

Fig. 5. Time history of species volatilisation (3)

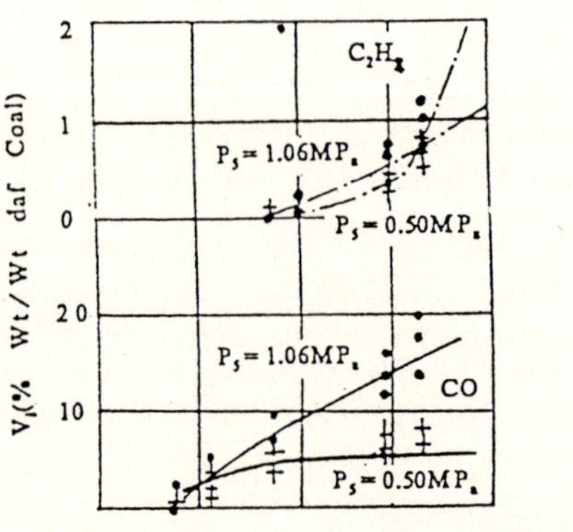

Fig. 6. Influence of pressure on volatile concentrations

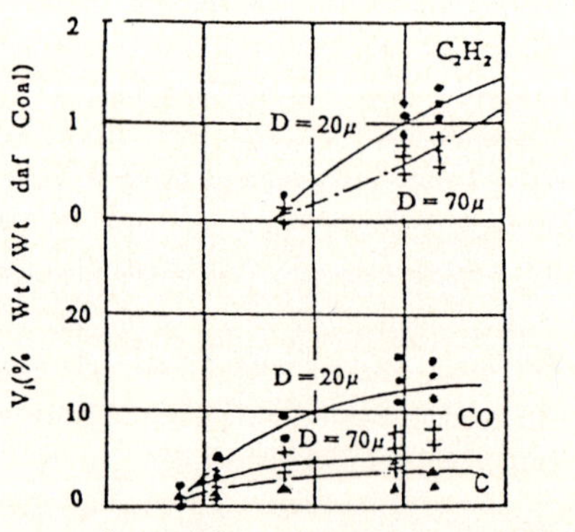

Fig. 7. Influence of fineness on volatile concentrations

Table 3. Characteristic constants by the simple model

	CC-01			CC-02			CC-03		
	V(%)	K (1/s)	E (J/mol)	V(%)	K (1/s)	E (J/mol)	V(%)	K (1/s)	E (J/mol)
CO	19.95	1.58E+5	6.51E+4	16.62	1.77E+5	6.91E+4	29.78	1.19E+5	5.59E+4
CO_2	7.35	4.91E+4	4.88E+4	6.12	5.10E+5	5.18E+4	10.97	4.36E+5	4.19E+4
H_2	0.42	8.82E+5	5.70E+4	0.35	9.51E+4	6.05E+4	0.63	7.12E+4	4.89E+4
CH_4	0.69	1.91E+5	9.97E+4	0.58	2.47E+5	1.06E+5	1.03	1.01E+5	8.56E+4
C_2H_2	0.52	1.98E+5	1.32E+5	0.43	2.97E+5	1.40E+5	0.78	7.53E+4	1.13E+5
C_2H_4	1.53	5.58E+4	8.26E+4	1.28	6.11E+4	8.63E+4	2.29	3.25E+4	6.98E+4
C_2H_6	0.32E-3	5.52E+4	8.25E+4	2.65E-3	1.33E+5	9.71E+4	0.47E-3	6.09E+4	7.86E+4
C_3H_6	0.36	1.18E+4	6.11E+4	0.30	1.29E+4	6.47E+4	0.55	9.28E+3	5.24E+4
C_4H_8	0.31E-3	2.45E+4	7.12E+4	0.26E-3	2.81E+4	7.55E+4	0.47E-3	1.73E+4	6.11E+4
$n-C_6^o$	0.63	2.96E+5	1.05E+5	0.52	3.94E+5	1.12E+5	0.94	1.46E+5	9.08E+4

Fig.6 shows the CO yield at two different fineness values. Fig.7 shows the influence of pressure on the product yield.

The above results are consistent in the main with many preceding investigations (Frieske et al. 1981; Doolan et al. 1981, 1982; Szydlowski et al. 1981). With the increase of temperature, at first CO, CO_2 separated out. Then the great majority of the alkanes and alkene appears at about 1200 K. At about 1400 K, all gaseous products, excepting a few alkenes (such as C_3H_6), arise with the increase of pyrolysis temperature and residence time. However the increment range is lower than what expected from the simple model. It is evident that the Arrhenius first rate model, which considers the simple volatile process, can be accepted only in the starting period of the pyrolysis process. Beyond the proper temperature, because of the appreciable decomposition of the volatiles taking place and of the complexity of the decomposition process, the simplified model is not representative of the practical physical and chemical process. Therefore a method, by which the decomposition kinetic data of a specified volatile (for example C_4H_8) were obtained from the sole shock tube thermal decomposition experiment, was suggested to relate the volatile and decomposition process of the coal dust pyrolysis.

Fig.6 represents the strong influence of pressure on volatile concentrations. These phenomena have been also observed in many other investigations (Frieske et al. 1981; Kusmierz et al. 1987). Therefore, for coal dust pyrolysis under a shock wave heat pulse, the gaseous hydrocarbon products not only depend on the heated rate but also on the instantaneous pressure level. Because the mechanisms affecting the pyrolysis process are different, so the temperature and pressure could hardly be related by an appropriate parameter. However, it is interesting that, regulating the pressure in region "5" and the dust fineness, the shock tube can be used in studying pyrolysis processes where the heating rate is about $10^4 - 10^5$ K/s and the pyrolysis pressure 0.1 MPa, i.e. the same as in an industrial dust boiler. Curve C in Fig.7 represents the experimental result where the pyrolysis pressure is 0.11 MPa and the fineness is 140 μm.

Acknowledgments

Acknowledgment is made to the donors of the Chinese National Natural Science Foundation and China Coal Scientific Research Institute. The author is indebted to Professor Hong-Ru Yu and Professor Chun-Lai Wu for many supports and to our Colleagues Shi Miu Liu, Chen Hong and Xue Zheng Yuan for many helpful comments during the research.

References

Doolan KR, Mackie JC (1981) In: Treanor CE, Hall JG (eds) Proc. 13th Intl. Symp. on Shock Tubes and Waves, pp 641-647

Doolan KR, Mackie JC, Mulcahy MFR, Tyler RJ (1982) 19th Symp. on Combusion, pp 1131-1138

Frieske HJ, Seelbach E, Adomeit G (1981) In: Treanor CE, Hall JG (eds) Proc. 13th Intl. Symp. on Shock Tubes and Waves, pp 790-799

Kuśmierz A, Woliński M, Wolanski P (1987) In: Grönig H (ed) Proc. 17th Intl. Symp. on Shock Tubes and Waves, pp 811-817

Lowenstein AL, von Rosenberg Jr CW (1977) In: Ahlborn B, Hertzberg A, Russell D (eds) Proc. 11th Intl. Symp. on Shock Tubes and Waves, p 366

Szydlowski SL, Wegener DC, Merklin JF, Lester TW (1981) In: Treanor CE, Hall JG (eds) Proc. 13th Intl. Symp. on Shock Tubes and Waves, p 800

Part 5: Biological Aspects

Extra-Auditory Effects of Single and Multiple Blasts

P. Vassout, G. Parmentier, G. Evrard and A. Dancer
Group Physiology, French-German Research Institute F-68301, Saint Louis France

Abstract. Blast injuries occur only on gas-containing structures: auditory system, upper respiratory tract, lung and gastro-intestinal tract. The severity of the injuries depends upon the physical characteristics of the shock waves: peak pressure, rise time and A-duration, and upon the number of exposures. We present results obtained on swine and sheep concerning tracheal, pulmonary, gastrointestinal injuries, and a survival estimate for man.

Key words: Blast injury, Trachea, Lung, Intestine, Lethality

1. Introduction

Blast injury is a general term that refers to the bio-physical and patho-physiological events and clinical syndromes that occur when a living body is exposed to blast of any origin.

Blast-wave physical properties (peak pressure, A-duration and rise time) (Fig.1), the complexity of the wave form (free field or reflecting surface), body position, and the number of blast repetitions determine the potential for primary blast injury (PBI). PBI occurs when the blast wave strikes and compresses the body. Energy is transferred directly from the transmitting medium to the body surface. Damage is almost totally limited to the auditory system and gas-containing structures of the respiratory and gastro-intestinal tracts. Homogenous organs are never affected by the direct effect of shock wave. Death is commonly associated with the formation of arterial air and/or fibrinous emboli that are believed to originate from the lung. Rupture of the visceral pleura, formation of pneumothorax, haemothorax, gross haemorrhage, and pulmonary oedema are commonly observed (Stuhmiller et al. 1991). We will present injury curves determined for upper respiratory tract, lung, gastrointestinal tract, and for mortality, after exposures in free field with the incidence perpendicular to the body axis. The results from ISL and US research teams were obtained on swine and sheep (body weight from 25 to 40 kg) (Vassout et al. 1984). US authors (1966, 1968) estimate that results obtained on large animals can be extrapolated to human body with a maximum error of 15%. Injury scores are standardised for all organs like in this example:

Ranking scheme for pulmonary contusion

Injury score	Description
1- None(o)	One or more pethechia
2- Slight Injury(l)	Superficial focal haemorrhages
3- Moderate Injury(m)	Sub pleural ecchymotic haemorrhage with superficial involvement (<30% of lobe)
4- Severe Injury(s)	Diffuse ecchymotic haemorrhage extending into lung parenchyma (30-60% of lobe involvement)
5- Confluent	Ecchymotic haemorrhage extending into lung parenchyma (>60% of lobe involvement)

2. Upper respiratory tract

Tracheal or laryngeal injuries occurred alone or associated with intestine or pulmonary injuries (Vassout et al. 1984). US authors consider this organ as the most sensitive structure to repeated blasts and a good indicator of extra-auditory susceptibility to blast (Dodd et al. 1990). Fig.2 present the Bowen's curves estimate to man exposed with the axis of the body perpendicular to

Shock Waves @ Marseille III
Editors: R. Brun, L. Z. Dumitrescu

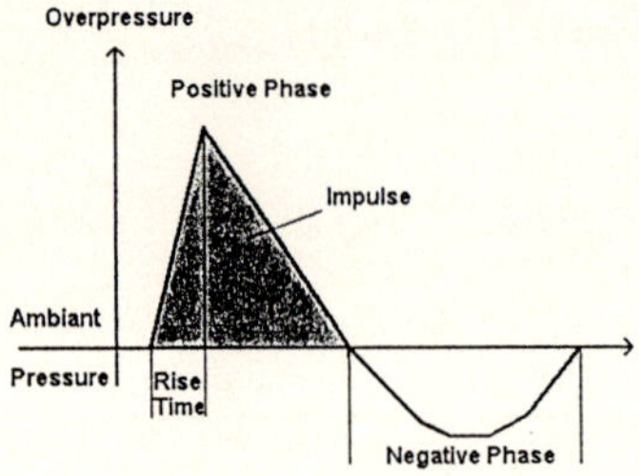

Fig. 1. Pressure-time pattern of an ideal blast wave

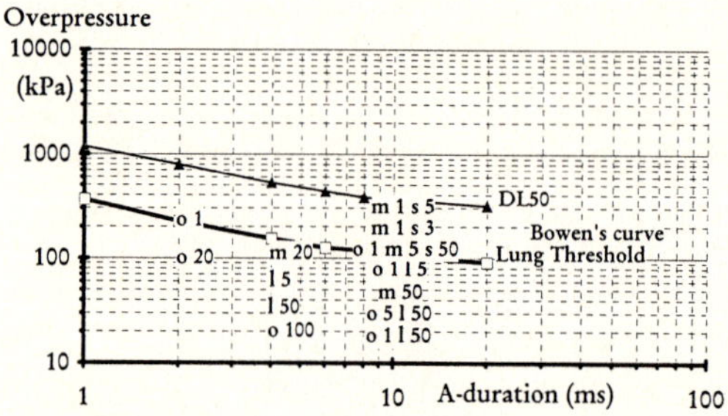

Fig. 2. Tracheal injuries

free field blast according to peak pressure (kPa) and A-duration (ms) (Bowen et al. 1968). We report the effect of single and multiple exposures obtained on swine and sheep. A-duration is an important parameter between 1 to 20 ms. For example, with the same peak pressure (100 kPa) and the same number of exposures (20), when the A-duration is divided by a factor 2 (from 4 to 2 ms), the laryngeal injuries change from moderate to none. For 4 ms, A-duration injuries occurred at a peak pressure level of 70 kPa for 5 exposures and 50 kPa for 100 exposures. For 10 ms A-duration 50 exposures induced injuries at a peak pressure of 40 kPa and 5 exposures at a peak pressure of 45 kPa. No injury is observed for 2 ms A-duration, 20 exposures at a peak pressure of 70 kPa, 4 ms A-duration, 100 exposures at a peak pressure of 55 kPa and 10 ms, 5 exposures at a peak pressure of 40 kPa. The same injury curves corresponding to the threshold for trivial laryngeal injury can be plotted according to the peak pressure and the A-duration (Fig.3) (Stuhmiller et al. 1991).

3. Pulmonary injuries

Closed chest trauma also occurs in non-combat situations due to falls or automobile accidents. Parameters of chest wall motion, intrathoracic pressure and rate of change of intrathoracic pressure have been suggested as important correlates of pulmonary injury in blast environments. Measurement of chest wall velocity during exposure to the blast indicate that it is a critical parameter for lung injury. No injury have ever been observed for a peak pressure lower than 100 kPa whatever the number of exposures and A-duration (64 rounds at 2 ms, 20 rounds at 4 ms and 50 rounds at 10 ms). For one exposure the threshold of trivial lung injury seems to be at peak pressure level of 200 kPa (Fig.4). At this pressure level (200 kPa), 2ms A-duration, doubling the number of exposures increases the level of lung injuries from slight to moderate and moderate to severe. Work by Vassout et al. (1981) has shown that severe pulmonary injuries on swine exposed to free field blast resolved in one month.

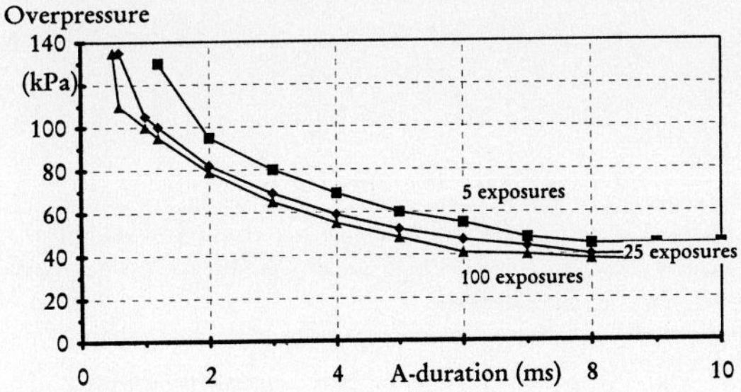

Fig. 3. Effect of number of exposures on tracheal injuries

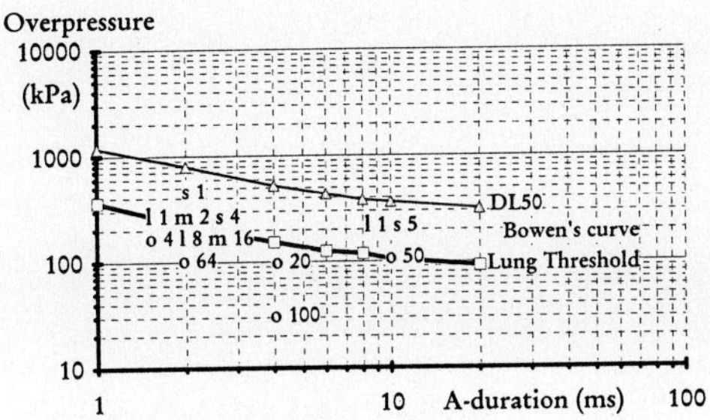

Fig. 4. Pulmonary injuries

4. Gastro-intestinal tract injuries

The mechanism of this kind of injury is different from the lung mechanism. The shock wave induces probably compressions and expansions of the gas phase, which damage the integrity of the intestine structure. It appears that intestine injury is observable at lower peak pressure levels than lung injury. The threshold for trivial intestine injury is situated at about 200 kPa, 2 ms A-duration and 100 kPa, 10 ms A-duration 20 exposure to 100 kPa, 4 ms A-duration induce severe injury. At 100 kPa, 10 ms A-duration one exposure induce threshold injury, 5 exposures moderate injury and 25 exposures severe intestine injury (Fig.5).

5. Mortality limits

We present Bowen's estimate for a man exposed to one blast, extrapolated from the first results obtained in Albuquerque on various species. We added recent results observed by Vassout et al. (1984) on sheep and swine exposed to multiple blasts in free field (Fig.6). It appears that these extrapolated curves are too conservative. 1% of the animals exposed to 400 kPa, 10 ms A-duration, 3 exposures, survived. 50% of the animals exposed to 250 kPa, 10 ms, 3 exposures survived and 99% of the animals exposed to 160 kPa, 10 ms, 5 exposures survived. It is likely that if an animal survives four hours after exposure it will probably survive.

6. Discussion

It is possible to estimate the potential risk of injury for people exposed to blast, explosive or strong impulse noise, from results obtained on large animals exposed in free field (long axis of the body perpendicular to the wave propagation). At low A-duration (< 20 ms) there is an interaction between peak pressure and duration. At durations longer than 20 ms (nuclear or very

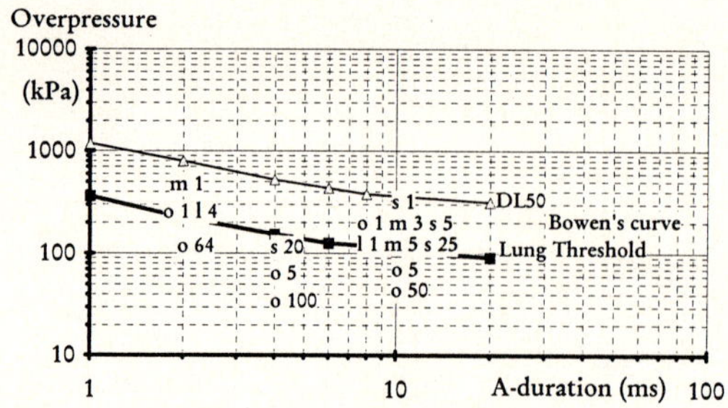

Fig. 5. Gastro-intestinal injuries

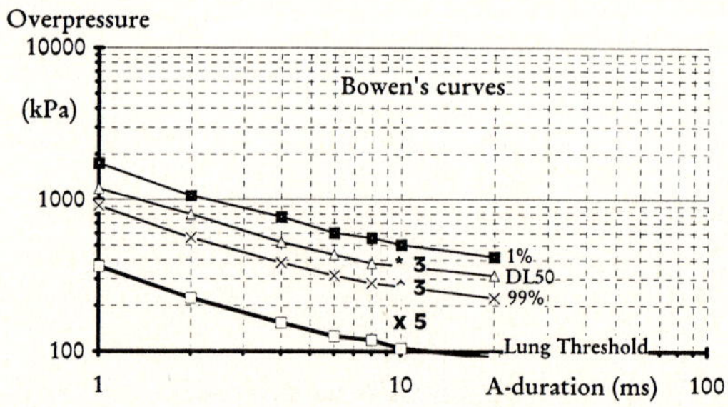

Fig. 6. Survival curves for man

large conventional explosive) only the peak pressure determines the injury level (Stuhmiller et al. 1991). The parameter *number of exposures* become very important now for the servants of heavy weapons (Richmond et al. 1981, Vassout et al. 1981). When the subject is exposed to shock waves with a long rise time (several ms, like people in enclosures) the severity of the injury is lower for the same peak pressure. The signature of the overpressure in this case depends of the reflecting surfaces and of the position of the subject in the enclosure.

References

Bowen I, Fletcher R, Richmond D (1968) Estimate of man's tolerance to direct effects of air blast. Report DASA 2113

Dodd K, Yelverton J, Richmond D, Morris J, Ripple G (1990) Nonauditory threshold for repeated intense freefield impulse noise. J. of Occup. Med.32: 260-266

Richmond D, Damon E, Fletcher E, Bowen I, White C (1966) The relationship between selected blast-wave parameters and the response of mammals exposed to air blast. Report DASA 1860, Defence Nuclear Agency

Richmond D, Yelverton J, Fletcher E (1981) The biological effects of repeated blasts. Report DNA 5842F

Stuhmiller J, Phillips Y, Richmond D (1991) The physics and mechanisms of primary blast injury. In Textbook of Military Medicine, Part 1 Vol. 5 Conventional Warfare: Ballistic, blast and burn injuries, Chap.7, pp. 241-270

Vassout P, Dancer A, Richmond D, Phillips Y (1984) Biological effects of shock waves: Importance of A-duration and multiple exposures. ISL report S-N 911/84

Vassout P, Parmentier G, Dancer A (1981) Effects of the number of blast exposures on swine. ISL report S-R 904/81.

Effects of Shock Waves on Living Tissues Using Shock Tubes for Bio-Tests

M. Tamagawa and T. Akamatsu
Department of Mechanical Engineering, Kyoto University, Kyoto 606-01, Japan

Abstract. Paper deals with the interaction between living tissue and plane shock waves. An experimental apparatus for the study of shock-induced injury of living tissue has been developed, and results of animal experiments using this apparatus are shown. The pressure wave propagating into living tissues and its deformation are analyzed numerically. To generate strong shock waves for experiments on injury of living tissues in water, a new diaphragmless device of a free-piston shock tube (Stalker Tube) is developed.

Key Words: ESWL, Free-piston shock tube, Diaphragmless shock tube

1. Introduction

Recently, Extracorporeal Shock Wave Lithotripsy (ESWL) is gaining widespread use, in which kidney stones are destroyed non-invasively by focusing shock waves (Chauchy 1982). During this treatment, the focused shock wave induces more or less inevitable damage on the living tissue. The reason why this damage occurs, the mechanisms and the dominant physical factors have not been elucidated. In medicine, the effects of shock waves on living tissue have been studied (Watanabe 1983). In this mechanism, there are many factors: (1) pressure, (2) gradient of pressure, (3) stress (velocity gradient), (4) energy impacted on the living tissue, but the dominant factor is not determined. Another factor is the mode of pressure wave, that is (1) compression wave, (2) expansion wave, (3) cavitation, induced by pressure waves into the tissue.

To examine simply and quantitatively the effects of plane shock waves on living tissues, Tamagawa and Akamatsu (1991) use a non-bursting diaphragm shock tube with a slit open to the atmosphere that generates a single positive saw-tooth pulse. Animal experiments proved that this shock tube with slit can produce damage on living tissues embedded in air, but not in water. That is to say, the non-bursting diaphragm type double cylindrical shock tube mentioned above cannot generate strong shock waves, sufficiently steep to produce effects on living tissues embedded in water.

On the other hand, free-piston shock tubes have been used for studies of strong shock waves, high enthalpy flows and hypersonic flows (Stalker 1967; Meyer 1965). We thought to use this free-piston shock tube for our bio-tests in the high pressure range. But on each shot, the diaphragm has to be replaced and the fragments of the broken diaphragm may induce damage to the test objects. However, these problems are overcome by replacing the diaphragm with a piston valve controlled passively by its back pressure.

To estimate the degree of injury in living tissue, it is necessary to analyse the flow around the living tissue and the pressure waves (stress waves) within the living tissue. It is difficult to measure directly the stresses in the living tissue. Thus we examine the pressure (stress) in the living tissue by experiment and simulation.

The pressure wave propagating in the living tissue is analysed in models: as a fluid layer (water) and an elastic fluid layer, whose acoustic impedance resembles that of the living tissue. Using a fluid-dynamical computation, the pressure wave analysis is carried out for the air-water, air-living tissue and air-water-living tissue models. Also, the decay of the pressure wave when the shock wave propagates into the living tissue is estimated. Then effect of the components (soft layer and rigid layer) in the living tissue is examined.

Shock Waves @ Marseille III
Editors: R. Brun, L. Z. Dumitrescu

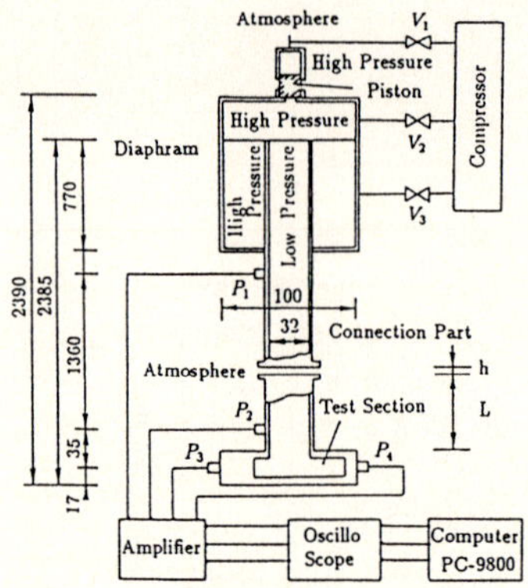

Fig. 1. Experimental apparatus

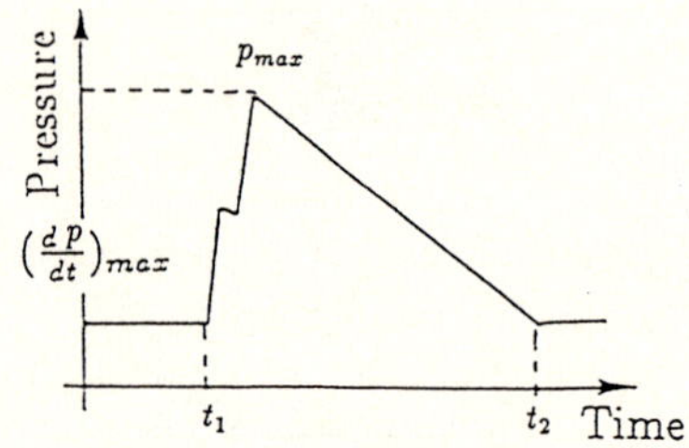

Fig. 2. Typical pressure history

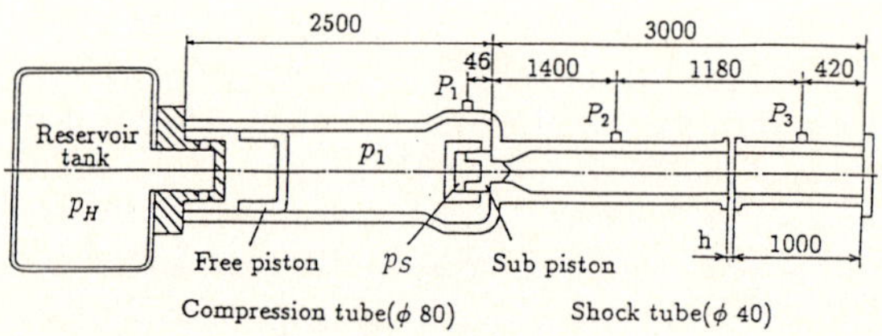

Fig.3a. Schematic of experimental apparatus

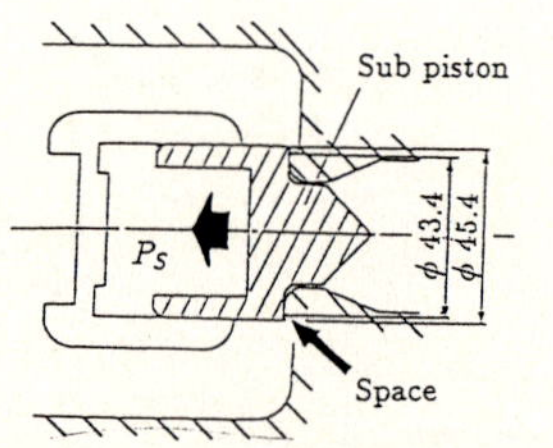

Fig.3b. Detail of sub-piston part

2. Experimental apparatus

2.1. Non-bursting diaphragm double cylindrical shock tube

To apply the pressure wave on living tissue, a special shock tube apparatus has been developed (Fig.1, Tamagawa and Akamatsu 1991). The shock tube is of the non-bursting diaphragm type with double-cylindrical chambers. The feature of this apparatus is that there is a slit to vent the pressure to the atmospheric, and that only a positive wave is generated in the test section. The experimental condition in this apparatus is such that a Mach number of 1.1 is produced in the region including the slit. Usually the high pressure in the driver section is up to 700 kPa. The slit width is adjusted for optimizing the pressure wave. A typical sketch of the pressure history at the position P_3 is shown in Fig.2. In this apparatus, the duration for bio-tests can be controlled by the length of the tube.

2.2. Free-piston shock tube for bio-tests

A schematic of the experimental apparatus is shown in Fig.3(a). The components of the apparatus consist of a high-pressure tank, compression tube, and shock tube. A piston valve device is located at the end of the compression tube

The details of the device are shown in Fig.3 b. The ambient pressure p_1 acts on the piston's right-end face (shown with an arrow in Fig.3 b). When the resultant force by this rising ambient pressure becomes larger than that maintained by the back pressure on the left face of the sub-piston, the sub-piston begins to move to the left and then the port opens. Subsequently the compressed gases flows into the low-pressure chamber to generate the shock wave.

After the shock wave propagates, an expansion wave runs behind the shock wave because the length of the driver section becomes short owing to the motion of the heavy piston.

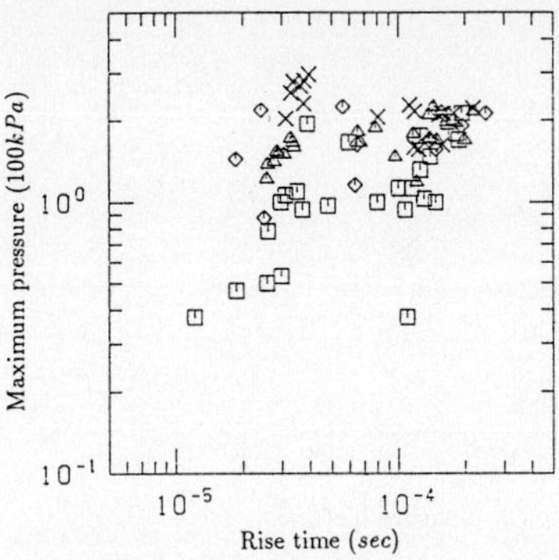

Fig. 4. Degree of injury (Maximum pressure and rise time)
□: no bio-effect; △: semi-damage;
×: bleeding; ◇: indistinguishable

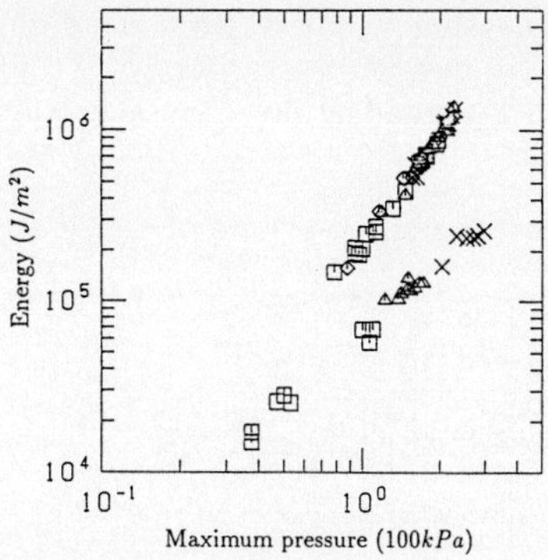

Fig. 5. Degree of injury
(Energy and maximum pressure)

The pressure in each chamber is defined as p_H, p_1 and p_S (Fig.3 a). Usually the initial working conditions are as follows; p_1 : $80 \cdots 100$ (kPa), p_S : $250 \cdots 450$(kPa). The pressures at P_1, P_S, P_2, P_3 are measured with PCB pressure transducers. Two kinds of driver gases (Air or Helium) are used to cover a wider operating range.

3. Animal experiments

In cooperation with Watanabe and his co-workers (Kyoto Prefectural University of Medicine, Japan), animal experiments using rats have been carried out in the non-bursting diaphragm-type double cylindrical shock tube. In these experiments, a living rat exposed to the incident shock wave is located at the test section (Fig.1). The degree of injury, especially on the kidneys, depends on dynamical quantities defined as follows:

$$\text{Risetime :} \quad \delta t = \frac{p_{max}}{\left(\frac{dp}{dt}\right)_{max}} \tag{1}$$

$$\text{Energy :} \quad J = \int_{t_1}^{t_2} p^2/Z dt \tag{2}$$

where p_{max}, $\left(\frac{dp}{dt}\right)_{max}$, t_1 and t_2 are defined in Fig.2; Z is the acoustic impedance of the surrounding fluid. Figs.4 and 5 show the relationship of δt, J and p_{max} with the degree of injury, that is expressed as □ (No bioeffect), × (Bleeding), △ (Semi-Damage: recovery after a quarter hour) and ◇ (Indistinguishable). In Figs.4 and 5, the kidney is damaged at an energy of $10^5 \cdots 10^6$ (J/m^3). In Fig.5, it is damaged at the maximum pressure of $200 \cdots 300$ kPa. It is remarkable that in the range of the rise time $4 \cdots 8 \times 10^{-5}$ (sec) there is a distinguishable degree of injury (Fig.4). The acoustic impedance of water being larger than that of air, Eq.2 indicates that the energy level is low for water compared with air at the same pressure level. Consequently, there is no damage in water. To overcome this problem, the free-piston shock tube for bio-tests has been developed, aiming to induce substantial damage to living tissues in water.

4. Computation of shock waves propagating in living tissue

4.1. Formulation

Air-water, air-living tissue and air-water-living tissue are modeled as fluids, including an interface between gas and liquid (the liquid layer simulates the living tissue),asssuming that the fluids are compressible and invisid. As plane shock waves are used in this study, computational objects are simplified as one-dimensional models (Fig.6). The saw-tooth type pressure wave shown in Fig.2 cannot be generated in these models because there is no open slit in this case.

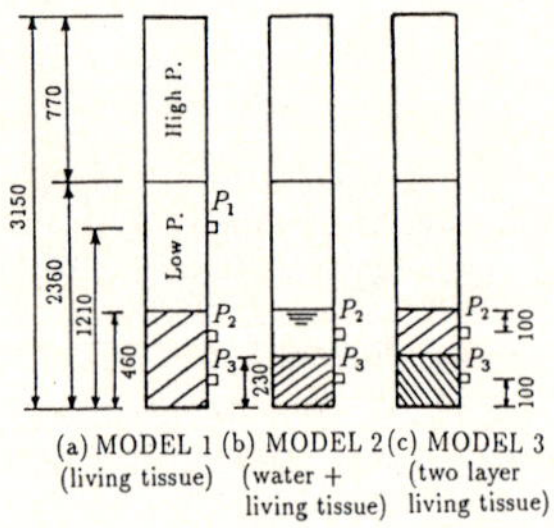

Fig. 6. Computational model of living tissue when the shock waves propagates in the shock tube

The governing equations are the conservation of mass, momentum and energy, while the state equation for air is taken as that of an ideal gas. They are written as follows:

For air (gas),

$$\frac{\partial}{\partial t}\begin{pmatrix}\rho_g\\ \rho_g u_g\\ \rho_g e_g\end{pmatrix} + \frac{\partial}{\partial x}\begin{pmatrix}\rho_g u_g\\ \rho_g u_g^2 + p_g\\ \rho_g u_g(e_g + p_g/\rho_g)\end{pmatrix} = 0 \tag{3}$$

For water,

$$\frac{\partial}{\partial t}\begin{pmatrix}\rho_l\\ \rho_l u_l\end{pmatrix} + \frac{\partial}{\partial x}\begin{pmatrix}\rho_l u_l\\ \rho_l u_l^2 + p_l\end{pmatrix} = 0 \tag{4}$$

where suffixes g or l indicate gas (air) or liquid (water).

Tait's equation, as the equation of state for water is expressed as follows:

$$p_l = (\rho_l/\rho_{l0})^n\,(p_{l0} + B) - B \tag{5}$$

where $n = 7.415, B = 2963 \times 100$ kPa.

The state equation in the living tissue is derived in the following, from the conservation of mass and momentum across the shock wave:

$$\left(1 - \frac{V_l}{V_{l0}}\right) = \frac{u_{lg}}{u_{ls}} \tag{6}$$

$$p_l - p_{l0} = \rho_{l0}\; u_{ls}\; u_{lg} \tag{7}$$

where suffixes l, g, s indicate liquid, gas and shock wave.

Assuming that the relation between u_s and u_g is linear,

$$u_{ls} = c_0 + s\; u_{lg} \tag{8}$$

where the derivative of K (bulk modulus) with respect to the pressure is expressed with the help of a parameter s as follows:

$$K' = \frac{dK}{dp} = 4s - 1 \tag{9}$$

Arranging the above Eqs. (6 - 9), a relation between pressure and density is obtained:

$$p_l = p_{l0} + \frac{c_0^2\,(1/\rho_{l0} - 1/\rho_l)}{[1/\rho_{l0} - s\,(1/\rho_{l0} - 1/\rho_l)]^2} \tag{10}$$

The parameter K is 2.5×10^8 (N/m^2) in the living tissue (Tamagawa and Akamatsu 1992), and 1.0×10^9 (N/m^2) in water. Also, K' is equal to 4.3. The above equations are discretized and solved using a finite-difference method, with 100 grid points.

4.2. Results and discussion about computations

4.2.1. Results for the case of air-living tissue The computation of a shock wave propagating into the living tissue through air corresponds to MODEL 1 in Fig.6. The pressure in the driven gas is atmospheric. Fig.7 shows the pressure history at P_2,P_3 in the living tissue in the pressure ratio range $r = 2 \cdots 6$. In this figure, there are oscillations produced by the pressure wave travelling between the end wall and the interface.This is in the conventional shock tube mode, not in the saw-tooth mode for bio-tests. So, we shall comment only on the incident pressure and its gradient.

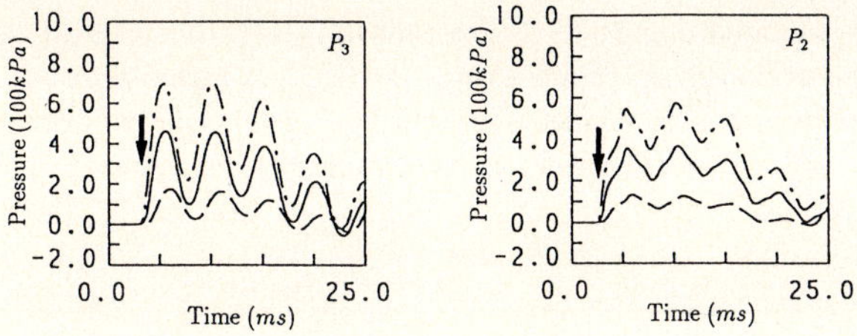

Fig. 7. Pressure history in MODEL 1 - - $r = 2$; — $r = 4$; - · - $r = 6$

Comparing the gradient of the incident pressure wave (marked in Fig.7 with an arrow) at P_2 with that at P_3 in the case $r = 4$, it is found that the gradient of the incident pressure wave at P_3 becomes smaller than that at P_2.

4.2.2. Results for the case of air-water-living tissue The computation of the shock wave propagating into the living tissue through water corresponds to MODEL 2 in Fig.6. Fig.8 shows the pressure history at P_2 and P_3 for the pressure ratio in the range $r = 2 \cdots 6$. Comparing the gradient of the incident pressure wave (marked with an arrow) at P_2 (in water) with that at P_3 (in the living tissue), the gradient of the incident pressure wave at P_3 becomes smaller than that at P_2. That is to say, the gradient of pressure decays because there is a difference between the bulk modulus of water (1.0×10^9 (N/m^2)) and that of living tissue (2.5×10^8 (N/m^2))

4.2.3. Results for the case of living tissue with the two-layer model The computation of the shock wave propagating into a living tissue consisting of two layers (rigid layer and soft layer), corresponds to MODEL 3 in Fig.6. For example, the component of kidney has two layers. Here, the bulk modulus K indicates the rigidity of the living tissue. That is to say, the softer the living tissue is, the smaller the bulk modulus K becomes. Fig.9 shows the pressure history at P_2 (upper layer) and P_3 (lower layer). In this case, the bulk moduli of the upper and lower layers are

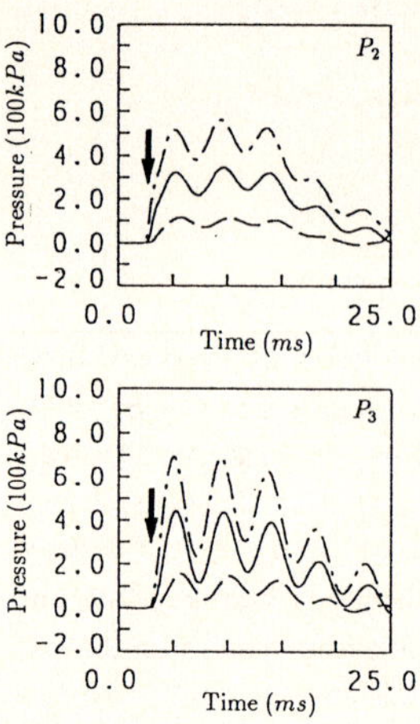

Fig. 8. Pressure history in MODEL 2
– – – $r = 2$; —— $r = 4$; – · – $r = 6$

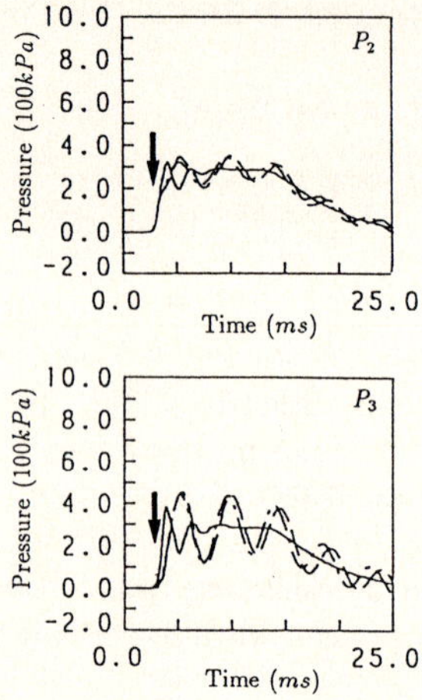

Fig. 9. Pressure history in MODEL 3
—— case 1 ($K_2 = 2.5 \times 10^8$, $K_3 = 2.5 \times 10^9$)
– – – case 2 ($K_2 = 2.5 \times 10^8$, $K_3 = 2.5 \times 10^8$)
– · – case 3 ($K_2 = 2.5 \times 10^9$, $K_3 = 2.5 \times 10^8$)

expressed as K_2, K_3. Comparison of the gradient of the incident pressure wave at P_2 with that at P_3 (marked with an arrow) shows the following; If K_2 is less than K_3, then the gradient of pressure at P_3 (lower layer) is as large as that at P_2 (upper layer). But if K_2 is higher than K_3, then the gradient of pressure at P_3 (lower layer) is smaller than that at P_2 (upper layer). Thus, the living tissue, whose upper layer is softer than the lower one, is considered to be exposed to a sharp gradient of the incident pressure wave.

5. Conclusions

(1) From the animal experiments, maximum pressure, rise time and energy are important factors for injury.

(2) The pressure wave analysis in the living tissue proves that the bulk modulus of the living tissue is an important factor.

Further development of this work has been made by using a free-piston shock tube for bio-tests.

Acknowledgement

The animal experiments have been carried out in cooperation with Prof. H. Watanabe, M. Uchida, Y. Imaide and co-workers. Also, this work was supported by a Grant-In-Aid for Scientific Research on Priority Areas, the Ministry of Education, Science and Culture, Japan.

References

Chauchy CH et al.(1982) Extracorporeal Shock Wave Lithotripsy. Karger

Meyer RF et al.(1965) The NAE hypersonic gun tunnel. Part I: The compression heater, LR-432

Stalker RJ (1967) A study of the free-piston shock tunnel. AIAA Journal, 5:12:2160

Tamagawa M, Akamatsu T (1991) Development of test apparatus for the shock induced injury of living tissue. In: Takayama K (ed) Proc. 18th Intl. Shock Wave Symp., Sendai, Japan, pp 1177-1180

Tamagawa M, Akamatsu T (1992) Effects of the shock wave on living tissue. 7th Intl. Conf. on Biomedical Engng., Singapore, p 219

Watanabe H et al.(1983) Micro-explosion cystolithotripsy. The Journal of Urology, 129:23

Extracorporeal Shock Wave Gallstone Lithotripsy

Hideo Ise*, **Osamu Kitayama***, **Seiki Matsuno***, **Kazuyoshi Takayama**†
and Toshiyuki Okawa‡
*First Department of Surgery, Tohoku University, School of Medicine, Sendai, Japan
†Shock Wave Research Center, Institute of Fluid Science, Tohoku University, Sendai, Japan
‡Medical Yachiyoda Co. LTD., Tokyo, Japan

Abstract. We have investigated 30 cases of gallbladder stone patients treated by extracorporeal microexplosive biliary lithotripsy. The lithotriptor used for tratment is a Yachyoda SZ-5000, using as the energy source microexplosive pellets (AgN_3).The stone-free rate was 12/30 (40%). Especially type Ia and Ib stones (ultrasonographic classification) were treated most effectively. The stone-free rate for types Ia and Ib was 4/6 (67%) and 5/5 (100%), while the side-effects and complications were mild. Extracorporeal microexplosive biliary lithotripsy was very safe and effective in patients whose ultrasonographic classification of the gallstone was type Ia or Ib.

Key words: ESWL, Gallstone, Micro-explosion, Shock wave, Ultrasonography

1. Introduction

Laparoscopic cholecystectomy (LC) has rapidly and radically changed the surgical treatment of gallstone disease. LC may have become the first choice of the treatment of cholecystolithiasis now in some hospitals. But extracorporeal shock wave lithotripsy (ESWL) is still now one of the treatments for selected cholesterol gallstones, because in most cases ESWL doesn't need general anesthesia, laparotomy, or admission. Careful patient selection provides considerable stone-free rates within a year. Side effects and complications are rare. So ESWL may be one of the established treatments of the gallstone disease. In this paper we report on 30 patients treated by microexplosive biliary lithotripter (SZ-5000).

2. Indication and exclusion criteria

In our indication criteria (Table 1), cholesterol gallstone patients will be the best candidates and patients are excluded when the stones are more than 30 mm in max. diameter and the number of gallstones are more than 3. Stone size was determined by ultrasonography. Radiolucent stone and functioning gallbladder will be suitable if posssible. The exclusion criteria are thickness of gallbladder wall, tumor in the gallbladder wall, haemorrhagic tendency, pregnancy, and a cardiac pacemaker. Informed consent is needed for all patients.

Table 1. Selection criteria

1	Cholesterol gallstone
2	Max. size of stone < 30 mm
3	Number of stones < 4
4	Symptomatic patient
5	Functioning gallbladder
6	No thickening of gallbladder wall
7	No tumor in the gallbladder wall
8	No haemorrhagic tendency, no pacemaker
9	Informed consent is needed

Shock Waves @ Marseille III
Editors: R. Brun, L. Z. Dumitrescu

3. Materials and methods

Patients were selected and treated with a microexplosive lithotripter "Yachiyoda SZ-5000", which is not a water-bath type and has a water cushion. Targeted stones are visualized by ultrasonography. The energy source of the shock waves is a microexplosive pellet. The pellet is 10 mg of silver azide (AgN_3) and it is about 0.8 mm cube in size. At the beginning the patient lies supine and during the treatment the patient lies in a 45 degrees head-up position. The gallstone is detected by an outside ultrasound probe viewing by a monitor and next the stone is targeted by an inside probe precisely. Moreover, during treatment, the progress of the fragmentation of the stones can be observed through the monitor.

The component and composition of the gallstones are predicted and diagnosed with the help of ultrasonographic classifications, as summarized by Tsuchiya (1986) and with the help of radiological findings of drip infusion cholangiography (DIC). The shock wave discharge is initiated by the operator, using electronic coupling to prevent discharge during the QRS complex of the heartbeat. Maximum number of shock waves in one session is limited to $\leq$ 500 following results of fundamental experiments using mongrel dogs and pigs in order to detect tissue damage, which were reported in the previous presentations at the 16th and 18th ISSW.

All patients had a complete history and physical examination before and after ESWL procedure. Gallbladder function was assessed by drip infusion cholangiography (DIC), the contraction of the gallbladder was induced by oral egg extract. The CBC, serum concentration of AST, ALT, alkaline phosphatase, CPK, amylase, BUN, creatinine, and amylase activity were determined before and after ESWL.

4. Results

Within a period of 30 to 120 minutes shock waves were generated extracorporeally by micro-explosions and focused by a semiellipsoidal reflector. Max. total number of focused shocks given to one patient was 101-3260 (mean: 773). Most patients were treated in several sessions; the mean number of treatments was 2.5. Most patients need no anesthesia and are treated in outpatient clinic.

Of the 30 patients treated by SZ-5000, gallstones were successfully disintegrated in 18 (60%). Under the influence of adjunctive therapy with ursodeoxycholic acid (UDCA) at the dose of 600 mg/day and a traditional Chinese medicine; Dai-saiko-to (Da-Chai-Hu-Tang: which is known to prevent gallstone formation) at the dose of 7.5 g/day, fragments completely disappeared in 12 patients (40%) (Table 2). In Tsuchiya's gallstone classification type Ia and Ib[1)] are the best candidates to be easily disintegrated and the complete clearance rate of those types of stones is very high. In our study the complete clearance rates of type Ia and Ib was 80% (4/5), and 100% (5/5) (Table 3). The group of single stones had 12 complete disappearance cases (Table 4). "Fair" cases means the patients those fragments were disintegrated below 4 mm, "poor" cases means the patients those fragments were more than 4 mm, and "ineffective" cases are when the stones were not disintegrated much.

Using SZ-5000 we could also fragment stones of 20-30 mm. But in some cases of small stones (< 10 mm), we might have misdiagnosed them for cholesterol stones. Most of them may be black pigment stones (Table 5). In our in vitro experiments, black pigment stones are not easily disintegrated into tiny pieces.

Side effects were minor and limited to laboratory evidence of obstruction of the biliary systems, as well as most cases of microhematuria and subcutaneous petechiae. And we have experienced 3 mild cases of acute cholecystitis, 1 mild case of acute pancreatitis. But there were no major complications associated with ESWL. In laboratory data, leucocytosis is found in 4 cases, elevation of AST and ALT is detected in 6 cases, elevation of LDH, ALP is found in 2 and 1 cases each. Elevation of serum amylase is found in only 2 cases.

Table 2. Results of ESWL

Good	18 (60%)
Complete disappearance	12 (40%)
Poor & Ineffective	12 (40%)

Table 3. Results of ESWL (US classification)

Gallstones	Good	Complete disappearance
Ia (n=6)	5 (83%)	4 (67%)
Ib (n=5)	5 (100%)	5 (100%)
IIa (n=15)	8 (53%)	3 (20%)
III (n=4)	1 (25%)	0

Table 4. Results of ESWL (number of gallstones)

Number of gallstones	Complete disappearance	Fair	Poor	Ineffective
1	12	4	3	4
2	0	2	3	1
>2	0	0	1	0

Table 5. Results of ESWL (size of gallstones)

Size of gallstones	Complete disappearance	Fair	Poor	Ineffective
< 10 mm	1	3	3	2
11 · · · 20 mm	8	2	3	2
> 21 mm	3	1	1	1

5. Discussion

Recently LC has gained wide spread around the world. LC may have become the first choice for the treatment of cholecystolithiasis now. But careful patient selection provides considerable stone-free rates within a short period. Side effects and complications are mild. So ESWL may be one of the established treatments of the gallstone disease.

We have surveyed 30 patients treated by microexplosive biliary lithotripter (SZ-5000). Of the 30 patients, gallstones were successfully disintegrated in 18 (60%). Under the influence of adjunctive therapy with UDCA and Dai-saiko-to, fragments completely disappeared in 12 patients (40%). We are now expecting Dai-saiko-to to prevent gallstone formation and to increase bile flow to take precautionary measures against bile stasis.

In Tsuchiya's gallstone classification types Ia and Ib[1)] are the best candidates to be easily disintegrated and the complete clearance rate of these is very high. Exact patient selection provides the complete clearance of gallstones. Side effects and complications are rare. So ESWL may be one of the established treatments of the gallstone disease.

ESWL will continue to be a successful technique for the treatment of thoroughly selected gallstone patients. And ESWL is a non-invasive method to get rid of the stones, but not of their gallbladder. We have not experienced recurrent stone cases yet, but about 10-20% of gallstone recurrence rate which is reported in several papers will continue to be the most important problem for biliary ESWL.

References

Tsuchiya Y et al. (1986) Ultrasonic properties of gallstones: Differentiation between cholesterol stones and pigment stones. Tan to Sui, pp 1483-1491 (in Japanese)

Underwater Shock Wave Propagation and Focusing in Inhomogeneous Media

H. Nagoya*, T. Obara† and K. Takayama*
* Shock Wave Research Center, Institute of Fluid Science,Tohoku University, 2-1-1 Katahira, Aoba, Sendai, Japan
†Faculty of Engineering, Saitama University, Shimo'okubo, Urawa, Japan

Abstract. It is important to clarify the mechanism of human tissue damage which occurs associated with Extracorporeal Shock Wave Lithotripsy (ESWL) clinical treatments. Human tissue being non-homogeneous, shock wave propagation in it involves complicated wave interactions, such as reflection, refraction and diffraction. Therefore, in order to understand the dynamic behavior of human tissue exposed to shock waves, it is necessary to determine experimentally the equation of state (EOS) of human tissue. This paper reports the experimentally obtained Hugoniot curve of human blood and other model tissue. Using these experimental results, a numerical simulation of shock propagation in inhomogeneous model tissue was carried out using a TVD finite difference scheme.

Key words: ESWL, Inhomogeneous media, Human tissue, EOS

1. Introduction

Extracorporeal Shock Wave Lithotripsy (ESWL) is one of most important clinical treatments used to disintegrate and remove kidney stones and gall bladder stones by using underwater shock wave focusing (Chaussy 1982; 1983). It has been reported that human tissue damage occurred near the focal region (Kuwahara 1992; Kandel 1992). In conjunction with the ESWL, the acoustic properties of human tissue are similar to those of water. Therefore, in the past human tissue has been assumed to be a homogeneous medium (Goss 1978). However, human tissue being in reality inhomogeneous, complicated shock wave interaction, reflection, refraction and diffraction would occur at various tissue interfaces. It is very important for conducting safe ESWL treatments to clarify analytically and experimentally shock wave reflection, refraction and diffraction both at the tissue interfaces and inside the tissue. First, this paper reports experimental determination of the pressure-density diagram (Hugoniot curve) of human blood and 5% and 10% gelatines which have been used for model tissue. Equations of state (EOS) of these specimens are determined by using a high-pressure test vessel. Second, using these measured data, numerical simulations have been carried out using a TVD finite difference scheme on a Cray Y-MP8/8128. Shock reflection, refraction and diffraction in inhomogeneous model tissue are clarified. A significant pressure difference was found between the model tissue and pure water.

2. Experiments

2.1. Measurement of the Hugoniot curve

Fig.1 shows the experimental set-up for measuring the Hugoniot curve of various specimens (Bridgman 1935). The experimental set-up consists of a high pressure pump, a pressure vessel, indicator of volume change, pressure gauge and water bath which keeps the test section at a specified temperature, which is, in this case, body temperature. The pressure vessel is filled with silicone oil and a special test tube in which different specimens are placed is inserted. This test tube is in a vertical position along which a piston can slide (Bridgman, 1949). The displacement of the piston movement represents the volume change of the test tube. The relation between the piston movement and volume change is calibrated by using mercury as a test liquid whose volume

Shock Waves @ Marseille III
Editors: R. Brun, L. Z. Dumitrescu

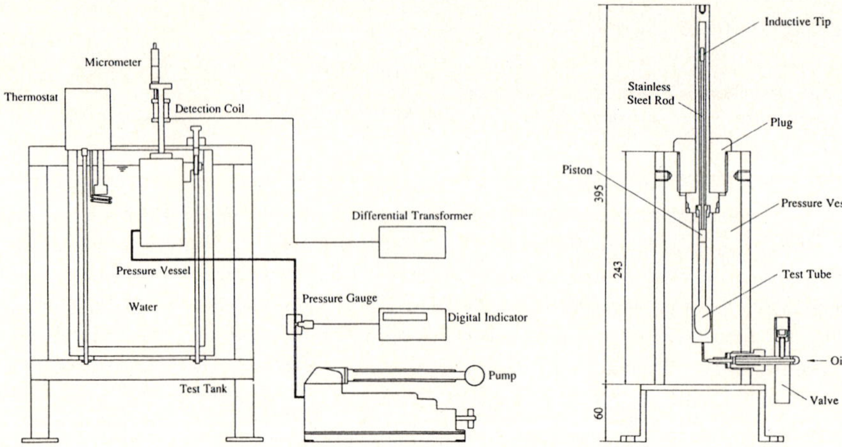

Fig. 1. Experimental set-up for measuring the Hugoniot curve of various specimens

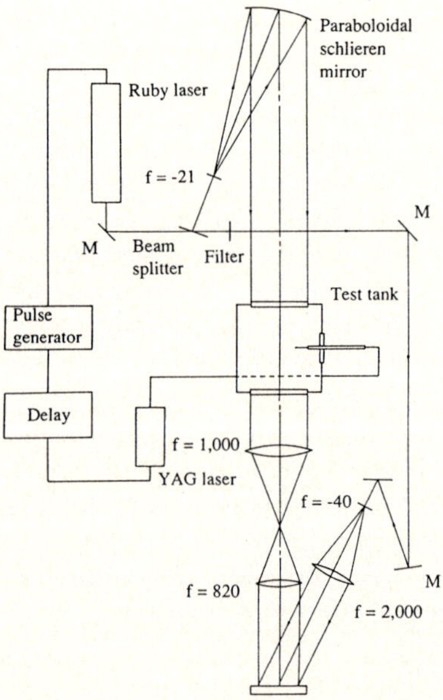

Fig. 2. Experimental set-up for holographic interferometry

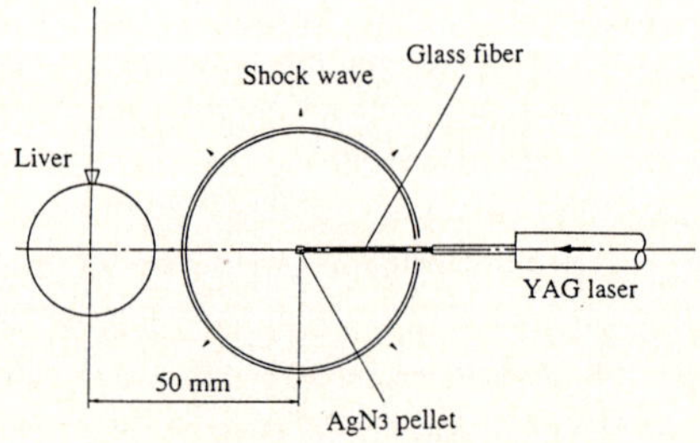

Fig. 3. Schematic diagram of flow visualisation

change against pressure is known. The volume change can be measured up to 0.12 mm^3, which is sufficiently accurate for our experimental purposes. The volume change of specimens was precisely measured at pressures ranging from 0.1 to 500 MPa using the standard measuring technique of volume change against high pressure (Bridgman 1950).

2.2. Holographic interferometry

Fig.2 shows the experimental set-up for holographic interferometry (Takayama et al. 1990). The light source is a holographic ruby laser (Apollo Lasers Inc., 25 nsec pulse duration, 2 J/pulse, 694.3 nm wave length). The laser beam is split into reference and object beams. The intensity ratio of the object to reference beams is adjusted between 2 : 1 to 3 : 1. In order to conduct a quantitative flow visualization in liquids, double exposure holographic interferometry is an appropriate method. The first exposure is done before the event and the second exposure is triggered by the phenomenon. The holograms record the information of phase change induced by the change of refractive index which occurs due to density change. The information of phase changes, that is, density change, is restored in the reconstruction process in the form of interference fringes. Fig.3 shows the experimental set-up. Shock wave loading on a pig liver, which is wrapped in 20 μm thick rubber film and positioned 50 mm away from the center of a 10 mg AgN_3 pellet, was visualized.

3. Numerical simulation

The governing equations of shock propagation in liquids are the two-dimensional unsteady Euler equations in axisymmetric polar coordinates. They were solved numerically using a second order accuracy upwind TVD finite difference scheme (Harten 1983). Since underwater shock waves are generated by microexplosions of 10 mg AgN_3 pellets, the explosion product gas is replaced with a high-pressure water sphere whose total energy is equivalent to that of the effective explosive pellet. The high-pressure water sphere is assumed to be released instantaneously to generate a spherical shock wave.

4. Results

4.1. Hugoniot curve

Measured relationships between the pressure ratio and the density ratio of human blood, and of 5% and 10% gelatines are shown in Figs.4(a)-(c). P_0 and ρ_0 denote the initial pressure and density and the dashed line shows the Hugoniot curve of water. The density of human blood increases 10.3% at 450 MPa and an inflection point appears in the neighborhood of 50 and 60 MPa. A hysteresis effect is found between the data obtained when the pressure is increasing and those obtained when the pressure decreased. When the pressure was decreased to ambient pressure, density differed from the initial value. Experimental results of 5% and 10% gelatines appeared to be close to that of water and did not show any significant difference. On the other hand, a formulation similar to Tait's equation, the empirical EOS of water is adopted to the EOS of the specimens, that is,

$$\frac{P+B}{P_0+B} = \left(\frac{\rho}{\rho_0}\right)^n \tag{1}$$

where n and B are constants (Glass and Heuckroth 1960; Zel'dovich and Raizer 1967). n and B in the EOS of the specimens are determined by fitting the measured data to Eq.1. The obtained values are n=5.527, B=614.6 MPa at 309 K for human blood, and n=7.415, B=296.3 MPa at 293 K for water (Table 1).

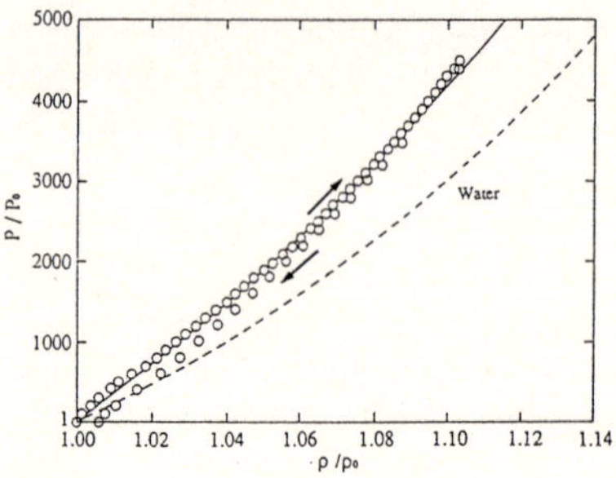

(a) Human blood

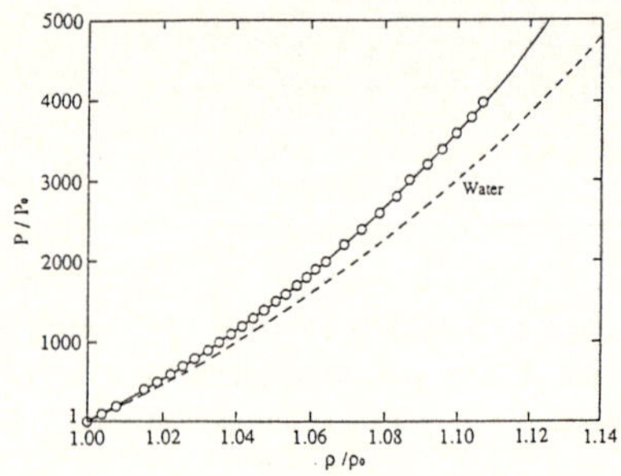

(b) 5 % gelatine

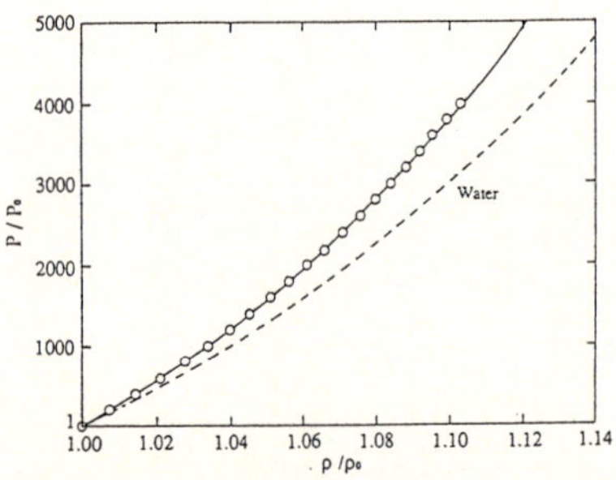

(c) 10 % gelatine

Fig. 4. Measured relationship between pressure ratio and density ratio

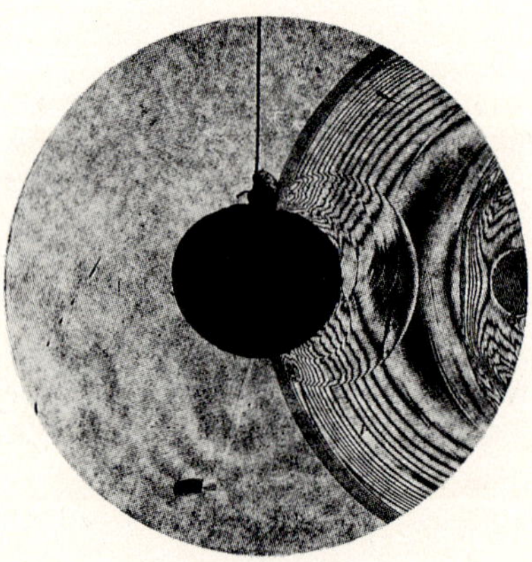

(a) 35.4 μsec

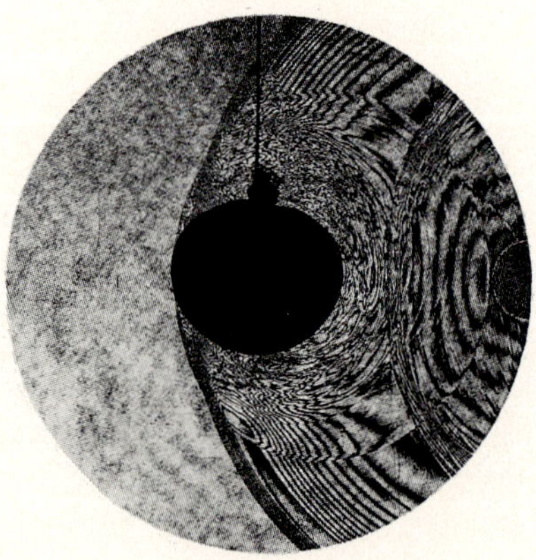

(b) 45.6 μsec

Fig. 5. Holographic interferogram

Table 1. n and B in the EOS of the specimens

	n	b (MPa)
Water	7.415	296.3
Human blood	5.527	614.6
10% Gelatine	8.549	302.7
5% Gelatine	8.525	288.3

4.2. Holographic interferograms

Fig.5 shows a holographic interferogram of a shock propagating through a pig liver (the model tissue). Part of the spherical shock wave was reflected at the model tissue interface and the shock wave propagates in the model tissue slightly faster than in water because the sound speed of the model tissue is higher than that of water. Shock wave refraction at the model tissue interface and associated complicated wave interactions can be clearly seen.

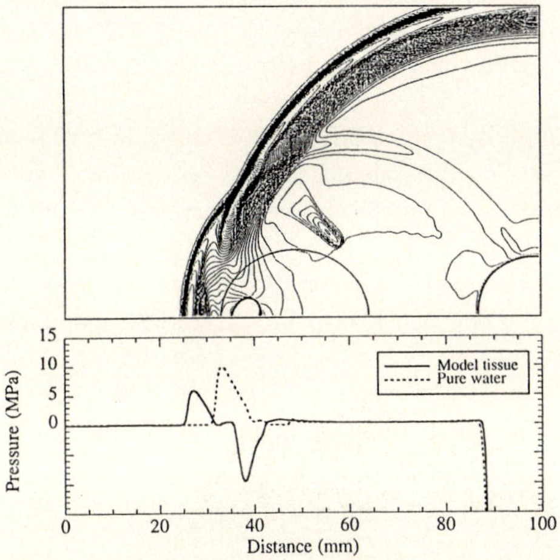

Fig. 6. Numerical results of the interaction of a spherical shock wave with the model tissue assuming the EOS of human blood

4.3. Numerical simulation

Fig.6 shows the numerical result of the interaction of a spherical shock wave with the model tissue assuming the EOS of human blood. The shock wave propagates in the model tissue faster. This trend agrees well with the experimental results. After the shock wave loading on the model tissue, an expansion wave was reflected from the model tissue interface. In the region exposed to negative pressure, the generation of a cavitation bubble is predicted.

5. Conclusions

In order to predict shock wave propagation in inhomogeneous media, an experimental and numerical study was carried out. The obtained results are summarized as follows:

(1) By measuring the volume change of human blood, and of 5% and 10% gelatines at pressures ranging from 0.1 to 500 MPa, the EOS of these specimens were determined.

(2) Numerical simulation, using these EOS, clearly showed shock wave interaction with the model tissue. The numerical results agree with holographic interferometric observation.

6. References

Bridgman PW (1935) The pressure-volume-temperature relations of the liquid, and the phase diagram of heavy water. J. Chem. Phys. 3, 10: 597-605

Bridgman PW (1949) Linear compressions to 30,000 kg/cm^2, including relatively incompressible substances. Proc. American Academy of Arts and Sciences 77, 6: 187-234

Bridgman PW (1950) Physics above 20,000 kg/cm^2. Proc. Royal Society A 203: 1-17

Chaussy Ch (1983) Shock wave treatment for stones in the upper urinary tract. The Urologic Clinics of North America 10, 4: 743-750

Chaussy Ch (1982) Extracorporeal Shock Wave Lithotripsy. Karger

Crum L (1988) Cavitation microjet as a contributory mechanism for renal calculi disintegration in ESWL. The Journal of Urology 140: 1587-1590

Glass II, Heuckroth LE (1960) An experimental investigation of the head-on collision of spherical shock waves. UTIAS Report No. 59, pp. 88-91

Goss SA, Johnston RL, Dunn F (1978) Comprehensive compilation of empirical ultrasonic properties of mammalian tissues. J. Acoust. Soc. America 64, 2: 423-456

Harten A (1983) High resolution scheme for hyperbolic conservation laws. J. Comp. Phys. 49: 357-393

Ioritani N et al. (1989) Renal tissue damage induced by focused shock waves. In: Kim YW (ed) Proc. 17th Intl. Symp. on Shock Waves and Shock Tubes, pp 185-190

Kandel LB (1992) Research issues of concern: Scientific and medical. In: Takayama K (ed) Proc. 18th Intl. Symp. on Shock Waves, Sendai, pp. 17-18

Kuwahara M (1992) Effects of focused shock waves on biological tissues: Cavitation in the prefocal region and the fluid flow in the focal region. In: Takayama K (ed) Proc. 18th Intl. Symp. on Shock Waves, Sendai, pp. 41-48

Takayama K et al. (1990) Holographic interferometric observation of shock wave focusing to extracorporeal shock wave lithotripsy. Proc. 19th Intl. Congr. on High-Speed Photography and Photonics, SPIE, Vol. 135 Part 2, pp 1180-1190

Zel'dovich YB, Raizer YP (1967) Physics of Shock Waves and High-Temperature Hydrodynamic Phenomena. Vol. 2, pp 709-711

Nonlinear Transmission of Focused Shock Waves in Nondegassed Water

Č. Štuka*, **P. Šunka**† **and J. Benes***
*Centre of Lithotripsy, Charles University, Prague, Czech Republic
†Institute of Plasma Physics, Czech Academy of Sciences, Prague, Czech Republic

Abstract. The aim of this contribution is to test the possibility of changing the ratio of the positive and the negative peak pressures of focused shock waves by transmission through nondegassed water.

Key words: Shock focusing, Lithotripsy

1. Introduction

Applying shock waves is a wide-spread method for destroying kidney stones and gallstones in clinical practice. However, the mechanisms of stone fragmentation and tissue damage have not yet been explained satisfactorily. It is generally admitted that the disintegration of stones is caused by compressive and tensile forces induced by the reflection of the positive pressure part of the shock wave within the stone and on its surface. This process can be reinforced by cavitation effects on the stone surface. On the other hand, while using limited pressures (not higher than 500 bars), which seems to be a safe limit (Folberth 1990), most adverse side-effects seem to be associated with cavitation effects, and hence with the negative pressure region of the shock wave. The goal of this paper is an attempt to test the possibility of controlling the ratio P^+/P^- of the positive (P^+) and negative (P^-) pressure peaks in the focused shock wave.

2. Experiment

The experiments were carried out in a specially-designed water tank to which the shock wave generator (lithotripter head) was attached. The shock waves were produced in degassed and nondegassed water by high-voltage spark discharges, and focused by a semiellipsoidal metallic reflector (with semi-axes a = 138 mm and b = 78 mm). The compact coaxial arrangement of the discharge circuit had an inductance of 67 nH with a capacity of 1 μF. All measurements were performed at a voltage of 7 kV with the distance of electrodes set to 1 mm. To achieve significant negative pressures it was necessary to adjust slightly the arrangement of the electrode holders. A hydrophone was fixed to the holder, adjustable in three axes and situated at the second focal point of the shock wave generator. It was not easy to choose a convenient hydrophone for exact measurements of shock wave pressure waveforms because it should have a very fast response time, a small size of the effective area and a satisfactorily long lifetime. For this reason we decided to perform the measurements with a broad-band PVDF needle hydrophone IMOTEC which had been devised specially for shock wave measurements. The effective area was 0.5 mm^2 with constant sensitivity up to 5 MHz (Müller 1985), and the rise time was approximately 65 ns. The electrical signals from the hydrophone were recorded on a digital storage oscilloscope Tektronix TDS 520 and the data were computer-processed.

Shock Waves @ Marseille III
Editors: R. Brun, L. Z. Dumitrescu

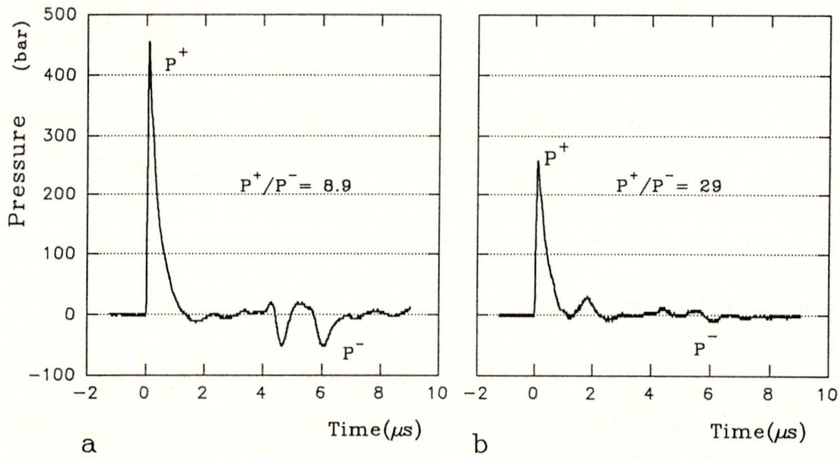

Fig. 1. Pressure waveforms measured with a PVDF needle hydrophone at the focus of the shock wave generator in a) degassed; and b) nondegassed water

3. Results

Our measurements were aimed at comparing the shock-wave propagation through degassed and nondegassed tap water. The positive and negative peaks of the pressure waveforms measured in the secondary focus were compared. Each measurement was repeated ten times and Fig.1 shows the mean of these ten waveforms. This figure shows clearly that the shock waves generated and transmitted in degassed (a) and plain tap water (b) differ substantially. According to our assumption in nondegassed water the negative peak pressure P^- is attenuated to a greater extent than its positive counterpart P^+ in comparison with the results of measurements in degassed water. The mean positive peak pressure P^+ in degassed water was found to be 454 bars, in nondegassed water it amounted to 258 bars, while the negative peak pressure P^- amounted to 51 bars and 9 bars, respectively. In nondegassed water P^+ was attenuated about 1.8 times and P^- was reduced 5.7 times as compared with peak pressures in degassed water. The pressure at the leading edge of both waveforms was ascending to peak values at the same rate, though we realize that the real rise time of the shock wave may have been shorter than the value measured. In the waveforms generated in nondegassed water we observed significant variations. Whereas the waveforms measured in degassed water showed equal patterns both in shapes and peak values (their variations were less than 10%), in the tap water we observed an irregular second pressure peak (as shown in Fig. 2) with a delay of about 1 μs after the main pulse. Its amplitude and position (time delay) vary from shot to shot. The cause of the irregular second peak has not been explained satisfactorily yet.

4. Discussion

In our opinion, the different degree of attenuation of the positive and negative pressures observed in tap water in comparison with the degassed water is associated closely with different cavitation thresholds in degassed and nondegassed water. The threshold for acoustic cavitation in water depends on the content of dissolved gas. In degassed water the cavitation threshold is about 0.5 MPa (5 bars) (Coleman 1987) whereas in nondegassed water we assume it to be lower. In tap water part of the dissolved gas takes the shape of microbubbles. According to Church's theoretical study (Church 1989) we can imagine clearly how these microbubbles influence the propagation of shock waves through the liquid. In the ideally degassed liquid low amplitude pressures are transmitted without any distortion. In the liquid with microbubbles the shock waves propagate with weak attenuation of the positive pressure, partly because the compressive pressure changes the radius of the microbubbles only within the range of one order of magnitude. However, the transmission of negative pressures can cause the volume of microbubbles to grow in the range of six orders of magnitude, not only around the secondary focus but mainly in the region of the primary focus. As a consequence of the amount of energy consumed for the enlargement of microbubbles the negative pressure part loses much of its energy and its peak is strongly attenuated. This should enable us to control the transmission of negative pressures by means of the concentration of dissolved gas. This method can help to minimize adverse effects of lithotripsy if necessary.

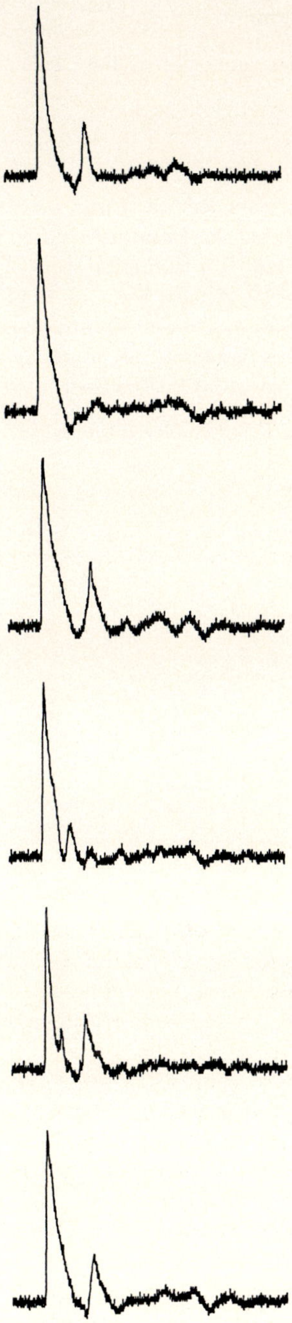

Fig.2. Variations in waveforms generated in nondegassed water

Acknowledgments

This work was supported by the Czech Ministry of Health Care, Grants no. 1133-3 and no. Z 293-3.

References

Coleman AJ, Saunders JE, Crum LA, Dyson M (1987) Acoustic cavitation generated by an extracorporeal shockwave lithotripter. Ultrasound Med. Biol. 13:69-76

Church ChC (1989) A theoretical study of cavitation generated by an extracorporeal shock wave lithotripter. J. Acoust. Soc. Am. 86:215-227

Folberth W (1990) Pressure-optimized lithotripsy with the Siemens Lithostar: successful and tissue-protecting treatment of urinary stones. European Urology 17:51-57

Müller M et al. (1985) Einsatz einer breitbandigen Piezodrucksonde auf PVDF-Basis zur Untersuchung konvergierender Stosswellen in Wasser. Acustica 58: 215-222.

Soft Flash X-ray System for Shock Wave Research

E. Sato*, K. Takahashi*, M. Sagae*, A. Shikoda*, T. Oizumi*, H. Ojima†, K. Takayama†, K. Sasaki‡, Y. Tamakawa‡ and T. Yanagisawa§

* Department of Physics, Iwate Medical University, Morioka 020, Japan
†Shock Wave Research Center, Institute of Fluid Science, Tohoku University, Sendai 980, Japan
‡Center for Radiological Science, Iwate Medical University, Morioka 020, Japan
§Department of Radiology, Iwate Medical University, Morioka 020, Japan

Abstract. A description is given of a soft flash X-ray system designed primarily for observing cavitation bubble clouds which are generated in Extracorporeal Shock Wave Lithotripsy (ESWL). This system consisted of a spectrum and image simulation system and two types of high-intensity single flash X-ray generators. The shock waves for ESWL were generated inside an ellipsoidal reflector by detonating a pellet of silver azide (AgN_3). High-speed radiography was performed by means of delayed radiography, and the spatial and temporal distribution of the bubble clouds generated by underwater shock wave focusing were observed in pseudo-human tissue.

Key words: Flash X-rays, High-speed radiography, Shock wave lithotripsy

1. Introduction

Together with advances in high-voltage pulse technology, various kinds of flash X-ray generators have been developed corresponding to specific radiographic objectives. Although most of these generators have been effectively applied to military and diffraction studies (Germer 1979), soft generators with photon energies of less than 150 keV have been designed to perform soft high-speed radiography including biomedical applications (Sato 1992). To be exact, a high-power pulse laser produces plasma flash X (water-window) rays (Tomie 1991) for performing microscopy, yet these plasma X-rays (Sato 1992) with high-intensity characteristic emission can be produced by means of a vacuum discharge achieved with a large-capacity condenser. In addition, when a high-speed image intensifier tube is employed, high-repetition-rate generators (Sato 1991) having hot-cathode radiation tubes are useful for performing high-speed cineradiography. In conjunction with the ESWL (Takayama 1992), there is an urgent need for visualizing shock wave-induced cavitation in living tissue. Although the occurrence of cavitation bubbles in human bodies during the ESWL treatment has been clearly demonstrated by ultrasound diagnostic devices, the spatial and temporal distributions of the bubbles have not been observed in vivo. For this purpose, we developed two types of high-intensity flash X-ray generators in conjunction with an image-simulation system and performed high-speed delayed radiography to clearly image cavitation bubble clouds in ESWL experiments.

2. Flash X-ray system

2.1. Low-vacuum flash X-ray generator

Fig.1 shows the block diagram (including the electric circuit) of a high-intensity flash X-ray generator (Type A). This generator consists of the following essential components: a high-voltage power supply, a high-voltage pulser having a four-cable transmission line and a 150 nF coaxial oil condenser, a mechanical-booster pump, a Krytron pulser as a trigger device, and a low-vacuum diode. The main condenser was charged from 50 to 100 kV, and the main discharge for generating the flash X-rays was produced after closing a gap switch by a Krytron pulser. The radiation diode was specially designed in order to produce stable vacuum discharges and was evacuated by a mechanical booster pump with a constant pressure of 1.7 Pa. Compared to the capacity of the

Shock Waves @ Marseille III
Editors: R. Brun, L. Z. Dumitrescu

cable transmission line, because the capacity of the main condenser is quite large, the ideal pulse output $V_{oi}(t)$ under the condition of no energy loss can be represented by the following Fourier series:

$$V_{oi}(t) = V_c - \sum_{m=1}^{\infty} \frac{4V_c}{m\pi} \cdot \sin\frac{m\pi}{2} \cdot \cos\frac{m\pi\nu t}{2l}, \tag{1}$$

where V_c is the charge voltage, l is the cable length, and ν is the propagation velocity of the coaxial cables (transmission line). Ideally speaking, the output voltage from this high-voltage pulser is double the charge voltage of the condenser.

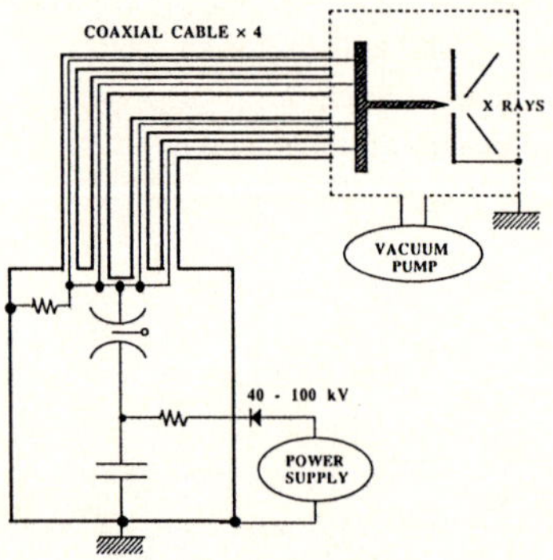

Fig. 1. Block diagram including the electric circuit of a high-intensity flash X-ray generator having a low-vacuum diode

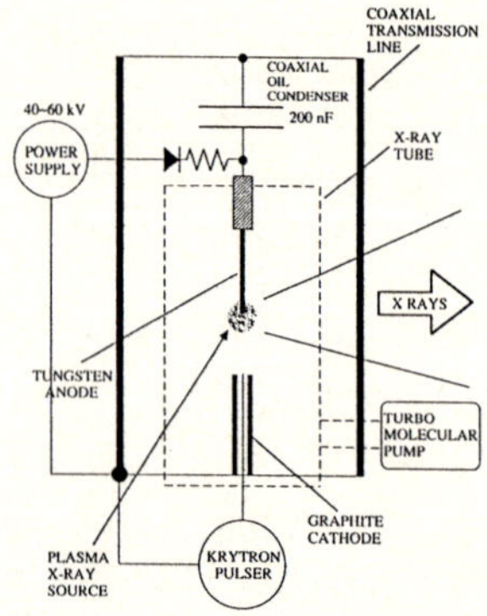

Fig. 2. Main transmission line of a high-intensity plasma flash X-ray generator

2.2. Plasma flash X-ray generator

The plasma generator (Type B) also employed a coaxial transmission line with a 200 nF large-capacity coaxial condenser (see Fig.2). The main condenser was charged from 40 to 80 kV, and the electric charges in the condenser were discharged to an X-ray tube by a thyristor trigger device. The radiation triode was connected to a turbo-molecular pump which allowed operation at a pressure of about 1×10^{-3} Pa. Because this tube employs a long tungsten anode tip, the plasma X-ray source which consists of tungsten ions and electrons can be easily produced. Using this generator, the maximum tube voltage was approximately equivalent to the initial charge voltage of the condenser.

2.3. Image simulation system

In this experiment, we developed an image simulation system SPECTRA (see Fig.3) in order to experimentally obtain the optimum control of the flash X-ray spectrum. The Bremsstrahlung energy spectra are calculated on the basis of Kramers' equation (Kramers 1923). Generally, we assume that the X rays permeate through N kinds of absorbers. Then the permeating energy spectrum $I_p(E)$ is written as:

$$I_p(E) = K \cdot \exp\left[-\sum_{k=1}^{N} x_i \cdot \mu_i(E)\right] \int_0^{\infty} J(t)\,[E_0(t) - E]\,dt, \tag{2}$$

where $J(t)$ is the tube current, $E_0(t)$ is the maximum kinetic energy of an electron which is equivalent to the X-ray tube voltage, E is the photon energy, and K is some factor, and x_i and

$\mu_i(E)$ are the thickness and the linear attenuation coefficient of the i-th absorber ($i = 1, \ldots N$), respectively.

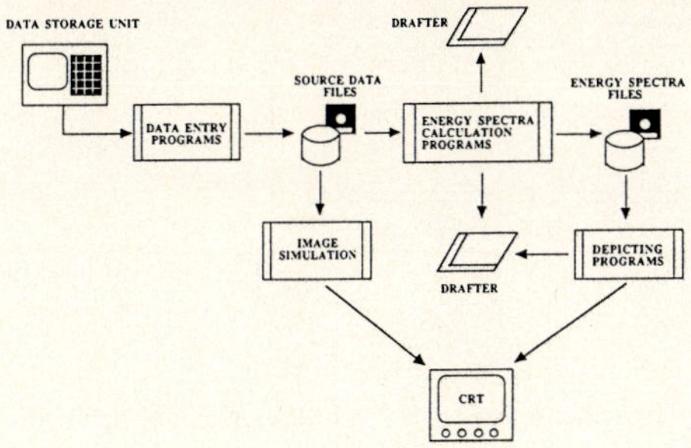

Fig. 3. Structure of the spectrum and image simulation system (SPECTRA)

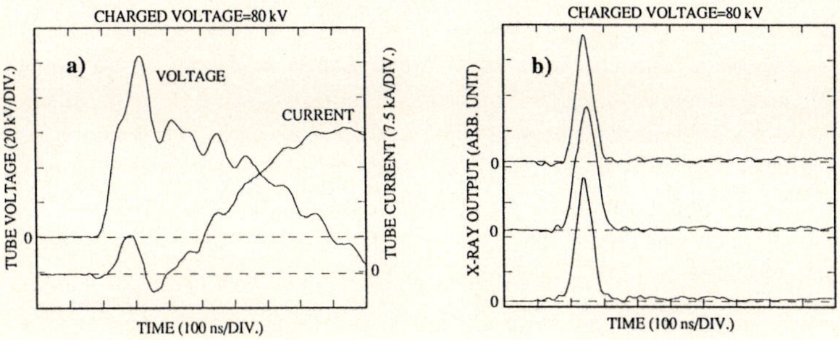

Fig. 4. Radiographic characteristics of the Type A generator with a charge voltage of 80 kV: (a) tube voltage and current; (b) flash X-ray outputs

3. Radiographic characteristics

3.1. Characteristics of the Type A generator

Fig.4 shows the radiographic characteristics of the tube voltage, current, and X-ray output achieved with the Type A generator at a charged main condenser voltage of 80 kV. Both the tube voltage and current displayed small-amplitude oscillations, because a cable transmission line was employed (see Fig.4(a)). The maximum tube voltage and current had values of 105 kV and 32 kA, respectively. Using this generator, stable X-ray outputs were obtained even when a low-vacuum diode was employed (see Fig.4(b)). In this condition, the pulse width of the flash X-rays was about 70 ns, and the X-ray intensities measured by an ionization chamber (Victoreen 660) had values of about 3.0 μC/kg at 1.0 m per pulse.

3.2. Characteristics of the Type B generator

The Type B generator produced extremely high-intensity L-series characteristic X-rays compared to those obtained by the Type A generator. Thus the simulation system SPECTRA cannot be used with this generator. In the present work, the tube voltage and current displayed damped oscillations, and the initial tube voltage before the discharging was equivalent to the initial charge voltage (see Fig.5(a)). Although the maximum tube current roughly increased as the charge voltage increased, the maximum current for a charge voltage of 60 kV had a value of about 23 kA.

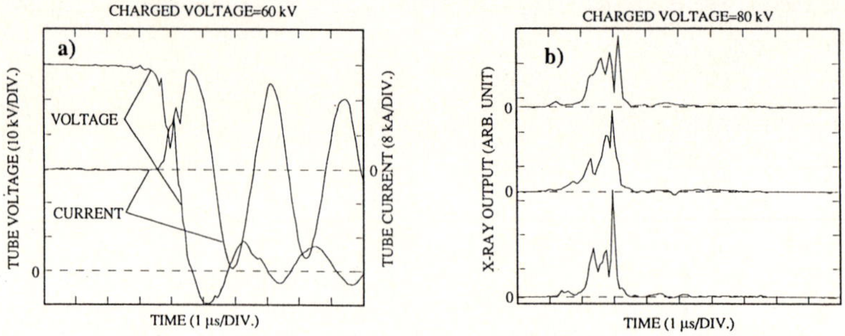

Fig. 5. Characteristics of Type B generator with a charge voltage of 60 kV: (a) tube voltage and current; (b) plasma flash X-ray outputs

In order to measure the X-ray intensity, including L-series spectra, we employed a thermoluminescence dosimeter (element: MSO-S). This generator produced comparatively long X-ray pulses of about 1 μs width (see Fig.5(b)); the X-ray intensity had a large value of about 12 μC/kg at 1.0 m per pulse.

3.3. Simulation images

SPECTRA provides various kinds of three-dimensional shaped models which can be used as objects in the radiographic image simulation. Fig.6 shows simulation results using the experimental data of the Type A generator with a charge voltage of 80 kV and a 1 m air path. In this simulation, it was assumed that a smaller concentric air cylinder was embedded in a larger water cylinder of 80 mm height (see Fig.6(a)). As the height of the air cylinder increased, the gray density slightly increased (see Figs. 6(b)-(d)). Thus the lower the height of water was, the higher image contrast of the bubble clouds was obtained.

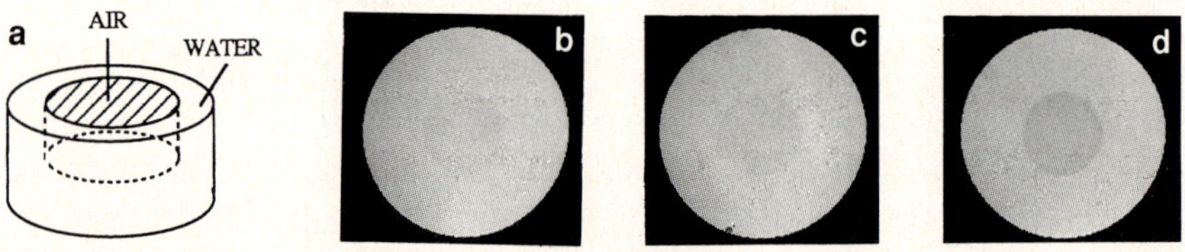

Fig. 6. Simulation results using an experimental result of the Type A generator: (a) concentric simulation model; (b) 5 mm air height; (c) 15 mm air height; and (d) 25 mm air height

4. Shock wave source for ESWL

Fig.7 shows a schematic drawing of the experimental set-up of the shock wave source for ESWL. Although both sides of the ellipsoidal cavity were cut to a width of 80 mm in order to clearly observe bubble clouds, the original major and minor diameters employed were 127.3 and 90.0 mm, respectively. The underwater shock waves were generated by means of a microexplosion produced by a 10 mg pellet of AgN_3; this pellet (which was located on a glass fiber tip) was spontaneously exploded by a Q-switched laser beam. The ignition time delay had an approximately constant value of about 200 μs.

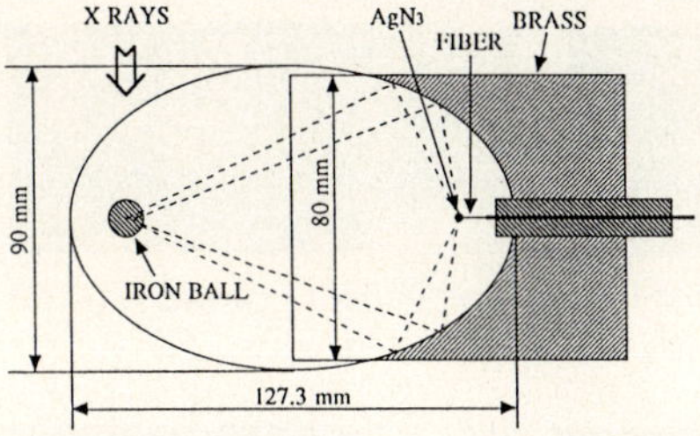

Fig. 7. Schematic drawing of an underwater shock wave source utilizing an ellipsoidal reflector

5. Radiography

The high-speed radiography achieved with the two types of generators and a computer radiography system (Sonoda 1983) was performed by means of delayed radiography. Fig.8 shows three radiographs of the gas bubbles produced by the detonation of AgN_3. These were obtained by the Type A generator at the indicated delay times (after the explosion). As the delay time increased, the size of the bubble increased, and increases in the water density were observed around the bubble. The radiographs in Fig.9 show the cavitation bubble clouds after underwater shock focusing. In this radiography, a steel ball was employed instead of a stone to clearly image the second focal point of the reflector. The dimension of the bubble cloud (dark spots on the surface of the ball facing the reflector) increased according to increases in the delay time. In addition, the bubble cloud from the edge of the reflector also grew with passing time. Finally, the cavitation bubble cloud in pseudo-human tissue cannot be easily observed by the optical visualization. As shown in Fig.10, both the cloud beside the ball and the turbulence pattern around the ball grew at the side facing the reflector as time passed.

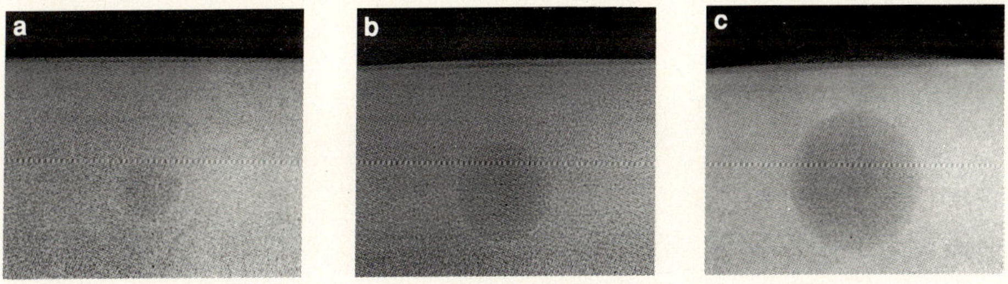

Fig. 8. Delayed radiographs of the gas bubble produced by the detonation of a AgN_3 pellet of 10 mg achieved with the Type A generator at the indicated delay times; (a) 100 μs; (b) 150 μs; and (c) 200 μs

6. Discussion

For the ESWL studies of underwater shock wave focusing, the Type A generator powerfully generated wide photon energy flash X-rays with pulse widths of less than 100 ns and it was easy to obtain complete stoppages by means of delayed radiography. Next, in order to image and to analyze the temporal distribution of the cavitation bubble cloud with high contrast, high-intensity soft flash X-rays produced from such a Type B generator are quite useful since harder X-rays produce X-ray scattering in the radiographic object; this scattering interferes with the high-contrast image production. Using these two flash X-ray generators in conjunction with a spectrum and image simulation system, the optimum use of the available radiographic conditions

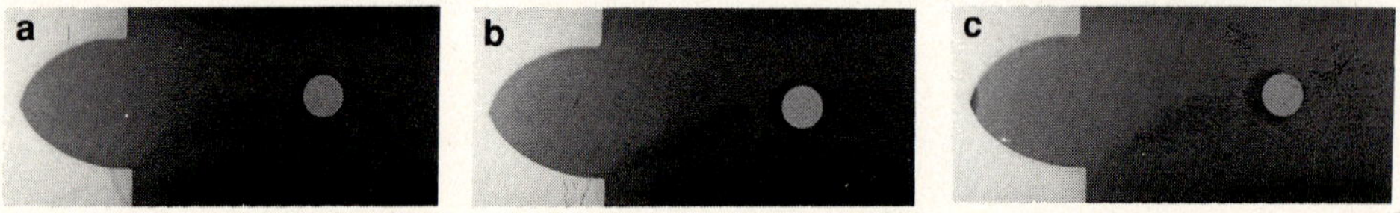

Fig. 9. Radiographs of cavitation bubble clouds induced by underwater shock wave focusing by a reflector: (a) 550 μs; (b) 650 μs; and (c) 750 μs

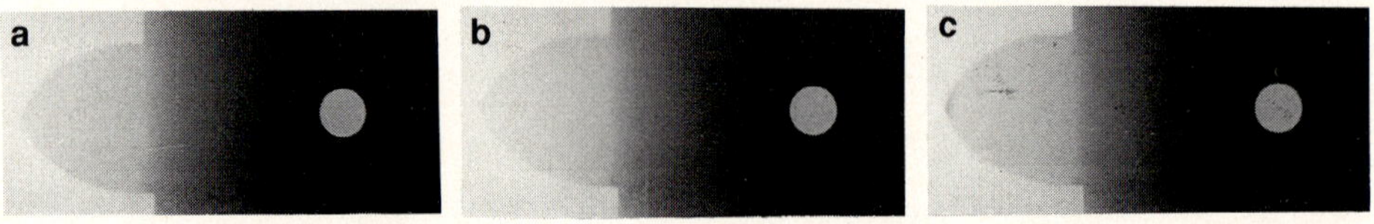

Fig. 10. Radiographs of cavitation bubble clouds in pseudo-human tissue: (a) 550 μs; (b) 650 μs; and (c) 750 μs

may be accomplished, so that this kind of flash X-ray system can be effectively applied to imaging shock wave phenomena in various kinds of media including living tissue.

Acknowledgment

This work was supported by Grants-in Aid for Scientific Research from the Iwate Medical University-Keiryokai Research Foundation, the Private School Promotion Foundation, and the Ministry of Education and Culture in Japan.

References

Germer R (1979) X-ray flash techniques. J. Phys. E: Sci. Instrum. 12:336

Kramers HA (1923) On the theory of X-ray absorption of the continuous X-ray spectrum. Philos. Mag. 46:836

Sato E, Shikoda A, Kimura S, Sagae M, Isobe H, Tamakawa Y, Yanagisawa T, Honda K, Yokota Y (1991) Kilohertz-range flash X-ray generator utilizing a triode in conjunction with an extremely hot cathode. Rev. Sci. Instrum. 62:2115

Sato E, Shikoda A, Kimura S, Sagae M, Oizumi T, Takahashi K, Hayasi Y, Shoji T, Shishido K, Tamakawa Y, Yanagisawa T (1992) Repetitive compact flash X-ray generators for soft radiography. SPIE 1801:628

Sato E, Sagae M, Kimura S, Oizumi T, Hayasi Y, Tamakawa Y, Ohta Y, Suzuki S, Obara H, Seino H, Yanagisawa T (1992) Conventional metal plasma X-ray-flash techniques using vacuum discharges. SPIE 1801:1001

Sonoda M, Takano M, Miyahara J, Kato H (1983) Computed radiography utilizing scanning laser stimulated luminescence. Radiology 148:833

Takayama K, Nagoya H, Obara T, Kuwahara M (1992) Application of holographic interferometric studies of underwater shock wave focusing to medicine. SPIE 1801:1029

Tomie T, Suzuki H, Majima T, Kanayama T, Kondo H, Yano M, Ono M (1991) Three-dimensional readout of flash X-ray images of living sperm in water by atomic-force microscopy. Science 252:691

New Discharge Circuit for Efficient Shock Wave Generation

P. Šunka*, V. Babický*, M. Člupek* and Č. Štuka †
* Institute of Plasma Physics, Prague
† 1st Department of Medecine, Charles University, Prague

Abstract. A new discharge circuit for the generation of shock waves in water was developed. The water spark gap operates in the self-breakdown regime. Measurements of electrical parameters of the discharge channel are presented. The discharge circuit was used for generating focused shock waves for ESWL.

Key words: Lithotripsy, Spark discharge

1. Introduction

Underwater spark discharges are used for the generation of shock waves in several commercially produced lithotripters. The discharge circuit is arranged there in such a way that a low-inductance condenser is charged up to 10-30 kV and then it is switched on the tip electrodes of an underwater spark gap by means of a gas-insulated switch (spark gap). In such an arrangement there are in fact two spark gaps in series. The gas switch not only increases the inductance of the discharge circuit but also it consumes a part of energy, which dissipates in it, thus weakening the shock generated in the water.

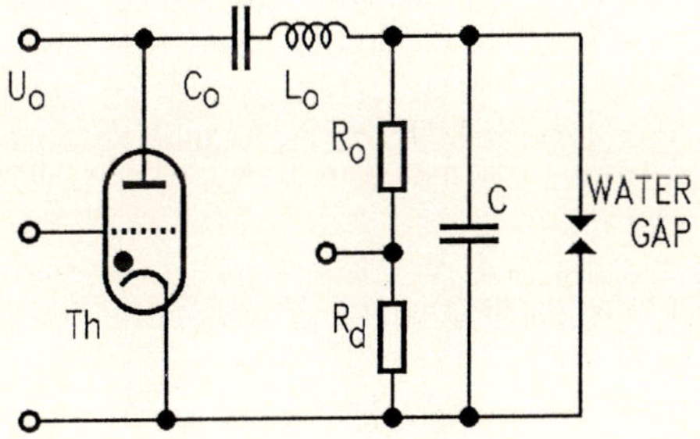

Fig. 1. Schematic diagram of the discharge circuit

2. Description of the discharge circuit

The main idea of our design is to operate the water spark gap in the self-breakdown regime avoiding thus the gas switch. The schematic diagram of such a circuit is shown in Fig.1. The electrodes of the water spark gap are directly connected to a low-inductance condenser C (0.2μF/20 kV, 60 nH) which is pulse-charged from the auxiliary condenser $C_0 = 0.25$ μF through the inductor $L_0 = 0.7$ mH and the thyratron Th. By switching the thyratron on, the charge from C_0 is almost totally transferred to the condenser C and the thyratron is automatically switched off. The gap between the electrodes is adjusted to such a value that breakdown occurs near the maximum

Shock Waves @ Marseille III
Editors: R. Brun, L. Z. Dumitrescu

voltage on the condenser C. Due to the dependence of the breakdown voltage in water on the duration of exposition to high voltage, the breakdown can occur also at later times during the exponential fall of the voltage on the condenser C (as shown in Fig.2). As the charging time of the condenser C is fixed ($t_0 = 28\ \mu$s), we use the duration of the time interval between the switching on the thyratron and the breakdown as a measure for feedback control of the water gap length. The compact design of the discharge circuit together with the coaxial construction of the water spark gap enable us to decrease the total inductance of the circuit L to about 110 nH.

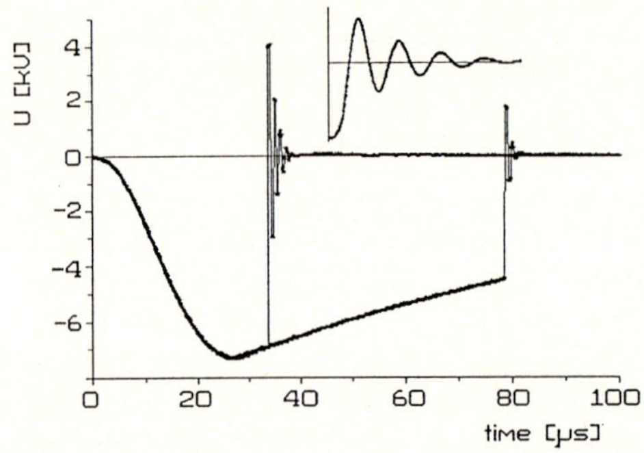

Fig. 2. Waveforms of the voltage on condenser C representing the charging phase and moment of discharge - two shot overlays. Insert: detail of the voltage during discharge

3. Estimates of the parameters of the discharge channel

As it follows from the theory and experiments of underwater sparks (Martin 1965, Okun 1969), during the initial stage of the discharge the resistance of the discharge channel falls exponentially with time to as low a value as $10^{-1}\Omega$ (the so called "resistive phase" t_r), remaining almost constant at later times. Most of the energy is transferred into the spark channel just during the resistive phase t_r. The following semiempirical formula was derived for the duration of t_r in water (Martin 1965):

$$t_r = \frac{232}{Z_0^{1/3}\ E^{4/3}} \qquad [\mathrm{s}, \Omega, \mathrm{V.m^{-1}}] \tag{1}$$

where E is the average electric field strength in the gap and Z_0 the characteristic impedance of the discharge circuit. The ratio G of the energy W dissipated inside the channel to the total energy $CU^2/2$ stored in the condenser C can be estimated as

$$G = \pi\, t_r\ /T \tag{2}$$

where T is the period of the discharge current. Thus, for an efficient shock wave generation we should minimize T and maximize t_r. For a given capacity of the condenser C this requires to minimize the circuit inductance L (shorter T) as well as the average electric field strength E (longer t_r). The minimum E is limited just by the self-breakdown voltage which is actually used. The pressure p inside the discharge channel can be estimated as (Vovtchenko 1978):

$$p = 8\,(P'/l)^{1/2} \qquad [\mathrm{P_a}, \mathrm{W.s^{-1}}, \mathrm{m}] \tag{3}$$

where P' is the time derivative of the power dissipated in the channel during the initial phase of the discharge, and l is the spark length.

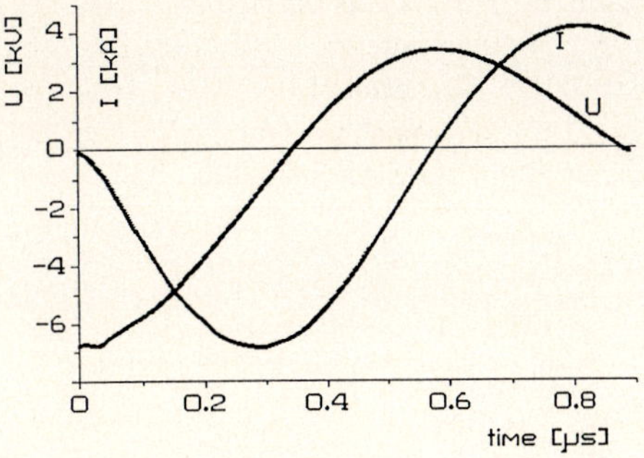

Fig. 3. Voltage and current waveforms during the discharge

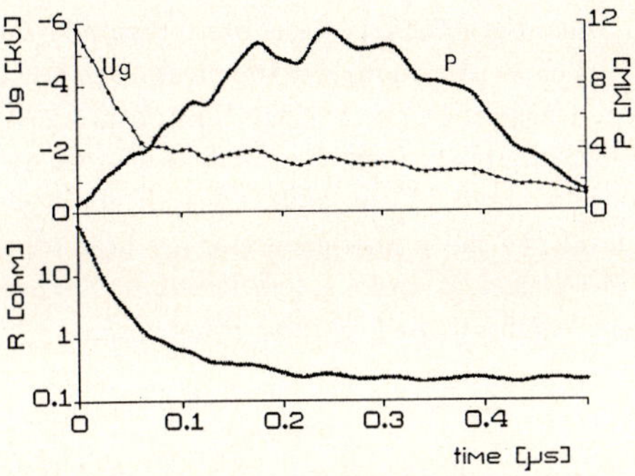

Fig. 4. Calculated voltage U_g across spark, power P dissipated in the discharge channel and spark resistance R (in logarithmic scale)

4. Experimental results

Using a resistive voltage divider, a Rogowski belt and two fast digital oscilloscopes, we have measured the voltage $U(t)$ on the condenser C and the discharge current $I(t)$. Results of these measurements, made at a charging voltage of 7 kV and inter electrode gap $l = 0.7$ mm are shown in Fig.3. Assuming that the circuit inductance L is constant (the time variation of the spark inductance is negligible), we may calculate the voltage on the spark:

$$U_g = U(t) - L \;\; dI/dt \tag{4}$$

the spark resistance:

$$R(t) = U_g(t) \; / \; I(t) \tag{5}$$

and the power dissipated inside the discharge channel:

$$P(t) = U_g(t) \cdot I(t) \tag{6}$$

Results of the computations are shown in Fig.4, where the spark resistance is plotted on a logarithmic scale. The duration of the resistive phase (rapid fall of U_g and R) is about 100 ns, in good agreement with Eq.1. The dissipated power reaches 10 MW during 200 ns ($P' = 5 \times 10^{13}$ W.s$^{-1}, l = 0.7$ mm). The pressure inside the channel, estimated from Eq.3 is $p = 2.13$ GPa (21300 bars).

The described discharge circuit was used for generating focused shock waves. The water spark gap was located in the primary focus of a semiellipsoidal cavity (semiaxes $a = 138$ mm, $b = 78$ mm)

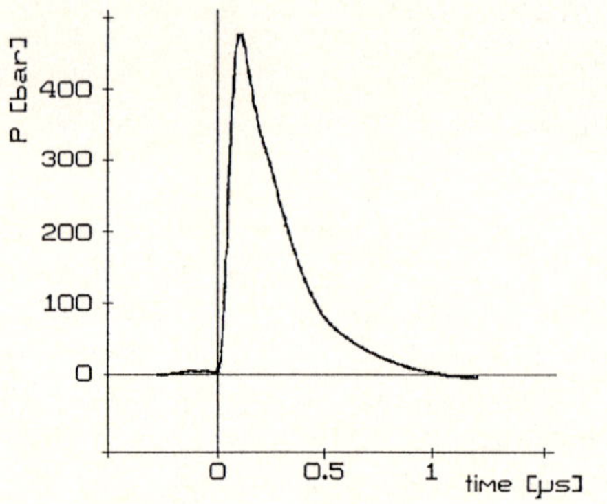

Fig. 5. Pressure pulse at the secondary focus. Charging voltage 14 kV

of the metallic reflector. The pressure distribution around the secondary focus was measured by a needle PVDF hydrophone IMOTEC with a rise time of 65 ns. An example of the pressure pulse in the secondary focus is shown in Fig. 5 for a charging voltage of 14 kV. The rise time of the shock wave is comparable with the time resolution of the hydrophone. Point by point measurements have shown that the diameter of the secondary focus is about 12 mm and its length about 45 mm (dimensions at half-maximum pressure). The developed shock wave generator has been used in the lithotripter Medilit produced by Medipo EM Co. Ltd. in Brno, Czech Republic. Several thousands patients with kidney stones have been successfully treated by these lithotripters.

References

Martin JC (1965): Duration of the resistive phase and inductance of spark channels. Res. Rep. Atomic Weapons Research Establishment Aldermaston No. SSWA/JCM/1065/25

Okun IZ (1969): Issledovaniye elektritcheskikh kharakteristik impulsnogo razryada v zhidkosti. J. Tech. Phys. 39: 850 (in Russian)

Vovtchenko AI et al.(1978): Elektricheskiye kharakteristiki podvodnogo iskrovogo razryada. In: Fizitcheskiye osnovy elektrogidravlitcheskoy obrabotki materialov, Kiev, "Naukova dumka", pp 14-25 (in Russian)

New Methods of Superpowerful Focused Shock Wave Generation for Medicine and Mineralogy

V. Veroman and O. Novikov
Medical Engineering Centre 'Hyperselective Effect', Saint Petersburg, Russia

Abstract. The disintegration process of solids by hyper-shock waves is successfully used in medicine for urinary-cholelithiasis treatment. The present paper considers problems concerned with further process development, brought about by new applications, both in medicine and industry, which require differentiation of disintegration modes and production of new equipment. Some other problems are also considered such as the interconnection between the speed of rift appearance in the solid body and the nature of disintegration. Besides, a new concept of the mechanical skin-effect is introduced, and a comparative-qualitative analysis of hyper-shock wave generators is given, which is based on different physical phenomena. This paper deals with such problems as the connection of acoustic pulse parameters with the pressure in the focal zone and solid body disintegration in a water medium, not degassed. Different reflectors and refractors have also been considered. A new type of toroidal reflector has been described and the problem of the accuracy of reflector focusing on the object is discussed. The possibility for disintegration of biomineralogical formations in a human body without knowing their spatial location by using an intercrystallite resonance phenomenon is shown, as well as the possibility for the process of selective disintegration of oncological cells against normal cells (work now in progress).

Key words: Rock disintegration, Hyper-shock wave, Lithotripsy

1. Introduction

The Dornier aircraft industry firm is known to have obtained excellent results in medicine having produced, in 1980, together with Dr. Chaussy, the first extracorporeal lithotriptor for the disintegration of kidney stones. At present, due to numerous works carried out in various countries, extracorporeal lithotripsy by hyper-shock waves has acquired a wide development in the theory and in engineering application, although, at the same time, more questions and problems than there were at the beginning have arisen. The following problems are among them: studying the physics of solid body disintegration by hyper-shock waves, in which case the process differs greatly from that produced by mechanical shocks; establishing the technology principles of extracorporeal disintegration; elaboration of the effect mode on the solid body, taking into consideration its physical properties and technological requirements. Contactless disintegration of solid bodies by shock waves has a number of peculiarities and advantages due, for instance,to the sterility of the process, to the possibility to put into effect a mode of superficial falling into dust-like fractions or its main disintegration into large fragments, etc. Currently there is a trend towards wide application of this process not only for various medical purposes, but also in industrial mineralogy in a wide sense of this term. As examples of problems encountered one should mention disintegration of ore materials, which requires producing new shock wave sources and special equipment. Even in medicine lithotriptors are nowadays differentiated into urolithotriptors, cholelithotriptors, salivolithotriptors, etc. which are essentially different both in the energetic characteristics of the shock waves and in their design peculiarities.

Shock Waves @ Marseille III
Editors: R. Brun, L. Z. Dumitrescu

2. Physical principles of solid disintegration

It is known that as a result of a pressure pulse applied to a solid body immersed in liquid there appears a stress on its surface, the value of which is proportional to the power being instantaneously supplied. The speed of rift development is proportional to the arising stress and the nature of solid body disintegration depends on the speed of rift development. So, the development of a superficial microrift embrio occurs at the initial loading by the pressure pulse with a speed of several mm per hour and is affected by the grouping of dislocations inside the crystal association and by subsequent plastic deformation. Further loading brings about spontaneous development of the main rift with a speed of several thousand meters per second, owing to the great number of single dislocations and crystallite displacements. Intensive reproduction of new microrifts takes place and the process of plastic deformation and saturation with dislocations loses its significance gradually. When the speed of rift development is 0.58-0.7 times the speed of transverse acoustic waves, main rift branching occurs, resulting in superficial falling of material. This phenomenon, discovered by us when carrying out experiments in extracorporeal lithotripsy of urinary stones was called a "mechanical skin-effect". At higher speeds of development of the rifts, in the range 2000-2200 msec, the pressure profile in the solid body develops a maximum concentration of stresses in front of the rift top, thus preventing development of a main rift. Besides, the mechanical properties of the destructed material have little effect on the destruction process. Therefore it is very important to produce hyper-shock wave generators of maximum instantaneous power.

3. Shock wave generators

3.1. Electrohydraulic generators

At first, these seemed to be the only ones providing maximum power of the acoustic pulse, because the growth speed of the pressure in the reflector internal focus is limited only by the inductance of external and internal wires, the inductance of the storage capacity and the reactive component of the complex electric resistance of the plasma channel. However further investigations of shock wave propagation processes in the area from the internal to the external focus showed that their characteristics change greatly as a result of their non-linear behaviour, as reported by us at the Symposium held at York University (Great Britain) in 1990. Besides, the supposed absolute advantages of electrohydraulic generators are not exceptional. We made attempts at increasing the frequency of pulse repetition from 5 to 20,000 Hz for an electrohydraulic generator, correspondingly increasing the power, as it was ascertained that the pulse power was the decisive factor of disintegration, rather than its energy. But this idea proved to be incorrect. First, we have experimentally determined the pulse energy threshold value, i.e. at a lower value there would be no disintegration even if the impulse power is high. Secondly, Dr.Coleman (St.Thomas Hospital, London) has convincingly proved that the pulse frequency is limited by the relaxation time of gas bubbles in the water after an acoustic pulse has passed, corresponding to units of Hz.

3.2. Electromagnetic generators

In spite of the high coil inductance affecting the membrane of the generator by its field and in spite of the mass inertia of the membrane itself, such generators produce acoustic pulses of sufficient power for extracorporeal chole-urinary lithotripsy, and were successfully applied by Siemens. One great advantage of the electromagnetic generators is the absence of zones of extreme energy density. Although it may seem unexplained, the energy distribution in the focal zone of electromagnetic generators is close to an elliptic form because of complex membrane oscillations which differ from radial ones.

3.3. Laser generator

This generator is based on optic-thermo-acoustic effects, providing the possibility to obtain short shock wave pulses. The limited energy value, however, and its losses in refractors do not give the possibility to obtain shock waves of desirable instantaneous power though disintegration of concretions into fragments by main rifts takes place after 10-30 pulses.

3.4. Piezoelectric generators

Such generators provide pressure pulses which can be compared with those generated by electrohydraulic and electromagnetic generators but, unlike them, having radial radiation, the latter being obviously an advantage of piezoelectric generators, as well as, in some cases, its small energy.

3.5. Thermoimpulse generators

In these generators a heated membrane radiates a shock wave. Thermoimpulse generators have the following advantages: membrane immobility and radiation strictly along the radius of curvature. Due to it, it becomes possible to obtain pressure pulses in the focal point with pressures up to 3 kbar with a duration of 50 microsecs.

3.6. Generators of distributed discharges

They are based on the smooth discharge which occurs between a set of electrodes and a focusing grid. These represent a further development of the electrohydraulic generator but unlike the latter they provide uniform radiation at the grid surface, no zones overloaded by energy and enable to multiply radiation energy within any limits correspondingly.

4. Focusing

When the first lithotriptors were developed, the main attention was paid to shock waves focusing on the disintegration object. The best results were obtaind by applying reflectors. Attempts to apply metal refractors which have corresponding acoustic contact with the radiator had no positive results. We have investigated various elliptic reflectors. It was shown that the energy distribution in the focal zone was non-linear and its characteristic form was elliptical. The zones of strongest concentration of energy appear as the excentricity is reduced, tending to 1. The high requirements put forward in the beginning were not satisfied, concerning the precision of shock wave location on the kidney stones, because of kidney excursions of about 2-4 cm, although the disintegration process of the stones was successful. Currently we are elaborating a disintegration process which does not need spatial locating at all and which is based on the selective effect of acoustic energy frequency. This important subject is unfortunately beyond the range of this paper. We have investigated in details the problem of shock focusing precision by elliptical reflectors while displacing the source relatively to the ellipsoid focus and displacing the disintegration object relatively to the reflector focus. The results of these investigations showed that the most critical was the source position relatively to reflector focus. It was also shown that the ellipsoid aperture limited focusing to only 10-15% of the total radiated energy.

5. Polypositional generators

The exhaustive use of a single shock wave generator with a view to increase the acoustic pulse power for some practical purposes (disintegration of coral-shaped kidney stones or disintegration of ore and rubber, etc.) brought about the necessity to elaborate polypositional shock wave generators. We have carried out experiments by increasing the number of electrohydraulic generators with elliptical reflectors adjusted to a common focal point and radiating shock waves in-phase. At first the experiments with two anti-directed generators showed that disintegration efficiency became tenfold higher, for constant pulse energy. This was the reason of producing the lithotriptor for coral-shaped stones in kidneys. The investigations of four generators connected opposite pairwise, at right angles between the pairs, showed that in this case the value of generated pressure was 3 times as high as that of a single generator and the required number of shocks hundredfold less for identical disintegration results.

The most interesting focusing system consists of 34 series-connected dischargers located on a ring in the focal points of a toroidal reflector the radial sections of which are ellipses, with the purpose of producing disintegration at their external focus. Such a system enables to generate pressures at the focus equal to 1000 MPa. Non-linear phenomena prevent to fully use concentrated energy of greater power. Therefore, in order to utilize powerful generators of shock waves it is necessary to apply a liquid denser than water. The part of energy which is used directly for disintegration is proved to be a rather small amount: 5-7% only. Impulse scanning with the help of Fourier integral into continuous spectrum of frequencies enabled us to put forward the hypothesis of intercrystallite resonance disintegration of biocrystal structures. Currently we are elaborating this hypothesis for the selective disintegration of oncocells.

6. Conclusions

Results obtained enable us to formulate the following conclusions:

1. Disintegration of materials by hyper-shock waves can be applied in various fields of medicine and industry.
2. It is necessary to produce special shock wave generators in order to solve different problems.
3. Special shock wave generators enable to obtain different modes of superficial disintegration, from mechanical skin-effect to main rifts, depending on the rift spreading speed.
4. Powerful generators of hyper-shock waves with narrow beams are preferable when electrohydraulic effect and elliptical reflectors are applied.
5. Powerful shock wave generators producing maximum pressure in a physical point must have a spherical radiating surface with no reflectors and refractors.
6. Polypositional generators represent a compromise providing maximum energy concentration in the focal point.
7. Without locating it is possible to disintegrate materials by applying intercrystallite resonance to selective effect on oncocells against the normal cells and to crystal structure disintegration.

Author Index (Volumes I - IV)

T

Contents - Volume I

Survey Paper

Plenary Lectures

Part 1: Hypersonic Flow and Aerospace Studies

Part 2: Combustion and Ram Accelerators

Part 3: Shock Tube Technology and Diagnostic Techniques

Part 4: Numerical Computations

Contents - Volume II

Survey Paper

Plenary Lectures

Part 1: Chemical Kinetics

Part 2: Combustion Kinetics

Part 3: Non-Equilibrium Flow

Part 4: Plasmas, Astrophysics

Contents - Volume IV

Survey Paper

Plenary Lectures

Part 1: Shock Structure. Propagation and Focusing

Part 2: Shock Reflection and Diffraction

Part 3: Shock - Interface Interactions

Part 4: Interactions Involving Shocks

Part 5: Explosions, Blast Waves, Detonations